ACS SYMPOSIUM SERIES **715**

Mineral–Water Interfacial Reactions

Kinetics and Mechanisms

Donald L. Sparks, EDITOR
University of Delaware

Timothy J. Grundl, EDITOR
University of Wisconsin

American Chemical Society, Washington, DC

Library of Congress Cataloging-in-Publication Data

Mineral–water interfacial reactions : kinetics and mechanisms / Donald L. Sparks, editor, Timothy J. Grundl, editor.

p. cm.—(ACS symposium series , ISSN 0097–6156 ; 715)

"Developed from a symposium sponsored by the Division of Geochemistry at the 213th National Meeting of the American Chemical Society, San Francisco, Calif., Apr. 13–17, 1997."

Includes bibliographical references and index.

ISBN 0–8412–3593–7

1. Water chemistry—Congresses. 2. Geochemistry—Congresses. 3. Surface chemistry—Congresses. 4. Solid–Liquid interfaces—Congresses.

I. Sparks, Donald L., Ph. D. II. Grundl, Timothy J., 1953– . III. American Chemical Society. Division of Geochemistry. IV. American Chemical Society. Meeting (213th : 1997 : San Francisco, Calif.) V. Series.

GB855.M55 1998
551.46—dc21

98–34534
CIP

The paper used in this publication meets the minimum requirements of American National Standard for Information Sciences—Permanence of Paper for Printed Library Materials, ANSI Z39.48–1984.

Distributed by Oxford University Press

PRINTED IN THE UNITED STATES OF AMERICA

Advisory Board

ACS Symposium Series

Foreword

The ACS Symposium Series was first published in 1974 to provide a mechanism for publishing symposia quickly in book form. The purpose of the series is to publish timely, comprehensive books developed from ACS-sponsored symposia based on current scientific research. Occasionally, books are developed from symposia sponsored by other organizations when the topic is of keen interest to the chemistry audience.

Before agreeing to publish a book, the proposed table of contents is reviewed for appropriate and comprehensive coverage and for interest to the audience. Some papers may be excluded in order to better focus the book; others may be added to provide comprehensiveness. When appropriate, overview or introductory chapters are added. Drafts of chapters are peer-reviewed prior to final acceptance or rejection, and manuscripts are prepared in camera-ready format.

As a rule, only original research papers and original review papers are included in the volumes. Verbatim reproductions of previously published papers are not accepted.

ACS Books Department

Contents

Preface....................ix

Introduction

1. **Kinetics and Mechanisms of Reactions at the Mineral–Water Interface: An Overview....................2**
Timothy J. Grundl and Donald L. Sparks

Spectroscopic–Microscopic Techniques

2. **Use of X-ray Absorption Spectroscopy To Study Reaction Mechanisms at Metal Oxide–Water Interfaces....................14**
Gordon E. Brown, Jr., George A. Parks, John R. Bargar, and Steven N. Towle

3. **New Directions in Mineral Surface Geochemical Research Using Scanning Probe Microscopes....................37**
M. F. Hochella, Jr., J. F. Rakovan, K. M. Rosso, B. R. Bickmore, and E. Rufe

4. **Atomic Force Microscopy as a Tool for Studying the Reactivities of Environmental Particles....................57**
Patricia A. Maurice

Digital Modeling

5. **From Molecular Structure to Ion Adsorption Modelling....................68**
W. H. VanRiemsdijk and T. Hiemstra

6. **Interlayer Molecular Structure and Dynamics in Li-, Na-, and K-Montmorillonite–Water Systems....................88**
Fang-Ru Chou Chang, N. T. Skipper, K. Refson, Jeffrey A. Greathouse, and Garrison Sposito

Sorption of Inorganic Species

7. **Kinetics and Mechanisms of Metal Sorption at the Mineral–Water Interface..108**
Donald L. Sparks, André M. Scheidegger, Daniel G. Strawn, and Kirk G. Scheckel

8. **Evaluation of Oxyanion Adsorption Mechanisms on Oxides Using FTIR Spectroscopy and Electrophoretic Mobility..................136**
D. L. Suarez, S. Goldberg, and C. Su

9. **Kinetics and Reversibility of Radiocesium Sorption on Illite and Sediments Containing Illite..179**
Rob N. J. Comans

Sorption of Organic Species

10. **A Revised Physical Concept of Natural Organic Matter as a Sorbent of Organic Compounds..204**
J. J. Pignatello

11. **A Three-Domain Model for Sorption and Desorption of Organic Contaminants by Soils and Sediments..........................222**
Walter J. Weber, Jr., Weilin Huang, and Eugene J. Leboeuf

Precipitation and Dissolution

12. **Interfacial Kinetics Through the Lens of Solution Chemistry: Hydrolytic Processes at Oxide Mineral Surfaces..........................244**
William H. Casey, Brian L. Phillips, and Jan Nordin

13. **Reactivity of Dissolved Mn(III) Complexes and Mn(IV) Species with Reductants: Mn Redox Chemistry Without a Dissolution Step?..265**
George W. Luther, III, David T. Ruppel, and Caroline Burkhard

Heterogeneous Electron Transfer

14. **Degradation of Tetraphenylboron at Hydrated Smectite Surfaces Studied by Time Resolved IR and X-ray Absorption Spectroscopies...282**
D. B. Hunter, W. P. Gates, P. M. Bertsch, and K. M. Kemner

15. **The Role of Oxides in Reduction Reactions at the Metal–Water Interface**........301
Michelle M. Scherer, Barbara A. Balko, and Paul G. Tratnyek

16. **The Reduction of Aqueous Metal Species on the Surfaces of Fe(II)-Containing Oxides: The Role of Surface Passivation**........323
Art F. White and Maria L. Peterson

17. **Pollutant Reduction in Heterogeneous Fe(II)–Fe(III) Systems**........342
Stefan B. Haderlein and Klaus Pecher

18. **Auto-Inhibition of Oxide Mineral Reductive Capacity Toward Co(II)EDTA**........358
Scott Fendorf, Phillip M. Jardine, David L. Taylor, Scott C. Brooks, and Elizabeth A. Rochette

Photochemical and Microbially Mediated Processes

19. **Adsorption and Sensitization Effects in Photocatalytic Degradation of Trace Contaminants**........374
T. David Waite, Stephan J. Hug, and Andrew J. Feitz

20. **Microbially Mediated Oxidative Precipitation Reactions**........393
B. M. Tebo and L. M. He

Author Index........415

Subject Index........416

Preface

Processes that occur at mineral–water interfaces, such as sorption, precipitation, dissolution, and electron transfer are central to the full understanding of the geochemistry of natural systems. Geochemical processes at the mineral–water interface also markedly affect the fate, mobility, speciation, and bioavailability of inorganic and organic contaminants in the environment. In the past, researchers focused mostly on equilibrium aspects of mineral–water interfacial reactions. However, it is increasingly clear that a knowledge of the kinetics and mechanisms of interfacial reactions is critical for a full understanding of natural systems and the prediction of contaminant fate in the environment.

Recently, significant advances have been realized in the employment of kinetic methods and molecular-scale spectroscopic and microscopic techniques to study mineral–water interfacial reactions in-situ. In the past decade, researchers have coupled the detailed understanding of surface structure derived from molecular-scale studies with experimental and modeling studies of macroscopic surface behavior. This synergism has resulted in a dramatic increase in the ability to decipher and predict both the kinetics and mechanisms of mineral–water interfacial reactions. It is these important aspects that form the central theme of this book.

The chapters in the book are based on invited papers presented at a major American Chemical Society (ACS), Division of Geochemistry Symposium, "Kinetics and Mechanisms of Reactions at the Mineral–Water Interface", held April 13–17, 1997, in San Francisco, California. We clearly recognized that the symposium topic was inherently multidisciplinary. Accordingly, the invited authors of the chapters, who are internationally recognized in their fields, encompass a wide spectrum of areas including environmental chemists, engineers, soil chemists, microbiologists, geochemists, hydrogeologists, limnologists, oceanographers, and spectroscopists–microscopists.

The last major compilation dealing primarily with interfacial processes in the environment was based on an ACS Symposium held six years ago. Reflective of the progress made in the intervening time, this volume contains research advances that cut across a much broader spectrum of phenomena.

The first section of the book deals specifically with spectroscopic–microscopic techniques that can be used in combination with macroscopic approaches to glean mechanistic information on mineral–water reactions and processes. The second section emphasizes computer models that are used to elucidate surface mediated reaction mechanisms. The remainder of the volume is organized around reaction type. Sections are included on sorption–desorption of inorganic species, sorption–desorption of organic species, precipitation–dissolution processes, heterogeneous electron transfer reactions, photochemically driven reactions, and microbially mediated reactions. The book should be of interest to professionals and students in chemistry, soil science, geochemistry, microbiology, environmental engineering, and marine studies.

Acknowledgments

The editors are grateful to the authors for their outstanding contributions, to the referees of the chapters for their thoughtful and comprehensive reviews, to George Luther for his support and encouragement, and to the editorial staff at ACS Books, especially Anne Wilson. The Geochemistry Division of the ACS and the Petroleum Research Fund (Grant 32047–SE) both provided partial support for the symposium from which this volume originated.

TIMOTHY J. GRUNDL
Geosciences Department
University of Wisconsin at Milwaukee
Milwaukee, WI 53201

DONALD L. SPARKS
Department of Plant and Soil Sciences
University of Delaware
Newark, DE 19717–1303

INTRODUCTION

Chapter 1

Kinetics and Mechanisms of Reactions at the Mineral–Water Interface: An Overview

Timothy J. Grundl[1] and Donald L. Sparks[2]

[1]Department of Geosciences and Center for Great Lakes Studies, University of Wisconsin at Milwaukee, Milwaukee, WI 53201
[2]Department of Plant and Soil Sciences, University of Delaware, Newark, DE 19717

The study of environmental aquatic chemistry has matured over the past 30 years and with it the understanding of natural aquatic systems has grown. One outgrowth of this more mature understanding is a realization that reactions occurring at mineral-water interfaces are central to many, perhaps most, processes of geochemical importance. The rates of many ecosystem scale or even global scale processes are controlled by mineral-water interfacial reactions. The centrality of reactions at mineral-water interfaces is not a new idea as evidenced by previous research compilations which have emphasized the mineral-water interface to varying degrees (*1-9*). Rapid advances were made in the 1980's with the advent of a variety of modern spectroscopic techniques that, for the first time, allowed the direct study of mineral surfaces (*10*). In the last decade, researchers have coupled the detailed knowledge of surface structure available from these spectroscopic techniques with experimental and modeling studies of macroscopic surface behavior. This synergism has resulted in a dramatic increase in the ability to decipher and predict both the mechanisms and kinetics of surface mediated reactions. It is these mechanisms and rates that form the central theme of this volume.

Underlying Themes

The underlying intent in the preparation of this volume was to provide a compilation of papers that focus on the involvement of mineral surfaces in geochemical processes as a topic in and of itself. There is a particular need for a compilation of this sort because the topic encompasses a wide spectrum of researchers including environmental engineers, environmental microbiologists, geochemists, hydrogeologists, limnologists, oceanographers, soil scientists and spectroscopists. The resulting literature is quite fragmented and the rapid pace of progress further exacerbates the problem. The last major compilation dealing primarily with interfacial processes in the environment was based on a symposium held 6 years ago (*2*). Reflective of the progress made in the intervening time, this volume contains work that cuts across a much broader spectrum of phenomena. Also indicative of the continuing evolution and maturation of this field is the incorporation of principles from completely unrelated fields of knowledge. Two clear examples of this include the use of the metallurgical corrosion literature to approach the problems inherent with zero valent iron remediation (*11*) and the use of the glassy-rubbery phase transition of the synthetic polymer literature to explain non-idealities in the sorption of organic contaminants to natural organic matter (NOM)(*12,13*).

The diversity of reactions that are considered to be surface mediated has also increased over the past decade. It is not only strict sorption/desorption and precipitation/dissolution processes that are important but also the surface mediation of reactions such as electron transfer (eg. *14-17*), hydrolysis (*18*) and various photochemical transformations. In addition certain solid phases, in particular metallic iron, iron oxides and smectitic clays, are capable of transferring electrons in and out of their bulk structure (eg. *19-23*). When viewed in this context, minerals should not be considered as passive solids, or even as simple sources of a reactive surface but must be considered as bulk reactants.

An additional emphasis that we hope becomes discernable to the reader is in the presentation of work that uses evolving spectroscopic techniques to complement either experimental data obtained from the macroscopic behavior of surfaces or modeling efforts aimed at a mechanistic understanding of surface mediated reactions. Much of the mechanistic detail that is needed to attain a predictive capability towards mineral surface behavior is derived via combined studies of this type.

Volume Overview

The first section deals specifically with the spectroscopic/ microscopic tools that can be used in concert with macroscopic techniques. The second section emphasizes computer models that are used to elucidate surface mediated reaction mechanisms. The remainder of the volume is organized around reaction type. Sections are included on sorption/desorption of inorganic species; sorption/desorption of organic species; precipitation/dissolution processes; heterogeneous electron transfer reactions; photochemically driven reactions; and microbially mediated reactions. What follows are a few highlights taken from the work presented in this volume.

Spectroscopic/Microscopic Tools. It has been long recognized qualitatively that natural surfaces are heterogeneous, however researchers can now explore surface structure, and the structures associated with phenomena such as sorption or precipitation and dissolution on a microscopic scale in real time. Equally important as real-time capability is the development of both spectrographic and microscopic techniques that allow the imaging of samples under *in situ* conditions (i.e. with water present). Environmental Scanning Electron Microscopy (ESEM), Atomic Force Microscopy (AFM), Scanning Tunneling Microscopy (STM), Electron Paramagnetic Resonance (EPR), Fourier Transform Infrared (FTIR), Nuclear Magnetic Resonance (NMR), Mossbauer spectroscopy, X-ray Adsorption Fine Structure (XAFS) and X-ray Adsorption Near Edge Structure (XANES) are all extant ambient techniques that are applicable to geochemical problems (for a description of these techniques see (*10*)). A spatially resolved XAFS technique (X-ray spectromicroscopy) has recently been developed that has wide application to geochemical problems (*24*). Numerous examples of studies using real-time, *in situ* techniques are presented in this volume (*25-29*).

Modeling of surface reactions has kept pace with the new spectroscopic views of mineral surface structures. Chang et al. (*30*) and Van Riemsdik and Hiemstra (31) present excellent examples of the incorporation of new surface structure paradigms into computer modeling studies.

These new paradigms can immediately be applied to the many geochemical processes that are in essence surface-mediated. This will result in an improved understanding of geochemical processes on all time scales. Over geologic time, the surface-controlled dissolution of silicates and the precipitation of metal oxides are important sinks and sources, respectively in the global proton balance of surface and groundwaters (*32*). On more human time scales, carbonate precipitation/dissolution and various adsorption reactions are important buffers. Soil scientists recognize the importance of surface binding to the availability of soil nutrients. Hochella et al. (*28*) address this issue by applying real-time AFM data to the problem of potassium release from phlogopite. Surface-mediated reactions have been invoked to remove or supply ionic constituents from or to water, poise redox levels, and catalyze photochemical transformations. The importance of heterogeneous electron transfer reactions to the poise of sediment/water systems is probably matched only by the importance of silicate and carbonate dissolution to the buffering of the same systems.

Sorption of Inorganic Species. The intricacies of the sorption of inorganic cations to mineral surfaces is directly addressed by Sparks et al. (*33*) who point out that "sorption" is a general term that describes the retention of an aqueous species on a mineral by any of a number of processes, only one of which is actual adsorption (i.e. the association of an aqueous species strictly to the surface). Adsorption itself is highly variable with adsorption of radiocaesium being highly selective towards ion exchange type reactions at the edge sites of illite (*34*) whereas the sorption of Ni^{2+} on clay minerals and aluminum oxides is shown to be a combination of adsorption and the surface precipitation of a mixed nickel-aluminum hydroxide (*33*). Surface precipitation occurs at Ni^{2+} concentrations well below the solubility limit of $(NiOH)_2$ and is associated with release of aluminum and silicon from the mineral itself. The formation of mixed cation oxide phases appears to be a substitution of trace metals of similar size as Al^{3+} and Si^{4+} for silicon or aluminum in the surface layers. Similar Fe^{2+}/Fe^{3+} mixed cation surface precipitates on the surface of magnetite are suggested by Haderlein and Pecher (*35*). Implicit in these findings is that the formation of mixed cation oxides at mineral surfaces may be quite prevalent in the environment. Furthermore, these precipitates seem very geochemically active as an efficient scavenging mechanism for the removal of trace metal ions and in the enhancement of clay and oxide mineral dissolution. These precipitates also appear quite effective in promoting heterogeneous electron transfer.

Suarez et al. (*36*) use a combination of FTIR spectroscopy, electrophoretic mobility and pH titration data to deduce the specific nature of anionic surface species sorbed to aluminum and silicon oxide minerals. Phosphate, carbonate, borate, selenate, selenite and molybdate data are reviewed and new data on arsenate and arsenite sorption are presented. In all cases the surface species formed are inner-sphere complexes, both monodentate and bidentate. Two step kinetics is typical with monodentate species forming during the initial, rapid sorption step. Subsequent slow sorption is presumed due to the formation of a bidentate surface complex, or in some cases to diffusion controlled sorption to internal sites on poorly crystalline solids.

Sorption of Organic Compounds. The classic view of the sorption of neutral organic contaminants to sediment holds that hydrophobic forces cause the contaminant to preferentially associate, or partition, to NOM (e.g. *37,38*). This linear partitioning model is attractively simple, and to a first approximation, accounts for the behavior of organic contaminant sorption. However the often observed presence of non-linear isotherms, sorption-desorption hysteresis, competitive effects, and slow kinetics indicates that this model does not accurately reflect the complete sorptive process. Weber et al. (*13*) and Pignatello (*12*) both address this problem and summarize a new model in which NOM can be viewed as having two physical states; glassy and rubbery. The rubbery state behaves as a classic "partitioning" type sorbent whereas the glassy state acts as a specific site-to-site type sorbent. The proportion of glassy to rubbery states in NOM is a function of the provenance and extent of weathering. This model is promising in that it may provide fundamental insights into the sorption of organic contaminants. This should allow a mechanistic understanding of such poorly understood phenomenon as the meaning of the exponential term in Freundlich isotherms, slow sorption kinetics, and hysteresis. This model may also provide a framework in which compositional differences in NOM and its effect on sorption can be classified.

Mineral precipitation and dissolution: The study of mineral precipitation and dissolution rates is advancing rapidly with the advent of modern spectroscopic techniques, in particular AFM (see (*26*)) and a review has recently been published that is devoted explicitly to the rates of mineral dissolution (*39*). Casey et al. (*40*) point out that in spite of the acknowledged complexity of reactive sites at mineral surfaces, much can be learned by comparing mineral dissolution and growth to the much simpler metal-ligand exchange reactions that occur in solution. This analogy is most useful when using multidentate ligands which more closely resemble the coordination conditions that exist at mineral surfaces. Luther et al. (*41*) report on a rapid disproportionation reaction of colloidal sized MnO2 particles when they are complexed to oxalate. The reaction yields Mn2+ and CO2. In this reaction, solid MnO2 is not only dissolved but also reduced and is therefore a potential oxidant in natural systems. This reaction is particularly interesting because the colloidal particles involved are only 50 nM in size and as such would be considered soluble in that they readily pass the 0.2 or 0.45 uM pore size filter that is typically used to differentiate soluble from insoluble constituents.

Heterogeneous electron transfer. It has long been obvious that redox conditions in heterogeneous systems are not adequately defined in terms of aqueous parameters (*42,43*). The underlying reason is that biologic control of redox reactions dominates in the natural environment causing widespread disequilibrium among redox active species. The geochemical importance of heterogeneous redox reactions is underscored by the fact that rapid bacterial reduction of several ferric oxides, manganese oxide, and iron bearing montmorillonite has been demonstrated (*44-49*).

In recent years, much progress has been made in the understanding of electron transfer across mineral surfaces. Taken as a whole, this work has greatly improved the understanding of redox geochemistry of environmental systems and of the specific processes in operation during the engineered remediation of groundwater. Reactions have been reported in which the bulk mineral is oxidized (eg. *50,51*), is reduced (e.g. *15,16,17,21*), and acts to mediate the transfer of electrons between two aqueous species (52). In the first two cases, where the mineral serves as a bulk source/sink of electrons, the structure of the actual surface and how it differs from the structure of the bulk mineral is key to the understanding of the actual electron transfer. Scherer et al. (*11*), working within the context of metallic iron oxidation, presents three models for the redox behavior of surface oxide layers. Evidence suggests that surface oxides can act as passivating layers, as semiconductors, or as sites for the formation of redox active surface complexes. These behaviors are not mutually exclusive and may often operate concurrently.

Surface passivating layers are further discussed by White and Peterson (*53*) in the oxidation of magnetite by Cr^{6+} and by Fendorf (*29*) in the reduction of pyrolusite by Co^{2+}-EDTA complexes. In both cases, surface passivation removes the bulk mineral from further reaction with the solution and significantly reduces the redox capacity of the mineral.

In addition to exchanging structural electrons, ferric oxyhydroxide minerals also act to mediate electron exchange from surface bound Fe^{2+} to several reducible pollutants of environmental concern (*52*). In this case, the redox capacity of the mineral is not limited by the formation of a passivating layer because the bulk reductant is aqueous ferrous iron and reactive surface sites are continually regenerated. Haderlein and Pecher (*35*) review the environmental factors affecting the reactivity of surface bound Fe(II) in heterogeneous systems.

In contrast to the behavior of oxide minerals, iron bearing montmorillonite is capable of being reversibly reduced (*21,54,27*). A variety of inorganic reductants have been shown to be capable of reducing montmorillonites including hydrazine (*55*), sodium sulfide (*56*), benzidine (*57,58*), dithionite (*59*), and tetraphenylboron (*27*). Exposure to oxygen re-oxidizes the clay. The mechanism of electron transfer with the iron, which is located in the inner octahedral layer of these clays, is not fully understood but apparently occurs across the outer layer of tetrahedrally coordinated silicate (*27,57,58,60*). Using XANES and EXAFS techniques, Hunter et al. (*27*) examine structural changes induced in the clay matrix by the reduction of tetrahedral iron.

Photochemical reactions. Photocatalyzed degradation of pollutants at semiconducting surfaces has been reported for a long time. The majority of this work has been done in simple laboratory systems. Waite et al. (*61*) present data on the photocatalyzed oxidation (on TiO_2) of a complex algal toxin in the presence of a large excess of other, poorly defined, algal exudates. Composite systems of this sort are more representative of natural systems in which poorly defined humic or fulvic acids and various microbial exudates are common. Waite at al. (*61*) suggest a conceptual model for this system in which the background organics (algal exudates) sorb to TiO_2 and form long-lived radicals which in turn oxidize the pollutant (algal toxin). Interestingly, the long-lived radicals appear to desorb and primarily react with the algal toxin in solution rather than on the surface.

Microbially Mediated Reactions. In many respects the true geochemists of the world are the prokaryotic bacteria. These microbes are universally abundant and catalyze so many geochemically important reactions that the distinction between "biotic" and "abiotic" reaction is often unclear. Biotic reaction mechanisms have been observed on scales ranging from electron transport across the fluid interface of mineral grains (*44,48,62*) to the microbial involvement in weathering and elemental cycling on a global scale (*63,64*). Tebo and He (*65*) provide an excellent overview of the diversity of microbes that are involved in the oxidation of Fe(II) and Mn(II). New insights into the mechanism of the microbial catalysis of Mn(II) oxidation are also presented. It seems possible that the enzymes responsible are relatively general and may be capable of oxidizing a variety of trace metal ions (*66*). While it is not always clear what metabolic benefit is derived from the oxidation of metal ions, it is clear that microbial cycling of Fe(II) and Mn(II) is central to many environmental processes.

Progress in this field is being driven by dramatic advances in molecular biology. It is now possible to study microbes at the molecular level. The ability to form genetic mutants and thereby manipulate internal metabolic processes means that a detailed understanding of the metabolic pathways is possible. This, in conjunction with the newest techniques in electron microscopy that allow imaging under ambient conditions (e.g. *66*), are allowing further insight into the mechanisms of microbial processing at mineral interfaces.

Conclusion

This volume summarizes an increasingly clear understanding of reaction mechanisms at mineral surfaces on the part of researchers in the field. Along with this understanding comes the attendant realization of the complexity of these reactions. Long standing impediments to the detailed understanding of these complex surface processes are falling as experimental techniques in spectroscopy/microscopy and molecular biology continue to evolve. We hope that as the reader progresses through this volume, new conceptualizations of mineral surfaces and their reactivity will arise and ultimately lead to new avenues of research into the effects of mineral surfaces on the aqueous environment at large.

Acknowledgments

The editors would like to acknowledge the participating authors for their excellent contributions as well as the reviewers for time spent providing anonymous and unheralded, yet constructive, criticisms of the manuscripts. The Geochemistry Division of the American Chemical Society and the Petroleum Research Fund (grant #32047-SE) both provided partial support of the symposium from which this volume originated. We would also like to thank George Luther for his help and encouragement along the way.

Literature Cited

(1) *Mineral-water Interface Geochemistry*; Hochella, M.F., White, A.F., Eds., Reviews in Mineralogy #23; Mineralogical Society of America, Washington, DC, **1990**.

(2) *Aquatic Chemistry Interfacial and Interspecies Processes*; Huang, C.P.; O'Melia, C.R.; Morgan, J.J., Eds.; Advances in Chemistry Series 244; American Chemical Society: Washington DC, **1995**.

(3) *Physics and Chemistry of Mineral Surfaces*; Brady, P.V., Ed.; CRC Series in Physics and Chemistry of Surfaces and Interfaces; CRC Press: Boca Raton, FL, **1996**.

(4) *Spectroscopic Methods in Mineralogy and Geology*; Hawthone, F.C., Ed.; Reviews in Mineralogy, vol. 18; Mineralogy Society of America: Washington DC, **1988**.

(5) *Perspectives in Environmental Chemistry*; Macalady, D.L., Ed.; Topics in Environmental Chemistry; Oxford University Press: New York, NY; **1998**.

(6) *Geochemical Processes at Mineral Surfaces*; Davis, J.A. and Hayes K.F., Eds. ACS Symposium Series 323; American Chemical Society: Washington DC, **1986**.

(7) Stumm, W. and Morgan, J.J. *Aquatic Chemistry, 3rd edition*; Wiley Interscience: New York, NY, **1996**.

(8) Stumm, W. *Chemistry of the Solid-Water Interface*, Wiley Interscience: New York, NY, **1992**.

(9) *Geomicrobiology: Interactions Between Microbes and Minerals*, Banfield, J.F; Nealson, K.H., Eds. Reviews in Mineralogy #35; Mineralogical Society of America, Washington, DC, **1997**.

(10) Brown, G.E. In *Mineral-water Interface Geochemistry*; Hochella, M.F., White, A.F., Eds., Reviews in Mineralogy #23; Mineralogical Society of America, Washington, DC, **1990**; pp 309-364.

(11) Scherer, M.M, Balko, B.A., Tratnyek, P.G. In *Kinetics and Mechanisms of Reactions at the Mineral/Water Interface*; Sparks, D.L. and Grundl, T.J., Eds.; American Chemical Society: Washington DC; **1998**; this volume.

(12) Pignatello, J.J. In *Kinetics and Mechanisms of Reactions at the Mineral/Water Interface*; Sparks, D.L. and Grundl, T.J., Eds.; American Chemical Society: Washington DC; **1998**; this volume.

(13) Weber, W.J., Huang, W. LeBoeuf J. In *Kinetics and Mechanisms of Reactions at the Mineral/Water Interface*; Sparks, D.L. and Grundl, T.J., Eds.; American Chemical Society: Washington DC; **1998**; this volume.

(14) Junta, J.L.; Hochella, M.F. *Geochim. Cosmochim. Acta*. **1994**, *58*, 4985-4999.

(15) Kriegman-King, M.R.; Reinhard, M. *Environ. Sci. Technol.* **1992,** *26*, 2198-2206.

(16) Kriegman-King, M.R.; Reinhard, M. *Environ. Sci. Technol.* **1994,** *28*, 692-700.

(17) Stone, A.T. In *Geochemical Processes at Mineral Surfaces*; Davis, J.A. and Hayes K.F., Eds. ACS Symposium Series 323; American Chemical Society: Washington DC, **1986**; pp 446-460.

(18) Stone, A.T. In *Perspectives in Environmental Chemistry*; Macalady, D.L., Ed.: Topics in Environmental Chemistry; Oxford University Press: New York, NY; **1997**, pp 75-93.

(19) Jolivet, J.P.; Tronc, E. *J. Colloid Interfacial Sci.* **1988**, *125*, 688-701.

(20) Stucki, J.W.; Bailey, G.W.; Gan, H. In *Metal Speciation and Contamination of Soil*; Allen, H.E.; Huang, C.P.; Bailey, G.W.; Bowers, A.R., Eds.; Lewis Publishers: Boca Raton, FL, **1995**; pp 113-181.

(21) Komadel, P.; Madejova, J.; Stucki, J.W. *Clays & Clay Minerals*, **1995**, *43*, 105-110.

(22) Matheson, L.J.; Tratnyek, P.G. *Environ. Sci. Technol.* **1994**, *28*, 2045-2053.

(23) Agarwal, A.; Tratnyek, P.G. *Environ. Sci. Technol.* **1996**, *30*, 153-160.

(24) Droubay, T.; Mursky, G.; Tonner, B.P. *J. Elec. Spectros. Rel. Phenom.* **1997,** *84*, 159-169.

(25) Brown, G.E.; Parks, G.A.; Bargar, J.R.; Towle, S.N. In *Kinetics and Mechanisms of Reactions at the Mineral/Water Interface*; Sparks, D.L. and Grundl, T.J., Eds.; American Chemical Society: Washington DC; **1998**; this volume.

(26) Hochella, M.F.; Rakovan, J.F.; Rosso, K.M.;Bickmore, B.R.; Rufe, E. In *Kinetics and Mechanisms of Reactions at the Mineral/Water Interface*; Sparks, D.L. and Grundl, T.J., Eds.; American Chemical Society: Washington DC; **1998**; this volume.

(27) Hunter, D.B.; Gates, W.P.; Bertsch, P.,M.; Kenner, K.M. In *Kinetics and Mechanisms of Reactions at the Mineral/Water Interface*; Sparks, D.L. and Grundl, T.J., Eds.; American Chemical Society: Washington DC; **1998**; this volume.

(28) Maurice, P.A. In *Kinetics and Mechanisms of Reactions at the Mineral/ Water Interface*; Sparks, D.L. and Grundl, T.J., Eds.; American Chemical Society: Washington DC; **1998**; this volume.

(29) Fendorf, S.; Jardine, P.M.; Taylor, D.L.; Brooks, S.C. In *Kinetics and Mechanisms of Reactions at the Mineral/Water Interface*; Sparks, D.L. and Grundl, T.J., Eds.; American Chemical Society: Washington DC; **1998**; this volume.

(30) Chang, F.C; Skipper, N.T.; Refson, K.; Greathouse, J.A.; Sposito, G. In *Kinetics and Mechanisms of Reactions at the Mineral/Water Interface*; Sparks, D.L. and Grundl, T.J., Eds.; American Chemical Society: Washington DC; **1998**; this volume.

(31) VanRiemsdijk, W.H.; Hiemstra, T. In *Kinetics and Mechanisms of Reactions at the Mineral/Water Interface*; Sparks, D.L. and Grundl, T.J., Eds.; American Chemical Society: Washington DC; **1998**; this volume.
(32) Brady, P.V.; Zachara, J.M. In *Physics and Chemistry of Mineral Surfaces*; Brady, P.V., Ed.; CRC Series in Physics and Chemistry of Surfaces and Interfaces; CRC Press: Boca Raton, FL, **1996**; pp 307-356.
(33) Sparks, D.L.; Scheidegger, A.M.; Strawn, D.G.; Scheckel, K.G. In *Kinetics and Mechanisms of Reactions at the Mineral/Water Interface*; Sparks, D.L. and Grundl, T.J., Eds.; American Chemical Society: Washington DC; **1998**; this volume.
(34) Comans, R.N.J. In *Kinetics and Mechanisms of Reactions at the Mineral/Water Interface*; Sparks, D.L. and Grundl, T.J., Eds.; American Chemical Society: Washington DC; **1998**; this volume.
(35) Haderlein, S.B.; Pecher, K. In *Kinetics and Mechanisms of Reactions at the Mineral/Water Interface*; Sparks, D.L. and Grundl, T.J., Eds.; American Chemical Society: Washington DC; **1998**; this volume.
(36) Suarez, D.L.; Goldberg, S.; Su, C. In *Kinetics and Mechanisms of Reactions at the Mineral/Water Interface*; Sparks, D.L. and Grundl, T.J., Eds.; American Chemical Society: Washington DC; **1998**; this volume.
(37) Karickhoff, S.W. *J. Hydraulic Engrg.* **1984**, *110*, 707-735.
(38) Chiou, C.T.; Porter, P.E.; Schmedding, D.W. *Environ. Sci. Technol.* **1983**, *17*, 227-297.
(39) *Chemical Weathering Rates of Silicate Minerals*, White, A.F. and Bantely, S.L., Eds. Reviews in Mineralogy #31; Mineralogical Society of America, Washington, DC, **1995**.
(40) Casey, W.H., Phillips, B.L., and Nordin, J. In *Kinetics and Mechanisms of Reactions at the Mineral/Water Interface*; Sparks, D.L. and Grundl, T.J., Eds.; American Chemical Society: Washington DC; **1998**; this volume.
(41) Luther, G.W., Ruppel, D.T. and Burkhard, C. In *Kinetics and Mechanisms of Reactions at the Mineral/Water Interface*; Sparks, D.L. and Grundl, T.J., Eds.; American Chemical Society: Washington DC; **1998**; this volume.
(42) Lindberg, S.E.; Runnells, D.D. *Science*, **1984**, *225*, 925-927.
(43) Grundl, T.J. *Chemosphere*, **1994**, *28*, 613-626.
(44) Kostka, J.E.; Nealson, K.H. *Environ. Sci. Technol.* **1995**, *29*, 2535-2540.
(45) Lovely, D.R. *Microbiol. Rev.* **1991**, *55*, 259-287.
(46) Lovely, D.R.; Phillips, E.F. *Appl. Environ. Microbiol.* **1988**, *51,* 683-689.
(47) Stucki, J.W.; Komadel, P.; Wilkinson, H.T. *Soil Sci Soc. Amer. J.* **1987**, *51*, 1663-1665.
(48) Wu, J; Roth, C.B.; Low, P.F. *Soil Sci. Soc. Amer. J.* **1988**, *52*, 295-296.
(49) Kostka, J.E.; Stucki, J.W.; Nealson, K.H.; Wu, J. *Clays & Clay Minerals,* **1996,** *44*, 522-529.
(50) Jolivet, J.P.; Tronc, E.; Barbe, C.; Livage, J. *J. Colloid Interface Sci.*, **1990,** *138,* 465-472.
(51) White, A.F.; Yee, A. *Geochim. Cosmochim. Acta,* **1985**, *49,* 1263-1275.
(52) Klausen, J.; Trober, S.P.; Haderlein, S.B.; Schwarzenbach, R.P. *Environ. Sci. Technol.*, **1995**, *29*, 2396-2404.

(53) White, A.F.; Peterson, M.L. In *Kinetics and Mechanisms of Reactions at the Mineral/Water Interface*; Sparks, D.L. and Grundl, T.J., Eds.; American Chemical Society: Washington DC; 1998; this volume.

(54) Lear, P.R.; Stucki, J.W. *Clays & Clay Minerals*, **1985**, *33,* 539-545.

(55) Rozenson, I; Heller-Kallai, L., *Clays & Clay Minerals*, **1976**, *24*, 2271-282.

(56) Rozenson, I; Heller-Kallai, L., *Clays & Clay Minerals*, **1976**, *24*, 283-288.

(57) Solomon, D.H.; Loft, B.C.; Swift, J.D. *Clay Minerals*, **1968**, *7*, 389-397.

(58) Tennakoon, D.T.B.; Thomas, J.M.; Tricker, M.J. *J. Chemical Soc. Dalton,* **1974**, 2211-2215.

(59) Stucki, J.W.; Golden, D.C.; Roth, C.B. *Clays & Clay Minerals*, **1984**, *32*, 191-197.

(60) Lear, P.R. Stucki, J.W. *Clays & Clay Minerals* **1985**, *33,* 539-545.

(61) Waite, T.D.; Hug, S.J.; Feitz, A.J. In *Kinetics and Mechanisms of Reactions at the Mineral/Water Interface*; Sparks, D.L. and Grundl, T.J., Eds.; American Chemical Society: Washington DC; **1998**; this volume.

(62) Heijman, C.G.; Grieder, E.; Holliiger, C.; Schwarzenbach, R.P. *Environ. Sci. & Technol.,* **1995**, *29,* 775-783.

(63) Barker, W.W.; Welch, S.A.; Banfield, J.F. In *Geomicrobiology: Interactions Between Microbes and Minerals;* Banfield, J.F. and Nealson, K.H., Eds.;Mineralogical Society of America: Washington, DC, **1997**; pp 391-428.

(64) DesMarais, D.J. In *Geomicrobiology: Interactions Between Microbes and Minerals;* Banfield, J.F. and Nealson, K.H., Eds.;Mineralogical Society of America: Washington, DC, **1997**; pp 429-448.

(65) Tebo, B.M.; He, L.M. In *Kinetics and Mechanisms of Reactions at the Mineral/Water Interface*; Sparks, D.L. and Grundl, T.J., Eds.; American Chemical Society: Washington DC; **1998**; this volume.

(66) Little, B.J.; Wagner, P.A.; Lewandowski, Z. In *Geomicrobiology: Interactions Between Microbes and Minerals;* Banfield, J.F. and Nealson, K.H., Eds.; Mineralogical Society of America: Washington, DC, **1997**; pp 123-160.

Spectroscopic–Microscopic Techniques

Chapter 2

Use of X-ray Absorption Spectroscopy To Study Reaction Mechanisms at Metal Oxide–Water Interfaces

Gordon E. Brown, Jr.[1,2], George A. Parks[1], John R. Bargar[2], and Steven N. Towle[1,3]

[1]Department of Geological and Environmental Sciences, Stanford University, Stanford, CA 94305–2115
[2]Stanford Synchrotron Radiation Laboratory, SLAC, MS 69, Stanford, CA 94309

Chemical reactions at mineral-water interfaces are responsible for the partitioning of metal ions from aqueous solutions to mineral surfaces as well as the dissolution of minerals in natural systems. These processes, in turn, affect the fate of environmental contaminants and the composition of natural waters. Little is known quantitatively about the mechanisms of these important reactions, the nature of the surface reaction products, or the nature of reactive surface sites on minerals in contact with bulk water. Changes in the molecular-level speciation of aqueous metal ions during these interfacial reactions, *e.g.*, formation of inner-sphere surface complexes of different sizes or three-dimensional precipitates and redox reactions of sorbate ions, can dramatically affect their solubilities and transport properties, thus their potential for environmental impact. As a consequence, this class of chemical reactions has received increasing attention in recent years. Macroscopic studies of surface reactions at mineral-water interfaces, including uptake measurements of metal ions as a function of pH, ionic strength, and metal-ion concentration, have been used to infer reaction stoichiometries and sorption complex types, leading to quasi-thermodynamic models of sorption that are not unique (**1-5**). As will be discussed in this paper, *in-situ* spectroscopic studies (*i.e.*, those carried out in the presence of bulk water under ambient conditions) have great potential for providing important constraints on these models (**6-12**). X-ray absorption fine structure (XAFS) spectroscopy, in particular, has proven to be well suited for this purpose.

During the past decade, synchrotron-based XAFS spectroscopy has been used in simple model sorption systems to provide detailed information on the molecular-scale speciation of metal ions sorbed at submonolayer coverages at metal oxide-aqueous solution interfaces [see references in (**9-14**)]. Intense, wavelength-tunable x-rays from synchrotron sources are essential for these studies because they allow one to examine sorbate concentrations as low as 10 ppm for a wide variety of sorbed elements even in natural systems where more than one sorbate element is often present. XAFS spectroscopy is an ideal probe of reaction products at mineral-water interfaces because the technique is element specific and does not require long-range order around the sorbate element (**7**). The ability to perform *in-situ* XAFS analysis on wet samples is particularly important because sample preparation such as drying may alter the state of the sorbate. XAFS analysis provides information on the distances to and identities of first-, second-, and, in favorable cases, more distant neighbors of the sorbate ion, thus can be used to help define reaction products. Careful analysis of second-neighbor atom contributions in the XAFS spectrum of a sorbate can detect the presence or absence of metal ions of the sorbent, indicating, if present, that the sorbate is bonded directly to the oxide surface as an inner-sphere complex, or if absent, that the sorbate may be present as an outer-sphere, non-specifically bonded surface complex or in the diffuse swarm of ions in the electrical double layer. If second-neighbor metal ions characteristic of the sorbate are detected, the formation of surface oligomers or precipitates is indicated. In addition, when a sorbate ion is bonded directly to oxygen or hydroxide sites of a mineral surface, XAFS can provide information on the composition and geometry of the reactive surface site. Thus, in many cases, XAFS spectroscopy is well suited for molecular-level studies of mineral-aqueous solution interfacial reaction products in model systems, which in turn is providing the basis for XAFS studies of sorbate species in real systems (**15-18**).

In some cases, however, XAFS analysis can lead to less than definitive results, such as when a mixture of inner-sphere and outer-sphere complexes is present on a mineral surface, when x-ray fluorescence from another element interferes with the XAFS signal from the sorbate (*e.g.*, when Co(II) is sorbed to an iron oxide, Fe Kα fluorescence will dominate the signal), or when the surface of a powdered sample has a number of different types of sites to which the sorbate bonds. In this last case, the XAFS signal represents the sum of all geometric configurations of the sorbate element, weighted by the abundance of each. Moreover, when the sorbate ion concentration is less than $\approx$ 50 ppm, the XAFS signal may be too weak to fully analyze, although this limitation should be lowered by about two orders of magnitude when third-generation synchrotron sources are combined with new high-throughput, multi-element, solid-state detectors. In spite of these limitations, XAFS spectroscopy is one of the few *in-situ* spectroscopic methods that can provide relatively direct structural information on sorbates. Nonetheless, it is best to combine XAFS analysis with other characterization tools that provide complementary information about sorption reaction products. Furthermore, it is essential that the sorbent be as thoroughly characterized as possible and that the speciation of the metal ion of interest in solution be well known as a function of pH, ionic strength, and metal-ion concentration.

In this paper, we review recent XAFS work from our laboratory on sorption reactions of Co(II) and Pb(II) on several metal oxide surfaces with the aim of defining reaction products and mechanisms, including the types of surface sites involved in the reactions. Although the bulk of our studies have been performed using high surface area powdered sorbents, we have carried out some sorption experiments and XAFS studies on oriented, single-crystal sorbents in order to reduce the number of different types of sorption sites and sorbate geometries that may occur on the high surface area sorbents. In these cases, a variant of XAFS spectroscopy, known as grazing-incidence (GI) XAFS, was used to probe sorbate species. In selected cases, we have also used transmission electron microscopy combined with energy dispersive x-ray analysis to help characterize precipitate phases in sorption samples. We have also introduced bond-valence analysis of sorbate bonding to help constrain the types of surface sites to which a given sorbate species may bond (**19**), and have used this approach to help constrain reaction stoichiometries, including the number of protons released or consumed during a sorption reaction. The examples we have chosen include aqueous Co(II) sorption on high surface area γ-Al_2O_3, α-SiO_2 (quartz), and TiO_2 (rutile) powders (**20-22**), Pb(II) sorption on α-Al_2O_3 (corundum), α-Fe_2O_3 (hematite), and α-FeOOH (goethite) powders (**11,12**), and Co(II) and Pb(II) sorption on oriented α-Al_2O_3 (**19,23-25**) and TiO_2 single crystal surfaces (**26**). Finally, we discuss how XAFS results help constrain macroscopic surface complexation models of these sorption reactions.

Experimental Details

A summary of the procedures used in these studies to prepare sorption samples is presented in **Figure 1**. Experimental details can be found in (**11,19**). All sorbents used were synthetic and are available commercially or can be easily synthesized. The identity and phase purity of the powdered sorbents were determined by powder x-ray diffraction, and in selected cases by transmission electron microscopy coupled with energy dispersive x-ray analysis. Surface area was determined for powdered sorbents by N_2-BET measurements. Single-crystal samples were characterized prior to sorption experiments by x-ray reflectivity measurements or atomic force microscopy, which provided quantitative estimates of surface roughness. Most of the single-crystal samples used in the sorption experiments had rms roughnesses of 1-3 Å. Both powdered and single-crystal sorbents were placed in 0.01M $NaNO_3$ solutions and pH was allowed to equilibrate. In separate experiments, either $Co(NO_3)_2$ or $Pb(NO_3)_2$ was added to each solution in contact with the sorbent at initial aqueous concentrations ranging from 0.09 to 9.8 x 10^{-3} M. pH was adjusted by addition of NaOH to achieve a desired percentage uptake of Co(II) or Pb(II) onto each sorbent, which was verified by graphite furnace atomic absorption (GFAA) spectrophotometry analysis of the supernatent. The resulting sorption densities ranged from $\approx$ 0.5 to 23 μmoles/m^2. Each sample was allowed to "equilibrate" at the final pH for 24-36 hours, then prior to GFAA analysis, the sample was centrifuged and about 95% of the supernatent was removed. Except at the highest metal concentrations, the solutions were undersaturated with respect to the stable Co- or Pb-hydroxide or oxide, thus no $Co(OH)_2$, $Pb(OH)_2$, or PbO precipitate phase was expected to form. The remaining sorption sample in the form of a wet paste was then loaded into a Teflon sample holder with Mylar windows, and XAFS spectra were collected at the Stanford Synchrotron Radiation Laboratory (SSRL) within two days. In none of these experiments was evidence found for oxidation or reduction of the sorbate ion during or after XAFS data collection, as indicated by the lack of energy shifts in the Co or Pb edge positions, EXAFS-derived Pb-O and Co-O bond lengths, and XPS measurements. The XAFS data analysis procedures used for the powdered sorption samples are described in (**11**) and (**21**).

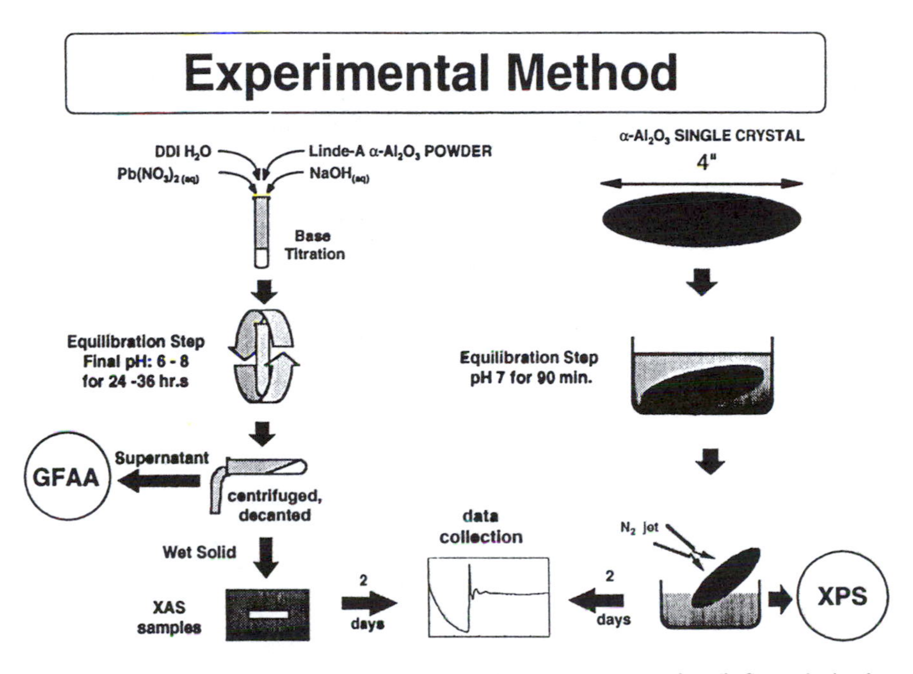

Figure 1. Summary of the procedures used to prepare powder (left) and single crystal (right) sorption samples for spectroscopic analysis.

In the case of single-crystal sorbent samples, following base titration and metal-ion uptake, the wafers were equilibrated in solutions containing no background electrolyte and were carefully removed from aqueous solution. During removal from solution, the solution was maintained at the meniscus using a gentle jet of dry N_2 gas. Thus, evaporation of solution on the surface, which could deposit unwanted sorbate ions from the supernatent, was avoided. A smaller wafer sample, which could be accommodated in the XPS sample chamber, was subjected to exactly the same procedure in the same batch experiment as the larger wafer and its surface was analyzed by x-ray photoelectron spectroscopy prior to GI-XAFS work to ensure that the sorbate coverage was known and sufficient to produce usable EXAFS. In practice, 0.5 μmoles/m^2 was the minimum coverage examined in our GI-XAFS experiments, which corresponds to less than 0.1 monolayer coverage. The α-Al_2O_3 and TiO_2 wafers were 4 inches and 2 inches in diameter, respectively. The goniometer used in the GI-XAFS experiments is described in (**19**). These experiments were run at SSRL within two days of this procedure, under either *in-situ* (bulk water present) or *ex-situ* (bulk water absent) conditions. Because of the relatively low energy of the Co-K absorption edge (7,709 eV) and the fact that x-rays of this energy are almost completely attenuated by the water path at grazing incidence, all GI-XAFS experiments on Co(II)/Al_2O_3 or Co(II)/TiO_2 samples were conducted *ex situ*. In contrast, the Pb-L_{III} absorption edge (13,055 eV) is high enough in energy to permit *in-situ* experiments. In all cases, even when *ex-situ* experiments were performed, the relative humidity was sufficiently high ($\approx$ 50%) to ensure that several monolayers of water were in contact with the wafer surface [see (**27**)]. The analysis procedure for the GI-XAFS data is described in (**19**).

Powder XAFS Studies of Sorption of Co(II) on γ-Al_2O_3, α-Al_2O_3, α-SiO_2, and TiO_2 (rutile)

Effect of sorption density on sorption complexes. A fundamental question in surface chemistry is how sorption density (Γ in units of μmole/m^2) of a particular sorbate affects the type of surface complex formed in a sorption reaction. To explore this effect, Chisholm-Brause *et al.* (**20**) and Towle *et al.* (**22**) used XAFS measurements to examine the types of Co(II) complexes formed on high surface area γ-Al_2O_3 and α-Al_2O_3 as a function of Co(II) sorption density, which ranged from $\Gamma \approx$ 0.1 to $\approx$ 23 μmoles/m^2. Monolayer coverage by Co(II) on α-Al_2O_3 corresponds to $\Gamma \approx$ 12 μmoles/m^2 on the (1-102) surface assuming bidentate bonding to the surface and that this surface is a perfect termination of the bulk structure (**12**). The EXAFS spectra and corresponding Fourier transforms for the Co(II)/γ-Al_2O_3 samples are shown in **Fig. 2**. Fits of these data and of similar data for Co(II)/α-Al_2O_3 suggest that Co(II) forms inner-sphere complexes on γ-Al_2O_3, as indicated by the presence of Co-Al correlations at 3.1 Å, and that no second-neighbor cobalt is present at the lowest Co(II) sorption density, indicating that Co(II) forms monomers on the γ-Al_2O_3 surface. With increased sorption density, the second-neighbor feature at $\approx$ 3.1 Å increases in magnitude, and fits of the data indicate that this feature is due to a mixture of cobalt and aluminum atoms. The presence of second-neighbor cobalt atoms indicate oxo- or hydroxo-bridged dimers or larger oligomers. At the highest sorption density, the cobalt K-EXAFS spectrum strongly resembles the spectrum of freshly

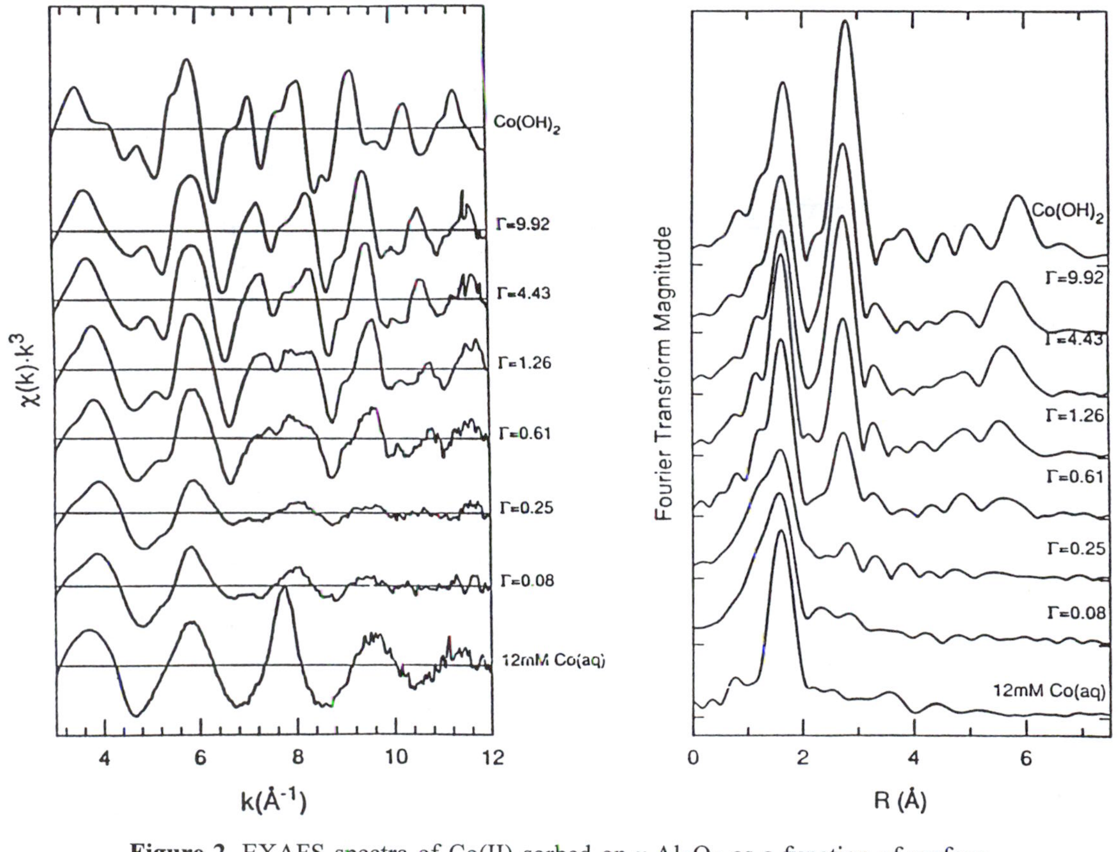

Figure 2. EXAFS spectra of Co(II) sorbed on γ-Al_2O_3 as a function of surface coverage (in μmole/m^2) and their corresponding Fourier transforms [from ref (**38**)].

precipitated $Co(OH)_2$, indicating that a precipitate phase similar to $Co(OH)_2$ formed. The FT feature at ≈ 6.1 Å (distance corrected for phase shift) for samples with $\Gamma \geq 1.3$ μmoles/m^2 is particularly useful as an indicator of a precipitate phase. This feature is due to multiple scattering among second-neighbor cobalt atoms arranged around a central cobalt atom such that each Co-O-Co(2) linkage forms a linear scattering path (**28**). Linear or near-linear Co-O-Co(2) linkages are characteristic of the $Co(OH)_2$ type solid. Co-O-Co(2) angles of less than about 170° cause a marked reduction in the 6.1 Å correlation (**28**). More recent XAFS and TEM work by Towle *et al.* (**22**) has shown that the precipitate phase formed in these systems is actually a Co(II)-Al(III) hydroxide. The results of these studies as well as those from a similar XAFS study of Ni(II) sorption on pyrophyllite as a function of surface loading (**29**) indicate that (1) sorption complexes formed at the lowest sorption densities are dominantly monomers, (2) as sorption density increases, the metal ion sorbate is dominantly in the form of oxo- or hydroxo-bridged oligomers, and (3) at the highest sorption densities below monolayer coverage metal hydroxide precipitates form. These studies cannot rule out the presence of monomers, dimers, and small oligomers when precipitates are present nor the presence of outer-sphere complexes of Co(II) or Ni(II) in these sorption systems, although they are expected to be minority species. XAFS results for Co(II) sorption on γ-Al_2O_3 will be used to constrain surface complexation models of Co(II) uptake on γ-Al_2O_3 in a later section of this paper.

Effect of sorbent type on sorption complexes. Another fundamental question is how sorbent type affects the types of surface complexes formed. O'Day *et al.* (**21**) studied Co(II) sorption on powdered α-SiO_2 and TiO_2 (rutile) at similar sorption densities and solution ionic strengths. When combined with similar studies of Co(II) sorption on γ-Al_2O_3 (**20**) and α-Al_2O_3 (**22**), this work showed that large multinuclear surface complexes or precipitate phases similar to $Co(OH)_2$ form on α–SiO_2 at the lowest sorption densities considered (Γ = 0.77 μmoles/m^2), whereas there is no evidence for the formation of a precipitate phase on TiO_2 until sorption densities are 2 to 3 times higher (Γ > 1.3 μmoles/m^2). At lower sorption densities, the EXAFS analysis suggests that Co(II) substitutes for Ti(IV) in octahedral positions on the rutile surface, although such substitution must be limited because of local charge balance problems created by this substitution. The findings for γ-Al_2O_3 and α-Al_2O_3 appear to be intermediate between those for quartz and rutile with no evidence for a precipitate at the lowest sorption densities but clear evidence from EXAFS analysis and high-resolution transmission electron microscopy that such phases are present in samples with sorption densities ≥ 1.6 μmoles/m^2 (**22**).

These solids differ in several important ways, including pH_{pznpc}, solubility, crystal structure, and types of surface sites available for sorption reactions. The pH_{pznpc} value is lowest for quartz (≈ 2.9), intermediate for rutile (≈ 5.8), and highest for the aluminas (≈ 9) (**30,31**). Other factors being equal, one might anticipate that inner-sphere Co(II) sorption on quartz would be greatest, whereas Co(II) sorption on alumina would be lowest at the pH values of the sorption experiments (pH = 6-10). However, this is not the case. The reasoning leading to this incorrect conclusion is as follows. The surface of quartz should be negatively charged over this pH range, thus there should be a strong electrostatic attraction between the quartz surface and charged hexaaquo-Co(II) solution complexes. In contrast, the surface of alumina should be positively charged over most of this pH range, resulting in electrostatic repulsion of hexaaquo-Co(II) solution complexes. Sorption of Co(II) on rutile should be intermediate relative to quartz and alumina. Macroscopic uptake measurements of Co(II) on these surfaces as a function of pH clearly show that pH_{ads} for quartz (defined as the pH value at 50% metal ion uptake; pH_{ads} (SiO_2) ≈ 6.0, dependent upon solid:liquid ratio, ionic strength, and $[Me]_T$) is well above the pH_{pznpc} of quartz, whereas pH_{ads} of alumina (≈ 7.0) is well below the pH_{pznpc} of alumina. The pH_{ads} value for rutile (≈ 5.5) corresponds approximately to its pH_{pznpc} value. These results suggest that Co(II) forms strong chemical bonds with the alumina surface at pH values lower than its pH_{pznpc} where the surface charge is positive and electrostatic repulsion must be overcome. In contrast, Co(II) does not bond strongly to the quartz surface until the surface charge is quite negative, suggesting that bonding in this case is dominated by electrostatic forces. Co(II) sorption on rutile behaves approximately as expected, with uptake beginning below the pH_{pznpc} and complete just above the pH_{pznpc} of rutile.

A relatively simple interpretation of these results is offered based on differences in crystal structure and bond-valence sums at surface oxygens for these sorbents. The quartz structure consists of tetrahedral SiO_4^{4-} units which are linked through their corners (oxygen atoms) to other SiO_4 tetrahedra. The bond-valence sum at an oxygen that bridges between two linked tetrahedra in bulk quartz is 2.0 valence units (v.u.), which satisfies Pauling's electrostatic valence principle (see (**19**) for discussion). Thus surface oxygens of the bridging type should not be particularly reactive to Co(II) in solution unless the bridging bond is broken. In this case, a non-bridging oxygen (bonded to only one Si in tetrahedral coordination, resulting in a bond-valence sum of 1.0 at the non-bridging oxygen) could also bond to two octahedrally coordinated Co(II), assuming that steric effects do not prevent this. Alternatively, a non-bridging oxygen could bond to one octahedrally coordinated Co(II) and a proton (**19**). Both of these arrangements would result in a bond-valence sum close to the ideal value of 2.0 v.u.. If the quartz surfaces in contact with a Co(II)-containing aqueous solution have mostly bridging oxygens, then bonding of $Co(OH_2)_6$ to the surface would not be energetically favored. O'Day *et al.* (**21**) used a somewhat different argument involving the dissimilarity of SiO_4 and $Co(OH_2)_6$ polyhedral dimensions to rationalize the fact that $Co(OH)_2$-like precipitates form on quartz even at the lowest sorption densities considered. They suggest that polyhedral edge sharing between an SiO_4 tetrahedron and a $Co(OH_2)_6$ octahedron is not as favorable as edge sharing between two $Co(OH_2)_6$ octahedra due to dimensional mistmatch of the tetrahedral and octahedral edges. In contrast, edge sharing between two $Co(OH_2)_6$ octahedra would not result in a strain energy. Moreover, electrostatic repulsion between divalent Co(II) ions in two edge-sharing $Co(OH_2)_6$ octahedra would be less than the repulsion between a divalent Co and a tetravalent Si in an edge-shared SiO_4 tetrahedron and $Co(OH_2)_6$ octahedron. Although this reasoning based on dimensional mismatch and relative electrostatic repulsion between edge-sharing octahedra and tetrahedra is attractive due to its simplicity, it must be used with caution because shared edges between divalent cation-containing octahedra and Si-containing tetrahedra are common in rock-forming silicate minerals such as olivines and pyroxenes. Also, these arrangements address only one type of contribution to the total free energy of the complex.

Turning to sorption of Co(II) on alumina surfaces, these surfaces in contact with bulk water should be dominated by $Al(OH,OH_2)_6$ octahedra (**19**), which would favor polyhedral edge sharing between an Al-containing octahedron and a Co-containing octahedron. A bond-valence analysis of sorption of Co(II) on alumina surfaces was carried out by Bargar *et al.* (**19**) with the assumption that a particular alumina surface is a perfect termination of the bulk structure. Although this assumption is probably not correct in detail for alumina surfaces in contact with bulk water, it is likely to approximate the polyhedral topology of the wet α-Al_2O_3 surface as argued by Bargar *et al.*. This analysis suggests that Co(II) can bond to a variety of surface oxygen sites on alumina. Thus, larger Co(II)-containing oligomers are predicted to be more stable on alumina surfaces than on silica surfaces.

Co(II) sorption complexes on rutile surfaces appear to form extensions of the bulk structure, with EXAFS-derived Co-Ti distances very similar to Ti-Ti distances in rutile. The small number of Ti second neighbors detected by EXAFS spectroscopy indicates that Co occurs dominantly in Ti-equivalent sites at the rutile surface (**21**). A bond-valence analysis of Co(II) sorption on rutile also indicates that Co(II) can bond stably to a variety of surface oxygen sites (**26**). When the sorption density of Co is high on rutile, O'Day *et al.* (**21**) found Co-Ti distances characteristic of anatase. The observed differences in the style of Co sorption on quartz, rutile, and alumina under similar conditions and sorption densities are consistent with structural differences among the three sorbents that result in different reactivities of available surface sites.

Formation of Co-Al hydroxide precipitates on alumina. Precipitation of solids is another means of metal ion sorption at high surface loadings of the metal ion, as shown, for example, in past studies of Co(II) sorption on kaolinite (**33**). Until recently, it was commonly assumed that such precipitates in simplified model sorption systems were hydroxides of the sorbing metal ion. However, such precipitates have been observed to form at $[Me(II)]_T$ values well below saturation with respect to the $Me(OH)_2$ solid [*e.g.*, ref. (**33**)]. An explanation for this observation was recently provided by several XAFS studies of Co(II) and Ni(II) sorption products on various Al- or Si-containing solids (**22,34,36**). For example, our XAFS study of aqueous Co(II) sorption on alumina powders (pH $\approx$ 8, sorption density = 0.28-23.1 μmoles/m^2; $[Co(II)]_T$ = 100 μM - 12.6 mM), coupled with high resolution transmission electron microscopy (HRTEM) observations of the dried sorption samples following spectroscopic examination, provides strong evidence for a nanoscale precipitate phase consisting of a mixed Co-Al hydroxide at Co(II) sorption densities above about 3 μmoles/m^2 (**22**). Under these solution conditions, the concentration of Co(II) is well below the solubility of solid Co-hydroxide; however, sufficient Al(III) dissolves from the alumina surface to combine with Co(II) (in bulk solution or at the alumina-water interface), resulting in the formation of this mixed-metal hydroxide phase, which has a hydrotalcite-like structure (**22,34,36**). Small monomeric and multimeric surface complexes of Co(II) form at lower sorption densities, even though the solution is free from multinuclear species, indicating that the alumina surface causes polymerization of the Co(II) surface species. A similar Co(II)-Al(III)-hydroxide precipitate was found by d'Espinose de la Caillerie *et al.* (**34**) in the Co(II)/alumina system and by Thompson *et al.* (**35**) in the Co(II)/kaolinite system. In addition, evidence for Ni(II)-Al(III)-hydroxide precipitates was found in several sorption systems by Scheidegger *et al.*

(**36**). These results have important implications for surface complexation and reactive transport models as co-precipitation of nanoscale multicomponent phases may be a significant mode of sorption, particularly under conditions where the concentration of metal ions in the aqueous solution is fairly high. Such phases also provide new surfaces on which further sorption of aqueous ions can occur. However, detection of these phases and distinguishing them from the sorbate metal hydroxide phase requires very careful EXAFS analysis and HRTEM studies of sorption samples.

Powder XAFS studies of Pb(II) sorption on α-Al_2O_3 and α-Fe_2O_3

We have also used XAFS spectroscopy to study sorption reactions of aqueous Pb(II) on high surface area powdered samples of α-Al_2O_3 (**11**) and iron oxides (**12**). A major objective of these studies was to determine if Pb(II) sorption complexes on isostructural α-Al_2O_3 and α-Fe_2O_3 are the same or different. On α-Al_2O_3, Pb(II) ions adsorb preferentially to the edges of $Al(O,OH)_6$ octahedra as mononuclear bidentate complexes at pH 6-7 and Pb sorption densities of 0.5 to 5.2 μmoles/m^2 (**11**). Above a sorption density of 1.5 μmoles/m^2, a second Pb(II) surface species was observed, which is bonded only to corners of $Al(O,OH)_6$ octahedra in a monodenate fashion. A smaller number of dimeric Pb(II) complexes was observed at sorption densities of 3.4 μmoles/m^2 and above. No evidence was found for $Pb_4(OH)_4^{4+}$ surface complexes, a stable solution species under the conditions of our experiments.

Aqueous Pb(II) sorption on hematite and goethite powders at room temperature was studied using XAFS spectroscopy as a function of pH (6-8), sorption density (2-10 μmoles/m^2), and total Pb concentration (0.2 μM to 1.2 mM) in 0.1 M $NaNO_3$ electrolyte solution (**12**). The Pb(II) ions were found to be hydrolyzed and adsorbed as mononuclear bidentate complexes to edges of FeO_6 octahedra on both goethite and hematite under all experimental conditions studied. Application of the bond-valence model described above, combined with the XAFS results, suggests that Pb(II) adsorption occurs primarily at unprotonated surface sites. Hydrolysis of Pb(II) appears to be the primary source of proton release during Pb(II) sorption. The absence of dimeric Pb(II) complexes on iron oxides and their presence on α-Al_2O_3 may be related to edge-length differences of $Al(O,OH)_6$ and FeO_6 octahedra and to differences in bond-valence sums at the surface oxygens to which Pb(II) bonds.

We have also carried out an XAFS study of aqueous Pb(II) sorption on goethite and alumina surfaces in the presence of Cl^- (pH = 5-7, sorption density = 1.2-5 μmoles/m^2, $[Pb]_T$ = 0.42 - 4.5 mM) (**32**). At pH 7, the Pb(II)/goethite and Pb(II)/alumina spectra are similar to those in the absence of Cl^-. However, at pH $\leq$ 6, Pb(II)-chloro ternary complexes were observed on goethite. The Pb is attached to the goethite surface through Pb-O_{sfc} and Cl^--O_{sfc} bonds simultaneously. These results suggest that Cl^--O_{sfc} bonding should be taken into account when describing Pb sorption on Fe-(hydr-)oxides in the presence of Cl^-. In contrast, no Pb-Cl^--Al_{sfc} ternary complexes were observed on alumina. This behavior parallels that of Cl^- aqueous complexation to Fe(III) [moderately strong, *i.e.*, log $K_{fm,\ FeCl^{2+}(aq)}$ = 1.5 (**37**)].

Grazing-Incidence XAFS Studies of Co(II) and Pb(II) Sorption on Single Crystal α-Al_2O_3 and TiO_2

Two questions that can not be easily answered by XAFS spectroscopic studies of sorption complexes on high surface area powdered sorbents are (1) What specific surface sites are involved in sorption? and (2) Do different crystallographic surfaces of the same sorbent cause different types of sorption complexes? These questions can be addressed, however, by XAFS studies of sorption complexes on oriented single crystal sorbents, which should minimize the number of different types of surface sites available for sorption, thus simplifying interpretation of the data. In addition, to increased surface sensitivity, one can perform XAFS measurements at an incident x-ray angle less than the critical angle for solid sorbents, resulting in total external reflection of the x-rays. For α-Al_2O_3, this angle is of the order of 0.2°.

We have examined aqueous sorption complexes of Co(II) and Pb(II) on α-Al_2O_3 (0001) and(1-102) surfaces and on the (110) and (001) surfaces of TiO_2 using grazing-incidence fluorescence XAFS spectroscopy. Pb(II) was found to form outer-sphere complexes on α-Al_2O_3 (0001) but inner-sphere complexes on α-Al_2O_3 (1-102). XPS analysis of the samples showed that an order of magnitude more Pb(II) is present on the (1-102) surface than on the (0001) surface following uptake from solution. In contrast, Co(II) forms inner-sphere complexes at about equal concentrations on both surfaces of α-Al_2O_3, adsorbing dominantly to tridentate sites on the (0001) surface and dominantly to tetradentate sites on the (1-102) surface. These differences can be explained using a bond-valence model which takes into account the structural differences on the two alumina surfaces and the differences in Co(II) and Pb(II) coordination complexes (**19**). This surface bonding model quantitatively accounts for O-H bonds as well as hydrogen bonds from solvating water molecues using a bond valence-bond length curve for O-H bonds derived from neutron diffraction studies of hydrated solids (**19**). These results provide the first direct structural evidence we are aware of that crystallographically different surfaces of a given oxide can adsorb aqueous cations in quite different modes. They also have implications for the mechanisms of compositional sector zoning in crystals grown from aqueous solution. The observed Pb(II)-surface oxygen distance for the outer-sphere Pb(II) complex on the α-Al_2O_3 (0001) surface can be used to constrain the capacitance of water between the oxide surface and the outer Helmholz plane of the diffuse double layer (**24**). In our GI-XAFS study of Co(II) on the (110) and (001) surfaces of TiO_2, we found that Co adsorbs at surface sites corresponding to Ti-equivalent positions in an extension of the rutile structure (**26**). Even though these two surfaces of TiO_2 have different reactivities to H_2O in UHV systems, we find that aqueous Co(II) sorbs in a similar manner on both.

Surface Complexation Modeling: Co(II) on γ-Al_2O_3

We have argued that molecular-scale understanding of the structure and composition of mineral surface-aqueous solution sorption complexes is vital to development of robust reactive contaminant transport models. This argument suggests that significant error may be expected in predicted environmental behavior in the absence of such knowledge. In an attempt to test this claim, we have used the results of our XAFS studies of Co(II) sorbed by alumina to refine a quasi-thermodynamic uptake model, then used the model to evaluate the sensitivity of predicted Co(II) partition coefficients to the choice of reactions included in the sorption model.

Our uptake data (**Fig. 3**) and XAFS results (**Fig. 4**) for sorption of Co(II) on γ-Al_2O_3 (**20,38**), together with limitations suggested by coordination chemistry and bond-valence theory (**19**) impose significant constraints on the types of surface

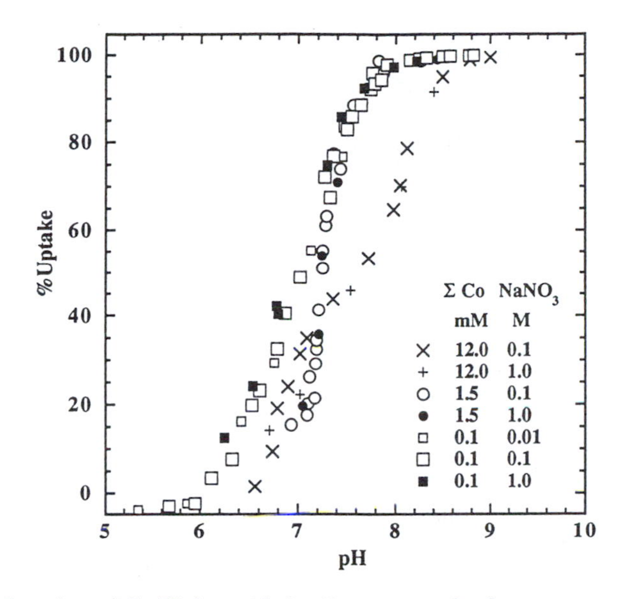

Figure 3. Sorption of Co(II) by γ-Al_2O_3: Percent uptake from aqueous solution as a function of pH, ΣCo, $NaNO_3$ background electrolyte concentration (**38**). Notice the lack of $NaNO_3$ concentration dependence.

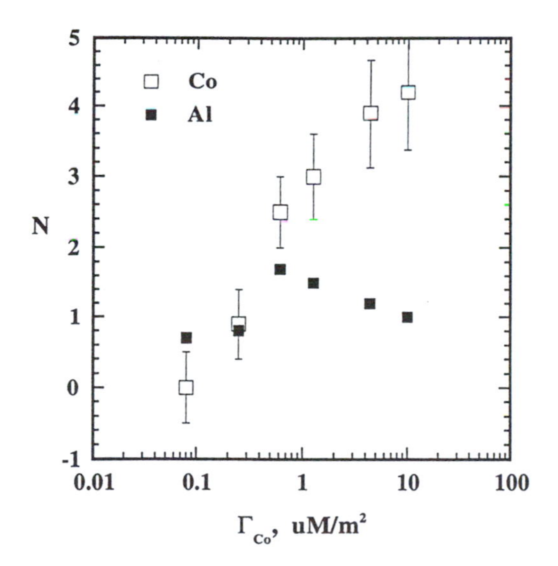

Figure 4. Sorption of Co(II) by γ-Al_2O_3: Average number of Co and Al atoms, N_{Co} and N_{Al}, among near neighbors of each sorbed Co as derived from XAFS (**38**).

complexes formed during sorption, thus on the reactions used in simulation models. Hayes and Katz (**10**) described a "state-of-the-art" approach to the simulation of similar uptake and XAFS data for Co(II) on α–Al_2O_3 and found it necessary to use monomeric and oligomeric sorption complexes and a hydroxide precipitate of variable thermodynamic activity in a surface complexation, triple layer model, derived from HYDRAQL (**39**). We have developed a similar HYDRAQL-based sorption model for our data and used it, in turn, to assess the impact that reaction selection might have on the partition coefficients computed for use in coupled reaction/transport codes. Coupled codes using HYDRAQL-like thermodynamic and sorption models are available (**40, 41**) and are proving useful in simulating the behavior of contaminant metals in, *e.g.*, acid mine drainage.

The model. In this surface complexation model (SCM), sorption by the alumina is attributed to a single set of surface hydroxide functional groups, AlOH. Protonation and deprotonation of these sites and supporting electrolyte adsorption are represented by reactions coded 100, 102, 105, and 106 in **Table I**. Lacking the data needed for more sophisticated fitting, we used the equilibrium constants derived for these reactions by James and Parks (**42**) from data by Huang (**43, 44**) for γ-Al_2O_3. This requires: (1) accepting the AlOH site density (N_S = 8 sites/nm^2) and triple layer model (TLM) capacitances associated with the outer Helmholz plane (OHP) region (C_2 = 0.9 F/m^2) and the near-surface region (C_1 = 0.2 F/m^2) on γ-Al_2O_3 used by James and Parks; and (2), inasmuch as our supporting electrolyte was $NaNO_3$ and Huang's was NaCl, accepting James' chloride binding constant as an approximation of the nitrate binding constant.

Cobalt sorption reactions. Our uptake and XAFS results appear to be in good qualitative agreement with the monomer/multimer/precipitation sorption sequence proposed by earlier investigators. This sequence implies a gradual increase in N_{Co}, from zero to perhaps as high as 6.0 if ordered $Co(OH)_2$ precipitates at the highest pH and Γ_{Co}. The XAFS-derived variation in N_{Co} with Γ_{Co} shown in **Figure 4** supports this scenario. For simulation purposes, then, we must identify sorption reactions producing mononuclear and multinuclear complexes and/or a precipitate, and estimate equilibrium constants for these and necessary protonation/deprotonation derivatives. We did this by assuming that the 0.1 mM ΣCo uptake isotherm represents mononuclear sorption complexes only, and that the high-pH, high uptake branch of the 12 mM uptake isotherm represents precipitation. Multinuclear complexes were added if mononuclear reactions and precipitation failed to account for uptake and N_{Co} data.

Weak ionic strength dependence in uptake results suggests that sorbed Co is bound in inner-sphere complexes (**3**). XAFS N_{Co} results (**Fig. 4**), Co-Al distances, and N_{Al} confirm inner-sphere bonding and suggest bidentate stoichiometry. We assume, therefore, that any Co atom bonded to the surface is bound in an inner sphere, bidentate mode, probably sharing two oxygen atoms in the edge of the CoO_6 coordination octahedron with two adjacent AlOH sites.

Mononuclear reactions. Adsorption producing mononuclear surface complexes, accompanied by deprotonation or hydrolysis of AlOH and/or Co(II), is represented by the generic reaction:

$$2AlOH + Co(H_2O)_6^{2+} = [(AlOH)_{2-p}(AlO)_pCo(OH)_q(H_2O)_{4-q}]^{6-p-q} + (p+q)H^+$$

Reactions were chosen from this set by trial and error fitting to the 0.1 mM ΣCo uptake isotherm. Reactions 203 and 204 (**Table I**) proved necessary and adequate, while 202 produced unacceptably low uptake curve slopes. The solution is not unique. Other combinations of reactions and equilibrium constants (K) can fit the 0.1 mM data equally well, but K-values significantly different from the selected range degrade the fit at higher ΣCo.

Table I. SCM/TLM Sorption Reactions and Equilibrium Constants

	ID[a]	*Log K*	
Model Number →		*M1*	*M9*
Surface site protonation/deprotonation: Protons located in the surface at x = 0			
$AlOH + H^+ = AlOH_2^+$	102	5.2	5.2
$AlOH = AlO^- + H^+$	100	-11.8	-11.8
Background electrolyte adsorption: Adion bonding outer-sphere, located at x = β			
$AlOH + H^+ + NO_3^- = AlOH_2–NO_3$	106	7.9	7.9
$AlOH + Na^+ = AlO–Na + H^+$	105	-9.2	-9.2
Co(II) adsorption: Co bonding assumed inner-sphere and bidentate			
Monomeric sorption complexes—Co and protons at x = 0			
$2AlOH + Co^{2+} = [(AlOH)_2Co]^{2+}$	202	5.7	--
$2AlOH + Co^{2+} + H_2O = [(AlOH)_2CoOH]^+ + H^+$	203	-0.65	-0.6
$2AlOH + Co^{2+} + 2H_2O = [(AlOH)_2Co(OH)_2]^0 + 2H^+$	204	-9.8	-11.0
Hydroxide and AlO^- bridged dimer—all at x = 0			
$3AlOH + 2Co^{2+} + H_2O = [(AlOH)_2AlOCo_2OH]^{2+} + 2H^+$	221	--	0.3
$Co(OH)_2$ fragment nonamer, 3Co at x = 0			
$8AlOH + 9Co^{2+} + 18H_2O = [(AlOH)_4(AlO)_4Co_9(OH)_{14}]^0 + 18H^+$	902	--	-86.5
Precipitation			
$Co^{2+} + 2H_2O = Co(OH)_2\ (ppt) + 2H^+$	--	-13.2	-13.2

[a]ID = Model reaction identification numbers

Site saturation and effective surface area. Abrupt changes in the slope of the 0.012 M Co uptake curve suggest that sorption here involves two processes. In other systems, the first stage has been interpreted as evidence of saturation of a subset of sorption sites in a heterogeneous surface (**45**) Alternatively, all sites may saturate completely, followed by a different, site independent process, perhaps precipitation nucleated on the surface but not constrained to two dimensional growth. We chose to retain the original site density, $N_S = 8$, and to mimic saturation of a subset of sites by reducing the total amount of surface area used in the simulation enough to produce the apparent sorption saturation suggested by the 12 mM Co uptake data.

Precipitation and multinuclear complexes. Both precipitation and multinuclear sorption reactions probably contribute to uptake at the highest Co concentrations and pH investigated. If we interpret the second, highest pH stage of the 12 mM isotherm as precipitation, then the first, lower pH branch of the isotherm cannot represent precipitation of the same compound. Similarly, no part of the 1.5 mM isotherm can represent precipitation because, since the total Co concentration is lower throughout the pH range, solutions in this series would all be subsaturated relative to the precipitate formed at the higher ΣCo. Since the 1.5 mM curve is much steeper than the 0.1 mM curve, yet apparently does not represent precipitation—and XAFS results require $N_{Co}>0$—sorption in this range must produce a mixture of mononuclear complexes and a precipitate, or multinuclear surface complexes. Since N_{Co} never exceeds 4.5, while $N_{Co}= 6$ for $Co(OH)_2$, we have at least two alternatives: either the precipitate may be a hydrotalcite-like phase (an "HTC") with $N_{Co}\leq 4.5$ or mono- and/or low-N_{Co} multinuclear species coexist with $Co(OH)_2$ in Co sorbed at high pH at 12 mM such that the average $N_{Co} \leq 4.5$.

We chose the second alternative for two reasons. XAFS spectra of sorption samples in this range are similar to that of homogeneously precipitated $Co(OH)_2$.. Furthermore, early trials showed that $Co(OH)_2$ and an HTC were equally successful in simulating uptake, $Co(OH)_2$ is the simpler and better known choice. Accordingly, if precipitation dominates uptake in the extreme high uptake, high pH range, the second step in our fitting strategy was to allow precipitation of $Co(OH)_2$, adjusting the solubility product ($^*K_{s0}$, (**46**)) to fit the 12 mM total-Co uptake data for pH>7.5. $Co(OH)_2$ precipitation accounts well for the steep slope in this range if log $^*K_{s0}$ = 13.2, close to the solubility product of "active pink" $Co(OH)_2$.

Model M1: monomeric complexes and precipitation alone. Our first cobalt sorption model ("M1" in **Table I**) was constructed to test the necessity of multinuclear complexes, so assumed that mononuclear complexes and precipitation of $Co(OH)_2$ alone contribute to sorption. Equilibrium constants were optimized by trial and error. The best fit achieved with this model underestimates uptake for 7.5<pH<8 and N_{Co} over much of the Γ-range of interest. Uptake cannot be explained on the basis of mononuclear complexes and precipitation alone.

Multinuclear sorption: By analogy with the hydrolysis and polymerization of Co(II) in solution, Katz and Hayes (**4**) suggested multimeric surface complexes related to the Co_2OH^{3+} and $Co_4(OH)_4^{4+}$ inferred from potentiometric titration (**46**). A dimer alone, with $N_{Co} = 1$, is too small to account for the observed N_{Co} over any significant pH/concentration range. If the tetramer has a structure similar to that proposed for $Pb_4(OH)_4^{4+}$, as has been proposed (**46**), then $N_{Co} = 3$, and a mixture of monomers, the tetramer, and a hydroxide precipitate might explain the XAFS results. This structure, however, brings Co atoms closer together than the XAFS-derived Co-Co distances, and we consider it unlikely. The "hydroxide-like" structure indicated by XAFS suggests complexes built by bonding $Co(OH)_2$ units onto a sorbed Co through hydroxide bridges. By trial and error, a 3 by 3 "nonameric" complex, essentially a fragment of $Co(OH)_2$, bonded in inner-sphere mode along one edge only, *i.e.*, $(AlOH)_4Co_9(OH)_{18}{\cdot}4AlOH_{masked}$ was selected. This nonamer reaction is listed as 902 **Table I** with a trial and error fit equilibrium constant. Early attempts with models incorporating the nonamer as well as monomers and precipitation were reasonably successful in simulating uptake data, but produced values of N_{Co} lower than observed in some ranges of sorption density. To improve simulation of N_{Co}, a dimer (reaction 221) was added as an intermediate between monomers and the nonamer—and all equilibrium constants were refit.

Model M9: Mono- and multinuclear complexes and precipitation. The best model yet achieved comprises reactions 203, 204, 221, 902, and precipitation. Simulations of uptake data and N_{Co} are shown in **Figures 5A,B, 5C, and 6.** Model M9 fits both uptake and N_{Co} data well. The fitting exercise reinforces our inferences and the conclusion offered by Hayes and Katz (**10**), namely that uptake data alone are inadequate to derive a robust sorption model. Spectroscopic information is needed to support selection of reactions. Future models must account for the mixed-cation hydroxide precipitates observed by Scheidegger *et al.* (**36**), Towle *et al.* (**22**), and Thompson *et al.* (**35**), but we found it unnecessary to invoke a co-precipitate or variable activity precipitate. Obviously, M9 is not a unique solution. A continuum of multimeric species seems more likely than a single species. More data are needed to test for protonation/deprotonation of all proposed surface complexes and precipitates, and sorption by the precipitates themselves.

Significance. Reactive chemical transport models used to simulate behavior of contaminants in ground water often use a partition coefficient, K_d, to represent sorption processes. The K_d may be assumed constant or it may be computed with an equilibrium or rate controlled subroutine. For the Co(II)/Al_2O_3 sorption system, $K_d = (\Sigma Co_{ads}/\Sigma Co_{aq})$. **Figure 7** shows the variation of K_d with initial total cobalt concentration as predicted by the M9 model and two simpler models. The figure demonstrates that K_d is not a constant but varies by at least two orders of magnitude in a concentration range that might be observed between the near and far-field environments of a breached hazardous waste site. It also shows that a simple, monomeric model may account for sorption if order-of-magnitude accuracy is adequate, but, in contrast, that neglect of monomeric species is entirely inadequate at low concentrations. Accurate sorption models used to predict K_d must be robust in the sense that they represent the chemical processes actually responsible for partitioning. Neglect of a single reaction may lead to large errors. The challenge is to identify all significant reactions involved in partitioning, then to develop a scheme for reliable simplification of models to minimize parametric information required and conserve computation time.

Conclusions

(1) X-ray absorption fine structure spectroscopy has become one of the key molecular-scale methods in studies of reactions mechanisms of aqueous cations and anions at environmental interfaces.

(2) Different types of oxide and oxyhydroxide surfaces can have a large effect on the types of surface complexes formed at a given surface coverage, due in part to surface structure and bonding.

(3) Specific types of surface sites to which metal ions bind can be characterized using grazing-incidence XAFS.

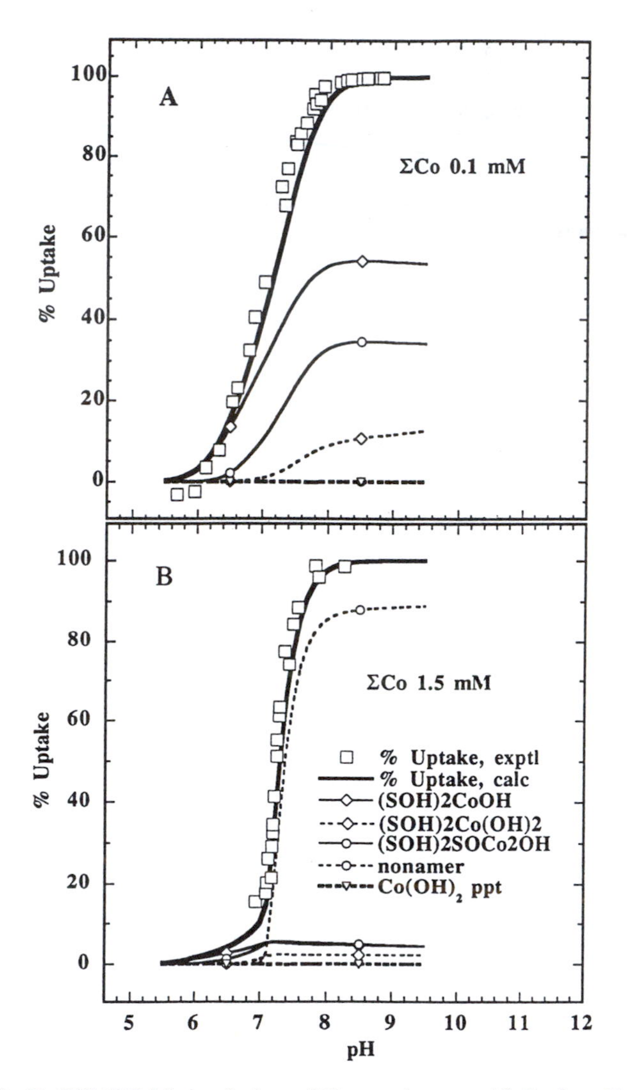

Figure 5 A, B. SCM/TLM simulation of Co uptake on γ-Al_2O_3 for ΣCo = 0.1 and 1.5 mM in 0.1 M $NaNO_3$, using model M9, including mononuclear, dimeric, one nonameric sorption complex, and precipitation. Experimental uptake, simulated uptake, and contributions of individual surface complexes are shown. The legend in 5B applies for figures 5, A, B, and C. In formulas, SOH represents AlOH.

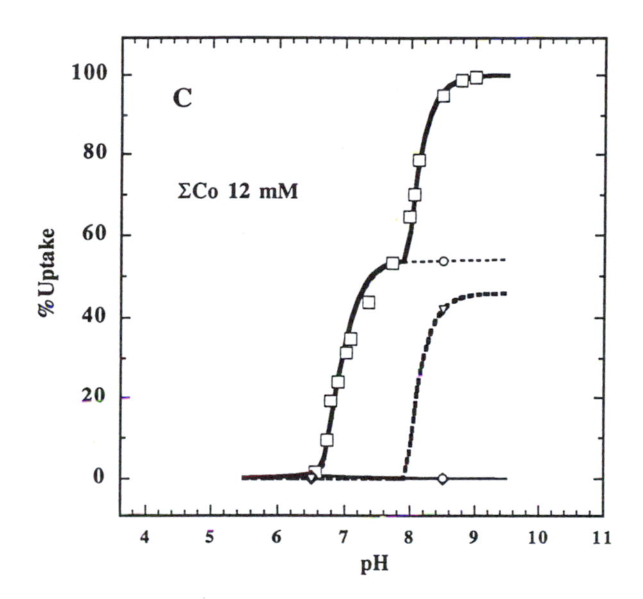

Figure 5C. SCM/TLM simulation of Co uptake on γ-Al_2O_3 for ΣCo= 12mM in 0.1 M $NaNO_3$, using model M9. The legend in 5B applies for 5C

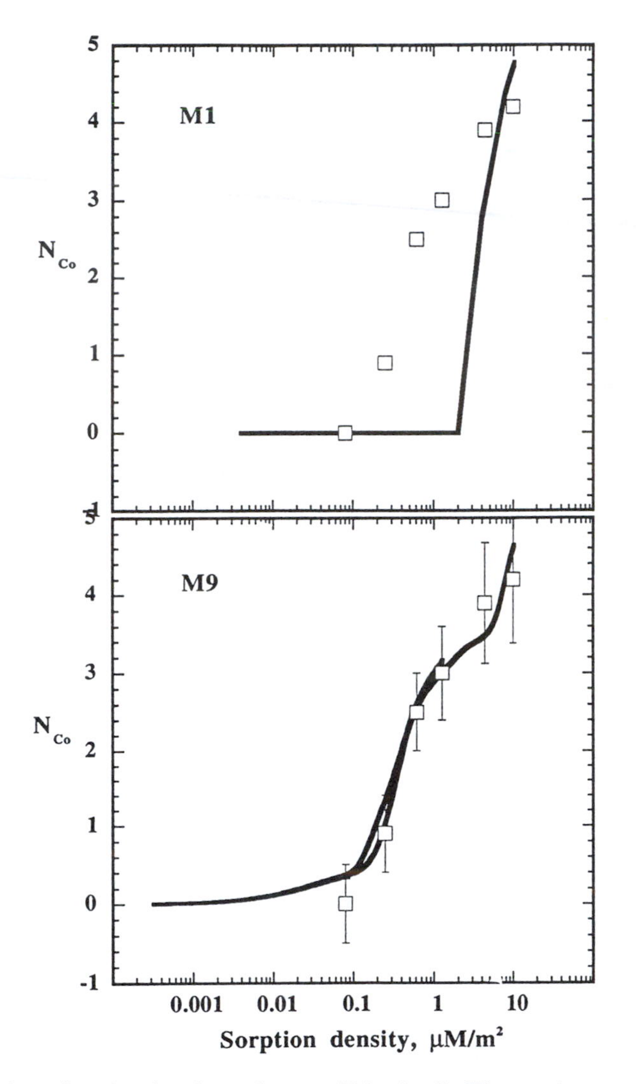

Figure 6. Sorption density dependence of N_{Co} in Co(II) sorption complexes on γ-Al_2O_3 as predicted by models M1 and M9 with HYDRAQL. M9 includes monomeric and multimeric sorption complexes and precipitation. M1 neglects multimeric complexes.

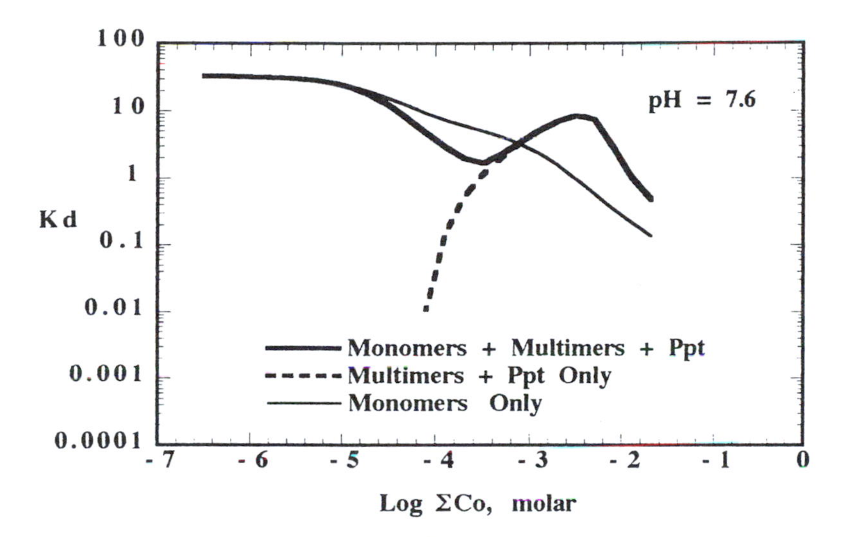

Figure 7. Partition coefficient, K_d, for Co(II) on γ-Al_2O_3 computed as a function of ΣCo for pH 7.6. Values of K_d computed with the complete model M9 are compared with values computed neglecting monomeric sorption species or multimers and precipitates.

(4) Different surfaces of the same sorbent can cause different types of sorption (outer- vs. inner-sphere) and sorption complexes.

(5) Surface complexation modeling reactions can be significantly constrained using XAFS-derived information on surface reaction products.

(6) Ignorance or neglect of individual sorption reactions or variables may lead to order-of-magnitude error in calculated aqueous solution-solid phase partition coefficients.

Acknowledgments

We thank D. L. Sparks for inviting us to participate in the ACS Symposium that resulted in this special volume and for his patience in waiting for our manuscript, and Samuel Traina for thoughtful review of the manuscript. We also thank the Department of Energy (OBES) for supporting this work through grant DE-FG03-93ER14347-A006. The synchrotron work reported in this paper was carried out at the Stanford Synchrotron Radiation Laboratory which is supported by the Office of Basic Energy Sciences of the Department of Energy, the DOE Office of Biological and Environmental Research, and the National Institutes of Health. We also gratefully acknowledge the support of the SSRL staff, particularly Dr. Britt Hedman.

Literature Cited

(1) Davis, J.A. and Kent, D.B. In: *Mineral-Water Interface Geochemistry,* Hochella, M.F., Jr. and White, A.F., Eds; Reviews in Mineralogy; Mineralogical Society of America: Washington, DC, 1990, Vol. 23; pp. 177-260. (and references therein).
(2) Schindler, P.W. and Stumm, W. In *Aquatic Surface Chemistry*, Stumm, W., Ed.; Wiley-Interscience: New York, 1987; pp. 83-110.
(3) Hayes, K.F. and Leckie, J.O. *J. Coll. Interf.Sci* **1987**, *115*, pp. 564-572.
(4) Katz, L.E. and Hayes, K.F. *J. Coll. Interf. Sci.* **1994**, *170*, pp. 477-490.
(5) Katz, L.E. and Hayes, K.F. *J. Coll. Interf. Sci.* **1994**, *170*, pp. 491-501.
(6) Brown, G.E., Jr., Parks, G.A., and Chisholm-Brause, C.J. *Chimia* **1989**, *43*, pp. 248-256.
(7) Brown, G.E., Jr. In: *Mineral-Water Interface Geochemistry*; Hochella, M.F., Jr. and White, A.F., Eds.; Reviews in Mineralogy; Mineralogical Society of America: Washington, DC, 1990, Vol. 23; pp. 309-363.
(8) Manceau, A., Charlet, L., Boisset, M.C., Didier, B., and Spadini, L. *Applied Clay Sci.* **1992**, *7*, pp. 201-223.
(9) Brown, G.E., Jr., Parks, G.A., and O'Day, P.A. In *Mineral Surfaces*, Vaughan, D.J. and Pattrick, R.A.D., Eds.; Chapman & Hall: London, 1995; pp. 129-183.
(10) Hayes, K.F. and Katz, L.E. In *Physics and Chemistry of Mineral Surfaces*; Brady, P.V., Ed.; CRC Series in Chemistry and Physics of Surfaces and Interfaces; CRC Press: Boca Raton, Florida, 1996; pp. 147-223.
(11) Bargar, J.R., Brown, G.E., Jr., and Parks, G.A. *Geochim. Cosmochim. Acta* **1997**, *61*, pp. 2617-2637.
(12) Bargar, J.R., Brown, G.E., Jr., and Parks, G.A. *Geochim. Cosmochim. Acta* **1997**, *61*, pp. 2639-2652.

(13) Scheidegger, A.M., Lamble, G.M., and Sparks, D.L. *Environ. Sci. Tech.* **1996**, *30*, 548-554.
(14) Scheidegger, A.M. and Sparks, D.L. *Soil Sci.* **1996**, *161*, pp. 813-831.
(15) Pickering, I.J., Brown, G.E., Jr., and Tokunaga, T.K. *Environ. Sci. Tech.* **1995**, *29*, pp. 2456-2459.
(16) Manceau, A., Boisset, M.C., Sarbet, G., Hazemann, J., Mench, M., Cambier, P., and Prost, R. *Environ. Sci. Tech.* **1996**, *30*, pp. 1540-1552.
(17) Peterson, M.L., Brown, G.E., Jr., Parks, G.A., and Stein, C.L. *Geochim. Cosmochim. Acta* **1997**, *61*, pp. 3399-3412.
(18) Foster, A.L., Brown, G.E., Jr., Tingle, T.N., and Parks, G.A. *Amer. Mineral.* **1998** (in press).
(19) Bargar, J.R., Towle, S.N., Brown, G.E., Jr., and Parks, G.A. *J. Coll. Interf. Sci.* **1997**, *85*, pp. 473-493.
(20) Chisholm-Brause, C.J., Brown, G.E., Jr., and Parks, G.A. In: *XAFS VI, Sixth Internat. Conf. on X-ray Absorption Fine Structure*, Hasnain, S.S., Ed.; Ellis Horwood Ltd, London, 1991; pp. 263-265.
(21) O'Day, P.A., Chisholm-Brause, C.J., Towle, S.N., Parks, G.A., and Brown, G.E., Jr. *Geochim. Cosmochim. Acta.* **1996**, *60*, pp. 2515-2532.
(22) Towle, S.N., Bargar, J.R., Brown, G.E., Jr., and Parks, G.A. *J. Coll. Interf. Sci.* **1997**, *187*, pp. 62-82.
(23) Towle, S.N., Bargar, J.R., Brown, G.E., Jr., Parks, G.A., and Barbee, T.W., Jr. In *Structure and Properties of Interfaces in Ceramics*, Bonnell, D.A., Chowdhry, U., and Rühle, M., Eds.; Materials Research Society Symposium Proceedings, Materials Research Society: Pittsburgh, PA, 1995, Vol. 357; pp. 23-28.
(24) Bargar, J.R., Towle, S.N., Brown, G.E., Jr., and Parks, G.A. *Geochim. Cosmochim. Acta* **1996**, *60*, pp.3541-3547.
(25) Towle, S.N., Brown, G.E., Jr., and Parks, G.A. *J. Coll. Interf. Sci.* (submitted).
(26) Towle, S.N., Brown, G.E., Jr., and Parks, G.A. *J. Coll. Interf. Sci.* (submitted).
(27) Yan, B., Meilink, S.L., Warren, G.W., and Wunblatt, P. IEEE Trans. Compon., Hybrids, Manuf. Technol. 1987, *CHMT-10*, pp. 247-256.
(28) O'Day, P.A., Rehr, J.J., Zabinsky, S.I., and Brown, G.E., Jr. *J. Amer. Chem. Soc.* **1994**, *116*, pp. 2938-2949.
(29) Scheidegger, A.M., Lamble, G.M., and Sparks, D.L. *Environ. Sci. Tech.* **1996**, *30*, pp. 548-554.
(30) Parks, G.A. *Chem. Rev.* **1965**, *65*, pp. 177-198.
(31) Sverjensky, D.A. *Geochim. Cosmochim. Acta* **1994**, *58*, pp. 3123-3129.
(32) Bargar, J.R., Brown, G.E., Jr., and Parks, G.A. *Geochim. Cosmochim. Acta* **1998** (in press).
(33) O'Day, P.A., Brown, G.E., Jr., and Parks, G.A. *J. Colloid Interface Sci.* **1994**, *165*, 269-289.
(34) d'Espinose de la Caillerie, J.-B., Kermarec, M., and Clause, O. *J. Amer. Chem. Soc.* **1995**, *117*, pp. 11471-11481.
(35) Thompson, H.A., Parks, G.A., and Brown, G.E., Jr. *Geochim. Cosmochim. Acta* (submitted).
(36) Scheidegger, A.M., Lamble, G.M., and Sparks, D.L. *J. Coll. Interf. Sci.* **1997**, *186*, pp. 118-128.

(37) Smith, R.M. and Martell, A.E. Critical Stability Constants , Vol. 4, Inorganic Complexes, Okenum Press, 1976.
(38) Chisholm-Brause, C. J. *Spectroscopic and Equilibrium Study of Cobalt(II) Sorption Complexes at Oxide/Water Interfaces*; Stanford University: Stanford, California, **1991**, pp 118.
(39) Papelis, C.; Hayes, K. F.; Leckie, J. O. "*HYDRAQL: A Program for the Computation of Chemical Equilibrium Composition of Aqueous Batch Systems Including Surface-Complexation Modeling of Ion Adsorption at the Oxide/Solution Interface*," Stanford University, Dept. Civil Engineering, Tech. Rept. 306, 1988.
(40) Yeh, G.-T.; Tripathi, V. S. *Water Resources Res.* **1991**, *27*, 3075-3094.
(41) Walter, A. L.; Frind, E. O.; Blowes, D. W.; Ptacek, C. J.; Molson, J. W. *Water Resources Res.* **1994**, *30*, 3137-3148.
(42) James, R. O.; Parks, G. A. In *Surface and Colloid Science*; Matijevic, E., Ed.; Plenum Press; New York, 1982; Vol. 12, pp 119-216.
(43) Huang, C. P. *The Chemistry of the Aluminum Oxide-Electrolyte Interface*; Harvard University: Cambridge, MA, 1971.
(44) Huang, C. P.; Stumm, W. *J. Colloid Interface Sci.* **1973**, *43*, 409-420.
(45) Schindler, P. W.; Liechti, P.; Westall, J. C. *Neth. J. Agric. Sci.* **1987**, *35*, 219-230.
(46) Baes, C. F., Jr.; Mesmer, R. E. *The Hydrolysis of Cations*; John Wiley and Sons: New York, 1986, 489 pp.

Chapter 3

New Directions in Mineral Surface Geochemical Research Using Scanning Probe Microscopes

M. F. Hochella, Jr., J. F. Rakovan[1], K. M. Rosso, B. R. Bickmore, and E. Rufe

Department of Geological Sciences, Virginia Polytechnic Institute and State University, Blacksburg, VA 24061–0420

Long-term studies in new areas of geochemical research that would not be possible without scanning probe microscopes are now well underway. Several examples are given in this chapter, including atomic force microscopy-based research that shows 1) that the rate of aqueous dissolution at specific reactive sites is slower in the presence of a thin water film relative to bulk water, 2) that clay-size phlogopite can rapidly delaminate and/or re-combine while establishing an equilibrium thickness, and 3) that aqueous Mn(II) sorption mechanisms vary with each mineral tested. The last example in this chapter involves ultra-high vacuum scanning tunneling microscopy. Using this technique, we show that low energy helium ion sputter cleaning and low temperature annealing of natural pyrite growth surfaces does not affect the original surface microtopography. This is important when using these natural growth surfaces for laboratory-based oxidation studies.

Scanning probe microscopes (SPM's) have been utilized in mineralogical and geochemical studies since 1989 (*1*). These microscopes have done for mineralogy and geochemistry what they have done for all other areas of science in which they have been used; they provide dramatic and direct views (images) of surface structure and microtopography, as well as reactions occurring at solid/fluid interfaces, often in real-time, and at unprecedented scales. Perhaps more than anything else, they have allowed us to observe, appreciate, and characterize the *local* nature of heterogeneous processes.

The original SPM, introduced in 1982 (*2,3*), was the scanning tunneling microscope (STM) built for use in ultra-high vacuum (UHV). The design was

[1]Current address: Department of Geology, Miami University, Oxford, OH 45056.

ingenious and unique at that time, but the development and refinement of SPM's since then have continued to come at a staggering pace. We estimate that there are a few dozen types of SPM's in widespread use today, although they are hard to count because many share very similar components and functions. Nevertheless, by far the most commonly used SPM today in geochemistry, and all other major fields where SPM's are used, is the atomic force microscope (AFM, also know as the scanning force microscope or SFM). This is due to a combination of factors. First, AFM has remarkable versatility, in that it can be used in air, solution, or vacuum, it can image conducting or non-conducting surfaces, and it can image rigid or soft surfaces including living cells. Second, it is relatively easy to use and relatively inexpensive to purchase. Finally, it is fairly straightforward to interpret AFM images as long as critical phenomena such as tip-surface shape convolution are properly accounted for. Yet it is the original SPM, the STM, that can measure the local (atomic scale) electronic structure of conducting and semi-conducting surfaces. This is why Gerd Binnig and Henrich Rohrer, the inventors of the STM, won the Nobel Prize in physics in 1986. This measurement is the single most remarkable achievement of the first and all subsequent SPM's, a dramatic breakthrough in science. Although not nearly as versatile as the AFM, it is this capability that assures the STM's long-term use and importance, even in the geological sciences. Fortunately, there are a few semi-conducting minerals such as hematite, goethite, and pyrite that have enormous relevance to geochemical and environmental research.

When SPM research in the earth sciences first started, the objective of most studies was to explore the capabilities of AFM and STM on mineral surfaces. It was important to determine what mineralogical or geochemical insight might be achievable with these techniques that were not obtainable in other ways. As both AFM and STM continue to mature, this exploration is still important. Today, however, most SPM research in the earth sciences is directed towards a specific research objective which requires the unique capabilities of SPM's. The purpose of this paper is not to review these capabilities or objectives, but to briefly describe some of the new directions in SPM research in geochemistry today. In each case presented here, the SPM is allowing new research areas to open up that would not otherwise be possible. These directions include the following: 1) using the AFM to quantitatively measure dissolution rates of individual features on mineral surfaces in the presence of bulk water and thin films of water, 2) using the AFM to measure the thickness of clay particles reacting with solutions in real-time, 3) using the AFM to observe and characterize specific sorption processes on mineral surfaces, and 4) using the STM in UHV for research relevant to geochemistry. We believe that each of these may develop into subfields of their own, although admittedly at the present time they are each in their infancy.

The Application of AFM to the Study of Dissolution Rates and Mechanisms

The study of chemical weathering, and more specifically determining mineral dissolution rates and mechanisms, has been a topic of great interest to geochemists for over 100 years. An excellent book which covers the field has been published recently (*4*).

Background. AFM may be used to 1) quantitatively measure the dissolution rate of specific features on mineral surfaces, and 2) to do this in the presence of both bulk water *and* confined thin films of water. The justification of the former is that generally, dissolution rates are measured by observing changes in solution composition with time, but they can also be studied by looking at the other side of the interface. The dissolution rate is the sum of many simultaneous reactions proceeding over the reacting surface, which on mineral grains is very complicated. So another way to understand dissolution mechanisms, and ultimately rates, is to directly observe the dissolution of individual features on a mineral surface with a local imaging probe, in this case AFM, and to measure the reaction rate at these features by observing the change in them with time. This may involve the observation of the size and/or shape of features, or their position in the case of steps. The justification for making observations of this type both in the presence of bulk water and confined thin films of water has been explored in detail elsewhere (*5*). Briefly, a significant portion of the silicate dissolution that occurs in natural soils and rocks happens in the presence of thin films of water, for example within microcracks and micropores, and along grain boundaries. Here, the properties of aqueous solutions (e.g. atomic structure, solvent properties, viscosity, dielectric constant) can be quite different than they are in bulk solution, and the dominant transport mechanism shifts from fluid flow (uni-directional and relatively fast) to diffusion (multi-directional and relatively slow).

Technique. We have devised a scheme which allows us to accomplish the two tasks described above in a relatively simple way (*6*). This scheme involves using sheet silicates because their cleavage surfaces are typically atomically flat over large areas. Perfect flatness is necessary in order to achieve a consistent thin film of water, down to the nanometer level, trapped between two mineral grains. However, the basal plane of sheet silicates is relatively unreactive in aqueous solution despite recent publications to the contrary (*7,8*), so our scheme involves controlled activation of this surface so that it becomes relatively reactive, promoting dissolution, in circumneutral aqueous solutions. We achieve this by reacting the sheet silicate in HF to create shallow etch pits. This, in effect, generates steps that are much more susceptible to dissolution processes than atomically flat terraces. For this work, we have initially chosen phlogopite, etched in concentrated (49%) HF for one second at room temperature, and then rinsed well. This results in a surface that does not have modified surface chemistry (as measured by XPS), and has easily defined and relatively consistent shallow pitting, and as a result, surface step morphology (Figure 1). Then the pretreated surface can either be placed "as is" in a batch or flow-through reaction vessel for dissolution in the presence of bulk solution, or wetted and clamped together with an untreated sheet and then placed in the reaction vessel for dissolution in the presence of the confined thin film of solution. In both cases, the surfaces are characterized with AFM before and after reaction. To make this comparison exact, the same etch pits are found after the reaction as characterized before the reaction.

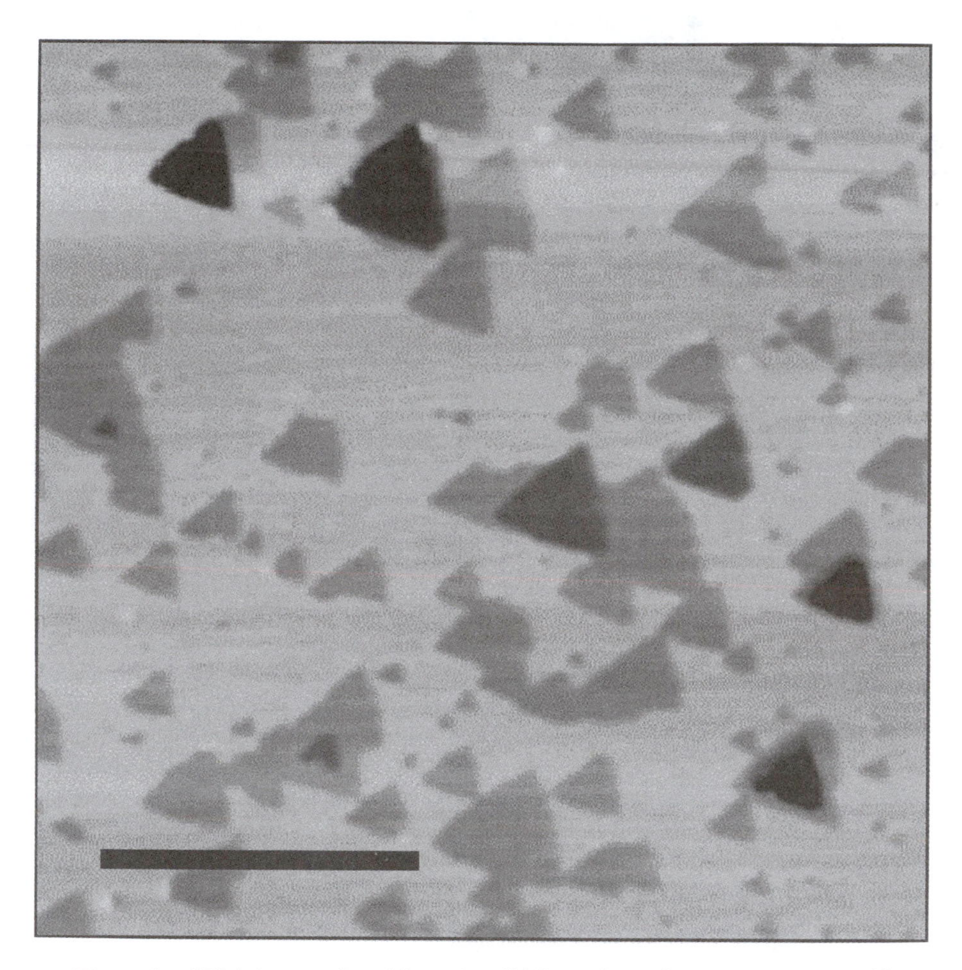

Figure 1. AFM image of a phlogopite (001) surface after pretreatment acid etch and rinse. The light gray triangular pits are 1 nm deep, equivalent to one unit cell along [001]. The black pits are deeper, typically between 2 and 20 nm. The scale bar is 1 μm long.

Results and Discussion. An example of one of the experiments just described is shown in Figure 2. A phlogopite (001) surface after a 1 second etching in HF is shown in the top portion of the figure. The image shows shallow triangular etch pits which are one unit cell (approximately 1 nm) deep, as well as a few deeper pits typically between several and 20 nm deep. The same area, after exposure to distilled water equilibrated with atmospheric CO_2 (pH 5.7) in a stirred batch reactor for 24 hours at room temperature, is shown in the lower portion. Note that the two images are rotated by approximately 30 degrees with respect to one another (it is very difficult to find the same area and achieve perfect alignment), but that these images are in fact taken from the same area. The 1 nm deep triangular-shaped pits are clearly enlarging after 24 hours. One can most easily visualize this by looking at the extent of their coalescence between the two images. We are currently refining methods that take the raw image data and quantify the amount of the top nanometer layer that is consumed between images (Rufe, E.; Hochella, M. F., Jr., manuscript in preparation). Here, we simply report the changes in the deeper triangular pits. During dissolution, these pits do not get deeper, but they do get larger laterally. We know that they do not increase in depth because we can identify and keep track of the uppermost sheet silicate layer as it slowly recedes, thus giving us a height reference. However, like the shallow pits, the deeper ones have increased laterally, and in this case by an average of 10%. When normalized to the surface area of the starting pit perimeter (the "reactive" surface area in this case), we calculate a dissolution rate for this feature of $5.1 \pm 2.1 \times 10^{-10}$ mol m^{-2} sec^{-1} based on Si release and the formula unit with (O,OH,F) = 24. When the experiment is repeated, except that the fluid above a pre-etched phlogopite surface is confined by clamping an untreated phlogopite sheet over it to generate a confined fluid film approximately 20-30 microns thick, the calculated dissolution rate is $1.1 \pm 0.6 \times 10^{-10}$ mol m^{-2} sec^{-1} based on Si release and (O,OH,F) = 24. As we discuss in length elsewhere (*5*), there are many competing effects in the confined space, some of which will accelerate and others of which will decelerate dissolution in the presence of a confined fluid relative to a bulk fluid. However, several of these factors only become important as the confined fluid approaches water monolayer thickness, a dimension not approached in these experiments. Nevertheless, the confined fluid in our experiments is stagnant, and therefore transport of components to and from the dissolving surface is subject to diffusion-limited transport. This alone may be enough to slow the dissolution rate considerably, and therefore the difference in measured dissolution rate in this study is not surprising. Differences in dissolution rates may be common in nature when comparing silicates dissolving in the presence of flowing solutions, or in a typical laboratory dissolution experiment, versus silicates dissolving in the presence of stagnant fluids found along internal weathering fronts within mineral grains, or intragranular surfaces in a vadose zone. Such differences may also contribute to the variation between dissolution rates measured in the field and in the lab.

It is also interesting to compare our measured dissolution rates with the dissolution rate of phlogopite measured conventionally by solution analysis under circumneutral pH conditions and room temperature. Lin and Clemency (*9*) measured this rate to be 3.8×10^{-13} mol m^{-2} sec^{-1}. This rate is normalized to the BET measured surface area of the starting mineral powder, and like our measurements is

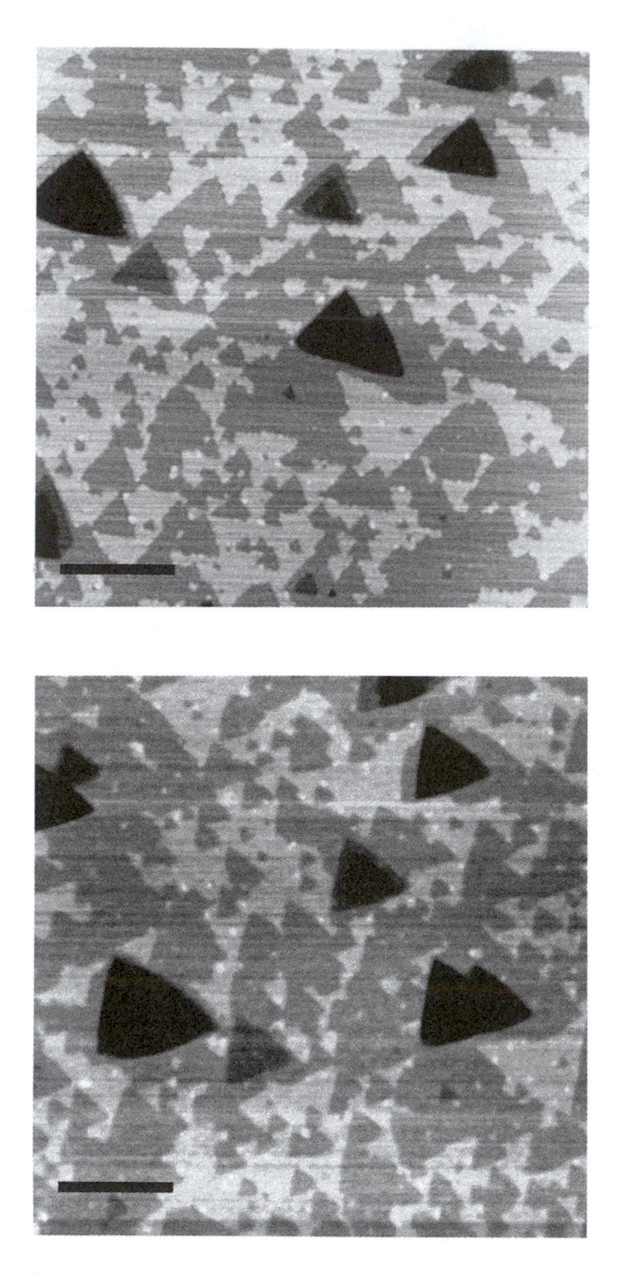

Figure 2. AFM images, approximately 2.5 μm on a side, of phlogopite (001) surfaces after pretreatment acid etch and rinse (top), and after pretreatment acid etch, rinse, and 24 hours in DI water equilibrated with air (bottom). Both images are taken from the same area of the same sample, although the bottom image is rotated about 30° counterclockwise from the top image. The scale bars in both images are 0.5 μm long. See text for additional details.

also based on Si release and a formula unit of (O,OH,F) = 24. The fact that our rates are a few orders of magnitude higher is not surprising, in that we are measuring the rate of a specific feature on the surface which constitutes a distinct portion of the total reactive surface area. Most of the surface area of these grains (the basal planes between steps) are relatively unreactive, but it is included in the overall measured bulk rate as reported by Lin and Clemency. The methods that we have used here have the potential of measuring the rates of all surface features, the area-weighted sum of which will equal the bulk dissolution rate.

Real Time Clay/Solution Interaction Studies Using AFM.

For both science and engineering applications, the AFM excels at measuring dimensions and dimensional changes at very small scales. In all of mineralogy and geochemistry, perhaps nowhere else is it more important to be able to observe dimensional change than for clay minerals. This section explores a new direction of research in this regard.

Background. We have become particularly interested in the weathering of micas in soils. This weathering process is geochemically and environmentally important because it is a major source of potassium in the soil system (and therefore of paramount importance to plants), and the weathering products typically include other sheet silicates with high cation exchange capacities (CEC). These secondary minerals play an important role in the mobility of toxic metals and nutrients in soils.

The first stage of mica weathering, for example phlogopite to vermiculite, is the exchange of interlayer potassium with a hydrated cation such as sodium. A number of groups have studied this important exchange process (see Norrish (*10*) and Nagy (*11*) for reviews and additional references). These studies have made it apparent that it is difficult to predict even this simple exchange behavior. It appears that the reaction must not only be considered from a chemical point of view, but also from a mechanical one (*12,13*). Progressive expansion of the interlayer space during exchange of potassium for a hydrated cation strains the particle. This strain should differ depending on the particle size and shape, and therefore particle dimensionality should play a role in the progress of the reaction. AFM therefore is one of the leading tools of the study of such reactions.

Atomic force microscopy has been used to image clays since its earliest use in the geological sciences (*14*). In addition, it has become commonplace to use AFM *in-situ*, that is in solution to observe mineral-solution interactions in real time. However, we are not aware of previous studies other than our own which image clay particles *in solution*. Why is imaging clays with AFM in solution important? Especially for the interlayer exchange reaction just discussed, observing the reactivity of individual particles in solution and in real time, allows one to draw direct conclusions about the role of dimensionality on the exchange process. Therefore, we set out to develop a technique which allows AFM observations of clay-size particles in solution. This is described next.

Technique. The methods and techniques used for AFM imaging in conjunction with a fluid cell have been reviewed by Dove and Chermak (*15*), and only a brief description will be given here. All commercial manufacturers of AFM's offer components which allow the cantilever with probe (i.e., the tip) to operate in a submersed state, although specific designs vary greatly. There are variables in how much fluid the cell can hold, whether the cell is opened or closed to the atmosphere, and whether the cell is of the static or flow-through type (the former providing no fluid movement, the latter providing fluid input and outflow). Other variables involve the different designs for visual access into the cell via an optical microscope, whether the tip scans across the sample or the sample is scanned across the tip, tip scanning capabilities (e.g. tapping), temperature control, sample mounting options, and the type of tips that are used (shapes, spring constants, resonant frequencies, etc.). Ideal set-up and/or configuration depends on specific application, but in every case, the AFM is readily adaptable to operation at the solid-fluid interface.

Developing ways in which individual clay particles can be affixed to a substrate for AFM imaging in the presence of water has been a challenge. When aqueous solutions or solvents are evaporated off a surface, there is always a thin (a few Angstroms to about a nanometer) residual layer that remains (*16*). This adventitious layer apparently acts as a kind of surface cement, in that even very fine fibers and clay-size particles can be imaged with contact mode AFM in air without displacing the sample. However, as soon as solution is added, the particles do not remain fixed and cannot be imaged with contact mode AFM, or even TappingMode™ AFM (TMAFM) which exerts very little lateral force on the imaged surface. Using cements or adhesives generally do not given satisfactory results. Although they may fix the particles, they may also either contaminate the solution or the tip. The tip commonly picks up minute amounts of cement, adversely affecting both its shape and its tracking (frictional) properties.

The most promising method that we have found which both holds clay particles during AFM imaging in solution *and* avoids the contamination problems works via the principle of surface electrostatic attraction (*17,18*). We use a polished sapphire single crystal as our substrate. This provides a flat, hard, and clean surface upon which to disperse the clay. Equally important, sapphire has a point of zero charge around pH 9. Therefore, in the circumneutral range in which we wish to conduct our experiments, the sapphire surface will be positively charged. Clays with interlayer cations, however, typically carry a net negative surface charge in the circumneutral pH range due to the loss of what were interlayer cations on their surfaces. Therefore, in solution, an electrostatic attraction should develop between sapphire and clay particles.

In a typical experiment, we prepare a dilute suspension of the clay in DI water, consisting of about 20 mL of water for every 1 mg of clay. We then sonicate the mixture so that the particles are well dispersed, and place a drop of it on a polished sapphire block that has been warmed to 90°C. After the drop has evaporated, one is ready to submerse the sapphire block into the fluid cell of the AFM.

To date, we have not been able to use this technique with conventional contact mode AFM. The lateral forces of the scanning tip apparently overcome the electrostatic attraction and the particles are dispersed. However, most particles remain fixed on the sapphire substrate when using TMAFM. This technique, where the tip oscillates at one of its resonant frequencies while it is rastered and tapped very lightly across a surface, minimizes lateral forces while maintaining excellent image quality.

Early Results and Comments. Figure 3 is a TMAFM image of a dispersed clay on a polished sapphire substrate in DI water. The particles are phlogopite from Ontario, Canada. Fine grains of phlogopite were obtained by grinding a sample on a fixed diamond wheel, from which the size fraction below two microns diameter was separated out using sedimentation techniques.

The height of individual particles in images like that shown in Figure 3 can be monitored in real time as exchangeable cations are added to the solution in the fluid cell. Early runs of this type have shown some surprises. Instead of just observing the particles increase in thickness as hydrated sodium (in these experiments) replaces interlayer potassium, we have observed particle delamination (where a particle on the substrate suddenly looses a significant portion of its thickness), and we have observed particle combination, where just the opposite happens. We have also observed slow particle expansion at a rate expected for interlayer cation exchange. The delamination process is not due to mechanical disturbance of the grains due to the tapping mode imaging process, as the same phlogopite grains are dimensionally stable when imaged in DI water. Clearly, particles are occasionally delaminating as part of the interlayer exchange process.

Taken together, we have observed that most particles stop increasing or decreasing in thickness when they reach a certain height, with an apparent dependence on particle width. This suggests that mechanical factors play a significant role in these exchange reactions, and also most likely in trioctahedral mica to vermiculite transformations.

Characterization of Sorption Processes Using AFM.

Sorption processes are the general category of heterogeneous reactions whereby free chemical species become associated in some manner with a surface. The field of sorptive geochemistry has mushroomed in the last several years largely because of the desire of scientists and engineers to study the mobility and fate of contaminants in the environment.

Background. A number of analytic breakthroughs have been made which have significantly advanced the field of sorptive geochemistry, most notably the development of X-ray absorption spectroscopic techniques (EXAFS, etc.) for the purpose of characterizing sorption reactions (e.g. *19*). In addition to all the spectroscopies that can now be used in studies of sorption mechanisms, scanning probe microscopes can be used to acquire previously unobtainable information on these processes. SPM's are the only high resolution microscopes ever developed that have the ability to image a surface while in contact with a fluid. This makes them ideally suited for sorptive geochemistry studies, especially when microprecipitation, which can be mistaken for adsorption reactions, is occurring. AFM, in particular, is

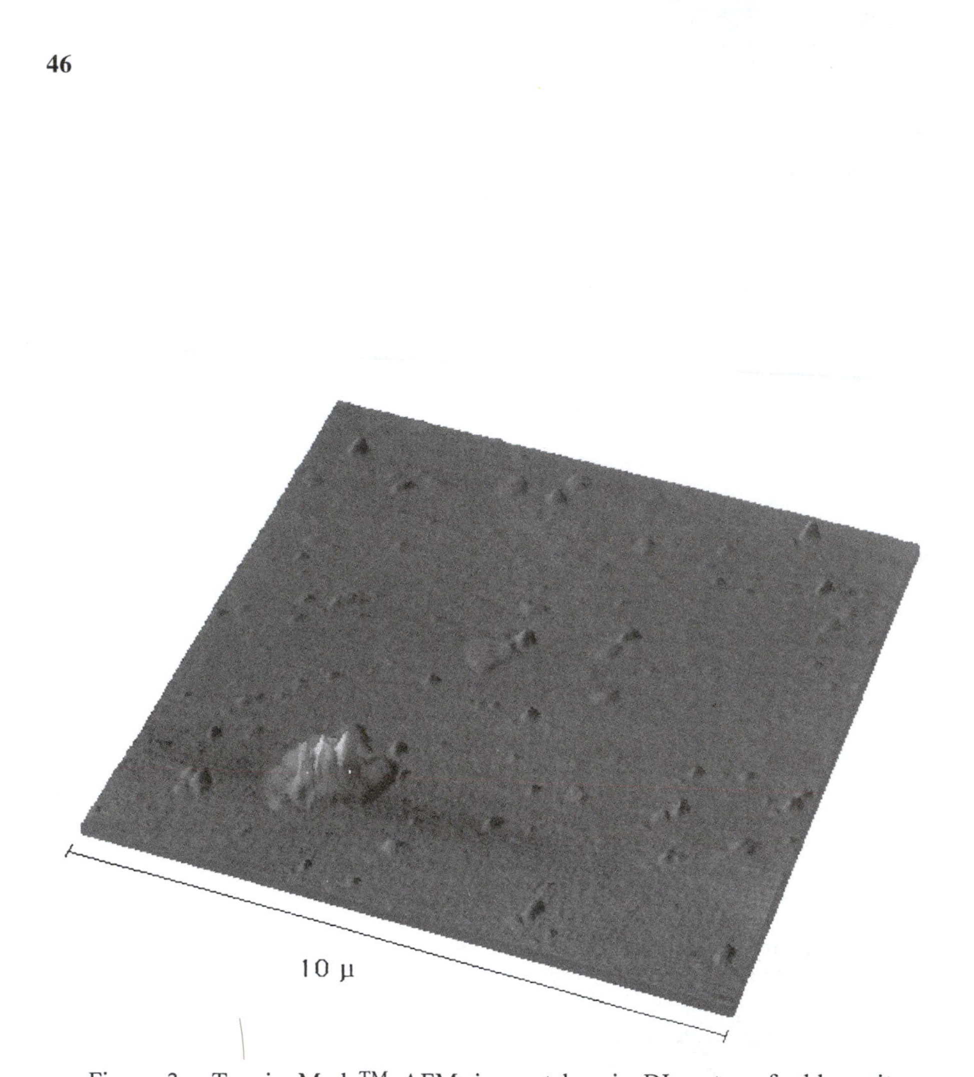

Figure 3. TappingMode™ AFM image taken in DI water of phlogopite particles (mostly submicron) on a sapphire substrate. The height of the small particles in the image are between 10 and 40 nm. The change in the particle heights, to within an Angstrom, are measured with time as the particles react with the solution.

appropriate for this type of study because of its ease of use in fluid cells, and its ability to image insulating mineral surfaces. However, at the present time, AFM remains relatively undeveloped and underutilized in this field of study. Below, we demonstrate how useful AFM can be in sorption research and discuss future directions.

Technique. AFM fluid cell research has already been briefly discussed above. In addition, we have already discussed in this chapter our electrostatic method of fixing clays so that they can be imaged in a fluid cell. For larger samples where individual pieces can be fixed as in the Mn sorption work described below, care must be taken to assure that the cement does not react with or contaminate the fluid. Both tip and solution contamination is a serious potential problem when using traditional cements such as epoxy or adhesive tape. These cements can cause artifacts that go unnoticed for what they are, or worse, interpreted for something that they are not. With samples that are big enough and when allowed by fluid cell/AFM design, mechanical non-reactive mounting devices, such as titanium clips, are always the safest choice.

Results and Comments. To date, our group has studied Mn sorption mechanisms on hematite, goethite, and albite mineral surfaces using AFM fluid cell techniques (*20-22;* Rakovan, J.; Hochella, M. F., Jr., manuscript in preparation). When aerated solutions having Mn(II) in the low ppm range are reacted with hematite and albite surfaces at pH's around 8, steps on both minerals are the most reactive site for the initial adsorption and oxidation of the Mn. Also, in both cases, Mn is oxidized to Mn(III), and micro-precipitation immediately begins to occur in the form of an oxyhydroxide (determined by XPS and XRD). But the development of the micro-precipitate is different on these surfaces. On hematite, the most reactive site of precipitate development is where the hematite, precipitate, and solution all meet, and as a result, the precipitate first grows in thin, nanometer thick films that eventually cover the surface. In the case of albite, the most reactive site for precipitate development is the precipitate itself, so that the precipitate grows up and away from the mineral surface, and does not grow over it. All of these observations were made in solution in real-time, and would not have been possible with any other technique.

A few of our latest AFM observations of similar reactions occurring on goethite surfaces are shown in Figure 4. The top image shows a freshly cleaved goethite (010) surface, the starting surface in these experiments. In the image, a 9.8 Å high step, the unit cell dimension along [010], is separated by atomically flat terraces. The bottom portion of Figure 4 shows an AFM fluid cell image of the same area after 5 minutes reaction with 5 ppm aqueous Mn(II) (added as a nitrate) at a pH of 7.8. The entire goethite surface is covered with islands of Mn-oxyhydroxide micro-precipitate averaging 4 to 5 nm in height. The reaction is much faster than that observed on hematite under similar conditions, and dramatically faster (and via different mechanism as discussed above) than that observed on albite.

The potential for AFM to significantly contribute further to sorptive geochemistry research is very high. There are many interesting and important systems to explore. Just as one example, Scheidegger and others (*23-25*) have performed some fascinating studies on Ni sorption on clays and aluminum oxides

a

Figure 4. (a) AFM contact mode image of a fracture-exposed (010) goethite surface taken in air. The atomically flat terraces are separated by a 9.8 Å high step. This is an example of a starting surface for an Mn sorption experiment. (b) AFM contact mode image taken in solution of the same area of the goethite surface shown above after a 5 minute exposure to 5 ppm aqueous Mn(II) solution adjusted to pH 7.8. The mounds average 4 to 5 nm in height. The scale bars in both images are 250 nm in length.

b

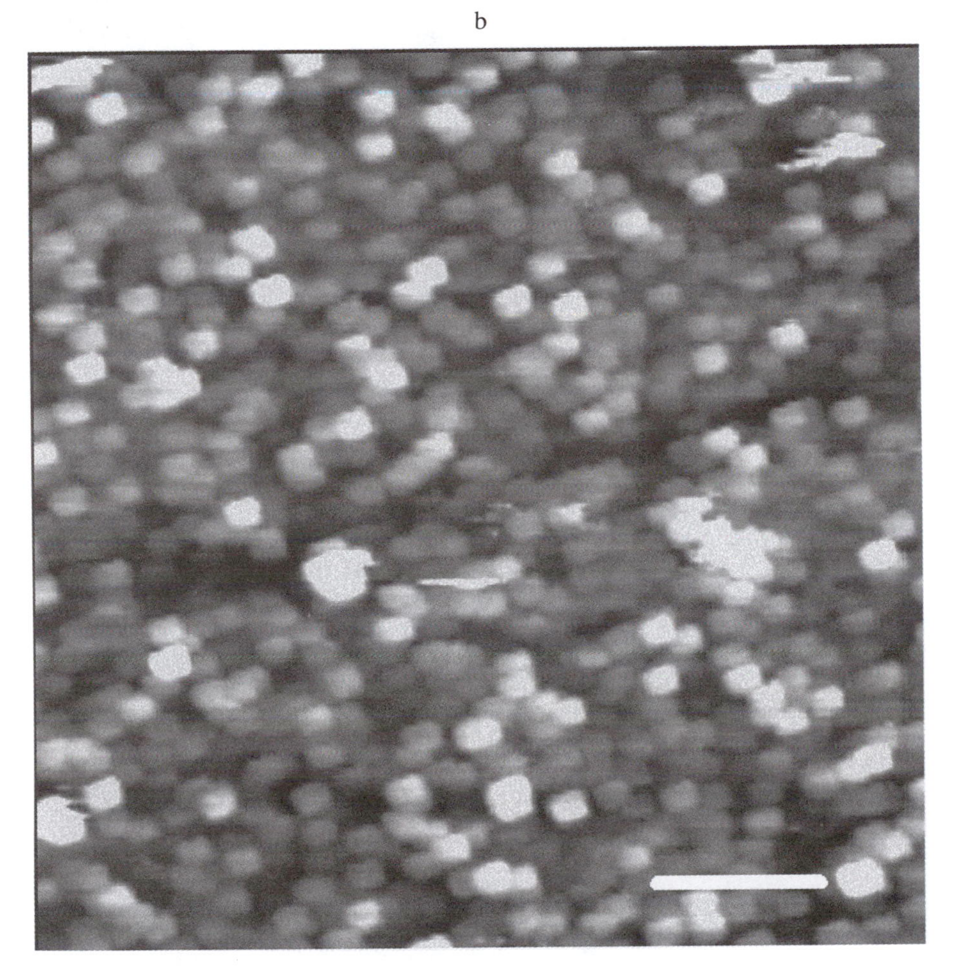

Figure 4. *Continued.*

using XAS techniques. In these studies, they have obtained spectroscopic evidence indicating that multinuclear metal complexes can form rapidly on mineral surfaces far below theoretical monolayer coverage and under solution conditions where precipitates are not expected to form. Such complexes, or micro-precipitates, may be visible with AFM. If so, the reactive site on the surface may be identifiable, as well as more information about the configurational nature of the multinuclear complexes.

Using UHV-based STM for Research Relevant to Geochemistry

As mentioned in the introduction of this chapter, the STM was first designed and built as a UHV device to be run typically in the pressure range of 10^{-9} and 10^{-12} torr. Soon thereafter, it was shown to operate successfully under non-vacuum conditions. Nevertheless, although ambient STM instruments are much less expensive and much easier to operate and maintain relative to UHV-based STM's, the latter has still been principally preferred by physicists and chemists. For geochemical applications, it would seem only appropriate to use STM in air or solution so that measurements are made on surfaces in "natural" conditions. This has been the case in most of the past STM research on mineral surfaces. However, there are compelling reasons to use UHV-based STM in geochemical applications for the same reasons that most physicists and chemist use it in UHV, and this represents a new direction in STM research in the earth sciences. We briefly explore these reasons and directions below.

Background. Here we list three general reasons why UHV-based STM is appropriate, and sometimes even preferred, in applications to geochemical research (*26*).

1) UHV provides for an ultra-clean environment in which the chemistry of a surface and the surface reactants, as well as surface reaction kinetics, are relatively controllable. Surfaces can be held in or reconditioned to a pristine state if needed, and the identity and purity of reactants can be assured. Reactions can be started and stopped (by controlling reactant concentrations in vacuum) to allow for the systematic study of reaction progress.

2) Spectroscopic measurements made with STM, i.e. scanning tunneling spectroscopy (STS), in non-vacuum conditions can easily be corrupted by unwanted adventitious material between the tip and the sample, or directly attached to the tip or the sample. STS in UHV conditions helps assure "clean" spectra, and is an important factor in helping to assure spectral reproducibility.

3) Atomically clean surfaces can be prepared in vacuum to allow for the fundamental characterization of intrinsic surface properties such as dangling bonds associated with coordinatively unsaturated surface atoms. Ultimately, both the local and long-range electronic structure of the surface can be investigated. It is this electronic structure which fundamentally controls the chemical reactivity of the surface, and of which we know relatively little.

In our group, the first mineral that we are studying extensively in vacuum with STM is pyrite (*27,28*). This seems a logical choice, in that it is a relatively narrow band gap semiconductor, making it easily accessible to STM/STS measurements, and it is relatively sensitive to air exposure. In addition, it is the most common sulfide in the crust, and it is also the single most influential mineral in acid mine drainage environments where acid is generated as a result of its oxidation. A number of excellent studies have addressed its oxidation behavior based on solution and isotopic chemistry (recent studies include *29-31*), scanning electron microscopy (*32*), x-ray photoelectron spectroscopy (e.g. *33-35*), and ambient STM (*36-39*).

Our new direction in pyrite UHV-based STM research includes working with growth surfaces as well as fracture surfaces. Although pyrite does not have any well defined cleavage directions, most STM work in the past has been done on surfaces that parallel {100} cubic growth surfaces as exposed by careful fracture. These surfaces, as shown in Figure 5, can be atomically flat over small areas, and commonly have steps whose height corresponds to the *a* cell edge (5.4 Å). We have obtained low energy electron diffraction (LEED) patterns of these surfaces which are consistent with {100} surface orientations. Nevertheless, surfaces of this type would rarely be observed under natural conditions. Much more common would be rounded faces on framboids, or crystallographically-controlled growth surfaces. Yet pyrite is a reactive mineral in air, and growth surfaces are not appropriate for studying the fundamental processes of pyrite oxidation reactions because they already have significant oxygen and water contamination, as well as other contaminants. Therefore, we use a technique that can clean a growth surface in vacuum without modifying the underlying pristine pyrite atomic structure, composition, or microtopography, all three of which control surface reactivity (*16*).

Technique. The principle of the technique that we use to "restore" pyrite growth surfaces in-vacuum is derived from the method published by Chaturvedi et al. (*40*). This method relies on ion bombardment to sputter the surface in vacuum to gently erode away the adventitious layers. To minimize differential sputtering and the disruption of the surface structure during bombardment (*41*), light ions (typically He^+) at low energy (a few hundred eV) are used. Even this exceptionally light bombardment will slightly disrupt the uppermost surface atomic structure, so in-vacuum annealing is also used. Temperatures are kept to a minimum, reaching no more than 300°C. The sputtering and annealing are performed in repeating cycles until XPS shows that the surface is clean (i.e. the surface has pyrite stoichiometry without adventitious contamination) and LEED shows that the surface is well-ordered with a pattern consistent with the termination of the bulk structure.

Early Results and Comments. Figure 6 shows a UHV STM image of a cleaned pyrite {100} growth face. This sample was cleaned using 10 cycles of He^+ ion bombardment at 200 eV (each bombardment lasting 10 minutes) and annealing to 300°C for 5 minutes in each cycle. The cleaning did not induce any microtopographic changes observable at this scale. After cleaning, LEED patterns of this surface showed those expected for the termination of the bulk structure, and XPS detected only Fe and S in the proportion expected for pyrite. This surface, therefore, should be representative of this mineral after growth was complete but before it was first exposed to an oxidizing environment. It is on such surfaces where oxidation

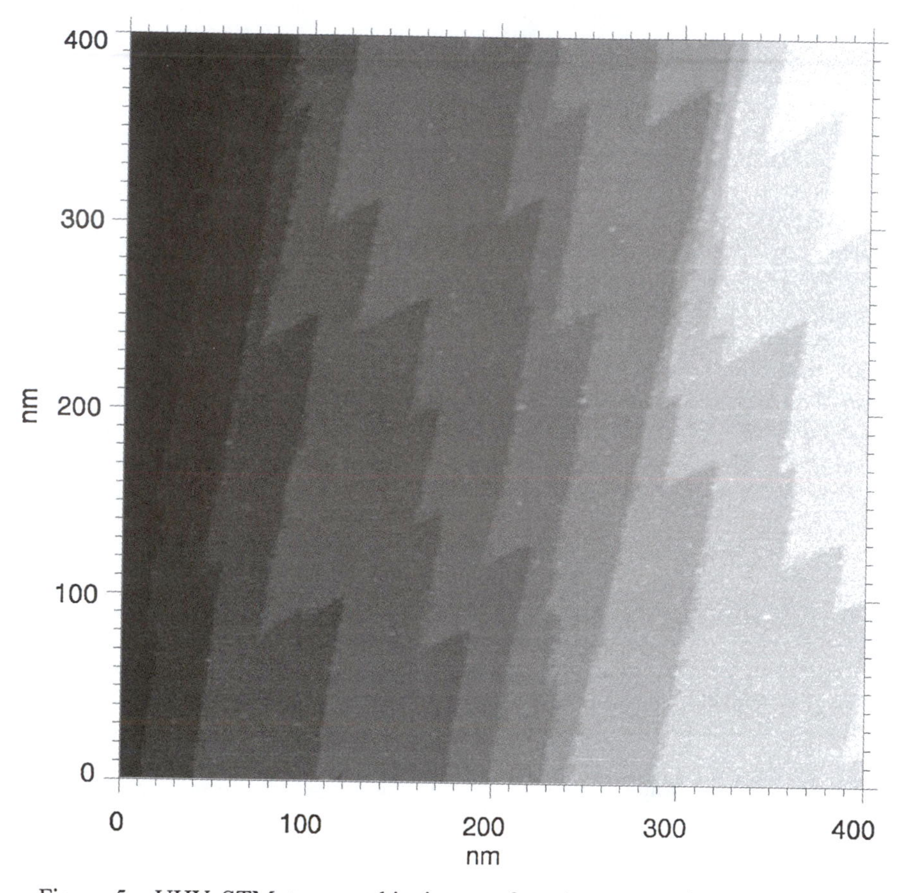

Figure 5. UHV STM topographic image of an in-vacuum fractured pyrite {100} surface. The steps are between 5 and 6 Å high. Tunneling conditions: +2 V bias and 1 nA tunneling current.

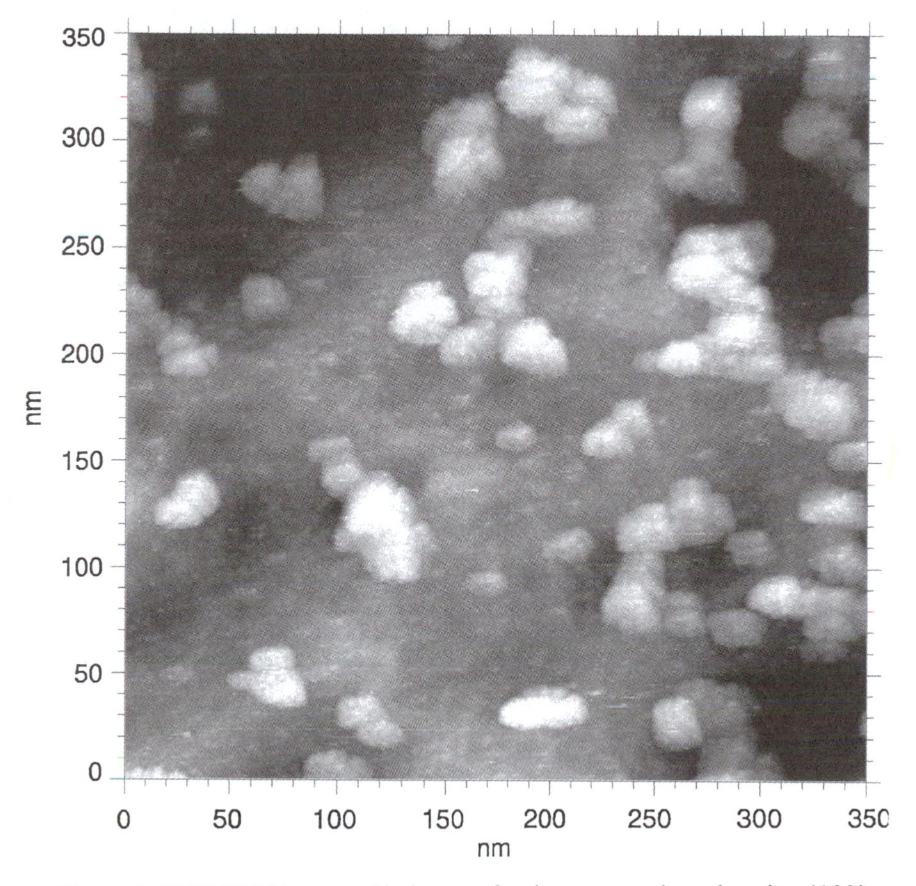

Figure 6. UHV STM topographic image of an in-vacuum cleaned pyrite {100} growth surface. The mounds on the surface are between 1.5 and 4 nm in height. Tunneling conditions: +1 V bias and 1 nA tunneling current.

would start in a natural environment, and therefore is of great interest to us in future studies.

Concluding Remarks

Like light and electron beam microscopies before them, scanning probe microscopies have become important techniques in geochemical research and they will continue to impact the field in the long term. As has been shown in this chapter, well established SPM's will find new uses within areas of geochemistry with long research histories as well as in areas that are just starting to be explored. In addition, new SPM's are being developed, and these may further support or help establish new areas of geochemical research. For example, two new SPM's are currently in the early to mid-development stages that may contribute to geochemical research. The scanning single-electron transistor microscope can map electric fields on surfaces with 100 nm resolution and detect surface charges as small as a fraction of a single electron charge (*42*). In addition, the magnetic resonance force microscope combines magnetic resonance imaging and AFM, a powerful combination which will find a considerable number of applications (*43*). If the past is any indication of the future, new SPM developments will continue at a brisk pace. New applications to geochemistry are essentially assured.

Acknowledgments

This work has been generously supported by the National Science Foundation (EAR-9527092 and EAR-9628023 to MFH, and a Graduate Student Fellowship to BRB) and the Petroleum Research Fund administered by the American Chemical Society (ACS-PRF 28720-AC2 and ACS-PRF 31598-AC2 to MFH). Our goethite and phlogopite specimens were kindly donated by the National Museum of Natural History, Smithsonian Institution, Washington, D.C. (NMNH #152060) and the Virginia Tech Mineralogical Museum (C3-204), respectively. We also acknowledge Jodi J. Rosso and Don Rimstidt for many stimulating discussions, and three anonymous reviewers for several helpful comments.

Literature Cited

1. Hochella, M. F. Jr.; Eggleston C. M.; Elings, V. B.; Parks, G. A.; Brown, G. E., Jr.; Wu, C. M.; Kjoller, K. *Amer. Mineral.* **1989,** *74*, 1235.
2. Binnig, G.;Rohrer H.; Gerber C.; Weibel E. *Appl. Phys. Lett.* **1982,** *40,* 178.
3. Binnig, G.;Rohrer H.; Gerber C.; Weibel E. *Appl. Phys. Lett.* **1982,** *49,* 57.
4. *Chemical Weathering Rates of Silicate Minerals;* White, A. F.; Brantley, S. L., Eds.; Reviews in Mineralogy; Mineralogical Society of America: Washington, D.C., **1995**; Vol. 31.
5. Hochella, M. F., Jr.; Banfield, J. F. In *Chemical Weathering Rates of Silicate Minerals;* White, A. F.; Brantley, S. L., Eds.; Reviews in Mineralogy; Mineralogical Society of America: Washington, D.C., **1995**; Vol. 31; pp 354-405.

6. Rufe, E.; Hochella, M. F., Jr. *213th Amer. Chem. Soc. Nat. Mtg. Book Abstr.* **1997,** GEOC 108.
7. Johnsson, P.A.; Hochella, M.F., Jr.; Parks, G.A.; Blum, A.E.; Sposito, G. In *Water-Rock Interaction*; Kharaka, Y.; Maest, A., Eds.; Balkema, Rotterdam, **1992**; Vol. *7*; pp 159-162.
8. Blum, A. In *Scanning Probe Microscopy of Clay Minerals*; Nagy, K.L.; Blum, A.E., Eds.; Clay Minerals Society: Boulder, Colorado, **1994**; Vol. 7; pp 171-202.
9. Lin, F. C.; Clemency, C. V. *Clays Clay Minerals* **1981,** *29,* 101.
10. Norrish, K. In *Proc. Int. Clay Conf.;* Madrid Eds.; Div. Ciencias C.S.I.C., **1973**; pp 417-432.
11. Nagy, K. L. In *Chemical Weathering Rates of Silicate Minerals;* White, A. F.; Brantley, S. L., Eds.; Reviews in Mineralogy; Mineralogical Society of America: Washington, D.C., **1995**; Vol. 31; pp 173-233.
12. Reichenbach H. G.; Rich, C.I. *Clays Clay Minerals* **1986,** *17,* 23.
13. Banfield J. F., Eggleton, R. A. *Clays Clay Minerals* **1988,** *36,* 47.
14. Hartman, H.; Sposito, G.; Yang, A.; Manne, S.; Gould, S. A. C.; Hansma, P. K. *Clays Clay Minerals* **1990,** *38,* 337.
15. Dove, P. M.; Chermak, J. A. In *Scanning Probe Microscopy of Clay Minerals;* Nagy, K. L.; Blum, A. E., Eds.; CMS Workshop Lectures; The Clay Minerals Society: Boulder, Colorado, **1994**; Vol. 7; pp 140-169.
16. Hochella, M. F., Jr. In *Mineral-Water Interface Geochemistry;* Hochella, M. F., Jr.; White, A. F., Eds.; Reviews in Mineralogy; Mineralogical Society of America: Washington, D.C., **1990;** Vol. 23; pp 87-132.
17. Bickmore, B. R.; Hochella, M. F., Jr. *Seventh Ann. V.M. Goldschmidt Conf.* **1997,** *LPI 921,* 27.
18. Bickmore, B. R.; Hochella, M. F., Jr. *Geol. Soc. Amer. Abstr. Prog.* **1997,** A-26.
19. Brown, G. E., Jr. In *Mineral-Water Interface Geochemistry;* Hochella, M. F., Jr.; White, A. F., Eds.; Reviews in Mineralogy; Mineralogical Society of America: Washington, D.C., **1990**; Vol. 23; pp 309-363.
20. Junta, J. L.; Hochella, M. F., Jr. *Geochim. Cosmochim. Acta* **1994,** *58,* 4985.
21. Junta-Rosso, J. L.; Hochella, M. F., Jr.; Rimstidt, J. D. *Geochim. Cosmochim. Acta* **1997,** *61,* 149.
22. Rakovan, J. F.; Hochella, M. F., Jr. *213th Amer. Chem. Soc. Nat. Mtg. Book Abstr.* **1997,** GEOC 70.
23. Scheidegger, A. M.; Lamble, A. M.; Sparks, D. L. *Environ. Sci. Technol.* **1996,** *30,* 548.
24. Scheidegger, A. M.; Lamble, A. M.; Sparks, D. L. *J. Coll. Inter. Sci.* **1997,** *186,* 118.
25. Scheidegger, A. M.; Strawn, D. G.; Lamble, A. M.; Sparks, D. L. (submitted).
26. Hochella, M. F., Jr. In *Mineral Surfaces;* Vaughan, D; Pattrick, R., Eds.; The Mineralogical Society Series; Chapman and Hall: London, England, **1994**; Vol. 5; pp 17-60.
27. Rosso, K. M.; Hochella, M. F., Jr *213th Amer. Chem. Soc. Nat. Mtg. Book Abstr.* **1997,** GEOC 107.
28. Rosso, K. M.; Hochella, M. F., Jr. *Geol. Soc. Amer. Abstr.* **1997,** in press.
29. Moses, C. O.; Herman, J. S. *Geochim. Cosmochim. Acta* **1991,** *55,* 471.

30. Reedy, B. J.; Beattie, J. K.; Lowson, R. T. *Geochim. Cosmochim. Acta* **1991,** *55,* 1609.
31. Williamson, M. A.; Rimstidt, J. D. *Geochim. Cosmochim. Acta* **1994,** *58,* 5443.
32. McKibben, M. A.; Barnes, H. L. *Geochim. Cosmochim. Acta* **1986,** *50,* 1509.
33. Buckley, A. N.; Woods, R. *Appl. Surf. Sci.* **1987,** *27,* 437.
34. Nesbitt, H. W.; Muir, I. J. *Geochim. Cosmochim. Acta* **1994,** *58,* 4667.
35. Guevremont, J. M.; Strongin, D. R.; Schoonen, M. A. A. *Surf. Sci.* **1997,** *391,* 109.
36. Fan, F. R.; Bard, A. J. *J. Phys. Chem.* **1991,** *95,* 1969.
37. Siebert, D.; Stocker, W. *Phys. Status Solidi (a)* **1992,** *134,* K17.
38. Eggleston, C. M.; Hochella, M. F., Jr. *Amer. Mineral.* **1992,** *77,* 221.
39. Eggleston, C. M.; Ehrhardt, J.; Stumm, W. *Amer. Mineral.* **1996,** *81,* 1036.
40. Chaturvedi, R.; Katz, R.; Guevremont, J.; Schoonen, M. A. A.; Strongin, D. R. *Amer. Mineral.* **1996,** *81,* 261.
41. Hochella, M. F., Jr.; Lindsay, J. R., Mossotti V. G.; Eggleston, C. M. *Amer. Mineral.* **1988,** *73,* 1449.
42. Yoo, M. J.; Fulton, T. A.; Hess, H. F.; Willett R. L.; Dunkleberger, L. N.; Chichester, R. J.; Pfeiffer, L. N.; West, K. W. *Science* **1997,** *276,* 579.
43. Jacoby, M. *Chem. Eng. News* **1997,** *75,* 35.

Chapter 4

Atomic Force Microscopy as a Tool for Studying the Reactivities of Environmental Particles

Patricia A. Maurice

Department of Geology, Kent State University, Kent, OH 44242

Atomic force microscopy has emerged as one of the premier techniques for studying the reactivities of environmental particles. This paper addresses some of the most important considerations involved in AFM applications to environmental particles, and presents several state-of-the-art examples of such applications.

Atomic force microscopy (AFM, also known as scanning force microscopy, STM) was developed in the mid-1980s (*1*), and it quickly emerged as an important techique for characterizing the structure, microtopography, and chemical reactivity of mineral surfaces. AFM allows for imaging of surfaces in air or immersed in aqueous solution, at extremely high resolution (nanometer to potentially atomic-scale under optimal conditions), and with minimal sample preparation (without surface coatings). While most AFM studies to date have been conducted on relatively flat, cm-scale, single-crystal surfaces, AFM is becoming an increasingly important technique for studying environmental particles such as clay minerals (e.g., *2, 3*), humic substances (e.g., *4*), and bacteria (e.g., *5-8*). Although application to particulates is challenging, recent advances are opening new avenues of research. The purpose of this paper is to describe some of the most important considerations that need to be addressed in applying AFM to environmental particles, and to discuss several recent advances that are facilitating particulate research.

Fundamental Principles of AFM. The principles of AFM as applied to geochemical systems have been described in detail in many previous publications (e.g., *9-13*). Briefly, AFM works by rastering a sample underneath a sharp tip that is attached to or part of a cantilever. A number of forces, including Van der Waals dispersion forces, the so-called 'atomic force' (related to the Pauli exclusion principle), electrostatic forces, and others, cause the tip to deflect as different surface features pass beneath it. By monitoring this deflection, a 3-dimensional map of the sample surface can be constructed. AFM is commonly operated in so-called contact mode, wherein the tip is in direct contact with the sample surface and repulsive forces are dominant. Non-contact mode imaging, which is conducted in an attractive regime, commonly in aqueous solution, also is possible and has resulted in more reliable and interpretable atomic-scale imaging (e.g., *14*).

A number of factors make AFM uniquely applicable to studies of environmental particle surfaces: (1) when used properly, AFM is for the most part nondestructive; (2)

under most operating conditions, micron- to nanometer-scale resolution is easily attainable; (3) under ideal operating conditions, molecular-to atomic-scale resolution sometimes may be achieved; (4) surfaces may be imaged in air or immersed in aqueous solutions; (5) a nanometer- to micron-scale portion of the surface may be imaged repeatedly such that reaction progress can be monitored on surfaces immersed in solution; and (6) sample preparation is generally minimal, and samples do not need to be coated prior to imaging.

Despite these advantages, application of AFM to studies of environmental particles is neither simple nor straightforward. Particulate material often is extremely difficult to image by AFM, for several reasons: (1) The finite size and shape of the tip can distort images. Blum (*2*) reported that for features less than about 30 nm high, the radius of curvature of the tip may broaden features. For features greater than 30 nm high, the shape of the tip (tip half angle) can interact with the shape of the sample, leading to an image that is at least in part a reverse image of tip shape. (2) AFM is by its very nature a surface technique, and thus cannot access internal pores or interlayers; moreover, if particle overlap occurs, underlying particles are obscured. (3) Lateral frictional forces during scanning may dislodge particles from underlying substrates, although recently developed tapping-mode (TMAFM, Digital Instruments, Inc.) technologies minimize such forces.

Additional difficulties include: (4) Because particulate material generally must be attached to a substrate for imaging, 3-dimensional conformations and interparticle relationships that normally are present when particles are suspended in solution may not be maintained. (5) As a microscopic technique, sampling statistics are extremely important; yet, the possibility of numerous artifacts precludes rigorous point-counting procedures and may result in significant sampling biases. (6) Although new AFM techniques have emerged which may allow materials with different electrostatic, adhesive, frictional, or other properties to be distinguished, AFM does not provide traditional chemical analysis. Hence, different particles in mixed samples must be distinguished based primarily on micromorphology or microtopography, alone. (7) Although atomic-scale AFM imaging may allow for structural determinations, most particulate imaging requires use of TMAFM, which has decreased resolution and does not permit atomic-scale analysis. The limit of TMAFM resolution is generally ~200 nm scan size.

AFM is best used as part of an integrated approach wherein a variety of techniques such as XRD, or SEM/EDX, TEM/ED, XPS, IR spectroscopy, Raman spectroscopy, etc. are used to determine the sample composition and to obtain images at a variety of scales. Because AFM is a high resolution technique, it is important to collect a large catalog of images of reaction 'blanks' and reacted samples to account for sample heterogeneity. Moreover, because AFM is a relatively new technique, it is essential to compare AFM images to images produced by the more established TEM and SEM techniques, to help determine imaging artifacts.

Tapping-Mode AFM and Phase Imaging. Most AFM imaging to date has been conducted in the contact mode, wherein the tip is in direct contact with the sample surface. During contact-mode imaging in air, a meniscus forms between water and sorbed contaminants on the surface of the sample and the tip. This meniscus results in strong attractive forces between the tip and the sample. Moreover, during the scan procedure, lateral frictional forces occur that can lead to substantial damage of the sample surface.

TappingMode AFM (TMAFM) was developed by Digital Instruments to overcome these problems (see http://www.di.com). In TMAFM, a piezoelectric driver is used to excite the cantilever into resonance oscillation. The resulting 'tapping' motion overcomes the meniscus forces and also results in decreased lateral frictional forces. During TMAFM imaging, microtopographic (height) data can be displayed side-by-side with amplitude data (Fig. 1). The amplitude data reflect the fact that the free oscillation amplitude of the tip is decreased as the tip contacts features of varying topography. The amplitude data give essentially the first derivative of the height data. Hence, amplitude images tend to highlight details, in a manner more easily interpretable by the human eye; however, amplitude data cannot be used to measure the height of a topographic feature, because heights are necessarily greatly distorted.

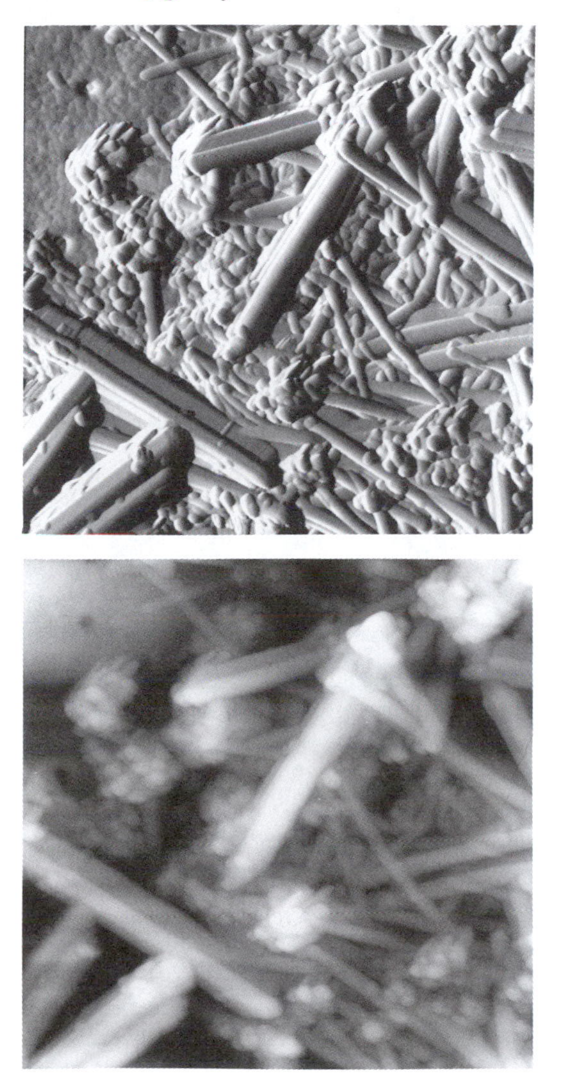

Figure 1. Height (top) and Amplitude (bottom) TMAFM images of HPY (needles) formed upon reaction of $[Pbaq]_i$ = 100 mg L^{-1} with HAP. Pc membrane appears in lower right-hand corner. Triangular pyramidal tip-sample interaction artifacts can be seen just to left of center. Maximum z-range (white to black) on Height image = 500 nm. Both images 2.3 μm on a side. Scales in μm.

Phase imaging is a recently developed extension of TMAFM which maps the phase lag of the cantilever oscillation; i.e. its actual phase relative to the signal originally sent to the cantilever's piezoelectric driver (*15*) (See Figure 2). The phase lag appears to be sensitive to variations in such physicochemical properties as composition, adhesion, friction, and viscoelastic properties. Phase imaging tends to highlight fine details of surface structure that are not apparent in other AFM images. At present, it is not fully understood how the different properties of the surface contribute to the overall phase image; nevertheless, this form of imaging provides nicely detailed images of many environmental particles.

Examples of applications

An integrated study of Pb uptake by hydroxylapatite. AFM can be used as an *ex-situ* technique for imaging particulates following sorption, growth, or dissolution experiments. The resulting micromorphological and microtopographical data may be used to help identify solid phases, to characterize associations between these phases, and to determine changes in microtopography that may provide clues to reaction mechanism (e.g., etch-pit formation). Because AFM does not provide chemical analysis of sample materials, mineral phases must be identified primarily based on morphology. In rare instances, atomic-scale surface structure may also be probed by AFM. However, TMAFM, which is most commonly used for particulates, does not provide atomic-scale resolution.

As part of a systematic approach to developing AFM applications to studies of reaction products, Lower et al. (*16*) applied TMAFM to study Pb uptake by reaction with the mineral, hydroxylapatite ($Ca_5(PO_4)_3OH$) (HAP). Ma et al. (*17-19*) and Xu and Schwartz (*20*) showed that HAP dissolution provides phosphate, which quickly reacts with Pb to form highly insoluble hydroxypyromorphite ($Pb_5(PO_4)_3OH$) (HPY), according to the following equations:

$$Ca_5(PO_4)_3OH + 7H^+ \xleftrightarrow{dis} 5Ca^{2+} + 3H_2PO_4^- + H_2O \quad (1)$$

$$5Pb^{2+} + 3H_2PO_4^- + H_2O \xleftrightarrow{ppt} Pb_5(PO_4)_3OH + 7H^+ \quad . \quad (2)$$

Lower et al (*21*) calculated a standard state Gibbs free energy change (G_r) of -137 kJ mol^{-1} for stoichiometric conversion of HAP to HPY, indicating that HPY formation is highly favorable. Lower et al's (*16*) experiments were designed to further test this proposed mechanism and to determine the conditions under which HPY nucleation occurred either heterogeneously on the surfaces of HAP particles or homogeneously, in solution.

Lower et al. (*16*) combined macroscopic measurements with *ex-situ* atomic force microscopy (AFM), transmission electron microscopy (TEM), scanning electron microscopy (SEM), energy dispersive spectroscopy (EDS), electron diffraction, and x-ray diffraction (XRD) to study the effects of initial saturation state on the formation of HPY from Pb(aq)-HAP interactions at pH 6, 22°C. Study results are summarized in Figure 2. Particle associations following reaction suggested that at high supersaturation (intial Pb concentration, $[Pb]_i$ = 500 mg L^{-1}), multiple nuclei formed simultaneously in solution, resulting in an intergrown network of small HPY crystals (generally $<<$ 1 µm in size). At somewhat lesser values of supersaturation ($[Pb]_i$ = 10 - 100 mg L^{-1}), nucleation again appeared to be largely homogeneous. Fewer nuclei formed and most Pb was consumed by crystal growth, leading to the formation of larger, euhedral needles of HPY (several µm in length) in solution. At the lowest initial saturation state ($[Pb]_i$ = 0.5, 1.0 mg L^{-1}), HPY or some other Pb-phosphate phase appeared to nucleate primarily heterogeneously on the surfaces of HAP crystals. These studies were conducted *ex situ*, by comparing reactants and products prior to and following reaction. Hence, interpretations were necessarily based on indirect evidence of particle morphologies and associations. The observed results agreed well with nucleation theory, as shown schematically in Figure 2.

Two significant problems were encountered during AFM analysis. First, the finite dimensions of the tip resulted in numerous tip-sample interaction artifacts, as shown in Fig. 1. Second, because AFM does not provide chemical information, identification of phases had to be based on morphological comparison to images of HAP and HPY sample blanks. Image interpretation was greatly facilitated by using a variety of complimentary techniques. XRD confirmed the presence of both HAP and HPY in the reacted samples. TEM/ED, and SEM/EDS demonstrated that the elongate needles were HPY, while the rounded clusters were most probably HAP.

Because these experiments were conducted *ex-situ*, on reactants and reaction products, evidence for heterogeneous versus homogeneous nucleation was necessarily indirect. The greatest strength of AFM imaging is the ability to study particles *in-situ*, over the course of reaction. Lower et al. (*21*) therefore expanded upon these *ex-situ* experiments, by conducting *in-situ* experiments on HAP particles at $[Pb]_i = 100$ mg L^{-1}. The *in-situ* AFM experiments were designed to further test the hypothesis that under these conditions, HPY formation is primarily through homogeneous precipitation in solution. This intial Pb concentration was chosen because Lower et al.'s (*16*) results suggested that at $[Pb]_i >> 100$ mg L^{-1}, the reaction would be so rapid that the system would become unstable with respect to *in-situ* AFM imaging, while at lower concentrations, reaction products might be too sparse to be detected by a high resolution microscopic technique. HPY was detected in the AFM fluid-cell effluent, in subsequent SEM images of reacted particles, and in SEM images of the integrated cantilever-tips used in the experiments. Using the AFM's binocular microscope attachment, HPY nucleation was observed in the environs of HAP particles. However, *in-situ* AFM imaging was unable to detect any HPY growth on HAP, despite imaging in numerous locations. Moreover, although HPY formed in the environs of HAP particles, the HPY needles were easily swept away as the sample was scanned under the tip, and did not appear to be attached to the HAP surfaces. Taken together, the various lines of evidence confirmed that HPY formed primarily by homogeneous nucleation in solution. Although SEM imaging of reacted samples showed that HPY needles were in close association with HAP surfaces, this was most probably due to diffusional controls on phosphate concentrations rather than to heterogeneous nucleation. Diffusion of phosphate away from dissolving HAP appeared to be the rate-limiting step in the overall reaction sequence.

Study results (*16, 21*) demonstrated the importance of applying a range of analytical techniques to a study of this type. On the one hand, AFM provided detailed images of particle surface microtopography, allowed for high-resolution analysis of particle associations, and permitted *in-situ* imaging of HAP particle surfaces over the course of reaction. On the other hand, SEM/EDS and TEM/ED both provided chemical means for identifying phases, along with information on larger scale features and associations of phases.

TMAFM Imaging of Fulvic acids. Humic and fulvic acids (HA, FA, respectively), are often the most abundant components of the natural organic matter pool in soils and natural waters, and they appear to play important roles in the transport of organic and inorganic pollutants, the coagulation of environmental colloids, and the growth and dissolution of minerals in soils and sediments. HAs and FAs contain numerous carboxylic and phenolic functional groups and are thus negatively charged except at very low pH, where they behave like uncharged polymers.

Schnitzer and his colleagues applied SEM and TEM to dried HA and FA samples, and they showed that whereas the pH, electrolyte concentration, and FA or HA concentration of the preparatory solution influenced the molecular conformations of

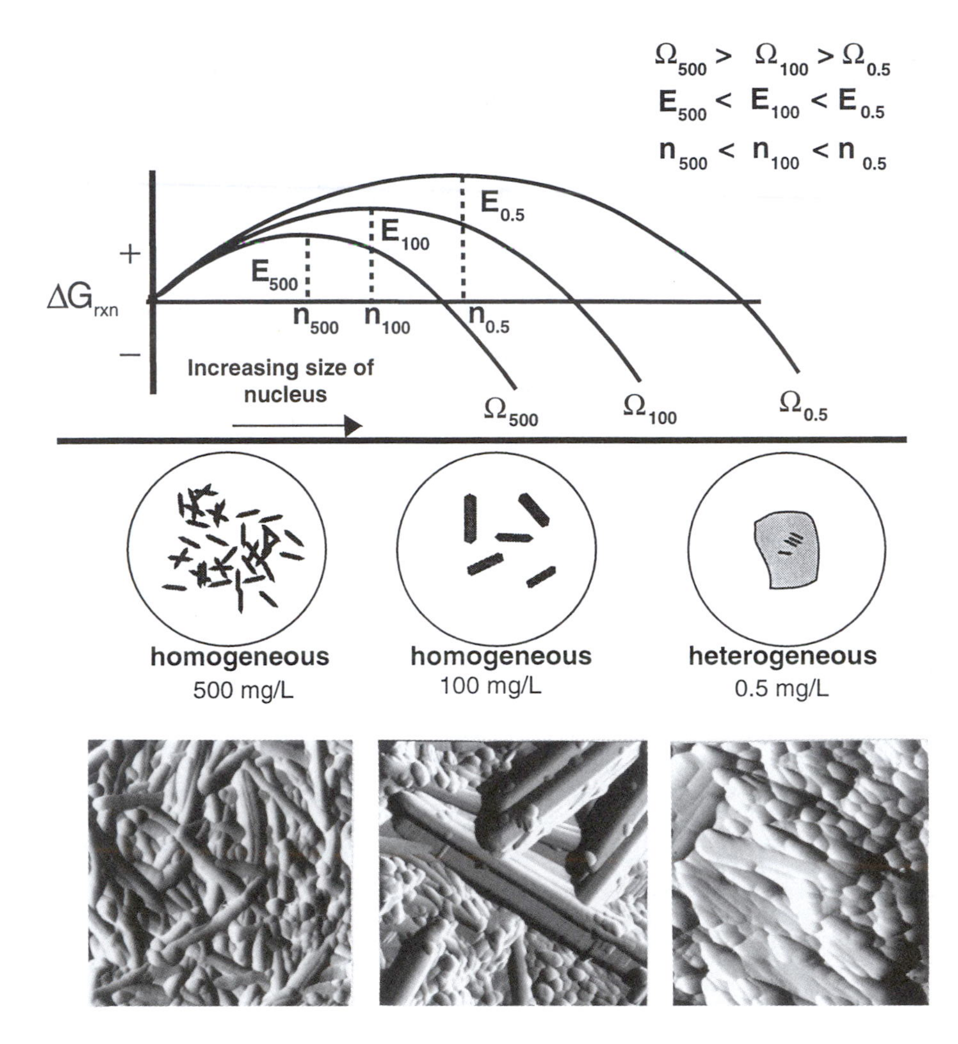

Figure 2. (top) Schematic illustration of relationship between saturation state (Ω), size of critical nucleus (n), and activation energy (E) (Adapted from *22*). Related to type of nucleation shown schematically (middle) and through AFM data (bottom). Dimensions of leftmost image = 1.5 mm on a side; center AFM images =1 μm on a side. Rightmost AFM image dimensions = 0.7 μm on a side. This figure modified from several figures in Lower et al. (*16*) as well as previously unpublished data.

FA molecules, the strongest influence appeared to be due to organic concentration (*23, 24*). Solution pH and ionic strength can, however, play important roles at low FA concentrations (*23*).

Namjesnik-Dejanovic and Maurice (*4*) compared AFM images of soil and stream FAs with previously reported TEM and SEM images (*23, 25*) of similar samples. To create samples similar to those used in TEM/SEM experiments, FAs were deposited on the basal-plane surface of muscovite mica, from solution, after which samples were allowed to dry, and imaged by TMAFM in air. This work was performed at pH~4 and ionic strength < 0.01, but at variable FA concentrations. Four main structures were observed on a sample of soil FA. At low concentrations, sponge-like structures consisting of rings (~15 nm in diameter) appeared, along with small spheres (10-50 nm). At higher concentrations, aggregates of spheres formed branches and chainlike assemblies. At very high surface coverage, perforated sheets were observed. On some samples, all of these structures were apparent, perhaps due to concentration gradients on drying (See figure 3). Surface-water samples from the Suwannee River, GA were only imaged at higher concentrations. Spheres, aggregated branches, and perforated sheets were apparent. Results of AFM imaging agreed well with previous work by Stevenson and Schnitzer (*24*), applying TEM to soil fulvic acids freeze-dried on muscovite. However, the TEM images did not detect the smaller spheres and sponge-like structures observed by AFM at low concentrations.

The environmental relevance of imaging dried samples is questionable because conformations may change significantly upon drying. For example, the perforated sheet structure that we observed at high concentrations is probably due to film rupture upon drying. Hence, imaging in solution should prove to be far more relevant to the study of how NOM behaves on surfaces. The key to such imaging most probably will be use of a strongly sorbing substrate, and solution conditions conducive to adsorption.

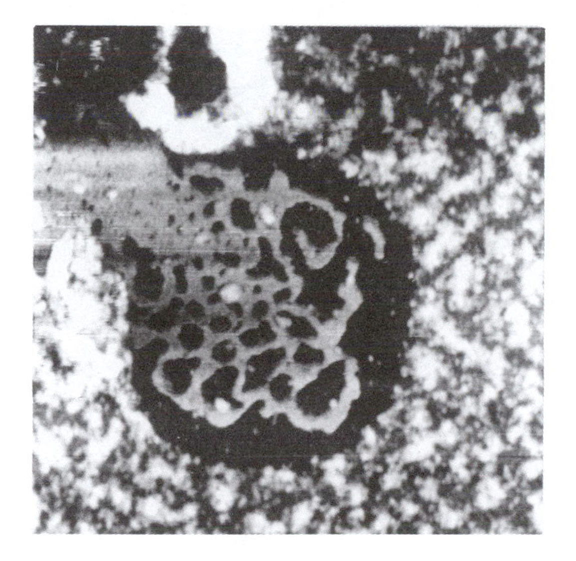

Figure 3. Fulvic acid on muscovite, imaged by TMAFM in air. Small spheres (bright), some of which have aggregated into a branched-chain structure, are apparent. Scan area = 1.25 μm on a side. Maximum relief = 22 nm.

The FA structures shown in Figure 3 most probably are subject to broadening due to the finite size and shape of the tip. Wilson et al. (*26*) describe how the finite size of the probe tip can result in laterally enlarged AFM images of biomolecules. These authors determined that in a typical imaging situation, the baseline dimensions of an imaged biomolecule may represent only 20% actual molecule and up to 80% tip structure. Using sharp tips can partially alleviate this problem. Morphological restoration procedures also may be employed, wherein a sample standard (e.g., a gold sphere) is scanned, the resulting AFM image is used to back out the shape of the tip, and the data on the tip shape are used to assess the effects of the tip on the AFM image of the biomolecule (*26*). While a variety of such procedures are being developed, they must be used with caution; for example, morphological restoration cannot compensate for imaging with a grossly distorted tip.

Phase imaging of microbial attachment features. Microbial attachment to mineral surfaces appears to be an important step in many forms of microbially mediated mineral dissolution (e.g., *27, 6, 7*). Although the detailed mechanisms of microbially enhanced dissolution are not widely understood, it has been hypothesized that attached microorganisms may create microenvironments conducive to dissolution, through release of extracellular material (e.g., *27*). Hence, characterization of microbial attachment features to mineral surfaces may provide valuable insight into the mechanisms of microbially mediated dissolution.

TMAFM and Phase Imaging are proving to be excellent methods for the characterization of microbial attachment features. Maurice et al. and Forsythe et al. (*5,*

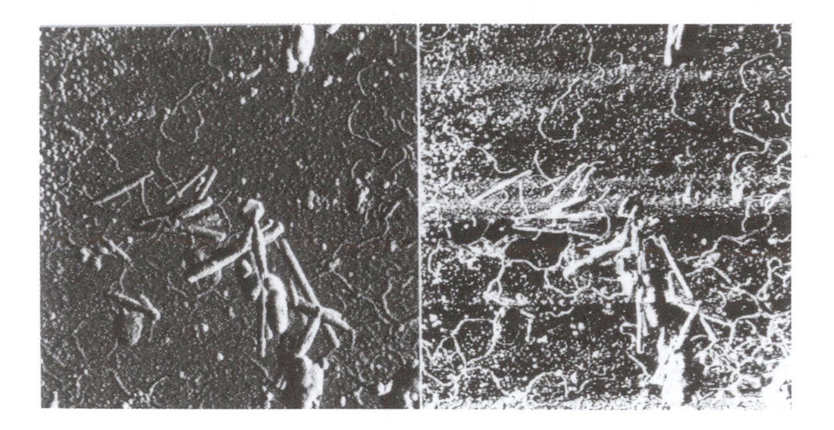

Figure 4a.(left) TMAFM amplitude image along the edge of a goethite-bacteria cluster. Figure 4b. (right) TMAFM Phase Image of the same area. This image was collected at 147 degrees phase offset. The fiber-like flagella and attachment features are highlighted in this image relative to the amplitude mode image. Goethite rods and oblong microbial bodies are also apparent. Horizontal striations are an imaging artifact. Both images 10 μm on a side.

8, 28) have studied the interactions between an obligate aerobic *Pseudomonas sp.* bacteria and hydrous Fe(III) oxide surfaces in the form of cm-scale specular hematite crystals and hematite and goethite particles. Microbial growth curves in Fe-limited media indicate that the bacteria are able to obtain Fe from the hydrous Fe(III) oxide particles. Results of TMAFM and Phase Imaging show that bacteria colonize some fraction of mineralogic aggregates, forming biofilms consisting of networks of fiber-like attachment features intertwined through films of organic material.

Figure 4a shows a TMAFM amplitude image collected along the edge of a goethite aggregate. This sample was prepared by culturing the bacteria with goethite particles for 8 days (See *8* for detailed discussion of experimental methods), removing a microbe-covered goethite cluster from the suspension, gently dropping onto a glass slide, and allowing to air dry prior to AFM imaging. This image shows acicular (needle-like) goethite crystals, oblong bacteria, and fibrous-like bacterial flagella and attachment features. As shown in Figures 4b, Phase Imaging can highlight details that are not readily apparent in Amplitude mode images. In Figure 4b, the fibrous attachment features are shown much more clearly than in Figure 4a. It should be noted that which details are highlighted (e.g., mineralogic particles, microbial bodies, or attachment features) appears to depend upon the phase offset angle used during imaging(additional imaging not shown here). Hence, this technique promises to provide clearer data, as well as potential information about physicochemical properties of particulate materials.

Acknowledgments

P. Maurice thanks the Petroleum Research Fund of the American Chemical Society, the National Science Foundation, the Department of Energy and the LACOR program of Los Alamos National Laboratories for funding various aspects of her AFM-related research. Bacterial interaction studies are being conducted in collaboration with J. Forsythe (KSU) and Drs. L. Hersman (LANL) and G. Sposito (U CA Berkeley); Pb-apatite studies are being conducted in collaboration with S. Lower, M. Manecki (both at KSU), and Drs. S. Traina (Ohio State Univ.) and E. Carlson (KSU); humic substance studies are being conducted in collaboration with K. Namjesnik-Dejanovic (KSU). P. Maurice wishes to thank these colleagues for their continued dedication to collaborative, interdisciplinary research.

References cited

(1) Binnig, G.; Quate, C. F.; Gerber, C. *Phys. Rev. Lett.* **1986**, 56, 930-933.
(2) Blum, A. E. In *Scanning probe microscopy of clay minerals* A. E. Blum and K. Nagy, Eds., Clay Minerals Society, Boulder, Colorado, 1994, pp. 172-202.
(3) Nagy, K. In *Scanning probe microscopy of clay minerals* A. E. Blum and K. Nagy, Eds., Clay Minerals Society, Boulder, Colorado, 1994, pp 203-239.
(4) Namjesnik-Dejanovic, K; Maurice, P.A. *Colloids SurfsA* **1997**, 120, 77-86.
(5) Maurice et al. (1996)
(6) Grantham, M.E. and P.M. Dove. *Geochim.Cosmochim. Acta* **1996**, 60, 2473-2480.
(7) Grantham
(8) Forsythe, J.H.; Maurice, P.A.; Hersman, L. In *Water-Rock Interaction International IX*, 1998, in press.
(9) Hochella, M. F., Jr. In *Mineral-water Interface Geochemistry*; M.F. Hochella, Jr. and A. F. White, Eds., Reviews in Mineralogy Vol. 23, Miner. Soc. Amer., Washington, D. C., 1990, pp. 87-132.
(10) Hochella, M. F., Jr. In *Mineral surfaces*, D. J. Vaughan and R. A. D. Pattrick, Eds., Chapman and Hall, New York, 1995, pp. 17-60.
(11) Eggleston, C. M. In *Scanning probe microscopy of clay minerals* A. E. Blum and K.Nagy, Eds., Clay Minerals Society, Boulder, Colorado, 1994, pp 1-90.
(12) Maurice, P.A. *Colloids Surf.A* **1996**, 107, 57-75.

(13) Maurice, P.A. and Lower, S.J. *Adv. Agron.* **1997**, 62, 1-43.
(14) Ohnesorge, F.; Binnig, G. *Science* **1993**, 260, 1451-1456.
(15) http://www.di.com
(16) Lower S.; Maurice, P.; Traina, S.; Carlson, E. *Amer. Min.* **1998**, in press.
(17) Ma, Q.Y., Traina, S.J., Logan, T.J., and Ryan, J.A. *ES&T* **1993**, 27, 1803-1810.
(18) Ma, Q.Y., Logan, T.J., Traina, S.J., and Ryan, J.A.*ES&T* **1994**, 28, 408-418.
(19) Ma, Q.Y., Logan, T.J., Traina, S.J., and Ryan, J.A.*ES&T* **1994**, 28, 1219-1228.
(20) Xu, Y. and Schwartz, F.W. *J. Contam. Hydro.* **1994**, 187-206.
(21) Lower et al., in review
(22) Berner, R.A. In Mineralogical Society of America Reviews in Mineralogy Vol. 8, Miner. Soc. Amer., Washington, D.C., 1981, pp. 111-134.
(23) Gosh, K.; Schnitzer, M. *Soil Sci.* **1982**, 129, 226.
(24) Stevenson, I.L.; Schnitzer, M. *Soil Sci.* **1982**, 133, 179-185.
(25) Chen, Y.; Schnitzer, M. *Soil Sci. Soc. Am. J.* **1976**, 40, 882.
(26) Wilson, D.L.; Dalal, P.; Kump, K.S.; Benard, W.; Xue, P.; Marchant, R.E.; Eppell, S.J. *J. Vac. Sci. Technol. B* **1996**, 14, 2407-2416.
(27) Hiebert, F.K.; Bennett, P.C. *Science* **1992**, 258, 278-281.
(28) Forsythe et al., in review

Digital Modeling

Chapter 5

From Molecular Structure to Ion Adsorption Modelling

W. H. VanRiemsdijk and T. Hiemstra

Department of Environmental Sciences, Laboratory of Soil Science and Plant Nutrition, Wageningen Agricultural University, P.O. Box 8005, NL 6700 EC Wageningen, Netherlands

Surface complexation has been studied quite independently from a variety of perspectives, like thermodynamics, colloid and interfacial chemistry, crystallography and mineralogy, and spectroscopy. The present challenge is to integrate the knowledge gained from the various disciplines as discussed.
The composition and the structure of the surface can be assessed by crystallography, which distinguishes different types of sites each with its own residual charge and affinity. The experimental charging behaviour of metal (hydr)oxides can be evaluated on the basis of such multiple site concept. The same crystallographic principles can be extended to ion binding, showing that the charge in surface complexes can be considered as spatially distributed. The concept of charge distribution can be incorporated in thermodynamic ion adsorption models, describing the macroscopic world.
Spectroscopy has provided a progressive flow of information concerning the binding mechanism(s) of ions and its complex structure. Qualitative and quantitative information from spectroscopy can be included in ion adsorption modelling, connecting molecular details to macroscopic measurements.
Various examples of the application of the charge distribution approach to the adsorption of cat- and anions are given.

Ion adsorption to mineral surfaces is an important phenomenon in many industrial applications as well as in the study of the natural environment. Ion exchange on constant charge minerals was initially the ion adsorption process that was studied most intensively both experimentally and theoretically. The success in this field was due to a combination of efforts. Mineralogists and crystallographers unravelled the detailed structure of the minerals after 1930, theoretical chemists developed the well known double layer theories around 1910-1925, and the thermodynamics of ion exchange was formulated around 1950. Colloid and soil chemists in the mean time developed the various applications of this work and it became widely used by practitioners.

Interest in the variable surface charge/potential of metal (hydr)oxides started somewhat later (*1-11*). Also here people from a very different background have contributed to the development of this field of science. The emphasis and the approach followed has differed depending on ones background and scientific interest. The topic has been studied from a thermodynamic, colloid chemical, modelling, mineralogical, spectroscopic, surface chemical, theoretical chemical or practical point of view. For simple chemical reactions there will be relatively little conflict between the various points of view. The dissociation reaction of acetic acid can be written as:

$$HAc \underset{\rightarrow}{\overset{\leftarrow}{}} Ac^{-} + H^{+} \qquad ;K_{HAc} \qquad (1)$$

This reaction can be characterized by one affinity constant (K_{HAc}). The molecular structure of acetic acid is well known and the origin of the proton dissociation is known to be the result of the dissociation of one proton from the carboxylic acid group in the molecule. The reaction can be followed by spectroscopy, can be interpreted from a theoretical chemical point of view, the thermodynamics are relatively straightforward and the modeler or practitioner will have little problem with this simple reaction. The different points of view will probably not lead to any serious conflict for such a simple case.
The situation for ion adsorption on mineral systems is much more complex and conflicts in interpretation of the data between different points of view can easily occur. For the description of ion binding on surfaces one will always need to formulate a series of reaction equations. Most approaches take the electrostatic interaction explicitly into account. The variable surface potential has a large contribution on the change of the Gibbs free energy of the reaction with increasing loading. The proton binding to a hypothetical homogeneous metal (hydr)oxide can be interpreted as being the result of a series of simple reactions of the type of equation 1 (*12*). This approach leads to a large apparent heterogeneity complicating further model development. Although most authors take electrostatics into account there are considerable differences in the approaches taken to model the effect of the potential near the solid-water interface (constant capacitance, purely diffuse, Stern-Gouy-Chapman, triple layer, three plane, multi layer model etc.).
The chemical composition of ideal crystal planes can be derived from crystallography. Most crystal planes have more than one type of surface group and the composition of different crystal planes on the same mineral may differ, affecting the reactivity of the particle (*13-17*). To determine the surface composition of mineral particles in aqueous systems is experimentally quite complex. The exact surface structure of mineral particles is in general ignored by most people who try to describe ion binding to these particles. It is common to assume that the surface is chemically homogeneous and to invoke a two step protonation reaction as the basis for the charging reaction with metal (hydr)oxides. As we will show below this is from a structural and mechanistic point of view not a very realistic situation. The primary models have been extended with adsorption reactions in order to describe the pH dependency of cations, oxyanions or weak organic acids. The same data set can normally be described with models that may deviate strongly from each other. The pH dependent binding of phosphate to goethite has been modelled by various researchers assuming the presence of completely different sets of surface species (*18-22*), ranging from one phosphate species (*18,21*) to five species (*19*).

However, the actual surface speciation of adsorbed phosphate on goethite has recently been determined by in situ spectroscopy. Two bidentate species and one monodentate species can be distinguished (*23*). The goal of the work of Hiemstra and VanRiemsdijk (*22*) was to describe a very extensive data set on phosphate binding taking into account available information on the surface structure of goethite and the experimentally determined surface speciation. The commonly used approaches are more or less successful in describing the pH dependent binding although the species used in the modelling have most probably little physical meaning. This situation is quite common in surface complexation modelling and is not very satisfactory from a spectroscopic or mechanistic point of view, because the actual surface species can nowadays be studied with various in situ spectroscopic techniques like EXAFS (*24-29*), NMR (*30*) or IR (*23*, *31-34*).
The pH dependence of the ion binding is from a thermodynamic point of view directly related to the exchange ratio between the ion that binds and the protons that are adsorbed or desorbed upon the ion adsorption (*35*,*36*). Apparently there are different ways in which this exchange process can be modelled. Fokkink et al.(*37*) have clearly shown that the proton-metal ion exchange ratio is influenced strongly by the distance of the adsorbed metal ion to the surface, the closer the distance the higher is the exchange ratio and thus also the pH dependence. The exchange ratio in the approach followed by Fokkink et al.(*37*) results solely from electrostatic interactions. It is clear that the formulation of a site binding reaction will in general also affect the exchange ratio that arises from the model (*38*). Most of the commonly used models even have great difficulty to give an accurate description of the measured exchange ratio. The description can probably easily be improved if more surface species are invoked (*36*) but such an approach gives no insight in the actual binding mechanisms.
The description of the ion binding in a way that is mechanistically reasonable is a great challenge. Whether the species invoked in the model have physical relevance is from a macroscopic thermodynamic point of view not very relevant. However the prediction of competitive binding of say phosphate and sulphate based on the separate singular monocompoment modelling of both types of ions is not necessarily successful from a thermodynamic point of view. The closer the model is to the physical reality the higher the chances are that the prediction of interaction for complex systems will be successful. Development of better mechanistic adsorption models that describe the binding with "real" species is therefore not only of theoretical both also potentially of great practical interest.

The Variable Charge Character of Metal (Hydr)Oxides

Metal (hydr)oxides are amphoteric minerals which can have a positive or negative surface charge/potential in the presence of so called "indifferent" electrolyte ions. The surface charge density is a function of the pH of the solution and the salt level. The pH at which the surface is uncharged, in the absence of specifically adsorbed ions other than H^+ and OH^-, is called the pristine point of zero charge (PPZC) (*9*) or simply the PZC. Interpretation of literature data of PZC values should be done with great care because a very small contamination of the sample with strongly adsorbing species can considerably affect the measured PZC. The PPZC can be quite different for different metal (hydr)oxides ranging from around 2 for silica to around 10 for gibbsite. Parks (*2-4*) was probably the first scientist who became intrigued by the difference in PZC for different minerals and developed a simple

model to estimate the PZC of a mineral. The model uses the charge and the radius of the cation and the coordination number of the metal ion. Later Yoon et al.(*39*) refined Parks model and used the Pauling bond valence of the metal-oxygen bond to estimate the PZC of metal (hydr)oxides.
The most widely used concept to describe the basic charging behaviour of metal (hydr)oxides is the two step protonation reaction which leads to a so called two-pK model for the description of the basic charging:

$$SO^{-1} + 2H^{+} \underset{\rightarrow}{\overset{K_1}{\leftarrow}} SOH^{0} + H^{+} \underset{\rightarrow}{\overset{K_2}{\leftarrow}} SOH_2^{+1} \quad (2)$$

This approach was also proposed by Parks (*4*). Stumm being influenced by Parks (*40*) used the same approach. It is obvious from equation 2 that the charge can become negative, uncharged or positive depending on the pH. In order to be able to describe the measured surface charge as a function of pH one has to extend equation 2 with equations that relate the electrostatic surface potential to the surface charge. The potential profile around the mineral particle is normally described by a combination of a purely diffuse double layer where the ions are treated as point charges and one or more electrostatic planes near the surface. In the purely diffuse Gouy-Chapman double layer model, as advocated by Dzombak and Morel (*41*), the head end of the diffuse double layer model coincides with the mineral-water interface. It is well known that the diffuse double layer model breaks down close the interface since hydrated counter ions have a finite size which means that there is a certain distance between the mineral surface and the centre of the counter ions that are closest to the surface, leading to charge separation. This was first realized by Stern (*42*). The most simple physical realistic electrostatic model is thus a Stern-Gouy-Chapman model. The combination of equation 2 with a Stern-Gouy-Chapman model can describe most charging curves of metal (hydr)oxides relatively easily. Parameters in the model are the two log$K$ values, the Stern layer capacitance and the reactive site density. Most authors find the parameter values by fitting the model to a data set. It is easy to show that the PZC equals $1/2(\log K_1 + \log K_2)$. The difference between the two log$K$ values is often referred to as the $\Delta \log K$ (or Δ pK). Using this model approach basically means that one assumes that the surface is chemically homogeneous and that the charging can be described by the two step charging reaction as given above.
Most authors pay very little attention to the real chemical nature of the reactive surface sites. An extreme approach in this respect is the approach followed by Fokkink et al.(*37*) who assume that the surface potential ψ_0 (V) behaves purely Nernstian :

$$\psi_0 = 0.059\,(\text{PZC-pH}) \quad (3)$$

The charging curves of goethite can for instance be described very well with this approach, in combination with a Stern-Gouy-Chapman model as has been shown by Venema et al.(*38*) and the only fitting parameter in this case is the Stern layer capacitance. For a better understanding of the basic charging behaviour of metal (hydr)oxides, and even more so for a sensible introduction of surface complexes as determined by spectroscopy, it is important to look in more detail to the possible reactive sites on metal (hydr)oxides.

Surface Structure and Sites. A surface of a mineral has so called "broken" bonds or "dangling" bonds which originate from the absence of coordinative bonds at the interface that are present in the bulk of the mineral. This leads to surface oxygens that are coordinated to less metal centres than in the solid. The charge of these surface oxygens is thus neutralized by charge attribution of one or more coordinating metal centres and by one or more protons that can bind with the oxygen. The most simple attribution of charge from a coordinating metal centre can be calculated using the Pauling bond valence ν (*43*). The Pauling bond valence expresses the available charge per bond and is simply the charge of the metal ion (z) divided by its coordination number (CN):

$$\nu = \frac{z}{CN} \tag{4}$$

All oxygens in the bulk of for instance silica (Figure 1) are coordinated by two bonds with Si and the negative charge (-2 v.u.) of the doubly coordinated oxygens is neutralised by two Si bonds in the structure, each contributing one valence unit since the Pauling bond valence of a Si-O bond in a tetrahedron equals one (4/4) as results from equation 4.
Based on the low reactive site density of silica, it is well known that the doubly coordinated oxygens ($=Si_2O$) can also occur at the surface of silica particles (*44,45*). This simple example already shows that the notion that a reactive surface oxygen has a charge of minus one as suggested by equation 2 is certainly incorrect for $=Si_2O$ surface groups. One could imagine that this surface group can react with a proton to form a positively charged Si_2OH^+ group:

$$Si_2O^0 + H^+ \overset{\leftarrow}{\underset{\rightarrow}{}} Si_2OH^{+1} \tag{5}$$

Assessment of the importance of this positively charged $=Si_2OH^+$ species in the interface requires the estimation of the proton affinity constant for the above reaction.
Another type of surface group in the silica/water interface is the singly coordinated surface oxygen that arises on silica due to the absence one of the two Si-O bonds (Figure 1). According to the Pauling bond valence concept the singly coordinated group has a charge of minus one ($\equiv SiO^-$). As indicated in Figure 1, this negative charge can be compensated by the adsorption of a proton ($\equiv SiOH^o$). The protonation of site SO^{-1} can be written with:

$$SO^{-1} + H^+ \overset{\leftarrow}{\underset{\rightarrow}{}} SOH^0 \tag{6}$$

Binding of a second proton results in a positive surface group ($\equiv SiOH_2^{+1}$). These reactions are exactly equivalent with the equations often used (equation 2). If this model is applied to the charging of silica particles it turns out that only the first protonation step is required to describe the proton titration data. The shape of the charging curve of silica using only one proton affinity constant is shown in Figure 2.

Note that the surface charge varies only relatively little at low pH values. On the basis of the value of the PZC one can estimate the value of the second protonation step. The log$K$ value of the first protonation step (equation 6) is found to be in the range of log$K$ = 7.5±1 (*7,46,47*). If the PZC value is around at 2 (*45,48*), it results in an estimated value of the second protonation constant in the range of around log$K$= -4. The calculated Δlog$K$ between the first and second protonation step is thus very large and in the order of 10 logK units. The very low value of the second protonation step, log$K$ around -4, indicates that the formation of Si_2OH^+ surface groups in aqueous systems is quantitatively insignificant under normal aqueous conditions. We conclude that the surface of a metal (hydr)oxide will in general have different types of surface groups that may show a large variation in reactivity. The structural approach followed in the case of silica, is the basis of the MUlti SIte Complexation (MUSIC) model (*49*) and can be applied to other metal (hydr)oxides. The hydroxyl groups in the bulk of the structure of $Al(OH)_3$ minerals like gibbsite and bayerite are all doubly coordinated with Al centres. The aluminium ions are in six fold coordination which means that the Pauling bond valence equals +0.5 (Figure 3).

The doubly coordinated hydroxyls in the structure are thus fully neutralized by coordination with aluminium as it should be. These doubly coordinated hydroxyls can also occur at the surface of $Al(OH)_3$ minerals (*50*). The doubly coordinated hydroxyl at the surface is thus uncharged and may associate and dissociate a proton, in accord with equation 2. In order to know whether these reactions are of practical relevance one needs to know the two log$K$ values. At the surface also singly coordinated hydroxyls can occur because of broken bonds. According to the Pauling bond valence concept, which is based on a bookkeeping of charges, the charge of this surface group (AlOH) is -0.5. The surface group can react with a proton to form a surface water molecule ($AlOH_2$) with a charge of +0.5. In principal dissociation of a proton from a surface hydroxyl could also occur leading to a singly coordinated oxygen with a charge of minus -1.5. The affinity of this highly negatively charged group for a proton is expected to be extremely high. That this configuration is very unlikely also follows from the speciation in solution. The various monomeric Al species have a variable number of OH and or H_2O in their first coordination shell and never an oxygen. Like for silica one can thus expect that the variable charge behaviour of $Al(OH)_3$ (s) can also be described with one affinity constant if we assume that the doubly coordinated groups are not reactive. These considerations have led in the past to the suggestion of the so called one-pK model which uses the following basic charging reaction (*50-52*):

$$SOH^{-1/2} + H^+ \overset{\leftarrow}{\underset{\rightarrow}{}} SOH_2^{+1/2} \tag{7}$$

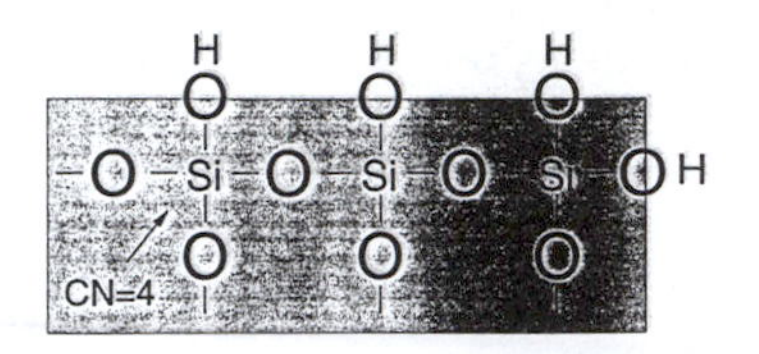

Figure 1. A schematic representation of the structure of silica. The Si^{4+} ion is fourfold coordinated with oxygens. The oxygens in the bulk are bound by two metal centres. In the interface singly and doubly coordinated oxygens can be found. The charge of the broken SiO bond can be compensated by the uptake of a proton: $SiO^{-} + H^{+} <=> SiOH^{0}$.

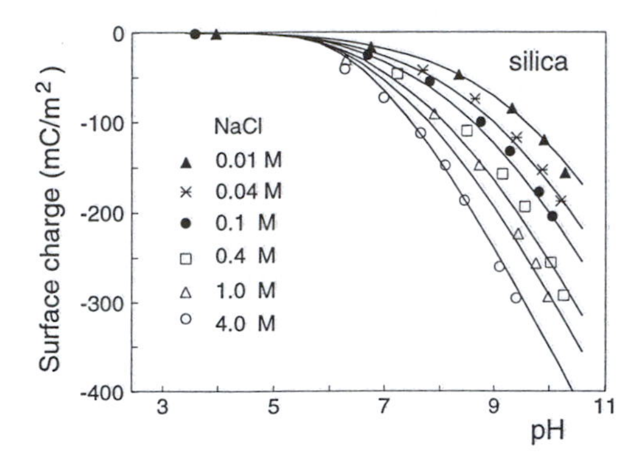

Figure 2. The charging of silica as function of pH for various concentrations of indifferent electrolyte ions. Data are from Bolt (1957). The lines are calculated using the one step surface protonation reaction $SiO^{-} + H^{+} <=> SiOH^{0}$ with $\log K=7.5$ and a site density $N_s=4.6$ nm^{-2} in combination with Gouy-Chapman-Stern double layer model with a capacitance $C=3.9$ F/m^2 (Hiemstra et al.1989b).

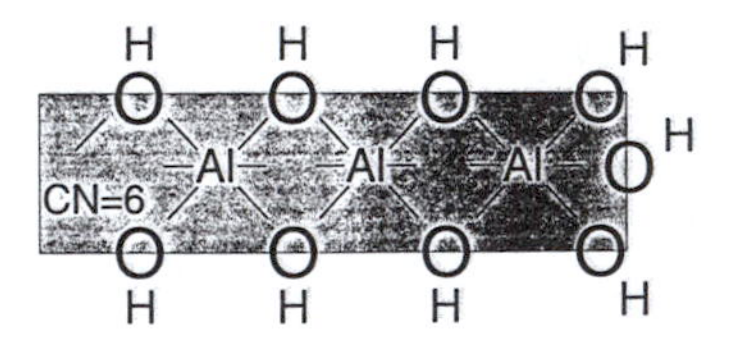

Figure 3. A schematic representation of the structure of Al hydroxide. The Al^{3+} ion is hexacoordinated with hydroxyls. The OH ions in the bulk are bound by two Al ions. In the interface singly and doubly coordinated hydroxyls are found. The charge of one broken bond of the singly coordinated group can be compensated by the uptake of a proton: $AlOH^{-1/2} + H^{+} <=> AlOH_2^{+1/2}$.

The log$K$ of this reaction follows directly from the PZC if this is supposed to be the only relevant charging reaction since the number of $SOH^{-1/2}$ and $SOH_2^{+1/2}$ groups should be equal at the PZC.
In the above examples of metal (hydr)oxides, $Al(OH)_3$ and SiO_2, the variation in surface charge mainly results from the presence of one type of reactive surface group, which reacts according to one single protonation step (equation 6 and 7). Nevertheless one observes experimentally a completely different charging behaviour, which is due to electrostatics. Equation 6 and 7 as such are mathematically equivalent descriptions of an one-step protonation reaction if electrostatics are ignored. However, introduction of the electrostatic contributions calculated on the basis of the charge attribution given in equation 6 and 7, leads to a different relationship between the surface speciation and the surface charge / potential. The different pH-charge-potential relation results in a completely different shape of the charging curve which is illustrated in Figure 4.
In this figure two calculated charging curves are compared for two hypothetical metal (hydr)oxides which have the same value of log$K$ (log$K$=8), the same reactive site density, the same Stern layer capacitance, but a different charge of the sites (-0.5 and +0.5 or -1 and 0). The one-pK approach, using equation 6 in combination with a relatively high reactive site density, will give very similar results for the charging behaviour as the Nernstian-Gouy-Chapman-Stern approach as has been explained by Venema et al. (*38*). The shape of experimental charging curves of metal (hydr)oxides of iron, aluminium and titanium corresponds to a -0.5/+0.5 charge transition whereas silica obviously corresponds to the transition -1/0 (*46*). The acid-base titration curves of most metal (hydr)oxides can be perfectly described by assuming a homogeneous surface with one log$K$ value for the protonation reaction in combination with a Stern-Gouy-Chapman model. The incorporation of weak outer sphere ion pair formation constants into the model may lead to an improved description of the data. The inclusion of the pair formation constants for the description of the curves of Al-hydroxides is even crucial (*50*). The formation of outer sphere complexes with simple monovalent electrolyte ions seems physically realistic (*53*). Formation of outer sphere complexes with electrolyte ions leads to an increased neutralisation of charge near the surface, diminishing the charge in the Gouy-Chapman diffuse double layer and may be therefore particularly important for the description of the salt dependence of the binding of other ions and will also be of importance for the understanding of the colloidal stability of the mineral colloids.

Proton Affinity. For a better understanding of the nature of the PZC and the resulting charging curves it is thus essential to have a better understanding of the factors that determine the proton affinity constants of the various types of reactive surface groups. The first reasonably successful attempt in this respect is the MUSIC model (*49*). The proton affinity of surface complexes is evaluated by applying the Pauling bond valence principal to the protonation reaction of the monomeric species in solution. The PZC of gibbsite is used as a calibration in the application of the model to surface species. One of the main findings of the model is that the Δlog$K$ of two consecutive protonation steps on the same oxygen is very large which means that in practice only one protonation step per type of reactive group is of importance as discussed before. The conclusion that the ΔpK between two consecutive protonation steps on one surface oxygen is expected to be extremely large also follows from the work of Borkovec (*54,55*), who used an Ising model approach for

the interaction between neighbouring sites. The MUSIC model also predicts that some surface groups are hardly reactive, like the doubly coordinated uncharged surface hydroxyls of trivalent iron and aluminium (hydr)oxides (Figure 3). The MUSIC model has been applied to various metal (hydr)oxides (*56*) like gibbsite, goethite, silica (*46*), rutile/anatase, (*46,57,58*), cerium oxide (*59*) and magnetite (*60*). Our initial results for goethite seemed quite satisfactory. However the crystal planes used in the calculations were not the correct ones as was pointed out to us by Schwertmann (personal communication). Using the most likely crystal planes led to the wrong prediction of the charging behaviour. This problem has stimulated us to develop a refined version of the MUSIC model. The refined version uses the actual bond valence instead of the Pauling bond valence. The bond valence is strongly related to the bond length (*61*). This means that asymmetry in the coordination structure will lead to an unequal distribution of the charge over the ligands. Relatively more charge will be attributed to the shorter bonds. This asymmetry is particularly important for the mineral goethite. All oxygens in the structure of goethite are triply coordinated with three iron centres.
The proton in the FeOOH structure is asymmetrically divided over the O and OH in the structure via hydrogen bonds. This is illustrated in figure 5. This observation reflects already the fact that both oxygens do not have the same proton affinity. It also shows that hydrogen bonding will play an important role in the interpretation of proton affinities as had been suggested before by Bleam (*62*). Hydrogen bonding is taken explicitly into account in the refined version of the MUSIC model. Some uncertainty exists with respect to the number of hydrogen bonds that should be accounted for, which may differ for monomeric species in solution compared to the surface because of steric reasons.
The refined MUSIC approach shows that undersaturation of the oxygen charge is the basis of the variation in proton affinity. The undersaturation depends not only on the number of coordinating metals, but also depends on metal-oxygen distance, as expressed in the actual bond valence and in addition is also affected by the presence of hydrogen bonds. Calculation of the charge undersaturation of the oxygen leads to the prediction of its proton affinity. Good predictions of the PPZC and the charging behaviour of all metal (hydr)oxides considered seems possible with the improved version of the MUSIC model (*63*). It is noted that the main conclusions of the MUSIC model as discussed above remain valid in the refined version.
Analysis of the surface structure of the main crystal faces of goethite (the dominant 110 and the 100 face) shows the presence of 1 singly, 1 doubly and 3 triply coordinated groups per unit cell. On the basis of our estimated proton affinities, it can be shown that apparently only 2 surface groups per unit cell will be proton active: one FeOH(H) and one $Fe_3O(H)$ (*22*). The Fe_2OH^0 surface group and the combination of one $Fe_3O^{-1/2}$ - $Fe_3OH^{+1/2}$ pair can be considered as inert. The apparent site density of the proton active groups is about 6 nm^{-2} and in agreement with the site density found by the extensive modelling of PO_4 adsorption (*22*).
Recently Rustad et al.(*64*) have followed a promising molecular modelling approach for estimation of proton binding constants for a few reactive groups on goethite. With the molecular modelling technique first the deprotonation energies of simple monomeric iron species in the gas phase are calculated. In a next step the interaction with water molecules is taken into account. The authors use empirically correlate gas phase energies and the measured protonation constants of aqueous monomeric iron species. Since the water molecule interaction can have

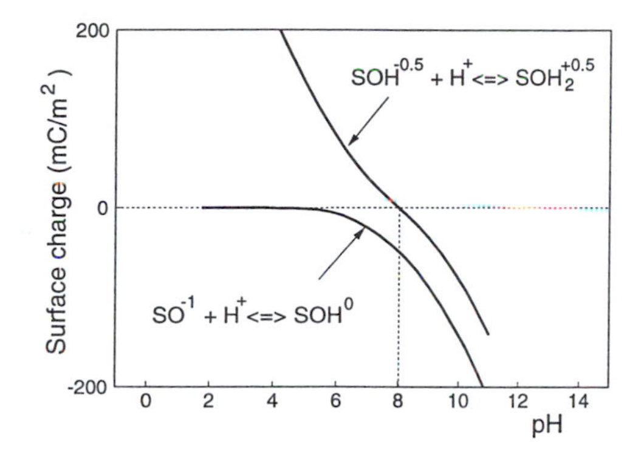

Figure 4. Comparison of the charging of two homogeneous metal (hydr)oxide surfaces, both having a one-step protonation reaction which only differs in charge attribution to the surface groups (-1/0 or -0.5/+0.5). The reactions are $SO^- + H^+$ <=> SOH^0 and $SOH^{-0.5} + H^+$ <=> $SOH_2^{+0.5}$. The difference in charge attribution is related to mineral structure of the metal (hydr)oxide, see text. The lines are calculated assuming for both reactions the same affinity log$K$=8, site density N_s=6 nm^{-2} and Stern layer capacitance C=2.5 F/m^2

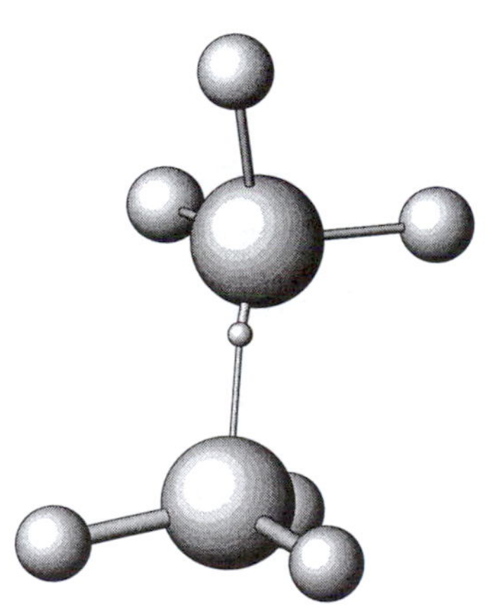

Figure 5. A schematic representation of the O and OH ions in the α-FeOOH structure. The O and OH are both triply coordinated (respectively bottom and upper half). The Fe-O distance is smaller than the Fe-OH distance, implying a relatively higher charge attribution of the coordinating Fe to the oxygen than to the hydroxyl.

a profound influence on the proton affinity as is clear from the refined MUSIC concept, application of such a correlation to interfaces remains rather uncertain unless the water structure and interaction with reactive groups are known. Further progress in this direction can be made by the development of a molecular modelling approach which accounts for the interactions with water in a less empirical way.

Ion Binding to Metal (Hydr)Oxides

Most cat- and anions have a chemical affinity for metal (hydr)oxides. Both inner sphere and outer sphere complexes can be formed. In case of an inner sphere complex one or more surface O or OH groups are shared between a metal centre of the metal (hydr)oxide and the central element (e.g. P or Cd) of the adsorbed species. The adsorption of a cation bound at a positively charged metal (hydr)oxide usually shows a linear isotherm at low pH and low cation ion concentrations. In a log-log plot of the adsorption linearity of the adsorption isotherm is reflected in the constant slope n with $n=1$ (figure 6).
The density of the surface complexes under these conditions is usually very low due to repulsive electrostatic interaction. The change in adsorption with increasing concentration in solution under these conditions leads to an increase in adsorption without a noticeable change in the surface potential which explains the linear behaviour under these limiting conditions. Metal ion binding close to the PZC may lead to a considerable adsorption already at very low metal ion concentrations in solution (figure 6). It results in a strongly nonlinear adsorption at constant pH due to the considerable change in surface charge and potential as a result of metal ion adsorption. The increase of surface/particle charge can be followed by measuring the proton - metal ion exchange ratio. This ratio for bivalent metal ions is typically below 2. For cadmium adsorption on goethite the ratio is approximately 1.5 (*65,66*).
The pH dependent metal ion binding can normally not be described satisfactorily assuming a homogeneous metal (hydr)oxide surface. A small number of high affinity sites is often invoked in order to be able to get a reasonable description in case of a wide range of concentrations and pH values (*41*). EXAFS measurements have shown for Cd (*27*) that these high affinity sites are related to particular surface configurations.
The adsorption of oxyanions and weak organic acids with more than one carboxylic group leads to high adsorption densities below the PZC for low concentrations in solution. This is due to the combination of a certain chemical affinity for the surface sites and a strong attractive electrostatic interaction. The phosphate adsorption at pH 4 on goethite is therefore still highly nonlinear at the very low concentration of 10^{-8} mol/l (*67*). It can be shown that non linearity is equivalent to a large variation in the overall Gibbs free energy of adsorption, i.e. a large variation in the apparent log$K$. In case of modelling without an electrostatic approach this will lead to the use of a very large series of apparent surface species. However models with the appropriate electrostatic model approach, will be able to describe the adsorption process using only the actual surface species observed by spectroscopy.
The description of the pH dependent ion binding is commonly done by extending a model used for the description of the pH dependent surface charge with a series of reactions that describe the formation of one or more surface complexes. The charge of the adsorbing species (for instance Cd^{2+}, PO_4^{3-}) is normally considered as a point charge and its charge is located in the surface plane or in an electrostatic plane at some short distance from the surface. It is clear from thermodynamics that

the pH dependence of the binding is due to the exchange between the adsorbing species and the protons which can be coadsorbed in case of anion adsorption or released in case of cation adsorption. This non stoichiometric exchange should be described by the model in order to get a description of the pH dependent ion binding behaviour. This exchange ratio is in the model influenced by a variety of factors. It is influenced by the type of species that is supposed to bind, e.g. Cd^{2+} or $CdOH^{+}$, it is influenced by the formulation of the binding reaction, whether it is formulated as a bidentate or a monodentate complex, and it is strongly influenced by the electrostatic interaction between the surface and the adsorbing species. The structure of the compact part of the double layer is a priori assumed by various authors, and in most cases one has either no choice where to place the adsorbing ion (purely diffuse, constant capacitance) or one chooses between two positions. The effect of the choice of a particular model and the positioning of adsorbing ions on the exchange ratio and the pH dependency has been discussed for data of cadmium adsorption on goethite (*38*). It is clear that the surface species that one needs in the various common approaches to describe the data depend strongly on the structure of the compact part of the double layer model that has been chosen. Considering that commonly little or no attention is paid to the real structure of the surface, that the reactions formulated to describe the proton binding are physically obscure, and that one uses the concept of point charges for inner sphere complexes, it is understandable that the surface species that are required to get a satisfactory description of the data, have normally only a vague resemblance with the surface species as determined by spectroscopy.
Spectroscopy is the only experimental technique able to assess the surface speciation. Todays challenge is still to determine the actual surface speciation unequivocally and quantitatively with sufficient detail. It should be noted however that even knowing the exact surface speciation does in itself not lead to a quantitative description of the ion binding of the species over a range of conditions. The reason for this is related to electrostatic and competitive interactions of adsorbed species. The affinity ($\log K$) of surface species established for a certain condition will change as function of pH, salt concentration, relative coverage and competition. A physical chemical adsorption model will have to be develloped to account for the change in overall affinity and its resulting actual speciation.

Interfacial Charge Distribution. Directly after the development of the MUSIC model (*49*) we tried to develop a new concept that would be able to describe the pH dependent ion binding using the surface structure and the surface species as derived from spectroscopy as a constraint. We decided to use phosphate adsorption on goethite as a test for our new approach since quite some information was available at the time for this system (*18,23,68*). The actual surface speciation had been determined semi-quantitatively as a function of pH and phosphate loading under aqueous conditions. The challenge turned out to be more difficult than we initially expected. One of the reasons was that we had at the time an incorrect interpretation of the dominant crystal planes present on the goethite as discussed before. It was essential to realize that there were at least two types of triply coordinated surface oxygens (figure 5) present at the surface of the important 110 face and 100 face, which differ considerably in proton affinity.
The newly developed approach for ion adsorption treats the charge of surface complexes no longer as point charges. The charge of the central atom in the adsorbing species (e.g. Cd^{2+} or P^{5+}) is distributed over the compact part of the double

layer. The model is a logical extension of the MUSIC approach. In case of inner sphere complex formation the common (surface) ligands (figure 7) are partly neutralised by the Me ions in the solid and part is neutralised by the central ion of the complex. The remaining charge of the central ion is attributed to the ligands that are more solution oriented, i.e. one assumes Charge Distribution (CD). We will refer to it as the CD-MUSIC model.

The charge distribution value is related to the surface complex structure. For instance a bidentate surface complex is expected to attribute more positive charge from the central ion to the solid than a monodentate complex since the latter one has only half as much bridging ligands with the surface than the first one. The CD value can be estimated with the Pauling bond valence concept in which an equal distribution of the charge over the ligands surrounding the central ion is assumed. A second principle for charge attribution is the condition of a minimum charge on the bridging ligands. The sum of the actual bond valences around an oxygen in the bulk of a mineral has been shown to be very close to zero (*61*). Since the surface is a charged interface, this is not necessarily true for the bridging oxygen in a surface complex. One nevertheless would expect a residual charge on the bridging oxygen or OH relatively close to zero. The actual CD value is found from the CD-MUSIC modelling. The charge distribution used in the CD-MUSIC model will influence the pH dependent binding as was discussed above. The correspondence between the charge distribution that is required in the model to get a good description of the measured data and the expected distribution can only be good if the model is a reasonable approximation of reality. It is noted that the structural interpretation of the CD value needs some precaution because the CD value is based on the overall interfacial change of charge accomplished from the adsorption of the species, which may comprise other (solute) structural contributions. Interpretation of the results for phosphate shows that the charge attribution that is obtained, is close to what is expected based upon estimates from the Pauling bond valence and the notion that the bridging oxygen should have a small residual charge.

The charge distribution concept can be applied to the adsorption of oxyanions as well as cations like Cd^{2+}. The surface speciation of cadmium on goethite has been estimated by EXAFS (*27*). Two main types of complexes are found to be present. At low Cd loading the Cd is mainly bound to hydroxyls forming the edges of octahedra of the solid. These sites are present at the head end of the goethite crystal (021 faces). At a higher Cd loading low affinity sites become occupied, where the Cd is bound to hydroxyl(s) present at the corner(s) of the Me octahedra of the solid. Both types of sites are used in the description of the Cd adsorption over a large concentration range. The modelling of an extended data set of Cd adsorption on goethite (*66*) revealed a CD value for the bidentate (double corner) complex of 0.3. This value is quite close to the value (2/6) expected on the basis of the Pauling bond valence for a Cd^{2+} octahedron, i.e. a symmetrical distribution of charge. The structure and the charge distribution for the bidentate Cd complex is schematically given in figure 9.

The discussed principle of a low charge on bridging surface ligands explains the formation of common hydroxyl ions in a Cd complex whereas in a complex like $Fe_2O_2PO_2$ (figure 8) oxygens are present as common ligand. The proton on the oxygen bridge in the $Fe_2(OH)_2Cd(OH_2)_4$ complex persists because removal creates a severe undersaturation of the valence, which is predicted to lead to an extremely high proton affinity, in the order of about $\log K$=20 (*63*). The same principle can also explain why PO_4 will not react with triply coordinated surface groups (Fe_3O)

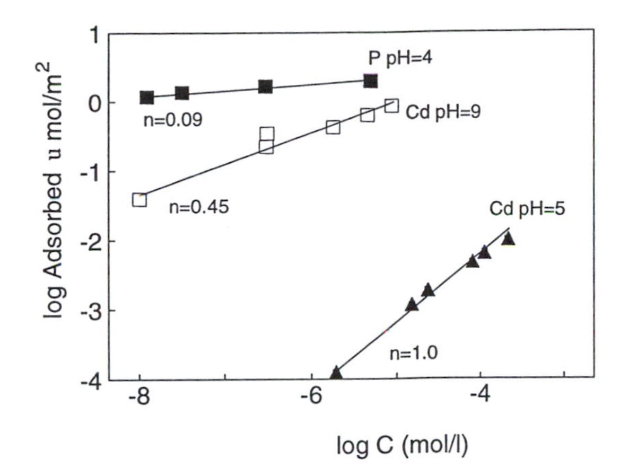

Figure 6. The adsorption isotherms of Cd and PO_4 plotted on a double logarithmic scale. The adsorption of Cd on goethite is linear at low Cd loading (low pH and concentration), as shown by the slope of the line (n=1), in contrast to the adsorption at high loading at pH=9 (n=0.45). The isotherm of PO_4 on goethite (pH=4, 0.01 M) is extremely non linear (n<0.1). The (non) linearity can be explained on the basis of electrostatics (see text). Data are taken from Venema et al.(1996b) and Geelhoed et al.(1997).

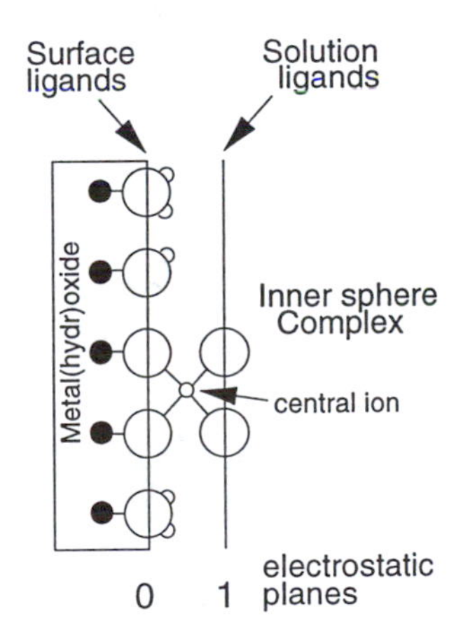

Figure 7. A schematic representation of the inner sphere complex formation at the surface of a metal (hydr)oxide.

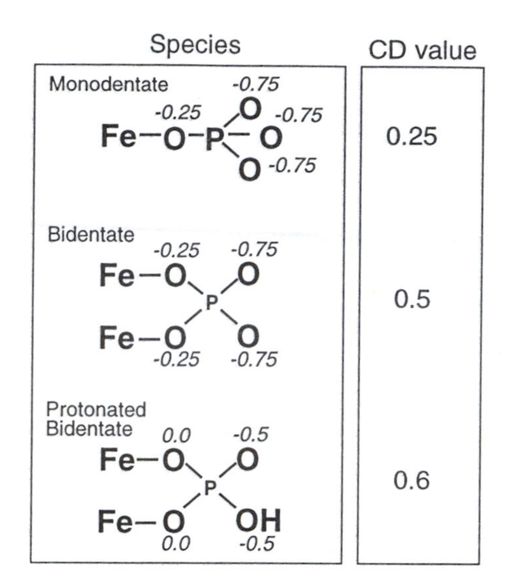

Figure 8. The phosphate surface complex structures on goethite in relation with the actual CD value found by modelling (Hiemstra and VanRiemsdijk 1996). The ligand charge is indicated on the various oxygens (italic).

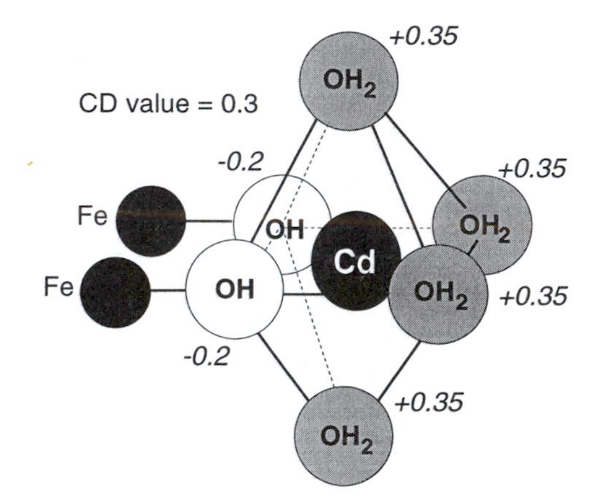

Figure 9. The Cd surface complexes structure on goethite found for low affinity sites (Spadini et al.1994) and the calculated charge on the ligands, based on the actual CD value found by modelling (Venema et al.1996b). The Cd surface complex has a hydroxyl as bridging ligand in contrast to the complexes of phosphate (figure 8).

because it would lead to a large over saturation of the valence of the common oxygen. The same concept has been applied to estimate the plausibility of the formation of surface complexes (*69*).
Recently the CD-MUSIC concept has also been applied to the description of the binding of weak organic acids like malonic and citric acid (*70*). The charge distribution concept plays also in this case a central role in the modelling and the interpretation of the surface complexes. Our knowledge of the actual surface species is at present still relatively limited and different interpretations can be found for the same system. This situation will probably improve the near future.

Multi Component Interaction. Application of an adsorption model in the natural environment is an appealing perspective and its success is built on the capability to describe the interaction of ions in multi component systems. This interaction can be synergistic, as found for Cd^{2+} and PO_4 or can be antagonistic like the competition between phosphate and sulphate.
The Cd^{2+}/PO_4 interaction has been studied by Venema et al.(*71*). An example is given in Figure 10. The data show an enhanced adsorption of Cd in the presence of PO_4 because adsorbed phosphate ions will decrease the positive charge, which diminishes the repulsive electrostatic contribution in the Gibbs free adsorption energy of cadmium. Although the adsorption of Cd in the single ion experiments could be modelled with the above mentioned species, the competition experiments revealed the presence of a low affinity monodentate cadmium species. The presence of this species is not necessarily in conflict with the EXAFS data. The modeling showed that this species is only dominantly present in the interface in the cadmium - phosphate systems (*71*). One of the reasons for this behaviour is related to the distribution of the charge. In monodentate complexes the positive charge is more solution oriented (Cd value=0.15) which enables a stronger electrostatic interaction with the negatively charged outer ligands of the PO_4 in the electrostatic 1 plane (figure 8).
The CD-MUSIC concept has been used to describe the pH dependent sulphate adsorption on goethite and to predict the competition between sulphate and phosphate (*67*). The predictions for the mixed systems are quite good over a very wide range of conditions. The very strong observed effect of the presence of sulphate on the phosphate concentration in solution at low pH and low phosphate loading can easily be understood. The goethite is still positively charged at low pH and very low phosphate concentration leading to a large positive effect of the electrostatic contribution to the Gibbs free energy of adsorption. The sulphate ion which has a relatively weak chemical affinity for the surface will bind relatively strong under these condition until the positive surface charge has approximately been neutralised. The sulphate adsorption thus reduces the positive contribution of the free energy of adsorption of phosphate compared to the situation in the absence of sulphate which explains the large effect on the equilibrium phosphate concentration for these conditions.
Both examples show that the modelling of the electrostatic interactions in a sensible way is essential for a proper prediction of the effect of interactions on the binding.

Conclusion We conclude that the CD-MUSIC model can be considered as a frame work, which is based as much as possible on structural information and well established concepts taken from mineralogy and colloid chemistry, and can be used

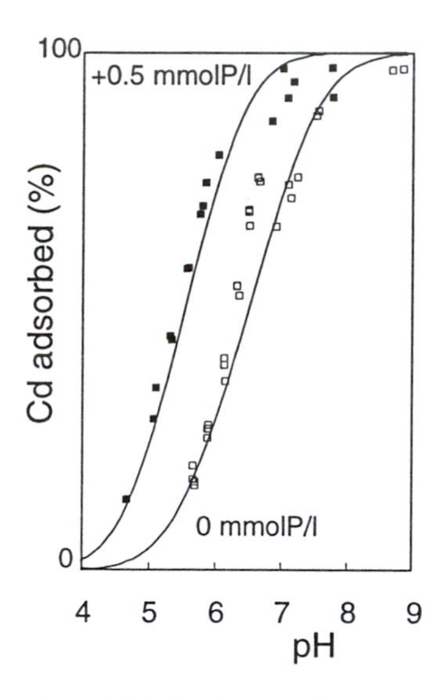

Figure 10. The adsorption of Cd (0.25 mmol/l) on goethite (570 m^2/l) in the presence of PO_4 (0.5 mmol/l) as function of pH (0.1 M $NaNO_3$). Data are taken Venema et al.(1997).

to describe adsorption data in a thermodynamically consistent way using physically realistic surface species. The results obtained so far for predictions in mixed systems are very encouraging.

References

1 Bolt,G.H. *J.Phys.Chem.* **1957**, *61*, 1166.
2 Parks,G.A.; DeBruyn,P.L. *J. Phys. Chem.* **1962**, *66*, 967.
3 Parks,G.A. *Chem. Rev.* **1965**, *65*, 177.
4 Parks,G.A. *Adv. Chem. Ser.* **1967**, *67*, 121.
5 Atkinson,R.J.; Posner,A.M.; Quirk,J.P. *J. Phys. Chem.* **1967**, *71*, 550.
6 Bérubé,Y.G.; DeBruyn,P.L. *J. Colloid Interface Sci.* **1968**, *27*, 305.
7 Schindler,P.W.; Kamper,H.R. *Helv. Chim. Acta* **1968**, *51*, 1781.
8 Stumm,W.; Huang,C.P.; Jenkins,S.R. *Croat. Chem. Acta* **1970**, *42*, 223.
9 Bowden,J.W.; Bolland,M.D.A.; Posner,A.M.; Quirk,J.P. *Nature Phys.Sci.* **1973**, *245*, 18.
10 Yates,D.E.; Levine,S.; Healy,T.W. *Chem. Soc. Faraday Trans.I* **1974**, *70*, 1807.
11 James,R.O., Davis,J.A.; Leckie,J.O. *J. Colloid Interface Sci.* **1978**, *65*, 331.
12 Borkovec,M; Rusch,U.; Koper,G.J.M.; Westall,J.C. *Colloids Surf. A* **1996**, *107*, 285.
13 Schwertmann,U. *Claymin.* **1984**, *19*, 9.
14 Torrent,J.; Barron,V.; Schwertmann,U. *Soil Sci. Soc. Am. J.* **1990**, *54*, 1007.
15 Schwertmann,U.; Cornell,R.M., *Iron Oxides in the Laboratory. Preparation and Characterization*; VCH verlag: Weinheim, Germany, 1991; Chap.5
16 Barron,V.; Torrent,J. *J. Colloid Interface Chem.* **1996**, *177*, 407.
17 Weidler,P.G.; Schwinn,T.; Gaub,H.E. *Clays Clay Min.* **1996**, *44(4)*, 437.
18 Bowden,J.W.; Nagarajah,S.; Barrow,N.J.; Posner,A.M.; Quirk,P.J. *Aust. J. Soil. Res.* **1980**, *18*, 49.
19 Sigg,L.; Stumm,W. *Colloid Surfaces* **1981**, *2*, 101.
20 Goldberg,S; Sposito,G. *Soil Sci. Soc. Am. J.* **1984**, *48*, 772.
21 VanRiemsdijk,W.H.; VanderZee,S.E.A.T.M. In *Interactions in the Soil Colloid - Soil Solution Interface*; Bolt,G.H.; DeBoodt,M.F.; Hayes,M.H.B.; McBride,M.B., Eds., NATO ASI Series (E); Kluwer Academic Publishers: Dordrecht, The Netherlands, 1991, Vol.190; Chap.8.
22 Hiemstra,T.; VanRiemsdijk,W.H. *J. Colloid Interface Sci.* **1996**, *179*, 488.
23 Tejedor-Tejedor,M.I.; Anderson,M.A. *Langmuir* **1990**, *6*, 602.
24 Hayes,K.F.; Roe,A.L.; Brown,G.E.; Hodgson,K.O.; Leckie,J.O., Parks,G.A. *Science* **1987**, *238*, 783.
25 Brown,G.E. In *Mineral-water Interface Geochemistry*; Hochella,Jr., M.F.; White,A.F., Eds., MAS: Reviews in Mineralogy, **1990**, *23*, 309-363
26 Waychunas,G.A; Rea,B.A.; Fuller, C.C.; Davis,J.A. *Geochim. Cosmochim. Acta* **1993**, *57*, 2251.
27 Spadini,L.; Manceau,A.; Schindler,P.W.; Charlet,L. *J. Colloid Interface Sci.* **1994**, *168*, 73.
28 Waite,T.C; Davis,J.A.; Payne,T.E., Waychunas,G.A., Xu,N. *Geochim. Cosmochim.Acta.* **1994**, *58*, 5465.
29 Scheidegger,A.M.; Sparks,D.L. *Soil Sci.* **1996**, *162*, 813.
30 Bleam,W.F. In *Advances in Agronomy* **1991**, *46*, 91-155.
31 Biber,M.V.; Stumm,W. 1994 *Environ. Sci. Technol.* **1994**, *28*, 763.

32 Sun,X., Doner,H. *Soil Sc.* **1996**, *161*, 865.
33 Hug,S. *J. Colloid Interface Sci* **1997**, *188*, 415.
34 Fendorf,S.; Eick, M.J.; Grossl,P., Sparks,D.L. *Environ. Sci. Technol.* **1997**, *31*, 315.
35 Perona,M.J.; Leckie,J.O. *J. Colloid Interface Sci.* **1985**, *106*, 64.
36 Cernic,M.; Borkovec,M.; Westall,J.C. *Langmuir.* **1996**, *12*, 6127.
37 Fokkink,L.G.J.; De Keizer,A.; Lyklema,J. *J. Colloid Interf.Sci.* **1989**, *127*, 116.
38 Venema,P.; Hiemstra,T.; VanRiemsdijk,W.H. *J. Colloid Interface Sci.* **1996a**, *181*, 45.
39 Yoon,R.H.; Salman,T.; Donnay,G. *J. Colloid Interface Sci.* **1979**, *70*, 483.
40 Stumm,W.; Morgan,J.J. *Aquatic Chemistry*; John Wiley & Sons: New York, 1970.
41 Dzombak,D.A.; Morel,F.M.M. *Surface complexation modelling: Hydrous Ferric Oxide*; John Wiley: New York, 1990.
42 Stern,O. *Z. Electrochem.* **1924**, *30*, 508.
43 Pauling,L. *J. Am. Chem. Soc.* **1929**, *51*, 1010.
44 Peri,J.B. *J. Phys. Chem.* **1965**, *4*, 211.
45 Iler,R.K. *The Chemistry of Silica*; Wiley: New York, 1979
46 Hiemstra,T.; De Wit,J.C.M.; VanRiemsdijk,W.H. *J. Colloid Interface Sci.* **1989b**, *133*, 105.
47 Sverjenski,D.A.; Sahai,N. *Geochim. Cosmochim. Acta.* **1996**, *60*, 3773.
48 Sposito,G. *The Surface Chemistry of Soils*; Oxford Univ.Press: New York, 1984.
49 Hiemstra,T.; VanRiemsdijk,W.H.; Bolt,G.H. *J. Colloid Interface Sci.* **1989a**, *133*, 91.
50 Hiemstra,T.; VanRiemsdijk,W.H.; Bruggenwert,M.G.M. *Neth. J. Agric. Sci.* **1987**, *35*, 281.
51 Bolt,G.H.; VanRiemsdijk,W.H. In *Soil Chemistry B Physio-chemical Models*; Bolt,G.H., Ed., Elsevier: Amsterdam, 1982, 2nd ed.; Chap.XIII.
52 VanRiemsdijk,W.H.; DeWitt,J.C.M.; Koopal,K.L.; Bolt,G.H. *J. Colloid Interface Sci.* **1987**, *116*, 511.
53 Sprycha,R., *J Colloid Interf. Sci.*, **1984**, *127*, 13.
54 Borkovec,M,; Koper,G.J.M. *Langmuir* **1997**, *13*, 2608.
55 Borkovec,M. *Langmuir* **1997**, *13*, 2608.
56 Jolivet,J.P. *De la Solution à l'Oxide*; CNRS Editions: Savoirs actuels Paris, 1994.
57 Giacomelli,C.E.; Avena,M.J.; DePauli,C.P. *Langmuir* **1995**, *11*, 3483.
58 Rodrigues,R.; Blesa,M.A.; Regazzoni,A.E. *J. Colloid Interface Sci.* **1996**, *177*, 122.
59 Nabavi,M.; Spalla,O.; Cabane,B. *J. Colloid Interface Sci.* **1993**, *160*, 459.
60 Vayssières,L. *PhD thesis*, Université Pierre et Marie Curie URA-CNRS 1466, Paris, 1966, 13.
61 Brown,I.D. *Chem. Soc. Rev.* **1978**, *7*, 359.
62 Bleam,W.F. *J.Colloid Interface Sci.* **1993**, *159*, 312.
63 Hiemstra,T.; Venema,P; VanRiemsdijk,W.H. *J. Colloid Interface Sci.* **1996**, *184*, 680.
64 Rustad,J.R.; Felmy,A.R.; Hay,B.P. *Geochim. Cosmochim. Acta* **1996**, *60(9)*, 1553.
65 Gunneriusson,L. *J. Colloid Interface Sci.* **1994**, *163*, 484.

66 Venema,P.; Hiemstra,T.; VanRiemsdijk,W.H. *J. Colloid Interface Sci.* **1996b**, *183*, 515.

67 Geelhoed,J.S.; Hiemstra,T.; VanRiemsdijk,W.H. *Geochim. Cosmochim. Acta* **1997**, *61* , 2398.

68 Barrow,N.J.; Bowden,J.W.; Posner,A.M.; Quirk,P.J. *Aust. J. Soil. Res.* **1980**, *18*, 395.

69 Bargar,J.R.; Brown,Jr,G.E.; Parks,G.A. *Geochim Cosmochim Acta* **1997**, *61*, 2639.

70 Filius,J.D.; Hiemstra,T.; VanRiemsdijk,W.H. *J. Colloid Interface Sci.* **1997**, *accepted.*

71 Venema,P.; Hiemstra,T.; VanRiemsdijk,W.H. *J. Colloid Interface Sci.* **1997**, *192*, 94.

Chapter 6

Interlayer Molecular Structure and Dynamics in Li-, Na-, and K-Montmorillonite–Water Systems

Fang-Ru Chou Chang[1], N. T. Skipper[2], K. Refson[3], Jeffrey A. Greathouse[4], and Garrison Sposito[5]

[1]Lamont-Doherty Earth Observatory of Columbia University, Palisades, NY 10964–8000
[2]Department of Physics and Astronomy, University College, Gower Street, London WC13 6BT, United Kingdom
[3]Department of Earth Sciences, University of Oxford, Parks Road, Oxford OX1 3PR, United Kingdom
[4]Department of Chemistry, University of the Incarnate Word, 4301 Broadway, San Antonio, TX 78209
[5]Earth Sciences Division, Mail Stop 90/1116, Lawrence Berkeley National Laboratory, Berkeley, CA 94720

The molecular structure of water and the distribution of counterion species at the montmorillonite-water interface are of abiding interest in surface chemical science. It is often hypothesized that the interlayer region of low-order montmorillonite hydrates (up to three adsorbed water layers) can be pictured as a two-dimensional ionic solution. Our recent Monte Carlo and molecular dynamics simulations of interlayer configurations in the low-order hydrates of Li-, Na-, and K-Wyoming montmorillonite are examined comparatively in order to assess the accuracy of the ionic solution picture. The results indicate that local ordering of water molecules around the interlayer cations is indeed like that in a concentrated ionic solution. However, competition between the cations and water protons for negatively-charged sites in the clay layer, and between the cations and clay structural hydroxyls for itinerant water oxygens, also contribute significant complexity to the molecular structure of water in montmorillonite interlayers.

Sposito and Prost (*1*) concluded 15 years ago from their review of the vibrationally-averaged structure (*V structure*, atomic motions averaged over many vibrational cycles) of water adsorbed by smectites that "the V structure of adsorbed water can be regarded principally as a result of superposed electric fields originating from the silicate surface and the exchangeable cations, when the water content is relatively low". The context in which this conclusion was drawn makes it clear that the influence of exchangeable cations (counterions) is paramount, with the charged sites in the smectite layers bordering the interlayer region passively donating electrons for H-bonding between adsorbed water molecules and the siloxane surfaces of the smectite layers. A decade later, Güven (*2*) found no reason to alter this "interlayer ionic solution" picture in his comprehensive examination of clay-water systems. He stated that "interlamellar hydration of clays at low relative water vapor pressures is primarily determined by the interactions of the interlayer cation with the water molecules. Additional contributions arise from the interactions of the clay surfaces with water dipoles and interlayer cations." Güven (*2*) added a corollary conclusion, following a concept developed by Parker (*3*), that, for given values of the structural surface charge density (σ_0) and counterion valence, the relative importance of water-siloxane surface interactions--as compared to water-exchangeable cation interactions--will grow as the radius of the exchangeable cation grows. For the structural surface charge density typical of smectites [$\sigma_0 \approx -0.1$ C m^{-2} (*4*)], approximately equal importance was assigned to the two kinds of interaction when the exchangeable cation is K^+ (*2, 3*). Thus, adsorbed water on K-smectites should exhibit a V structure that reflects more or less an equal influence of the counterions and the clay mineral surface.

These heuristic concepts of interlayer molecular structure for smectite-water systems in low-order hydration states [water activity ≤ 0.95 (*5*)] can be deepened by examining the chemical literature on cation hydration in aqueous solutions, particularly at high electrolyte concentrations (H_2O/cation molar ratio < 15). This literature has been reviewed by Enderby and Neilson (*6*) and, more recently, by Ohtaki and Radnai (*7*). Experimental data on the configurations of water molecules interacting with the Group IA metal cations are of particular quality and abundance, so as to permit the testing and refinement of computer simulation techniques designed to predict the structure and dynamics of monovalent cation hydration shells (*7*). These latter simulations, based on established Monte Carlo and molecular dynamics methods (*8*), have indeed led to deeper insight concerning the behavior of electrolyte solutions and to the design of more effective experiments to probe the coordination environments of hydrated ions (*6*).

The synergistic relationship between experiment and modeling that exists in the study of aqueous electrolyte solutions has not heretofore been possible in the study of smectite-water systems because of both theoretical and computational limitations (*9*). The latter difficulties have been nearly obviated by the availability of a generation of supercomputers exemplified in the Cray J90. The theoretical obstacles also have been largely removed by the emergence of reasonable models for water-smectite and cation-smectite potential functions based on quantum mechanical parameterization (*10-13*). Skipper et al. (*11, 12*) and Refson et al. (*13*) demonstrated the feasibility of these new models in their detailed Monte Carlo and molecular dynamics simulations of high-charge smectites bearing Na^+ or Mg^{2+} counterions and hydrated by either one or two layers of adsorbed water. These developments, in turn, have encouraged the undertaking of systematic computer simulation studies of low-charge smectite hydrates, particularly hydrated Wyoming-type montmorillonites (*14-21*). The results have generated a number of questions that require additional molecular experimentation (*22*), thus continuing the model-experiment spiral characteristic of liquid-state research (*8*).

The present chapter is an attempt at conceptual synthesis concerning smectite-water systems, designed to help sharpen the experimental issues that should be addressed in the next generation of laboratory investigations. The focus of our discussion is on Li-, Na-, and K-montmorillonite, based on the results of our previous simulation studies carried out separately for these homoionic smectites (*16-18*). The emphasis in our present analysis is on comparison with respect to the type of interlayer cation, in order to respond more definitively to the question of whether the aqueous phases in low-order montmorillonite hydrates are essentially "two-dimensional ionic solutions" (*1, 2*).

Simulation Methodology

Monte Carlo (MC) and molecular dynamics (MD) simulations are well-established and essential components of research in liquid-state physical chemistry (*8, 23*). The underlying paradigm in these simulations is to construct potential functions that represent all of the known interactions in a system of ions, atoms, and molecules, then devise a strategy for sampling the phase space of the interacting system in order to compute its properties. In a MC simulation, the configuration space of the system is sampled randomly under the guidance of an algorithm (Metropolis method) based in equilibrium statistical mechanics (*23, 24*). In a MD simulation, the phase space of the system is sampled through numerical integration (Beeman algorithm) of the Newton-Euler equations of motion for each molecular species, performed consistently with the suite of potential functions assumed (*8, 25*). Convergence of a MC simulation is monitored by examining the stability of calculated system properties (e.g., the average potential energy) as sampling proceeds, i.e., a convergence profile (*23*). In MD simulations, these same properties are monitored over time (*8, 23*). For the computer simulations described in the present chapter, the MC calculations were performed using the code MONTE (*24*), developed by N.T. Skipper and K. Refson, whereas the MD calculations utilized the code MOLDY (*25*) developed by K. Refson. Numerical calculations were carried out on Cray C90 or J90 supercomputers at the San Diego Supercomputer Center, the Pittsburg Supercomputing Center, or the National Energy Research Scientific Computing Center.

Physical Modeling. The results of molecular simulation cannot be more intrinsically accurate than the potential functions on which they are based (*purgamentum init, exit purgamentum*). A choice has been made in the present study to use water-water and cation-water potential functions based on *ab initio* (i.e., quantum mechanical) calculations, as opposed to optimized empirical models. The MCY potential function (*26*), developed from *ab initio* computations on water dimers in a variety of configurations, was chosen because its description of electron correlation and exchange interactions is more accurate than that in empirical models, and because its commitment to the tetrahedral bonding network in bulk liquid water is less strong than what occurs in empirical potential functions that have been optimized on the bulk water structure. This latter characteristic turns out to be important for accurate prediction of the properties of water molecules in the highly-constrained geometric environment offered by the interlayer region of low-order smectite hydrates. Comparative analyses and a full discussion of the water-water and cation-water potential functions are given by Skipper et al. (*11, 14*) and Chang et al. (*17*).

Although considerable progress has been made toward *ab initio* calculation of water-smectite interactions (*27-29*), the potential function used in the present study to represent these interactions was developed from a semi-empirical approach (*10, 11, 14*), in which siloxane O and octahedral-sheet OH are represented by MCY-type species, and the ions in the 2:1 clay layer are assigned effective charges, the latter being equal to the free-cation charge for octahedral Al^{3+}, Mg^{2+}, or Li^{+}, and equal to +1.2e for Si in each of the two tetrahedral sheets. This effective charge for Si lies between the Pauling estimate, +1.0e, and the "bond-covalency" estimate, +1.8e (*29*). Negative surface charge from isomorphic substitutions (*4*) is created by reducing the effective cation charge by 1.0e at any substitution site.

The representation of the smectite-water system by these model potential functions does involve significant approximation, but we have found that it identifies the most important features of the clay mineral-water-cation interaction leading to the experimentally observed V structure of adsorbed water in smectite interlayers (*10-18*).

Sampling Strategy. The simulation cell we used encloses about eight smectite unit cells, with boundaries cutting the *c*-axis of the 2:1 clay layer in the middle of its octahedral sheet. Negligible effects on MC convergence were observed after tripling this cell volume, or after altering the shape of its faces from rectangular (*14*). In order to avoid computational "edge effects", the cell is replicated periodically in three dimensions. The ions and molecules placed in each cell then interact mutually and with their periodic images. Artifacts related to the imposition of a periodic lattice are obviated by applying a real space cut-off to the short-range interactions (e.g., exponential terms in the MCY potential function), limiting their extent to < 9 Å in an "all image" convention (*14, 25*). This cut-off cannot be applied, however, to the coulomb interactions between ions without causing serious problems of inaccuracy for the simulation. Skipper et al. (*11, 14*) dealt with this important matter by using the Ewald sum method (*8, 25*) in conjunction with a reciprocal-space cutoff of $k < 3$ Å^{-1} to represent long-range coulomb interactions beyond the real-space cut-off distance. This choice of k results in < 0.1% inaccuracy when the Ewald-sum energy is calculated.

The MC simulations were performed at T = 300 K and with an applied stress σ_{zz} (normal to the clay layers) of 100 kPa (isothermal-isostress ensemble). The total interlayer potential energy, layer spacing, interlayer density profiles, orientation of water molecules, and radial distribution functions (RDF) were recorded every 500 steps. Among different simulations of the same hydrate, the results deemed "best" were those giving the lowest and most stable total interlayer potential energy and layer spacing as seen in convergence profiles (*14*). Equilibrium interlayer molecular configurations (with fixed layer spacings) as obtained from the final output of the "best" MC simulations were used as initial configurations for MD simulations (isothermal-isochoric ensemble). A typical MD simulation covered a period of 200 ps with 0.5 fs imposed as the timestep. This simulation time scale is appropriate to a realization of the V structure of adsorbed water (*1*). Intermediate configurations were saved every 100 timesteps in order to calculate self-diffusion coefficients (*D*), interlayer density profiles, orientation of water molecules, and trajectories of either water molecules or counterions.

Monte Carlo simulations should lead to the same thermodynamic properties when identical macroscopic parameters, but differing initial conditions, are specified. This ideal goal is achievable, however, only if the initial conditions are selected prudently and a large enough portion of the system phase space is explored in the simulations (*8*). We have found the selection of the initial configuration to be critical, not only to thermodynamic consistency, but also to system stability, especially for multilayer hydrates. An increasing number of water molecules in the interlayer region implies an increasing number of ways to distribute counterions initially among the water molecules. With more water molecules comes also a weakening of the screened coulomb forces that bind the system together and a growth in the number of "energy barriers" that can trap the system in local energy minima. These barriers make it difficult to equilibrate the system, or to sample a sufficient region of its phase space, within a feasible amount of computer time.

For example, we have found (*16, 17*) similar interlayer potential energy values among rather different cation configurations to be separated by "energy barriers", making it difficult to determine the correct "equilibrium" cation configuration accurately within an affordable simulation time. These "energy barriers" prevent counterions from passing through the water network and thereby make sampling all configurations an impossible task. This problem of a system becoming trapped in a too-small region of phase space means that MC simulations started with different initial configurations can result spuriously in different "equilibrium" interlayer molecular structures. Nonetheless, we have found (*16, 17*) that MC simulations started from different initial configurations were consistent in their thermodynamic properties (e.g., the mean interlayer potential energy varied by < 4%, and the layer spacing varied by < 5%) and in the positions of counterions bound directly to the clay mineral surface.

In three-layer hydrates, however, drift of the layer spacing was found if the simulation was started with counterions placed very near octahedral surface charge sites. This increase in layer spacing was usually accompanied by a partitioning of the midplane water molecules into networks. The MCY potential function used in the present study is known to have the problem of underestimating the mass density of bulk liquid water (*11*). This did not cause difficulty in simulating one- and two-layer hydrates, where coulomb forces still hold the system together. But, in the three-layer hydrate, this force no longer suffices if the clay layer charge is substantially neutralized by cations placed very near the mineral surface. Thus, it is necessary to place the counterions that are to be initially over octahedral charge sites in the middle of a simulation cell, instead of near the charge sites, to stabilize the interlayer water structure as well as the layer spacing variation.

These experiences led us to a phase-space sampling strategy that included runs with differing initial placement of the counterions and "optimizing runs" prior to a full MC simulation (*16, 17*). The latter were designed to provide the most reliable estimates of water molecule configurations subject to the subsequent movement of clay mineral layers and counterions. In the first "optimizing run" (10 - 100K steps), only water molecules were permitted to move, while the layer spacing and counterion positions remained fixed in their initial positions. In the second "optimizing run" (10 - 100K steps), water molecules moved and the layer spacing (under 100 kPa normal stress) was permitted to vary. Thereafter, all interlayer species moved and the upper clay mineral layer was allowed to move one step in any direction for approximately every five movements of interlayer water molecules (up to as many as 10M steps), until MC convergence.

Water structure network formation, an important mechanism to prevent counterions from moving away too quickly from their optimal positions, is particularly crucial in the early stages of MC simulation because the water molecules are arranged randomly at the outset. Our "optimizing runs", in which counterion movement is restricted, allowed the layer spacing to fluctuate and water molecules to equilibrate with the initial counterion configuration. If the initial counterion configuration is close to the equilibrium one, extensive "optimizing runs" are beneficial to ensure system stability. Thus, the true initial conditions for the MC simulations included the water molecule configurations and layer spacing achieved after the "optimizing runs" were completed.

Surface Complexes

Cations adsorbed on the basal planes of smectites can be immobilized in two kinds of surface complex (*4*). The surface complex is *inner-sphere* if the cation is bound directly to a cluster of surface oxygen ions, with no water molecules interposed, and it is *outer-sphere* if one or more water molecules is interposed between the cation and the siloxane surface to which it binds (*30*). Thus, adsorbed cations in outer-sphere surface complexes retain solvation shells, whereas those in inner-sphere surface complexes can at most be only partially solvated. Spectroscopic data (*31*) suggest that monovalent cations in inner-sphere surface complexes remain immobilized on a timescale of *ca.* 100 ps, whereas those in outer-sphere surface complexes are able to diffuse along the siloxane surface over this timescale.

Figures 1 and 2 are "snapshots" of equilibrium interlayer configurations, based on MC simulations (*21*), in Li-beidellite ($Li_{0.75}[Si_{7.25}Al_{0.75}]Al_4O_{20}(OH)_4$), which has isomorphic substitutions only in its tetrahedral sheets, and in Li-hectorite

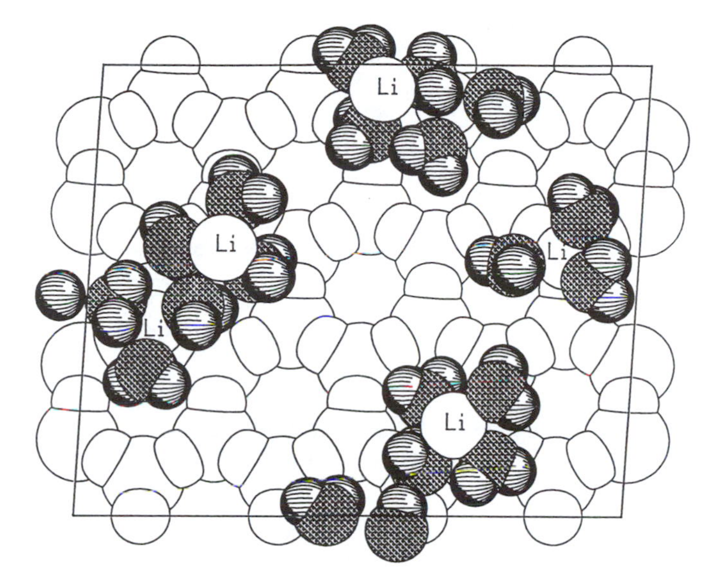

Figure 1. Plan view of the interlayer configuration in $Li(H_2O)_3$-beidellite, with the MC simulation cell outlined .

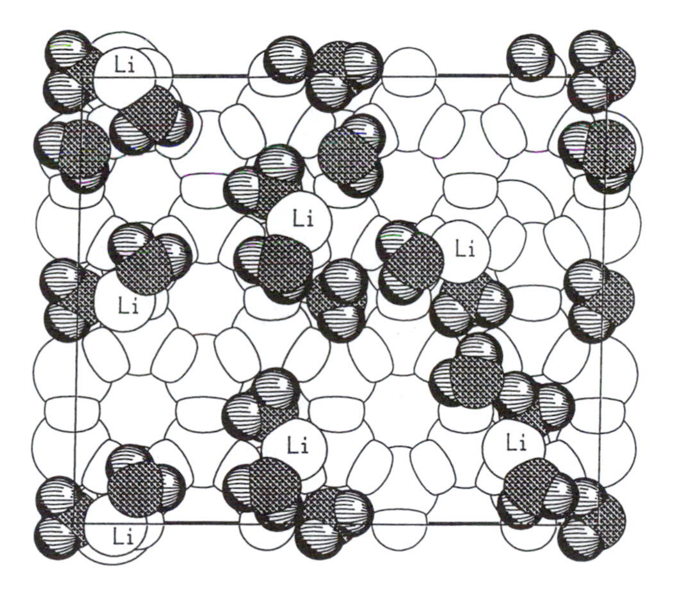

Figure 2. Plan view of the interlayer configuration in $Li(H_2O)_3$-hectorite, with the MC simulation cell outlined .

($Li_{0.75}[Si_8](Mg_{5.25}Li_{0.75})O_{20}(OH)_4$), which has substitutions only in its octahedral sheet (*4*). The view in each figure is along the *c*-axis, with only one of two opposing siloxane surfaces shown. All of the counterions and water molecules in a simulation cell (outlined) are depicted as prescribed in the MC output. The gravimetric water content in these simulations was fixed at *ca.* 5%, giving an average of three water molecules per interlayer Li^+ ($H_2O/Li = 3$).

All of the surface complexes shown in Figure 1 are inner-sphere, with Li^+ bound directly to a site comprising three surface oxygen ions. These oxygen ions form the base of a tetrahedron in which Al has been substituted for Si ("tetrahedral charge site"). Either two, three, or four water molecules solvate each Li^+. The counterions on the removed upper siloxane surface thus appear as unobscured discs in Figure 1, while those on the lower siloxane surface are obscured from view by their solvating water molecules. Figure 3 shows the trajectories of the solvating water molecules (gray) and the Li^+ they solvate (black) as observed in a MD simulation over 200 ps (*21*). The perspective is the same as in Figure 1, with the Li^+ trihydrate surface complex near the right edge of the simulation cell being the one selected for viewing. The hopping of the Li^+ around the three surface oxygen ions on which the layer charge deficit is localized (*4*) is apparent, as are the sympathetic translational motions of the three solvating water molecules that are located at the interlayer midplane.

By contrast, Figure 2 indicates only outer-sphere surface complexes have formed with the "octahedral charge sites"in Li-hectorite, most of them comprising just two solvating water molecules. The remaining water molecules have keyed themselves into the ditrigonal cavities of the siloxane surface (one such molecule is at the center of the simulation cell), evidently attracted by the proton in the structural OH group to be found at the bottom of each cavity (*4, 20*). In trioctahedral clay minerals, such as hectorite, these OH groups point directly toward the siloxane surface, whereas they point almost orthogonally to the *c*-axis of the 2:1 layer (i.e., toward the empty metal octahedral site) in dioctahedral clay minerals, such as beidellite (*4*). In the Li-hectorite system, it is water molecules that are bound directly to the clay mineral surface, while the counterions reside at the interlayer midplane--just the inverse of the interlayer configuration seen in the Li-beidellite system.

These two examples illustrate a trend, noted previously by Skipper et al. (*15*) in their extensive MC simulations of the interlayer configurations of 2:1 clay minerals bearing one monolayer of adsorbed water and Na^+ counterions: that monovalent adsorbed cations tend to form inner-sphere surface complexes with tetrahedral charge sites, but preferentially form outer-sphere surface complexes with octahedral charge sites. This trend can be associated with the smaller interaction distance between a counterion and a charged isomorphic substitution site--when considering a tetrahedral site as opposed to an octahedral site deeper in the clay layer. Close proximity of a charged site to the cation evidently is conducive to its desolvation during immobilization into a surface complex.

Counterion Solvation

Wyoming-type montmorillonite [layer charge x = 0.75, with 1/3 being tetrahedral charge sites (*4*)] bearing one monolayer of adsorbed water has a water/counterion molar ratio of 5 1/3, equivalent to a 10.4 *m* solution. For two or three water monolayers, the H_2O/counterion ratio increases proportionately and the equivalent solution molalities are 5.2 *m* or 3.46 *m*, respectively. Thus, from the perspective of counterion solvation, interlayer water on montmorillonite should be similar to a very concentrated aqueous solution.

The local coordination structure of water near Li^+, Na^+, and K^+ has been investigated experimentally by diffraction and spectroscopic methods (*6*), most

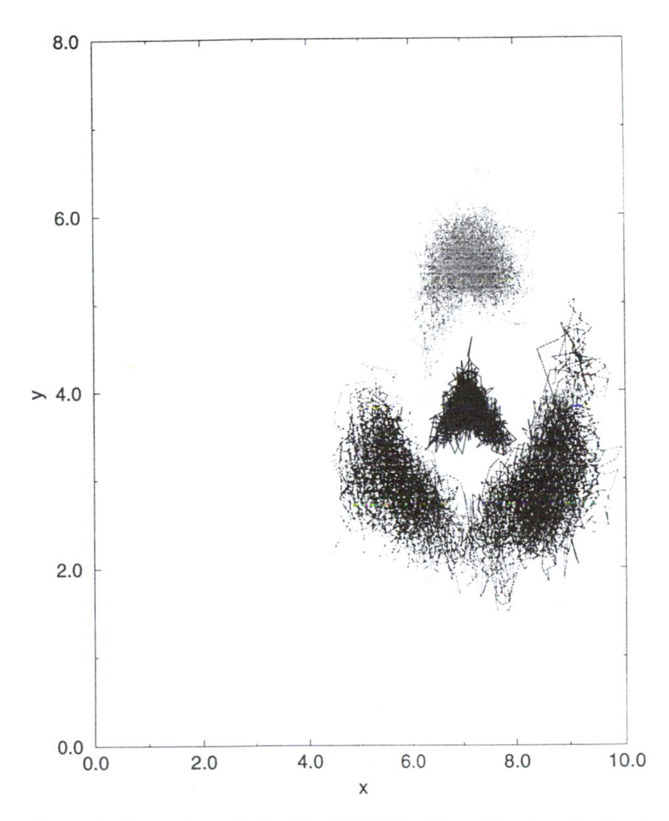

Figure 3. Trajectories of adsorbed Li^+ (black) and its threehydrating water molecules (gray), observed over 200 ps in a MD simulation of the interlayer region in $Li(H_2O)_3$-beidellite. This $Li(H_2O)_3^+$ species also appears at the right boundary of the MC simulation cell in Figure 1.

notably by X-ray and neutron diffraction and extended X-ray absorption fine structure spectroscopy (*7*). The cations Li^+ and K^+ can be examined in concentrated aqueous solutions with neutron isotopic difference diffraction, whereas Na^+ can be studied with X-ray isomorphic difference diffraction (*6*). The diffraction data analysis gives M^+-O (M = Li, Na, or K) distances of 1.95 ± 0.02, 2.4 ± 0.2 Å, and 2.8 ± 0.1 Å for Li^+, Na^+, and K^+, respectively--essentially the cation radius plus 1.4 Å for the radius of water O. Ohtaki and Radnai (*7*) have reviewed the published literature on the solvation structure near monovalent cations in concentrated aqueous solutions, noting that experimental M^+-O distances can often depend on the method of measurement (e.g., smaller distances inferred from neutron diffraction than from X-ray diffraction) and on solution concentration (e.g., smaller distances as the H_2O/M ratio drops below 5). The three M^+-O distances cited above are at the lower end of the range to be found in the published literature (*7*).

Figure 4 shows the M^+-O RDF for Li^+, Na^+, and K^+ in the one-layer hydrate of Wyoming-type montmorillonite, $M^+_{0.75}[Si_{7.75}Al_{0.25}]Al_{3.5}Mg_{0.5}O_{20}(OH)_4$, as computed by MC simulation (*16-18*). This RDF gives the average number of O atoms (i.e., water molecules) within an infinitesimal interval at a chosen radial distance from an adsorbed cation (Li^+, Na^+, or K^+) which has been placed at the origin of coordinates (*8, 22*). The first peak in the RDF gives the nominal nearest-neighbor M^+-O distance, which is 1.95 ± 0.05 Å (Li^+), 2.30 ± 0.05 Å (Na^+), or 2.8 Å (K^+) for the three adsorbed cations, in close agreement with the M^+-O distances cited above for concentrated aqueous solutions (*6, 7*). A second solvation shell for Li^+ at about 4.6 Å is indicated in Figure 4, which is close to the second-shell distance of 4.4 ± 0.1 Å inferred from diffraction measurements on aqueous solutions (*7*). Complex structure is seen in the RDF for Na^+-O, with peaks near 4.0 and 4.8 Å [MD simulations of Na^+ in aqueous solution also give second-shell distances ranging from 4.44 to 4.80 Å (*7*)]. The RDF for K^+-O beyond the first peak is essentially structureless, with a very weak maximum at 5.3 Å, the same "second-shell" distance as predicted by MC simulations of K^+ in aqueous solutions (*7*).

These features of the M^+-O RDF, which persist as well in the two- and three-layer hydrates (*16-18*), suggest that the near-neighbor coordination structure of water molecules about interlayer Li^+, Na^+, and K^+ is indeed very similar to that found in concentrated aqueous solutions, irrespective of the type of surface complex formed by the counterion. [Both types of surface complex are observed for monovalent counterions on montmorillonite, because the clay mineral has both tetrahedral and octahedral charge sites (*4, 16-20*).] What effect, then, does the clay mineral surface have? The answer to this question comes from an examination of cation mobility, as revealed by MD simulation (*17, 18*).

Table I shows self-diffusion coefficients (T = 300 K) for the three interlayer cations in low-order hydrates of montmorillonite, as calculated conventionally from the slopes of graphs of the (three-dimensional) mean-square cation displacement *versus* time (*8, 16-18*). Experimental values of the cation self-diffusion coefficients in aqueous solution also are listed (*32*). It is apparent that monovalent cation mobility in the one-layer hydrate is at best a few percent of that in bulk aqueous solution, and that the mobility increases significantly with increasing water content, to approach about 25% of the bulk-solution value in the three-layer hydrate. The constrained geometry and the charge sites on the clay mineral surface thus act to retard significantly the diffusive motions of interlayer cations through adsorbed water.

The restricted interlayer mobility of the cations can be seen qualitatively in MD trajectory plots projected onto the basal planes (YX plane) and onto the plane of the clay mineral *c*-axis (ZX plane). These are shown in Figures 5 and 6 for K^+ trajectories in the interlayer of Wyoming-type montmorillonite (*18*). Figure 5, for the one-layer hydrate, indicates sporadic (over 200 ps) entrapment of K^+ in the siloxane

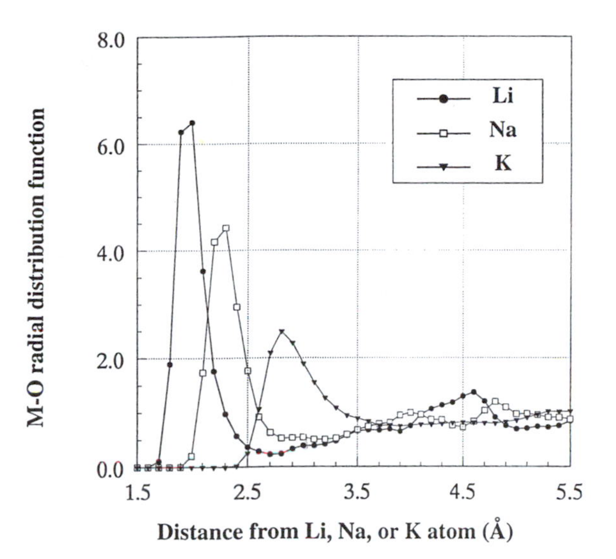

Figure 4. Radial distribution functions for M^+-O (M = Li, Na, or K) in one-layer hydrates of montmorillonite.

Table I. Self-Diffusion Coefficients for Monovalent Adsorbed Cations on Wyoming-Type Montmorillonite

Hydrate[a]	*Cation*	$D_{MD}(10^{-9}\ m^2\ s^{-1})$[b]
One-Layer	Li^+	0.011
	Na^+	0.015
	K^+	0.12
Two-Layer	Li^+	0.067
	Na^+	0.253
	K^+	0.25
Three-Layer	Li^+	0.43
	Na^+	0.11
	K^+	0.2-0.8[c]
Bulk Solution	Li^+	1.0
	Na^+	1.4
	K^+	2.0

[a] Montmorillonite hydrate D_{MD} values based on 200 ps MD simulations (*17, 18*); bulk solution values are experimental (*32*).

[b] The smallest value of D_{MD} calculable in a 200 ps MD simulation is about $10^{-11}\ m^2\ s^{-1}$. Typical experimental values of D for monovalent counterions range from 10^{-12} to $2 \times 10^{-10}\ m^2\ s^{-1}$ for one- to three-layer hydrates (*17, 18*).

[c] Three-layer hydrate of K-montmorillonite is hypothetical, with the value of D_{MD} sensitively dependent on the layer spacing (*18*).

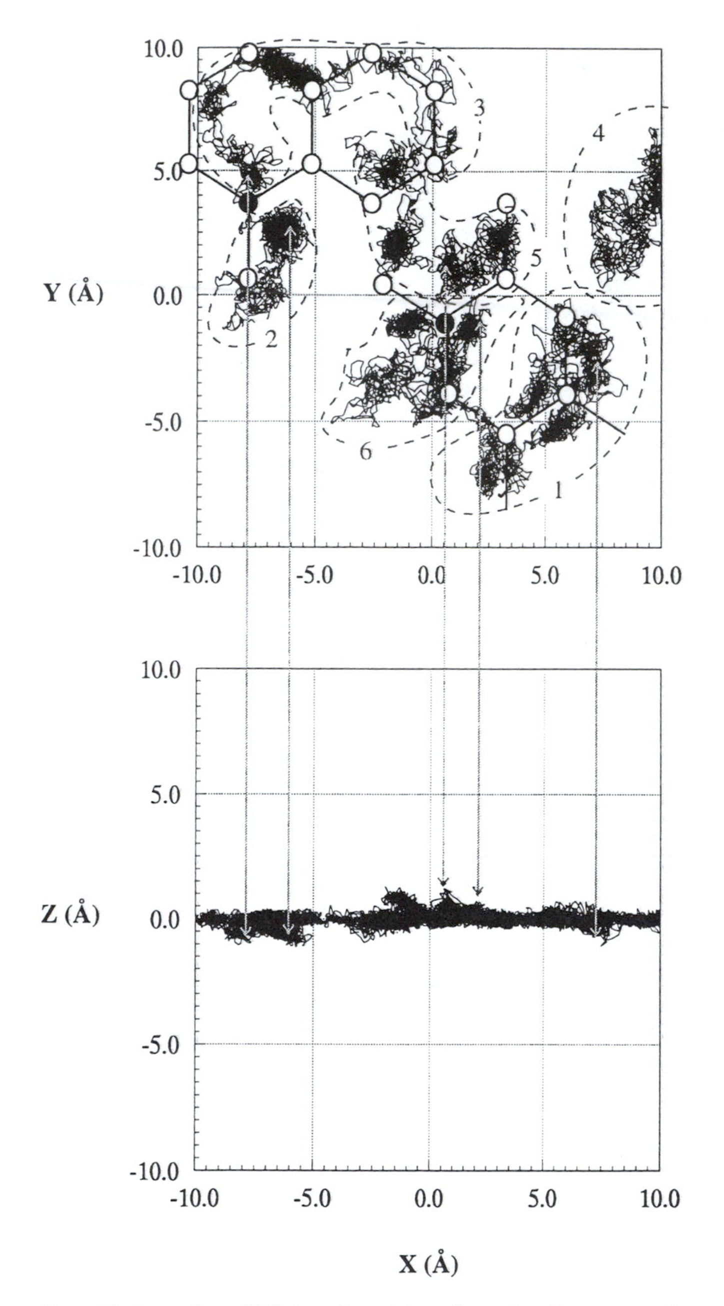

Figure 5. Trajectories of K^+ (numbered 1 to 6) on the siloxane surfaces of the one-layer hydrate of K-montmorillonite, observed during 200 ps in a MD simulation. Upper figure, plan view; lower figure, section view along the *a*-axis of the clay layer. Filled circles are tetrahedral charge sites. (Reproduced from reference 21. Copyright 1998 American Chemical Society.)

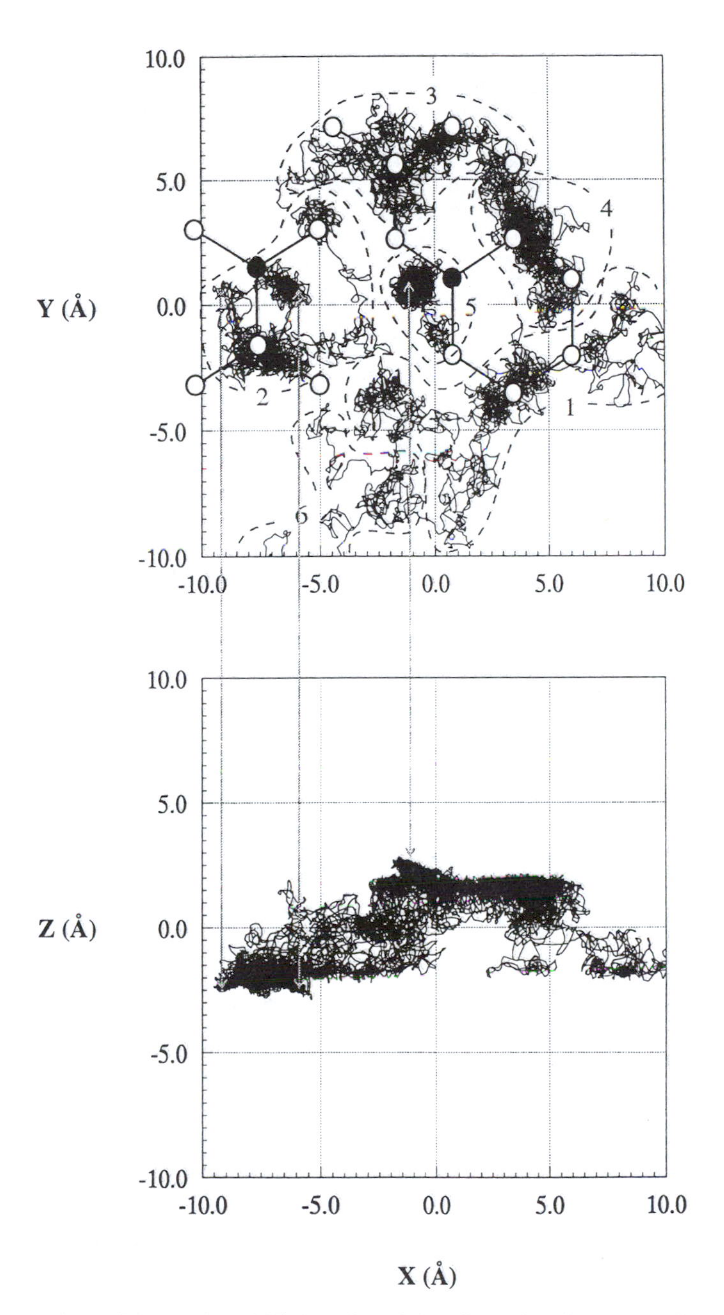

Figure 6. Trajectories of K^+ (numbered 1 to 6) on the siloxane surfaces of the two-layer hydrate of K-Montmorillonite, observed during 200 ps in a MD simulation. Upper figure, plan view; lower figure, section view along the *a*-axis of the clay layer. Filled circles are tetrahedral charge sites. (Reproduced from reference 21. Copyright 1998 American Chemical Society.)

ditrigonal cavities (V-shaped features in the ZX plane) near the two tetrahedral charge sites (denoted by filled circles in the YX plane). The rather limited excursion on the siloxane surface for the K^+ changes in the two-layer hydrate (Figure 6), as the K^+ begin to roam more widely on the surface and diffuse readily across the interlayer region. This surface-speciation-exchange behavior occurs more readily for K^+ (18) than for Li^+ (*17*), because the former cation interacts more weakly with water molecules (*6, 7*) and thus diffuses more easily through the adsorbed water network.

Interlayer Water Structure

Figure 7 shows the O-O RDF for monolayer hydrates of Li-, Na-, and K-montmorillonite, based on MC simulations (*16-18*). The first peak reflects the average number of water molecules that are nearest-neighbors of a water molecule placed at the origin of coordinates (*23*). This peak occurs at 2.8 Å for Li^+ and Na^+, whereas for K^+ it occurs at 2.9 Å. In bulk liquid water, the first peak is also found at 2.8 Å, both experimentally and theoretically (*23*). This peak position may be compared to that for the water dimer, 2.98 Å (*23*), to illustrate the more complex organization of water molecules in both adsorbed water and in the bulk liquid. The nearest-neighbor peak height is about 25% larger in MCY bulk water (*15*) than it is for Li^+ or Na^+ in Figure 7, and for K^+ it is about 70% larger. These differences indicate a less-ordered local structure around a water molecule in adsorbed water than in the bulk liquid, a characteristic that can be inferred also from the limited diffraction data available on montmorillonite hydrates (*22, 33, 34*). Evidently, the degree of structural disorder is substantially greater for water in the K-montmorillonite hydrate than in those of Li- or Na-montmorillonite.

Secondary peaks in the O-O RDF for adsorbed water appear near 4.6 Å and 5.4 Å, indicating the average position of next-nearest neighbors to the water molecule at the origin. In bulk liquid water, a broad next-nearest neighbor peak appears at 4.5 Å, followed by a low minimum in the O-O RDF at 5.5 Å (*23*). Moreover, the O-O RDF for liquid water has a much shallower minimum between 3.2 and 4.0 Å than what appears for the RDF in Figure 7 (*23, 35*). This "filling-in" of the region between the first and second peaks for bulk liquid water is associated with "interstitial" water molecules that are not H-bonded to the water molecule at the origin (*35*)--structural complexity that is not apparent for adsorbed water in the monolayer hydrates. On the other hand, the secondary peak near 5.4 Å, which is absent for bulk water, suggests that some water molecules in the clay mineral hydrates are arranged around the water molecule at the origin merely according to molecular size (i.e., 5.4 Å $\approx$ 2.8 Å + 2.8 Å), as would be the case predominately if H-bonding did not exist (*23*).

Further insight as to the structure of interlayer water in the one-layer hydrates can be obtained from Figure 8, which shows O-H RDF based on MC simulations (*16-18*). The first peak in these RDF represents water molecules that are H-bonded to a water proton not placed at the origin of coordinates (*15, 23*). This peak occurs at 1.8 Å for MCY water (*17*), and at 2.2 Å, 2.0 Å, or 2.1 Å for the monolayer hydrates of Li-, Na-, and K-montmorillonite, respectively. These significant differences point to correspondingly large differences in the average length of a H-bond, with those in the montmorillonite hydrates being systematically longer than those in liquid water. The second peak in the O-H RDF, representing the water proton not H-bonded to the water molecule at the origin of coordinates, would occur at 3.2 Å if strictly linear H-bonds formed (15). The second peak in fact does occur at about 3.2 Å for liquid water (*23*). Its occurrence at 3.4 Å instead for the three monolayer hydrates in Figure 8 indicates that the H-bonds in adsorbed water are bent significantly, by contrast with the bulk liquid (*23*).

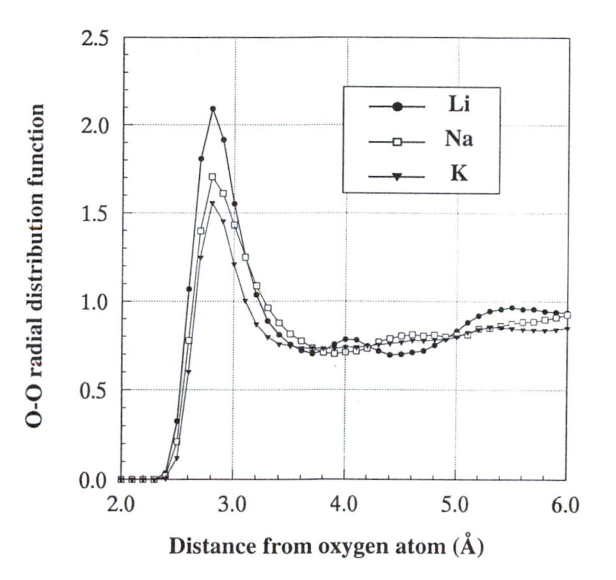

Figure 7. Radial distribution functions for O-O in one-layer hydrates of montmorillonite.

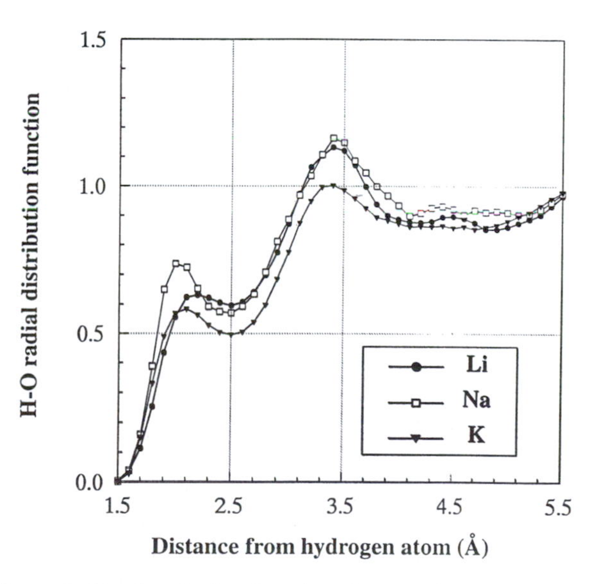

Figure 8. Radial distribution functions for O-H in one-layer hydrates of montmorillonite.

In the two-layer hydrates, the O-O and O-H RDF (Figures 9 and 10) change shape considerably. The first peak in the O-O RDF occurs at 2.8 Å in all three hydrates, with the peak height for Na-montmorillonite shifting downward relative to its value in the one-layer hydrate. There is greater depletion and less oscillation in the O-O RDF at distances beyond the nearest-neighbor position. Correspondingly, the two peaks in the O-H RDF occur at approximately the same distances in all three hydrates (1.9-2.0 Å and 3.3 Å, respectively). These results suggest that the major differences in structure between interlayer water in the two-layer hydrate and bulk liquid water are restricted to distances below about 3 Å around a water molecule. Powell et al. (*22*) have reached the same conclusion based on a very recent neutron diffraction study of the two-layer hydrate of Na-Wyoming montmorillonite.

These trends, which continue in the three-layer hydrates (*16-18*), suggest that, with increasing layer coverage, a more liquid-like structure is developing in adsorbed water on montmorillonite. However, the mobility of the water molecules remains much below that in bulk liquid water. Averaged for the three different clay mineral hydrates, the MD-calculated self-diffusion coefficients (D_w) of a water molecule are (at 300 K): $1.2 \pm 0.2 \times 10^{-10}$ m^2 s^{-1} (one-layer); $7 \pm 2 \times 10^{-10}$ m^2 s^{-1} (two-layer); and $12 \pm 4 \times 10^{-10}$ m^2 s^{-1} (three-layer), as compared to 23×10^{-10} m^2 s^{-1} for the bulk liquid (*17, 18, 36*). Water mobility increases by an order of magnitude between one and three monolayers, but still achieves only about 50% of the liquid water mobility. In terms of residence time, a water molecule in the one-layer hydrate oscillates in place before diffusing for about 100 ps [= $2a^2/3D_w$, where a = 1.44 Å is the van der Waals radius of a water molecule (*36*)]. This residence time is of the same order of magnitude as the residence time of an outer-sphere surface complex on a 2:1 clay mineral (*1*). The residence time of a water molecule in the bulk liquid is about 5 ps, and that of a water molecule in the first solvation shell of Li^+ in aqueous solution is about 30 ps (*7, 36*). These two latter values bracket the residence times for water molecules in the two- and three-layer hydrates of montmorillonite, 20 ps and 12 ps, respectively.

Concluding Remarks

Figures 11 to 13 are color snapshots of the MC equilibrium interlayer configurations for the one-layer hydrates of Li-, Na-, and K-montmorillonite (Wyoming-type), respectively. The view in each figure is along the *b*-axis of the montmorillonite unit cell. One simulation cell in the interlayer region is shown, so there are six counterions and 32 water molecules (*15*). The upper clay mineral layer has been removed for the sake of clarity. Prominent in the lower, visible clay mineral layer are its tetrahedral charge site (highlighted as a colored sphere with much lighter value than the others in the same sheet of tetrahedra) and the dioctahedral-sheet OH groups, the upper ones pointing into the plane of the figure, the lower ones out of the plane (OH proton in lighter value). The tetrahedral charge site in the upper, invisible clay layer is just above the cation [denoted by a blue (Li), yellow (Na), or pink (K) sphere] that is nearest to the right above the cation hovering over the tetrahedral charge site in the lower clay layer.

The complexity of adsorbed water molecule orientations is especially evident in Figure 11, which depicts the monolayer hydrate of Li-montmorillonite. Strong Li^+-water molecule interactions in this system produce a local hydrate structure like that in concentrated aqueous solutions of LiCl, but Li^+ interactions with tetrahedral charge sites also are strong and lead to inner-sphere surface complex formation with the oxygen ions of the clay mineral surface. As Figures 1 and 2 also illustrate, the configuration of water molecules (and counterions) differs substantially between inner-sphere and outer-sphere surface complexes. When these two surface complexes are forced to cohabit within the constrained spatial domain that exists in an interlayer region, disorder in the water network is likely, with distorted H-bonds and water

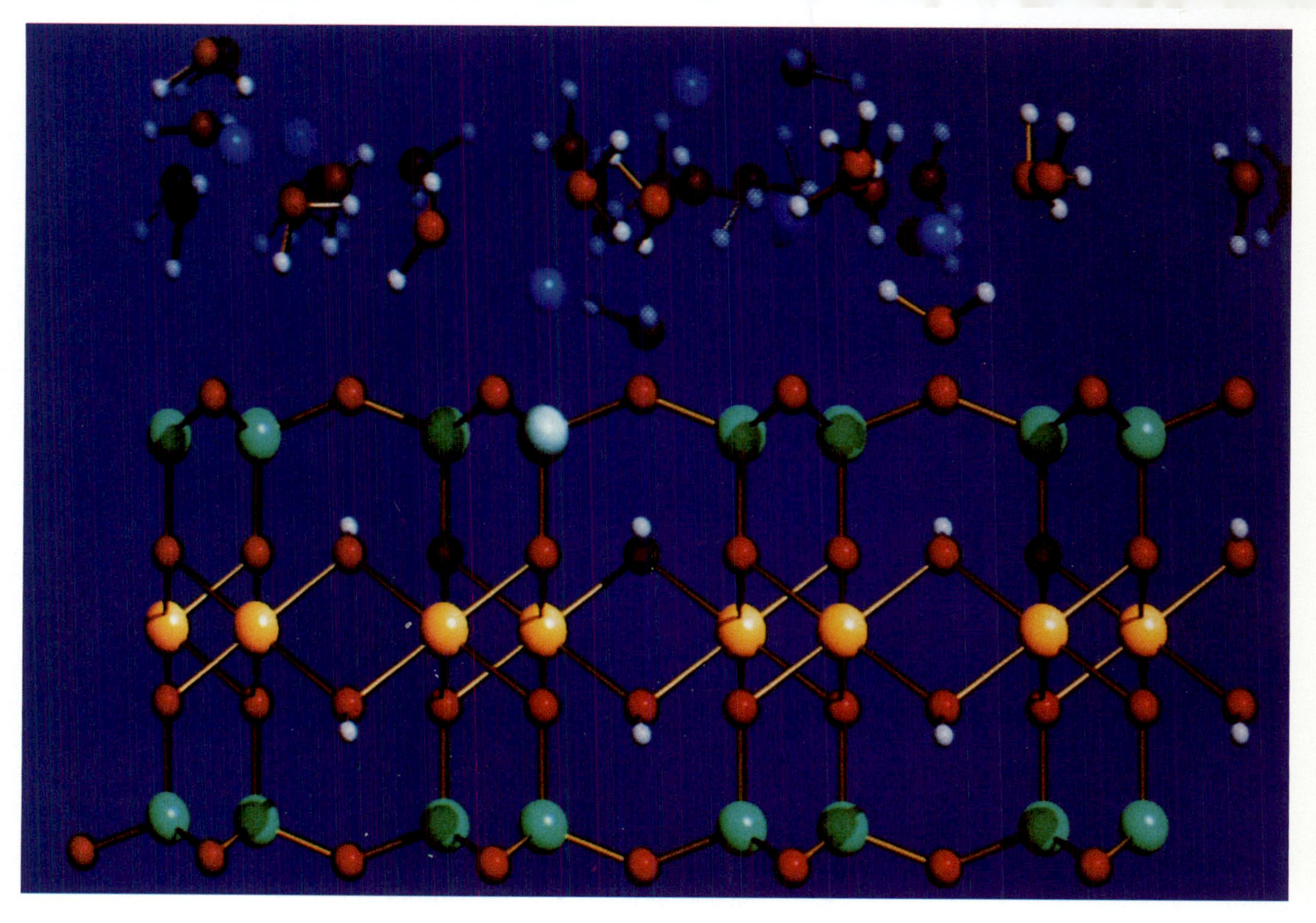

Figure 11. Section view, along the clay layer *b-axis*, of interlayer molecular structure in the one-layer hydrate of Li-montmorillonite.

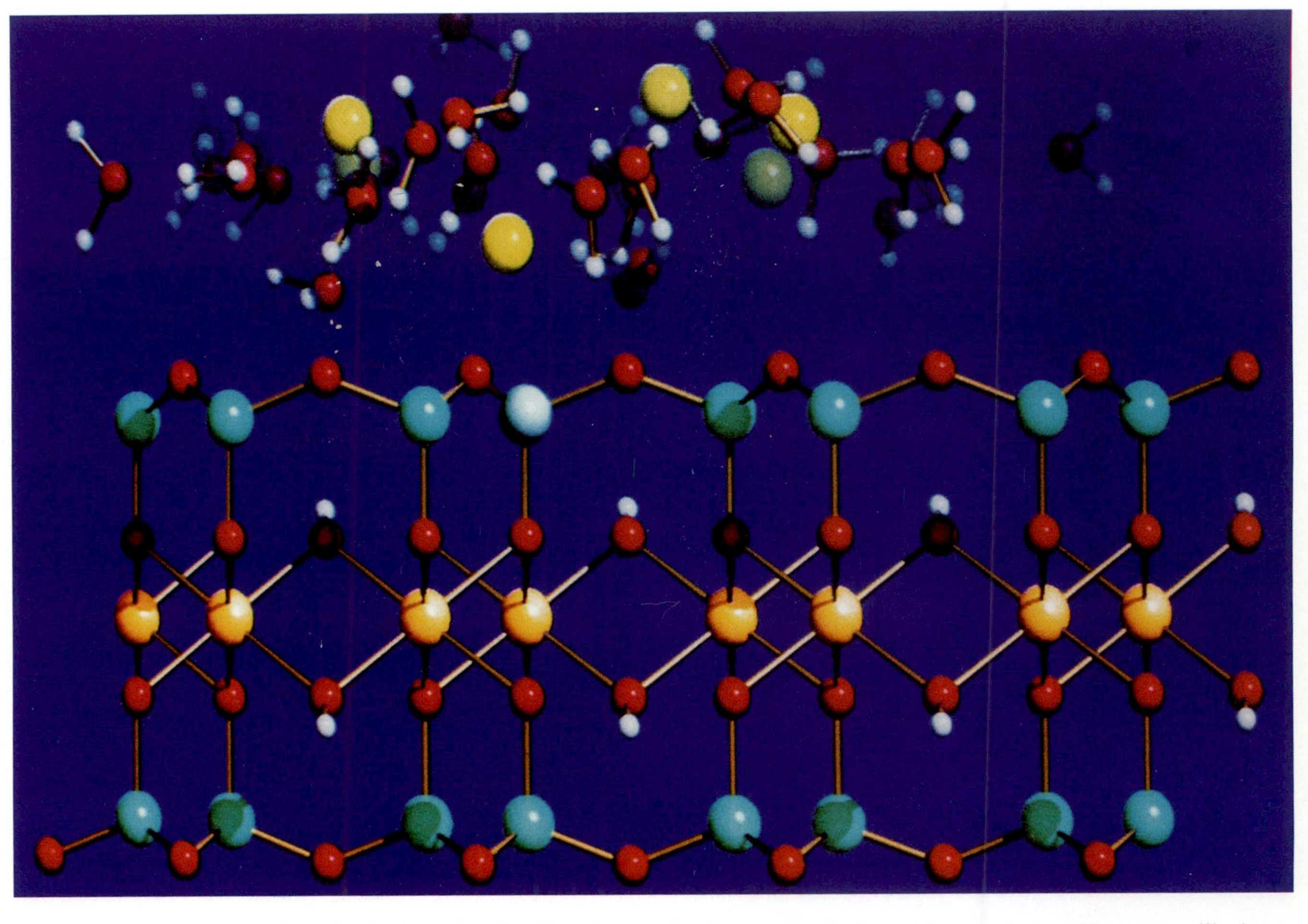

Figure 12. Section view, along the clay layer *b-axis*, of interlayer molecular structure in the one-layer hydrate of Na-montmorillonite.

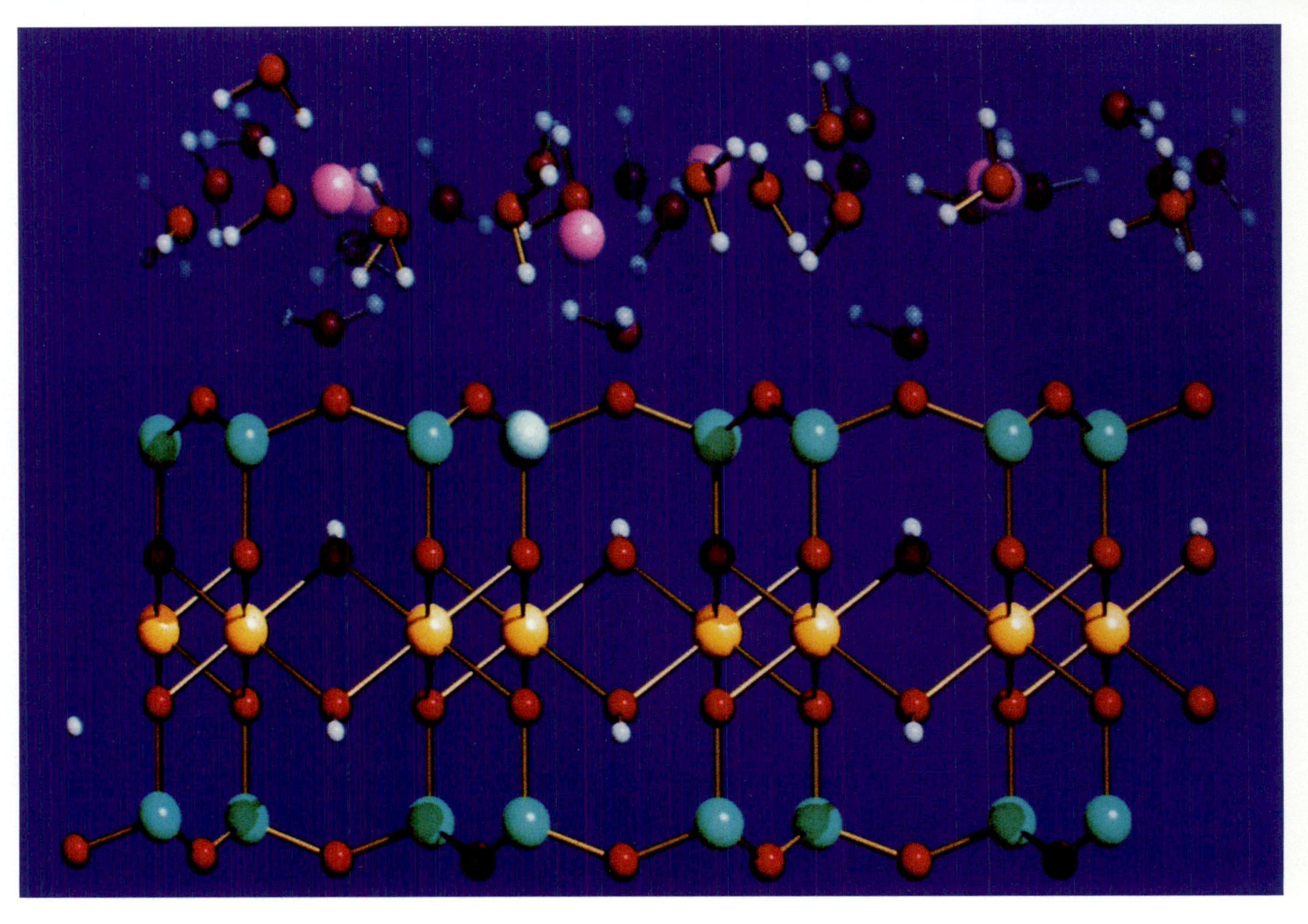

Figure 13. Section view, along the clay layer *b-axis*, of interlayer molecular structure in the one-layer hydrate of K-montmorillonite.

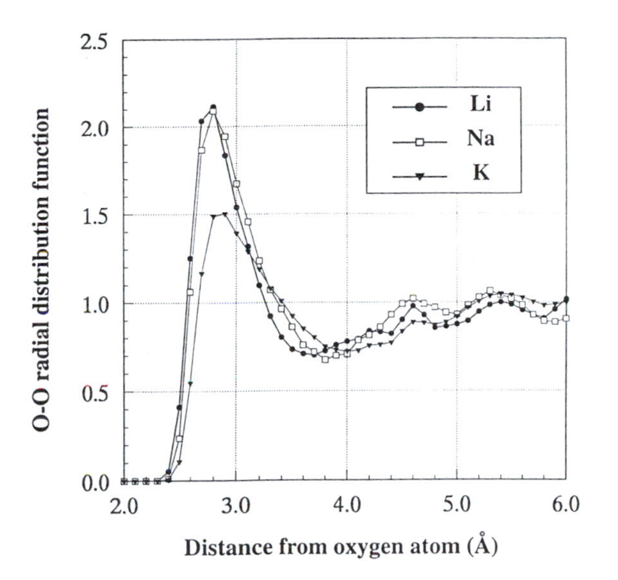

Figure 9. Radial distribution functions for O-O in two-layer hydrates of montmorillonite.

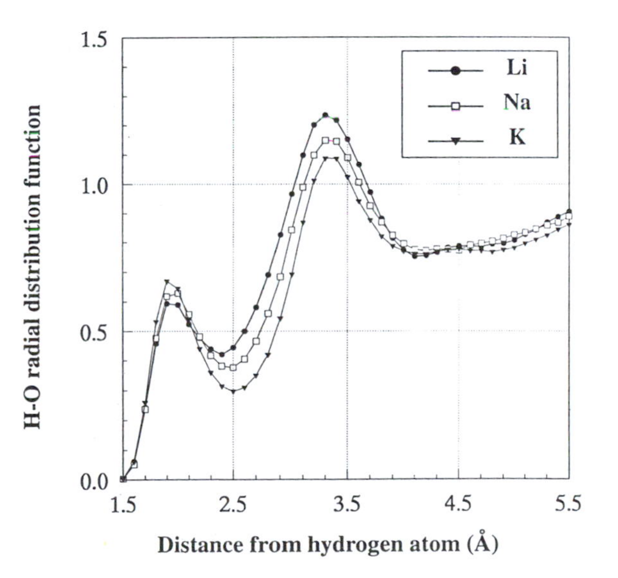

Figure 10. Radial distribution functions for O-H in one-layer hydrates of montmorillonite.

dipole orientations taking on almost every possible direction (*17*). This disorder is enhanced by the perverse attraction between the water dipoles and the protons of the structural OH groups in the octahedral sheet of the clay layer, a spinoff of the relative lability of outer-sphere complex formation (Figure 2). Two such "break-away" water molecules are observable in Figure 11, both of which are just below (but are not nearby) the Li^+ that is bound in an inner-sphere complex to the tetrahedral charge site in the lower clay layer.

The Na^+-water molecule interaction is not as strong as that of Li^+-water, with the result that the octahedral charge sites in the clay layer attract water protons [represented in the MCY model (*26*) as two positive charges, each equal to 0.7175e] competitively with Na^+ counterions. The final, compromise orientation adopted by many of the water molecules in the monolayer hydrate of Na-montmorillonite is apparent in Figure 12, and is typified by the leftmost water molecule in the snapshot. The water dipole points in a direction parallel to the clay mineral surfaces, with one hydroxyl group directed toward each of the latter (*16*), an orientation that should also occur if no counterions were present, because of the negative layer charge. Some water molecules are attracted to the proton in the structural OH groups in the clay layer, while the two Na^+ bound into inner-sphere surface complexes are located nearer to the interlayer midplane than were Li^+ in the same species. For K-montmorillonite (Figure 13), these last two features of the interlayer region are even more pronounced and, although not as sharply exemplified pictorially as for Na-montmorillonite, the tendency toward parallel alignment of the water dipoles with the clay mineral surface actually is even greater (*18*).

The three summarizing MC snapshots thus indicate structural complexity, despite the essential correctness of an "interlayer ionic solution" picture of the low-order hydrates of montmorillonite with monovalent counterions (*1-3*). Water molecules in the interlayers of these hydrates tend to organize in response to cation-water interactions, but the counterion spatial position is determined largely by how closely the counterion can approach an attracting charge site in the clay layer. If an inner-sphere surface complex is formed by a counterion, water molecules are relegated to the interlayer midplane. An outer-sphere surface complex, on the other hand, permits greater competition for negative layer-charge sites between water protons and counterions, as well as greater opportunities for water oxygens, instead of binding to the counterions, to seek out the protons in structural OH groups that point toward the interlayer region. This latter type of interaction appears to be just as important as H-bonding between water protons and the siloxane surface. Increasing monovalent cation size leads to an increasing tendency of water molecules, instead of organizing themselves simply into cation solvation shells, to interact with layer-charge sites and structural OH groups instead. This trend is accompanied by an increasing tendency of the counterions to diffuse readily along the water layers on the clay mineral surfaces (Table I).

Acknowledgments. The research reported in this paper was supported in part by NSF Grant EAR-9505629, and in part by the Lawrence Berkeley National Laboratory under its LDRD program. Without the unfailing support of Dr. John Maccini, U.S. National Science Foundation, the research described herein would not have been possible. The authors thank the San Diego Supercomputer Center, the Pittsburgh Supercomputer Center, and the National Energy Research Scientific Computing Center, for allocations of CPU time on Cray supercomputers. Thanks to Sung-Ho Park, Rebecca Sutton, and Tetsu Tokunaga for helpful comments on this chapter while in draft form, and to Angela Zabel for excellent preparation of the typescript.

Literature Cited

(1) Sposito, G. Prost, R. *Chem. Rev.* **1982**, *82*, 553.
(2) Güven, N. In *Clay-Water Interface and its Rheological Implications;* Güven, N., Pollastro, R.M., Eds.; The Clay Minerals Society: Boulder, CO, 1992; pp 1-79.
(3) Parker, J.C. In *Soil Physical Chemistry*; Sparks, D.L., Ed.; CRC Press: Boca Raton, FL, 1986; pp 209-296.
(4) Sposito, G. *The Surface Chemistry of Soils*; Oxford University Press: New York, 1984; pp 1-77.
(5) Bérend, I.; Cases, J.-M.; François, M.; Uriot, J.-P.; Michot, L.; Masion, A.; Thomas, F. *Clays Clay Miner.* **1995**, *43*, 324.
(6) (a) Enderby, J.E.; Neilson, G.W. *Rep. Prog. Phys.* **1981**, *44*, 593. (b) Neilson, G.W.; Enderby, J.E. *Adv. Inorg. Chem.* **1989**, *34*, 195.
(7) Ohtaki, H.; Radnai, T. *Chem. Rev.* **1993**, *93*, 1157.
(8) Allen, M.P.; Tildesley, D.J. *Computer Simulation of Liquids*; Clarendon Press, Oxford, 1987; pp 1-205.
(9) Bleam, W.F. *Rev. Geophys.* **1993**, *31*, 51.
(10) Skipper, N.T.; Refson, K.; McConnell, J.D.C. *Clay Miner.* **1989**, *24*, 411.
(11) Skipper, N.T.; Refson, K.; McConnell, J.D.C. *J. Chem. Phys.* **1991**, *94,* 7434.
(12) Skipper, N.T.; Refson, K.; McConnell, J.D.C. In *Geochemistry of Clay-Pore Fluid Interactions;* Manning, D.C., Hall, P.L., Hughs, C.R., Eds.; Chapman and Hall: London, 1993; p 40.
(13) Refson, K.; Skipper, N.T.; McConnell, J.D.C. In *Geochemistry of Clay-Pore Fluid Interactions*; Manning, D.C., Hall, P.L., Hughs, C.R., Eds.; Chapman and Hall: London, 1993; p 1.
(14) Skipper, N.T.; Chang, F.-R.C.; Sposito, G. *Clays Clay Miner.* **1995**, *43*, 285.
(15) Skipper, N.T.; Sposito, G.; Chang, F.-R.C. *Clays Clay Miner.* **1995**, *43*, 294.
(16) Chang, F.-R.C.; Skipper, N.T.; Sposito, G. *Langmuir* **1995**, *11*, 2734.
(17) Chang, F.-R.C.; Skipper, N.T.; Sposito, G. *Langmuir* **1997**, *13*, 2074.
(18) Chang, F.-R.C.; Skipper, N.T.; Sposito, G. *Langmuir* **1998**, *14*, 1201.
(19) Boek, E.S.; Coveney, P.V.; Skipper, N.T. *J. Am. Chem. Soc.* **1995**, *117*, 12608.
(20) Karaboni, S.; Smit, B.; Heidug, W.; Urai, J.; van Oort, E. *Science* **1996**, *271*, 1102.
(21) Greathouse, J.; Sposito, G. *J. Phys. Chem. B* **1998**, *102*, 2406.
(22) Powell, D.H., Tongkhao, K.; Kennedy, S.J.; Slade, P.G. *Clays Clay Miner.* **1997**, *45*, 290.
(23) (a) Beveridge, D.L.; Mezei, M.; Mehroortra, P.K.; Marchese, F.T.; Ravi-Shanker, G.; Vasu, T.; Swaminathan, S. In *Molecular-Based Study of Fluids*; Haile, J.M., Mansoori, G.A., Eds.; American Chemical Society: Washington, DC, 1983; pp 297-351. (b) Bopp, P. In *The Physics and Chemistry of Aqueous Ionic Solutions*; Bellissent-Funel, M.-C., Neilson, G.W., Eds.; D. Reidel: Boston, MA, 1987; pp 217-243.
(24) Skipper, N.T. *MONTE User's Manual*; Department of Physics and Astronomy, University College: London, 1996.
(25) Refson, K. *MOLDY User's Manual*; Department of Earth Sciences, University of Oxford: Oxford, UK, 1996. [www.earth.ox.ac.uk/%7Ekeith/moldy.html]
(26) Matsuoka, O.; Clementi, E.; Yoshimine, M. *J. Chem. Phys.* **1976**, *64*, 1351.
(27) Bridgeman, C.H.; Buckingham, A.D.; Skipper, N.T.; Payne, M.C. *Mol. Phys.* **1996**, *89*, 879.
(28) Bridgeman, C.H.; Skipper, N.T. *J. Phys. Condens. Matter* **1997**, *9*, 4081.
(29) Teppen, B.J.; Rasmussen, K.; Bertsch, P.M.; Miller, D.M.; Schäfer, L. *J. Phys. Chem. B*, **1997**, *101*, 1579.
(30) Sposito, G. *Soil Sci. Soc. Am. J.* **1981**, *45*, 292.
(31) Sposito, G. In *Ion Exchange and Solvent Extraction*; Marinsky, J.A.; Marcus, Y., Eds.; Marcel Dekker: New York, 1993, Vol 11; pp 211-236.

(32) Nye, P.H. *Adv. Agron.* **1979**, *31*, 225.
(33) Pezerat, H.; Méring, J. *C.R. Acad. Sci. Paris, Sér. D* **1967**, *265*, 529.
(34) Hawkins, R.K.; Egelstaff, P.A. *Clays Clay Miner.* **1980**, *28*, 9.
(35) Kusalik, P.; Svishchev, I.M. *Science* **1994**, *265*, 1219.
(36) Sposito, G. *J. Chem. Phys.* **1981**, *74*, 6943.

Sorption of Inorganic Species

Chapter 7

Kinetics and Mechanisms of Metal Sorption at the Mineral–Water Interface

Donald L. Sparks[1], André M. Scheidegger[2], Daniel G. Strawn[1], and Kirk G. Scheckel[1]

[1]**Department of Plant and Soil Sciences, University of Delaware, Newark, DE 19717–1303**
[2]**Waste Management Laboratory, Paul Scherrer Institut, CH-5232 Villigen, Switzerland**

The rates and mechanisms of sorption reactions at the mineral/water interface are critical in determining the mobility, speciation, and bioavailability of metals in aqueous and terrestrial environments. This chapter discusses nonequilibrium aspects of metal sorption at the mineral/water interface, with emphasis on confirmation of slow sorption mechanisms using molecular approaches. It is shown that there is often a continuum between sorption processes, viz, diffusion, sites of varying energy states, and nucleation of secondary phases. For example, recent molecular level in-situ studies have shown that metal adsorption and "surface" precipitation can occur simultaneously.

Contamination of aqueous and terrestrial environments with metals and semi-metals derived from industrial, municipal, and agricultural sources is a major concern worldwide. Industries and government are spending millions of dollars determining what pollutants are present in contaminated soils and ground water and implementing strategies to remediate subsurface environments. In many cases, however, speciation of the metals is inaccurate and difficult and the mechanisms by which metals are retained in soils and aquatic systems over long periods of time are not understood.

Dynamic reactions in the subsurface environment are critical in affecting the fate and transport of metals, as well as a number of other important processes (Figure 1). Soils and sediments have a remarkable ability to sorb metals. Sorption reactions at the mineral/water interface significantly affect the mobility, speciation, and bioavailability of trace metal ions in aquatic and soil environments. Therefore, one must precisely understand the kinetics and mechanisms of metal sorption on mineral surfaces to accurately predict the fate of such pollutants in subsurface environments and to facilitate effective environmental remediation procedures.

Adsorption can be defined as the accumulation of a substance or material at an interface between the solid surface and the bathing solution. It is strictly a two-dimensional process and does not include three-dimensional processes such as "surface" precipitation, and diffusion into the sorbent structure. Although in

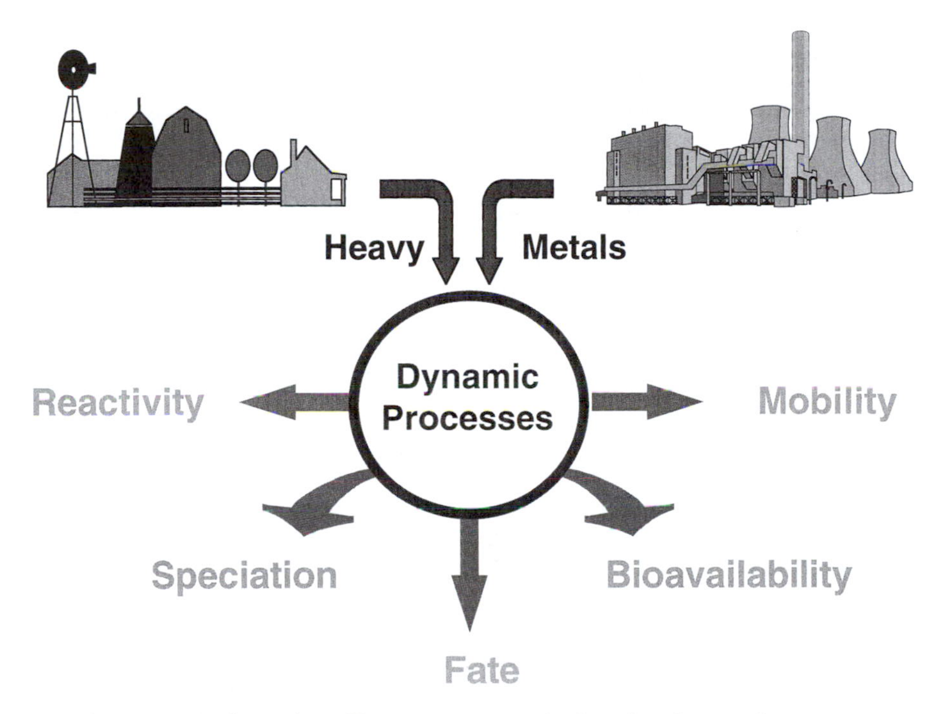

Figure 1. Metal reaction effects on processes in the subsurface environment.

many sorption studies spectroscopic techniques are used and reveal 3-dimensional products (as described later), it is often not clear whether a 3-dimensional process is actually occurring on a surface. To indicate this uncertainty, we will use the term "surface" precipitation. Adsorption, "surface" precipitation, and diffusion into the surface are collectively referred to as sorption, a general term that should be used when the retention mechanism at a surface is unknown. Although some scientists still use the term adsorption, sorption is a better term particularly when precipitation and absorption phenomena have not been eliminated as possible retention mechanisms.

The forces involved in adsorption can range from weak, physical, van der Waals forces (e.g., partitioning) and electrostatic outer-sphere complexes (e.g., ion exchange) to chemical interactions (Figure 2). Chemical interactions can include inner-sphere complexation that involves a ligand exchange mechanism, covalent bonding, hydrophobic bonding, hydrogen bonding, hydrogen bridges, and steric or orientation effects (*1,2*). Inner-sphere complexes can be either monodentate or bidentate (Figure 2).

With increasing levels of metal sorption, (i.e., increasing surface coverage) a "surface" precipitate can form. When the precipitate consists of chemical species derived from both the aqueous solution and dissolution of the mineral, it is referred to as a coprecipitate. There is often a continuum between surface complexation (adsorption) and "surface" precipitation.

Equilibrium Models

Most of the research on metal sorption at the mineral/water interface has dealt with equilibrium aspects. Numerous studies have used macroscopic approaches such as adsorption isotherms, empirical and semi-empirical equations (e.g., Freundlich, Langmuir), and surface complexation models (e.g., constant capacitance, triple layer) to describe adsorption, usually based on a 24 hour reaction time.

It should be recognized that adsorption isotherms are purely descriptions of macroscopic data and do not definitively prove a reaction mechanism. For example, the conformity of experimental adsorption data to a particular isotherm does not indicate that this is a unique description of the experimental data, and that only adsorption is operational. Thus, one cannot differentiate between adsorption and other sorption processes, such as "surface" precipitation, and diffusion using an adsorption isotherm, even though this has been done in the geochemistry literature.

Surface complexation models are chemical models that are based on molecular descriptions of the electric double layer using equilibrium derived adsorption data (*3*). Thus, no mechanistic information on sorption can be obtained. Surface complexation models often describe sorption data over a broad range of experimental conditions such as varying pH and ionic strength and have been used widely to describe metal cation and anion sorption reactions on metal (hydr)oxides, clays, and soils, and organic ligand and competitive sorption reactions on oxides. However, surface complexation models employ an array of

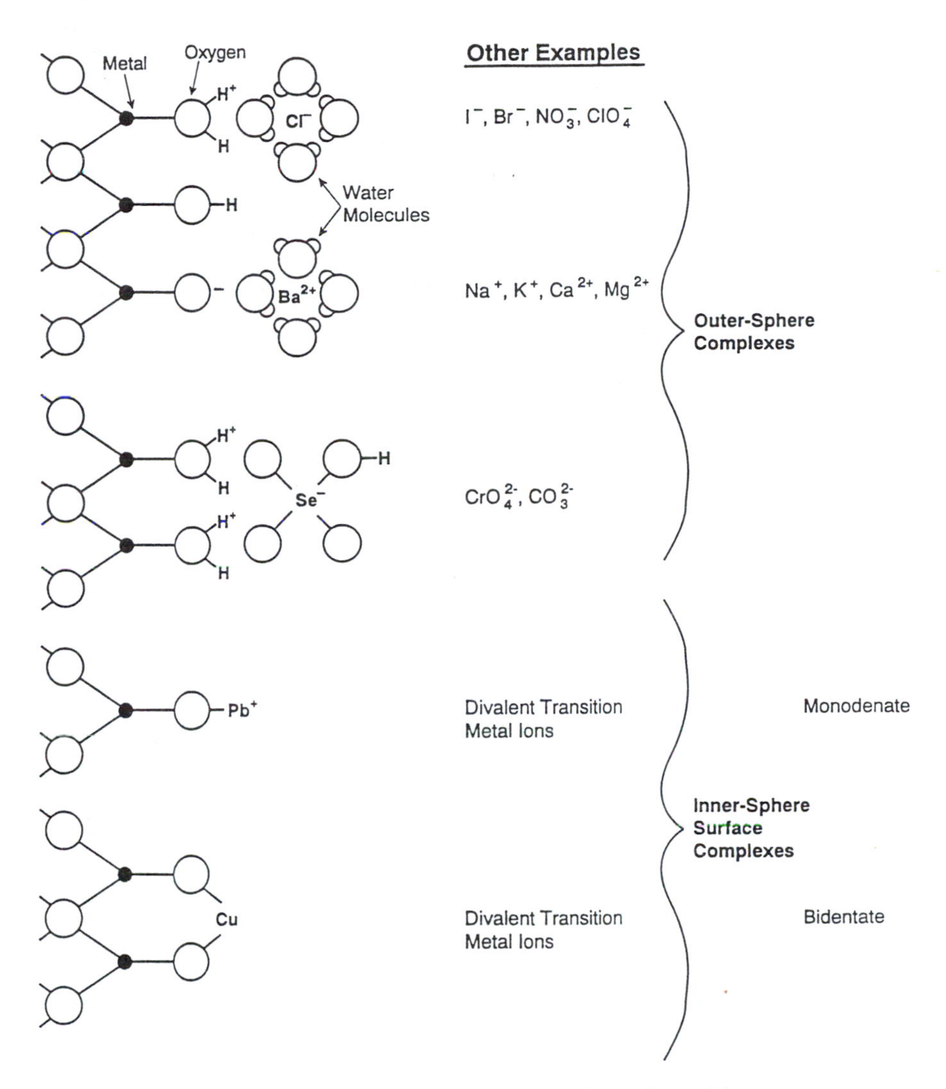

Figure 2. Schematic representation of surface complexes formed between inorganic ions and hydroxyl groups of an oxide surface. Modified from Hayes (*78*), with permission.

adjustable parameters to fit experimental data, and it has been shown that often several models may fit sorption data equally well (*4*).

Another major disadvantage of the commonly used surface complexation models, and of most equilibrium-based sorption models, is that three-dimensional surface products are not included as possible complexes. However, there are several exceptions. Farley et al. (*5*) and James and Healy (*6*) considered surface precipitation in successfully modeling sorption of hydrolyzable metal ions. Dzombak and Morel (*7*) modified the diffuse layer surface complexation model to include surface precipitation. However, these applications relied solely on macroscopic data without molecular-level identification of the sorption complex structure. Recently, Katz and Hayes (*8,9*) employed triple layer models, that included a surface solution model, a surface polymer model, and a surface continuum model to describe molecular level data for Co sorption on γ-Al_2O_3 over a wide range of surface coverages (0.1 to 100%).

Arguably, one of the major needs in modeling sorption on soils and natural materials is to include "surface" precipitation and other non-adsorption phenomena, based on molecular level data, as part of the model description and prediction. This is particularly important as recent research, based on in-situ spectroscopic analyses, indicates that metal-nucleation products form on an array of natural surfaces at low surface coverages and at relatively rapid time scales (*10-17*).

Kinetic and Molecular Approaches

In the past two decades, as concerns and interests about soil and water quality have increased, scientists and engineers have increasingly realized that reactions in subsurface environments are time-dependent. Kinetic studies can reveal something about reaction mechanisms at the mineral/water interface, particularly if energies of activation are calculated and stopped-flow or interruption techniques are employed. However, molecular and/or atomic resolution surface techniques should be employed to corroborate the proposed mechanisms hypothesized from equilibrium and kinetic studies. These techniques can be used either separately or, preferably, simultaneously with kinetic investigations (*2*).

There are two principal subdivisions in molecular spectroscopy: in-situ and non-in-situ methods (2,*18*). The principal invasive non-in-situ techniques used for soil and aquatic systems are x-ray photoelectron spectroscopy (XPS), auger electron spectroscopy (AES), and secondary mass spectroscopy (SIMS). Each of these techniques yields detailed information about the structure and bonding of minerals and the chemical species present on the mineral surfaces. XPS is one of the most widely used non-in-situ surface-sensitive techniques. It has been used to study sorption mechanisms of inorganic cations and anions such as Cu, Co, Ni, Cd, Cr, Fe, selenite, and uranyl in soil and aquatic systems (*19-28*). The disadvantage of invasive non-in-situ techniques is that they often must be performed under adverse experimental conditions, e.g., desiccation, high vacuum, heating, or particle bombardment. Such conditions may yield data that are misleading as a result of experimental artifacts (*2,29,30*). Review articles on XPS, AES, and SIMS are available (*29,31,32*).

In-situ methods require little or no alteration of the sample from its natural state (*2,18*). They can be applied to aqueous solutions or suspensions; most involve the input and detection of photons. Examples of in-situ techniques are electron paramagnetic resonance (EPR), Fourier-transform infrared (FTIR), nuclear magnetic resonance (NMR), x-ray absorption fine structure (XAFS), and Mössbauer spectroscopies. However, many other techniques are available (*2,33*).

XAFS has been heavily used to study the mechanisms of metal reactions at the mineral/water interface (*30, 33-37* and the chapter by Brown et al. in this volume for details on x-ray absorption spectroscopy and specific studies). While XAFS provides local chemical information, it provides no information on spatial resolution of surface species (*2*). Such information can only be obtained by microscopic methods. Scanning electron microscopy (SEM) and transmission electron microscopy (TEM or HRTEM - high resolution TEM) are well established methods for acquiring both chemical and micromorphological data on minerals and soils. TEM can provide information on spatial resolution of surface alterations and the amorphous nature or degree of crystallinity of sorbed species (ordering). It can also be combined with electron spectroscopies to determine elemental analysis (*2*). Scanning force microscopy (SFM) is also being increasingly used to study geochemical systems. SFM allows imaging of mineral surfaces in air or immersed in solution, and at subnanometer scale resolution (*40*). It has been applied to image mineral surfaces immersed in aqueous solutions, over the course of dissolution, precipitation, and nucleation reactions (*38-41* and the chapters by Hochella et al. and Maurice in this volume for detailed discussions).

Accordingly, to accurately predict the fate, mobility, speciation, and bioavailability of environmentally important metals and semi-metals in terrestrial and water environments, one must understand the kinetics and mechanisms of the reactions. This chapter focuses on nonequilibrium aspects of metal sorption/desorption and the confirmation of reaction mechanisms using in-situ atomic/molecular level spectroscopic and microscopic techniques.

Time Scales of Metal Sorption Reactions. Metal sorption reactions can occur over time scales ranging from milliseconds to several weeks depending on the sorbate/sorbent system. The type of sorbent can drastically affect the reaction rate. For example, sorption reactions are often more rapid on clay minerals such as kaolinite than on vermiculitic and micaceous minerals. This is in large part due to the availability of sites for sorption. Kaolinite has readily available planar external sites and sorption is often complete in minutes (*42*). Vermiculite and micas have multiple sites for retention of metals including planar, edge, and interlayer sites, with some of the latter sites being partially to totally collapsed. Consequently, sorption and desorption reactions on these sites can be slow. Often, an apparent equilibrium may not be reached even after several days or weeks. Thus, with vermiculite and mica, sorption can involve two to three different reaction rates - high rates on external sites, intermediate rates on edge sites, and low rates on interlayer sites (*43,44*).

The type of surface complex, i.e., outer-sphere versus inner-sphere (Figure

2), can also affect the rate and reversibility of metal sorption reactions. Outer-sphere complexation is usually rapid and reversible, whereas inner-sphere complexation is slower and may appear to be irreversible (*2*). Moreover, the rate of desorption from monodentate complexes may be higher than from bidentate complexes.

Metal sorption reactions on clay minerals, (hydr)oxides, humic substances and soils is usually characterized by a rapid, followed by a slow, reaction (Figure 3). The rapid reaction is ascribable to chemical reaction and film diffusion processes (*2,42*). For example, chemical reaction rates of metals on oxides occur on millisecond time scales (*45-47*). Half-times for divalent Cu, Pb, and Zn sorption on peat ranged from 5 to 15 seconds and were ascribed to film diffusion (*48*).

The slow metal sorption step on many minerals and soils occurs over time scales of days and longer. This slow sorption has been ascribed to several mechanisms including: interparticle or intraparticle diffusion in pores and solids, sites of low energy or reactivity, and "surface" precipitation/nucleation (*49-51*).

While it has been generally assumed that adsorption in comparison to "surface" precipitation is much more rapid, a recent study (*17*) has shown that "surface" precipitation processes can occur on time scales of minutes. The latter finding indicates that sorption and precipitation processes can occur simultaneously. However, in some cases, depending on reaction conditions and the metal involved, a particular sorption mechanism can dominate.

An important factor affecting the degree of slow sorption/desorption of metals is the time period the sorbate has been in contact with the sorbent (residence time). Ainsworth et al. (*52*) studied the sorption/desorption of Co, Cd, and Pb on hydrous ferric oxide (HFO) as a function of oxide aging and metal-oxide residence time. Oxide aging did not cause hysteresis of metal cation sorption-desorption. Increasing the residence time between the oxide and the metal cations resulted in hysteresis with Cd and Co but little hysteresis was observed with Pb. With Pb, between pH 3 and 5.5 there was slight hysteresis over a 21 week aging process (hysteresis varied from <2% difference between sorption and desorption to ≈ 10%). At pH 2.5 Pb desorption was complete within a 16 hour desorption period and was not affected by residence time (Figure 4). However, with Cd and Co (Figure 5), increasing hysteresis was observed as the metal/sorbent residence time increased from 2 to 16 weeks. After 16 weeks of aging 20% of the Cd and 53% of the Co was not desorbed, and even at pH 2.5, hysteresis was observed. The extent of reversibility with residence time for Co, Cd, and Pb was inversely proportional to the ionic radii of the ions, i.e., Co <Cd<Pb. Ainsworth et al. (*52*) attributed the hysteresis to Co and Cd incorporation into a recrystallizing solid (probably goethite) via isomorphic substitution and not to micropore diffusion.

Diffusion Mechanisms. Since minerals and soils are porous materials containing both macropores (>2nm) and micropores (<2nm) diffusion is a mechanism that can control the slow rate of metal sorption (Figure 6). These pores can be inter-particle (between aggregates) or intra-particle (within an individual particle).

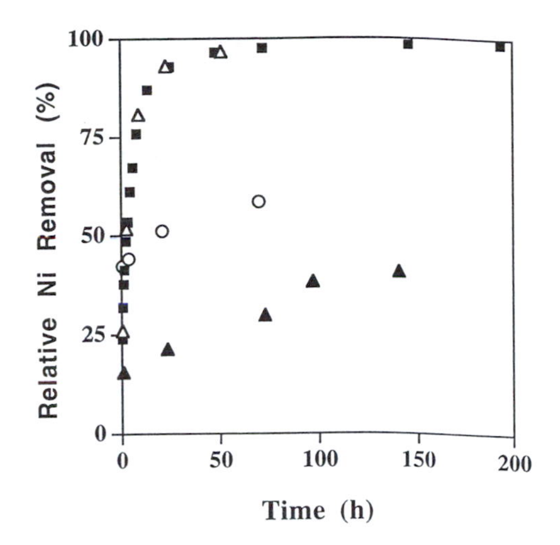

Figure 3. Kinetics of Ni sorption (%) on pyrophyllite (■), kaolinite (△), gibbsite (▲), and montmorillonite (o) from a 3 m*M* Ni solution at pH 7.5 and an ionic strength $I = 0.1$ *M* $NaNO_3$. The last sample of each experiment was collected and analyzed by XAFS. From Scheidegger et al. (*16*), with permission.

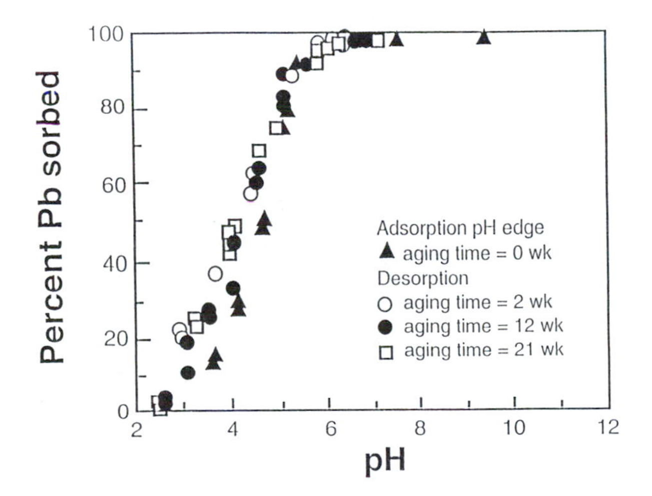

Figure 4. Fractional adsorption of Pb to hydrous Fe-oxide (HFO) as a function of pH and HFO-Pb aging time. From Ainsworth et al. (*52*), with permission.

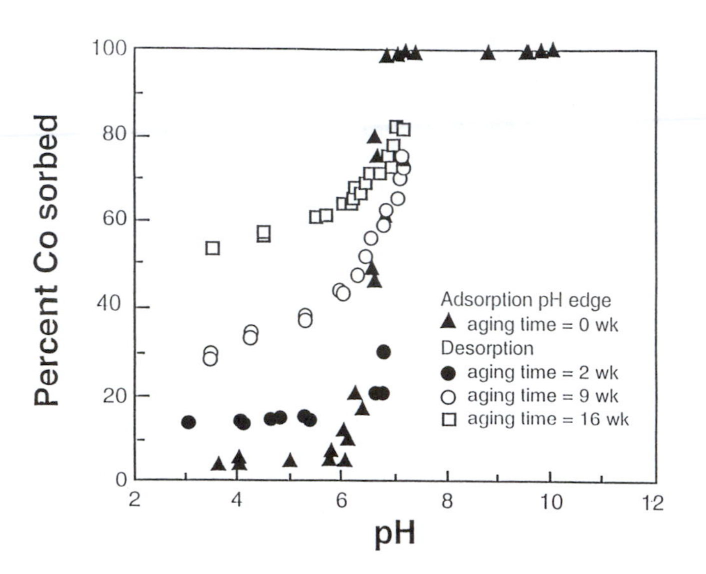

Figure 5. Fractional adsorption of Co to hydrous Fe-oxide (HFO) as a function of pH and HFO-Co aging time. From Ainsworth et al. (*52*), with permission.

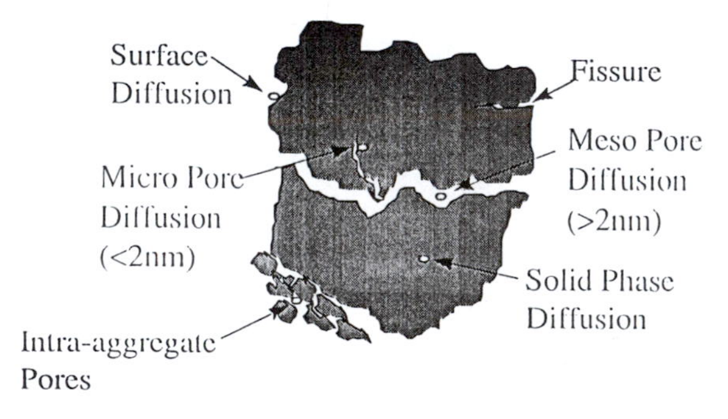

Figure 6. Diffusion processes in natural materials.

Intra-particle pores can form during weathering, upon solid formation, or may be partially collapsed interlayer space between mineral sheets, i.e., vermiculite and montmorillonite. The rate of diffusion through a pore is dependent on pore size, particle size, tortuosity, chemical interactions, chemical flux, and whether the pore is continuous or discontinuous. Besides pore diffusion, solid-phase diffusion is also a transport-limited process. Solid phase diffusion is dependent on the characteristics and interactions of the diffusant and the solid (*53*). Since there exists a range of diffusion rates in the soil, it follows that with increasing residence time the fraction of contaminants in the more remote areas of particles (accessible via slow diffusion) will increase. This slow sorption phenomenon is often the explanation researchers use to account for the slow continuous sorption and desorption observed between metals and natural materials (*42,50,54*).

Bruemmer et al. (*55*) studied Ni, Zn, and Cd sorption on goethite, a porous iron oxide known to have defects within the structure in which metals can be incorporated to satisfy charge imbalances. At pH 6, as reaction time increased from 2 hours to 42 days (at 293K), sorbed Ni increased from 12 to 70% of Ni removed from solution, and total increases in Zn and Cd sorption over this period increased 33 and 21%, respectively. The kinetics of Cd, Zn, and Ni were described well with a solution to Fick's second law (a linear relation with the square root of time). Bruemmer et al. (*55*) proposed that the uptake of the metal follows three-steps: (i) adsorption of metals on external surfaces; (ii) solid-state diffusion of metals from external to internal sites; and (iii) metal binding and fixation at positions inside the goethite particle. They suggest that the second step is the rate-limiting step. However, they did not conduct microscopic level experiments to confirm the proposed mechanism. In view of more recent studies, it is likely that the formation of metal-nucleation products could have caused the slow metal sorption reactions observed by Bruemmer et al. (*55*).

Similar observations on sorption of divalent metal ions were made by Coughlin and Stone (*56*). They hypothesized that the slow sorption and desorption on goethite was a result of slow pore diffusion. Axe and Anderson (*57*) also found that sorption of Cd and Sr could be characterized by a model which included two steps: a rapid reversible sorption step followed by a slow, rate-limiting process involving the diffusion of the cations through small pores existing along the surface.

While the above studies have suggested that diffusion is the rate-limiting step based on good model fits to data, macroscopic sorption experiments are not definitive proof of a mechanism (*50,58*). To give additional support to diffusion as a mechanism for sorption onto porous media, Papelis (*28*) measured surface coverages of Cd and selenite on porous aluminum oxides using XPS. He calculated the expected thickness of sorbed Cd and selenite from the total metal loss from solution using both external and internal surface areas. A good agreement was found between the calculated and the measured (using XPS) surface coverage thickness when the total surface area (i.e., internal and external surface area) was used. When the surface layer thickness was calculated without considering internal surface area, the calculated thickness exceeded the thickness observed using XPS. Therefore, the most likely sorption mechanisms were sorption to external sites, diffusion of Cd into the internal structure, and subsequent sorption. While Papelis (*28*) did not measure the kinetics of the reaction, it seems probable that the sorption rate on the interior sites was lower than on the exterior sites, and thus a slow kinetic sorption step would exist.

Sites of Varying Energy States. Rates of metal sorption are also affected by sites of varying energy states (*50*). McBride (*59*) used EPR to study Cu sorption on noncrystalline aluminum oxide. He found that sorption involved sites of varying reactivity. The first reaction step involved the rapid sorption of a low quantity of Cu on high energy sites. A second reaction occurred over several weeks and resulted in the uptake of a greater amount of Cu and EPR spectra distinct from the first reaction step. Hayes and Leckie (*46*) and Grossl et al. (*47*) used pressure-jump relaxation to measure the kinetics of Pb sorption on aluminum oxide and Cu(II) sorption on goethite, respectively. They found that the best fit to the data was obtained by using a kinetic model that included a transformation from outer-sphere to inner-sphere complexation. Their data suggested that sorption behavior was biphasic, which they explained by suggesting that the slower reaction was a result of sites with lower affinities or energies. Lehmann and Harter (*60*) measured the kinetics of chelate promoted Cu^{2+} release from a soil to assess the strength of the bond formed. Sorption/desorption was biphasic, which was attributed to high and low energy bonding sites. With increased residence time from 30 minutes to 24 hours, Lehmann and Harter (*60*) speculated that there was a transition of Cu from low energy sites to higher energy sites (as evaluated by release kinetics). Incubations for up to four days showed a continued uptake of Cu and a decrease in the fraction released within the first three minutes, which was referred to as the low energy sorbed fraction.

"Surface" Precipitation and Polynuclear "Surface" Complex Formation. Recent studies using surface spectroscopic and microscopic techniques such as XAFS, EPR, XPS, AES, TEM, SEM and SFM have shown that the formation of "surface" precipitates and polynuclear "surface" complexes on natural materials are important sorption mechanisms (*10-12,14,16,17,26,27,35,61*).

Nucleation products of Co, Cr(III), Cu, Ni and Pb on oxides and aluminosilicates have been observed (*10-14,16,17,61-68*). Such products have been observed at metal surface loadings far below a theoretical monolayer coverage, and in a pH range below the pH where the formation of metal hydroxide precipitates would be expected according to the thermodynamic solubility product (*10-12,14,16,17*).

Three different types of nucleation products have been proposed: formation or sorption of polymers (dimers, trimers, etc.) on the surface (polynuclear surface complexes); a solid solution or coprecipitate that involves co-ions dissolved from the adsorbent; and a precipitate formed on the surface composed of ions from the bulk solution, or their hydrolysis products (*5,15,33,62,69*). The two latter products are examples of "surface" precipitates. In case the nucleation product is associated with the surface (polynuclear surface complexes), the process leads to the saturation of sites, whereas precipitation mechanisms create new reactive surface area.

The formation of metal nucleation products could be a significant cause of slow metal sorption on mineral surfaces. For example, XAFS results for Co sorption on rutile (TiO_2) showed an increase in the number of backscattering Co atoms for residence times of one day to 11 days, suggesting an increase in the size of nucleation products formed (*61*). However, similar results were not seen for Co aging on quartz (α-SiO_2). Data analysis of the Co/quartz system showed that the reaction time had no effect on the Co coordination environment and revealed the presence of a Co hydroxide-like "surface" precipitate even at a low surface loading.

The authors hypothesized that the reason for the observed slow change in the "surface" precipitate on rutile and not on quartz was a result of the similar radii between $^{VI}Co^{2+}$ (0.75Å) and $^{VI}Ti^{4+}$ (0.61Å); [Table I]. As a result, Co sorption on rutile was consistent with the formation of a precipitate that had similar lattice dimensions as the surface, effectively extending the lattice structure of the bulk solid, i.e., an epitaxial growth.

On quartz, however, the formation of a precipitate that has similar lattice dimensions as the surface is not favorable unless Co undergoes a coordination change from octahedral to tetrahedral. Such a coordination change is unlikely for Co in aqueous solution (*70*) and the XAFS spectra of the Co/quartz sorption system showed no evidence of a change. Thus, the Co hydroxide-like precipitate that formed on quartz was attached only to corners of selected Si tetrahedra on the surface.

Ni Sorption on Clay Minerals: A Case Study. Initial research with Co/clay mineral systems demonstrated the formation of nucleation products using XAFS spectroscopy, but the structure was not strictly identified and was referred to as a Co hydroxide-like structure (*11,12*). Thus, the exact mechanism for "surface" precipitate formation remained unknown. Recent research in our laboratory and elsewhere suggests that during sorption of Ni and Co metal ions, dissolution of the clay mineral or aluminum oxide surface can lead to precipitation of mixed Ni/Al and Co/Al hydroxide phases at the mineral/water interface (*14,16,17,67,71*). This process could act as a significant sink for metals in soils. The following discussion focuses on some of the recent research of our group on the formation kinetics of mixed cation hydroxide phases, using a combination of macroscopic and molecular approaches (*14-17*).

Figure 3 shows the kinetics of Ni sorption on pyrophyllite, kaolinite, gibbsite, and montmorillonite from a 3 m*M* Ni solution at pH = 7.5 (*16*). For kaolinite and pyrophyllite relative Ni removal from solution follows a similar sorption trend with ≈90% Ni sorbed within the first 24 hours. At the end of the experiments, relative Ni removal from solution was almost complete (Ni/kaolinite system, 97% sorbed after 70 hours; Ni/pyrophyllite system, 98% sorbed after 200 hours). Nickel sorption on gibbsite and montmorillonite exhibited a fast initial step. Thereafter, relative Ni removal from solution distinctively slowed down. Relative Ni sorption increased from 42-58% for the Ni/montmorillonite system (time range 0.5-70 hours) and from 15-41% for the Ni/gibbsite system (time range

1-140 hours). Although not shown, experiments with gibbsite and montmorillonite were carried out over longer times (up to 6 weeks).

Figure 7 illustrates the kinetics of Ni sorption on pyrophyllite, along with dissolved Si data from the Ni-treated pyrophyllite and from an untreated pyrophyllite (*16*). The release of Si into solution shows a similar kinetic behavior as Ni sorption on pyrophyllite. When one compares the Si release rate with the dissolution rate of the clay alone, the Si release rate in the Ni-treated system is strongly enhanced as long as Ni removal from solution is pronounced (Figure 7). Although not shown, a similar correlation between Ni sorption and Si release was observed for the Ni/kaolinite system but not for the Ni/montmorillonite system. The dissolution rate of the Ni/gibbsite system was not determined since the [Al] in solution was too low (<50 ppb) to produce reliable ICP measurements.

Figure 8 shows normalized, background-subtracted and k-weighted XAFS spectra of Ni sorbed on pyrophyllite, kaolinite, gibbsite, and montmorillonite (*16*). The XAFS samples were collected at the end of each experiment shown in Figure 3. The corresponding surface sorption densities, Γ's, were 3.1 $\mu mol\ m^{-2}$ for pyrophyllite, 19.9 $\mu mol\ m^{-2}$ for kaolinite, 5.0 $\mu mol\ m^{-2}$ for gibbsite, and 0.35 $\mu mol\ m^{-2}$ for montmorillonite (Table II). The spectra of crystalline $Ni(OH)_2$ and takovite are shown for comparison. One can observe a strong XAFS signal out to higher energies, which indicates the presence of heavy back-scatterer elements such as Ni. There is an obvious similarity among the spectra of Ni sorption samples and the spectrum of takovite, a mixed Ni/Al hydroxide compound (Figure 8).

Figure 9 illustrates radial structure functions (RSFs) produced by forward Fourier transforms of the XAFS spectra represented in Figure 8 (*16*). The spectra were uncorrected for phase shift. All spectra showed a peak of $R \approx 1.8$Å, which represents the first coordination shell of Ni. A second peak representing the second Ni shell can be observed at $R \approx 2.8$Å in the spectra of the Ni sorption samples and takovite (Figure 9). These spectra also showed peaks beyond the second shell at $R \approx 5$-6Å (Figure 9); these peaks resulted from multiple scattering among Ni atoms (*12*).

The structural parameters derived from XAFS analysis are summarized in Table II (*16*). Least-square fits of filtered XAFS for the first RSF peak reveal that in the first coordination shell Ni is surrounded by six O atoms. This behavior indicates that Ni(II) is in an octahedral environment. The Ni-O distances for the Ni sorption samples are 2.02-2.03Å and are similar to those in takovite (2.03Å). The Ni-O distances in crystalline $Ni(OH)_2(s)$ are distinctly longer (2.06Å).

For the second shell, best fits were obtained by including both Ni and Si or Al as second-neighbor backscatterer atoms. Because Si and Al differ in atomic number by 1 ($Z = 14$ and 13, respectively), backscattering is similar. They cannot be easily distinguished from each other as second-neighbor backscatterers, especially in circumstances such as this where the contribution of both is small and cannot be resolved from each other in the Fourier transform.

The data analysis reveals 2.8 (montmorillonite) to 5.0 (gibbsite) Ni second-neighbor (N) atoms, indicative of the presence of Ni-nucleation products (see Table II). No correlation between the Ni surface sorption densities, Γs, and

Table I. Radii of Metal Cations[a]

Metal	Å
Al^{3+}	0.54[b]
Cd^{2+}	0.95
Co^{2+}	0.75
Cu^{2+}	0.73
Pb^{2+}	1.19
Ni^{2+}	0.69
Ti^{4+}	0.61
Zn^{2+}	0.74

[a] From Shannon (*77*)

[b] Radii for octahedral coordination

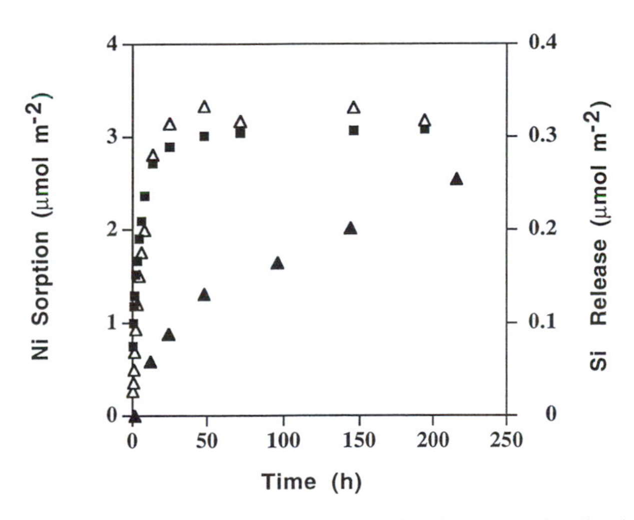

Figure 7. The kinetics of Ni sorption on pyrophyllite from a 3m*M* Ni solution at pH 7.5. (■) denotes the amount of sorbed Ni (μmol m^{-2}) and (△) the amount of simultaneous dissolved Si (μmol m^{-2}). The dissolution of untreated pyrophyllite at pH 7.5 is shown for comparison (▲). From Scheidegger et al. (*16*), with permission.

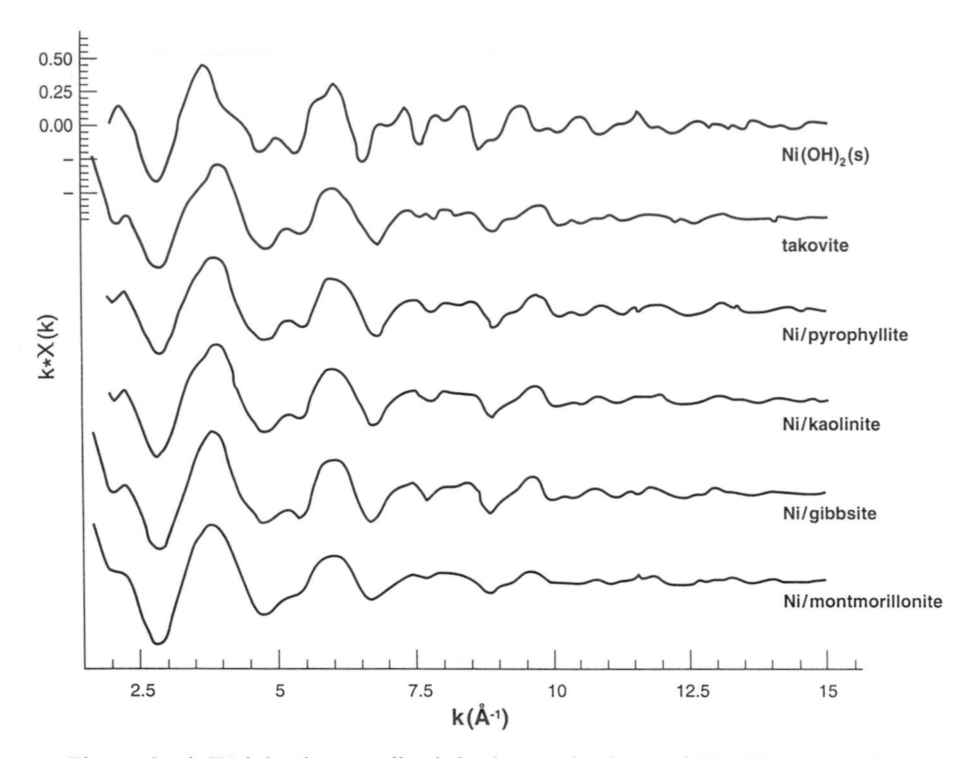

Figure 8. k-Weighted, normalized, background-subtracted XAFS spectra of Ni sorbed on pyrophyllite, kaolinite, gibbsite, and montmorillonite compared to the spectrum of crystalline $Ni(OH)_2(s)$ and takovite. Reaction conditions are those given in 7. The spectra are uncorrected for phase shift. From Scheidegger et al. (*16*), with permission.

Table II. Structural Information Derived From XAFS Analysis Using EXCURVE[a,b].

		Ni-O			Ni-Ni			Ni-Si/Al			
	Γ μmol/m^2	R(Å)	N	2σ^2	R(Å)	N	2σ^2	R(Å)	N	2σ^2	N(Ni)/N (Si/Al)
$Ni(OH)_2$		2.06	6.0	0.011	3.09	6.0	0.010				
Takovite		2.03	6.0	0.01	3.01	3.1	0.009	3.03	1.1	0.009	2.8
Pyrophyllite	3.1	2.02	6.1	0.01	3.00	4.8	0.009	3.02	2.7	0.009	1.8
Kaolinite	19.9	2.03	6.1	0.01	3.01	3.8	0.009	3.02	1.8	0.009	2.2
Gibbsite	5.0	2.03	6.5	0.01	3.02	5.0	0.009	3.05	1.8	0.09	2.7
Montmorillonite	0.35	2.03	6.3	0.01	3.03	2.8	0.011	3.07	2.0	0.015	1.4

[a] From Scheidegger et al. (*16*)

[b] Interatomic distances (R, Å), coordination numbers (N), and Debye-Waller factors (2σ^2, Å^2). The reported values are accurate to within R±0.02 Å, $N_{(Ni\text{-}O)}$ ±20%, $N_{(Ni\text{-}Ni)}$±40%, $N_{(Ni\text{-}Si/Al)}$±40%.

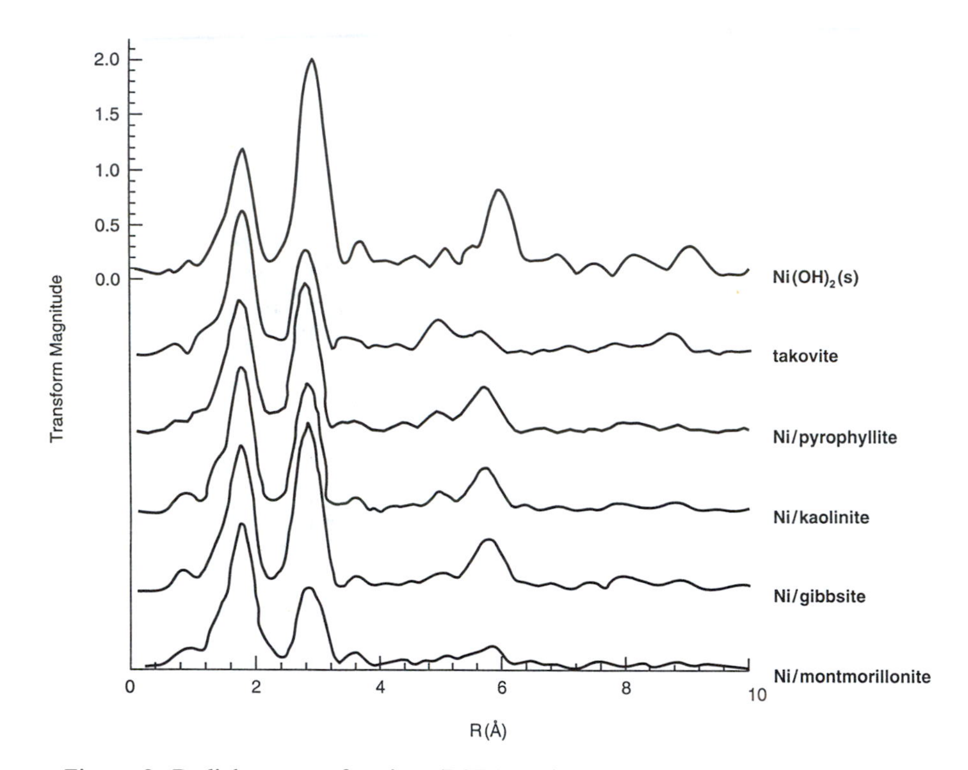

Figure 9. Radial structure functions (RSFs) produced by forward Fourier transforms of Ni sorbed on pyrophyllite, kaolinite, gibbsite, and montmorillonite compared to the spectrum of crystalline $Ni(OH)_2(s)$ and takovite. The spectra are uncorrected for phase shift. From Scheidegger et al. (*16*), with permission.

the Ns of the Ni sorption samples is evident. Observed Ni-Ni distances (3.00-3.03Å) are similar to those in takovite (3.01Å), but distinctly shorter (0.06-0.08Å) than those in crystalline $Ni(OH)_2(s)$ (3.09Å). XAFS data also reveal the presence of 1.8-2.7 Si/Al second-neighbor atoms at 3.02-3.07Å. Again, the bond distances are in good agreement with the Ni-Al distances observed in takovite (3.03Å).

Nickel sorption on pyrophyllite, kaolinite, gibbsite, and montmorillonite at pH 7.5 results in formation of Ni-nucleation products from solutions which are undersaturated with respect to the thermodynamic solubility product of $Ni(OH)_2(s)$. An important finding of the study of Scheidegger et al. (*16*) is that the structural environment of Ni in all Ni sorption samples is similar. There is also an obvious similarity among the spectra of the Ni sorption samples and the spectrum of takovite, suggesting the presence of Ni phases of similar structure (Table II).

The existence of mixed-cation hydroxide phases has been reported in the literature (*72,73*). These compounds consist of structures in which divalent and trivalent metal ions are randomly distributed within the same brucite-like octahedral hydroxide layer. The general chemical formula for the compounds is $[Me^{2+}_{1-x}\,Me^{3+}_{x}(OH)_2]^{+x}\bullet(x/n)A^{-n}\bullet mH_2O$, where, for example, Me^{2+} is Mg(II), Ni(II), Co(II), Zn(II), Mn(II), and Fe(II), and Me^{3+} is Al(III), Fe(III), and Cr(III). The compounds exhibit a net positive charge x per formula unit which is balanced by an equal negative charge from interlayer anions A^{-n} such as Cl^-, Br^-, I^-, NO^-_3, OH^-, ClO^-_4, and CO_3^{2-}; water molecules occupy the remaining interlayer space (*73*). The octahedral layers can be stacked with hexagonal symmetry and two layers per unit cell, with rhombohedral symmetry and three layers per unit cell, or with less symmetrical arrangements (*74*). Minerals with the chemical formula given above are classified as the pyroaurite-sjoegrenite group (*75*). Natural pyroaurite and sjoegrenite are polymorphs having the composition $Mg_6Fe_2(OH)_{16}CO_3\bullet 4H_2O$. The minerals takovite, $Ni_6Al_2(OH)_{16}CO_3\bullet H_2O$, and hydrotalcite, $Mg_6Al_2(OH)_{16}CO_3\bullet H_2O$, are among the most common natural mixed-cation hydroxide compounds containing Al (*16,73*).

The synthesis of mixed-cation hydroxide compounds can be performed by induced hydrolysis (*73*). When a suspension of a fully hydrolyzed cation is added to a solution of another cation, and the pH is maintained constant and slightly below the value at which the second cation hydroxide would precipitate, hydrolysis of the solution cation followed by the precipitation of a fully hydrolyzed mixed-cation hydroxide compound would occur (*16*). In the experiment of Scheidegger et al. (*16*), Al was the fully hydrolyzed cation (see discussion below), Ni was the second cation, and the pH was maintained constant with the pH-stat apparatus at about 0.5 pH units below the pH where the precipitation of $Ni(OH)_2$ would be expected in homogeneous solution.

In the literature mixed Ni/Al compounds have been synthesized containing Cl^-, Br^-, OH^-, ClO_4^-, and CO^{2-}_3 (*72,73*). In the systems of Scheidegger et al. (*16*) the anions present were NO^-_3 and OH^- (NaOH was used to maintain a constant pH). Dissolved Al could not be detected in the samples, a necessity for the formation of mixed Ni/Al compounds. Even so, Al could have been released into solution and incorporated into mixed (Al and Si) hydroxides (*16*). Indeed, the macroscopic data presented in Figure 7 suggest that surface complexes of Ni on pyrophyllite and kaolinite destabilize surface metal ions (Al and Si) relative to the bulk solution, and therefore lead to an enhanced dissolution of the clay. The association of Ni with Al could explain why the enhanced dissolution rate is only observable where Ni sorption is pronounced (see Figure 7).

The importance of Al in immobilizing heavy metals was recently pointed out by Lothenbach et al. (*76*). They studied the sorption kinetics of Cd, Cu, Pb, Ni, and Zn on montmorillonite, Al-montmorillonite and Al_{13}-montmorillonite. Addition of Al enhanced metal sorption of Ni and Zn and sorption increased with time, while Pb and Cd sorption were not affected by addition of Al. This finding agrees with the results of a recent XAFS study with Pb in our laboratory. No nucleation products seem to occur with Pb at surface loadings on clay minerals and γ-Al-oxide where nuclation products have been observed with smaller metals such as Co, Cu and Ni (Strawn, D. G., The University of Delaware, unpublished data). This sorption behaviour of Pb appears to be related to the mismatch in size between Pb^{2+} (1.19Å), and Al^{3+} (0.54Å) that is contained in the structure of the clay minerals and Al-oxide. The Pb ion is too large to fit into the mineral structure, while ions such as Ni^{2+} (0.69Å) and Co^{2+} (0.75Å) can fit into the structure (Table I).

To determine the formation kinetics of mixed Ni-Al hydroxide phases, Scheidegger et al. (*17*) used XAFS to spectroscopically monitor the sorption of Ni and formation of the phases with time on pyrophyllite. Figure 10 shows RSFs produced by forward Fourier transforms of the normalized, background-subtracted and k^3-weighted XAFS spectra of Ni sorbed on pyrophyllite for reaction times of 15 minutes, 75 minutes, 3 hours, 12 hours, 24 hours, and 3 months. The spectra are uncorrected for phase shift. All spectra show a peak of $R \approx 1.8$Å, representing the first coordination shell of Ni. A second peak representing the second Ni coordination shell can be observed at $R \approx 2.8$Å in the spectra of all Ni sorption samples. As reaction time progressed, and relative Ni removal from solution increased, the peak at $R \approx 2.8$Å in the RSFs increased in intensity. This finding suggests the formation of Ni-nucleation products increasing in size with increasing reaction time.

The structural parameters derived from XAFS analysis for Ni/pyrophyllite are shown in Table III (*17*). In the first coordination shell Ni is surrounded by 6 O atoms, indicating that Ni(II) is in an octahedral environment. The Ni-O bond distance (≈ 2.05Å) and coordination numbers were not affected by the reaction time.

Data analysis of the second coordination shell revealed that the number of second-neighbor Ni atoms (N_{Ni-Ni}) increased with increasing time for pyrophyllite (Table III). N_{Ni-Ni} increased from N = 1.4-4.5 as the reaction time increased from 15 minutes to 3 months while the number of second-neighbor Al atoms increased from 1.5 to 2.4. Data in Table III also further reveal that the Ni-Ni bond distances (2.99Å-3.03Å) were essentially unaffected by the reaction time.

The growth of mixed Ni/Al phases on mineral surfaces was investigated spatially using SFM (Figure 11). A one cm^3 cube of pyrophyllite was reacted with Ni for various times using a pH-stat batch reactor (pH ≈ 7.5). At each time, the

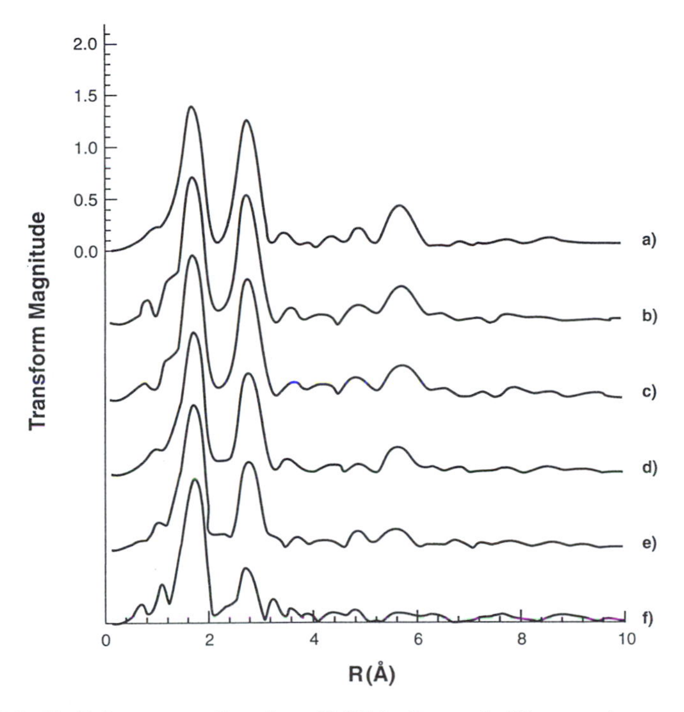

Figure 10. Radial structure functions (RSFs) of pyrophyllite samples reacted with Ni (reaction conditions are those given in Fig. 7) for a) 3 mo, b) 24 h, c) 12 h, d) 3 h, e) 75 min, and f) 15 min. The spectra are uncorrected for phase shift. Note the appearance of a peak at a R of about 2.8Å with increasing reaction time. From Scheidegger et al. (*17*), with permission.

Table III. Structural Information Derived From XAFS Analysis Using EXCURVE[a,b].

	relative Ni removal (%)	sorption density, Γ μmol/m²	Ni-O			Ni-Ni			Ni-Si/Al		
			R(Å)	N	$2\sigma^2$	R(Å)	N	$2\sigma^2$	R(Å)	N	$2\sigma^2$
15 min	28	0.7	2.03	6.0	0.009	2.99	1.4	0.007	2.96	1.5	0.004
75 min	44	1.1	2.03	6.4	0.01	3.01	3.1	0.009	3.04	1.6	0.01
3 h	67	1.7	2.02	6.2	0.009	3.03	3.4	0.01	3.07	1.6	0.011
12 h	93	2.4	2.03	6.4	0.01	3.02	4.1	0.009	3.07	2.1	0.009
24 h	97	2.6	2.03	6.5	0.01	3.02	4.2	0.01	3.07	1.8	0.013
3 mo		3.1	2.03	6.5	0.01	3.02	4.5	0.011	3.07	2.4	0.016

[a] From Scheidegger et al. (*17*)
[b] Interatomic distances (R, Å), coordination numbers (N), and Debye-Waller factors (σ^2, Å^2). The reported values are accurate to within R±0.02 Å, $N_{(Ni-O)}$±20%, $N_{(Ni-Ni)}$±20%, $N_{(Ni-Si/Al)}$±40%.

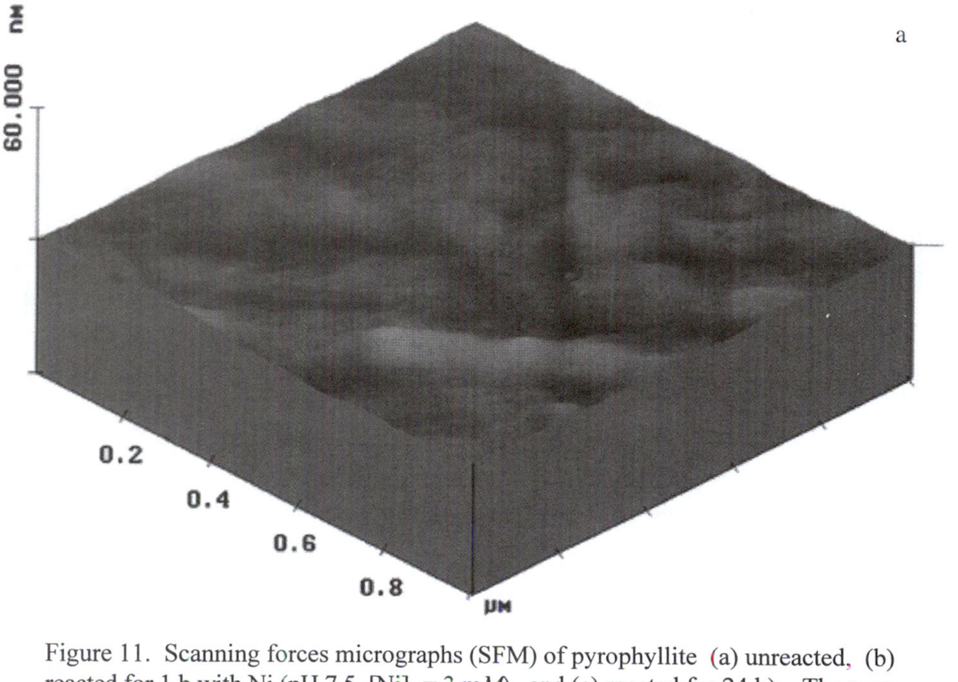

Figure 11. Scanning forces micrographs (SFM) of pyrophyllite (a) unreacted, (b) reacted for 1 h with Ni (pH 7.5, $[Ni]_o = 3$ m*M*) , and (c) reacted for 24 h). The scan size was 1 µm X 1 µm with a maximum Z-range of 60 nm. The micrographs were collected with Tapping Mode© SFM (Scheckel, K. G., The University of Delaware, unpublished data).

Continued on next page.

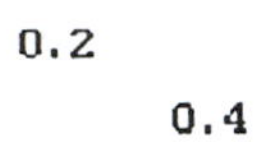

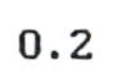

Figure 11. *Continued.*

sample was removed from the reactor and examined by SFM. The samples were scanned with a Digital Instruments Nanoscope IIIa SFM under TappingMode© AFM using multiple fresh tips. Each scan was run under the same parameters: the scan size for each image was 1 μm and the height range was 60 nm. Data in Figure 11 are presented in height mode only. Figure 11a shows the polished pyrophyllite surface prior to reaction with Ni. As the reaction progresses to one h, one begins to see the formation of nucleation products on the pyrophyllite surface (Figure 11b). When sorption is nearly completed at approximately 24 hours, large mountainous features are prevalent across the pyrophyllite interface, exhibiting the growth of the nucleation products with time (Figure 11c). Height and amplitude data were collected for each sample.

The XAFS and AFM data clearly show that mixed Ni-Al hydroxide phases form on relatively rapid time scales. After a reaction time of only 15 minutes, such phases appear on pyrophyllite. This suggests that adsorption and "surface" precipitation processes (mixed Ni/Al phase formation) can occur over similar time scales. It has traditionally been thought that precipitation processes are much slower than adsorption phenomena.

These above studies suggest that three phenomena occur at the mineral/liquid interface to cause formation of mixed-cation hydroxide phases: (1) non-specific and/or specific adsorption; (2) dissolution of Al where the dissolution rate is most probably dependent on the surface morphology and impurities present; and (3) precipitation of a mixed Ni/Al phase. The last step is rapid and proceeds until the cation concentrations correspond to the solubility product of the hydrotalcite-like phase. Dissolution of Al appears to be the rate-limiting step (*16,17*).

It has been proposed that the extent to which mixed-cation hydroxide compounds actually do form in aquatic and terrestrial environments is limited more by slow rates of soil mineral dissolution, a necessary preliminary step, than by lack of thermodynamic favorability (*51*). Because the dissolution rates of clays and oxide minerals are fairly slow, the possibility of mixed-cation hydroxide formation as a plausible "sorption mode" in 24 hour-based sorption experiments (and also most long-term studies) containing divalent metal ions such as Mg, Ni, Co, Zn, and Mn and Al(III)-, Fe(III)-, and Cr(III)-(hydr)oxide or silicate minerals has been ignored in the literature (*16,17*). This study and others recently published (*71*), however, suggests that metal sorption onto mineral surfaces can significantly destabilize surface metal ions (Al and Si) relative to the bulk solution, and therefore lead to an enhanced dissolution of the clay and oxide minerals. Thus, predictions on the rate and the extent of mixed-cation hydroxide formation in aquatic and terrestrial environments based on the dissolution rate of the mineral surface alone are not valid and underestimate the true values.

The studies of Scheidegger et al. (*16,17*) have emphasized the importance of combining time-dependent or kinetic studies with spectroscopic and microscopic investigations to better understand sorption processes at the soil mineral/water interface. Such studies can result in a detailed mechanistic understanding (e.g., distinguishing the rate of metal adsorption versus precipitation processes in sorption systems) which would be difficult to determine using a macroscopic approach alone.

Acknowledgments

We gratefully acknowledge the financial assistance of the DuPont Co., The State of Delaware, and the USDA (NRICGP), for their support of research reported in this review. Immense gratitude is extended to Dr. Noel C. Scrivner of the DuPont Co. for his unfailing support and encouragement during the past 10 years and to Dr. Robert G. Ford for his careful and thoughtful critique.

Literature Cited

1. Stumm, W.; Morgan, J. J. *Chemistry of the Solid-Water Interface*, 2nd ed.; Wiley: New York, NY, 1981.
2. Scheidegger, A.M.; Sparks, D.L. *Soil Sci.* **1996**, *161*, 813-831.
3. Goldberg, S. In *Advances in Agronomy*; Sparks, D. L., Ed.; Academic Press: San Diego, CA, 1992, Vol. 47; pp. 233-329.
4. Westall, J. C.; Hohl, H. *Adv. Colloid Interf. Sci.* **1980**, *12*, 265-294.
5. Farley, K. J.; Dzombak, D. A.; Morel, F. M. M. *J. Colloid Interf. Sci.* **1985**, *106*, 226-242.
6. James, R. O.; Healy, T. W. *J. Colloid Interf. Sci.* **1972**, *41*, 65-80.
7. Dzombak, D. A.; Morel, F. M. M. *Surface Complexation Modeling, Hydrous Ferric Oxide*, Wiley: New York, NY, 1990.
8. Katz, L. E.; Hayes, K. F. *J. Colloid Interf. Sci.* **1995**, *170*, 477-490.
9. Katz, L. E.; Hayes, K. F. *J. Colloid Interf. Sci.* **1995**, *170*, 491-501.
10. Fendorf, S. E.; Lamble, G. E.; Stapleton, M. G.; Kelley, M. J.; Sparks, D. L. *Environ. Sci. Technol.* **1994**, *28*, 284-289.
11. O'Day, P. A.; Brown, G. E., Jr.; Parks, G. A. *J. Colloid Interf. Sci.* **1994**, *165*, 269-289.
12. O'Day, P. A.; Parks, G. A.; Brown, G. E., Jr. *Clays Clay Miner.* **1994**, *42*, 337-355.
13. Papelis, C.; Hayes, K. F. *Coll. Surfaces.* **1996**, *107*, 89-96.
14. Scheidegger, A. M.; Lamble, G. M.; Sparks, D. L. *Environ. Sci. Technol.* **1996**, *30*, 548-554.
15. Scheidegger, A. M.; Fendorf, M.; Sparks, D. L. *Soil Sci. Soc. Am. J.* **1996**, *60*, 1763-1772.
16. Scheidegger, A. M.; Lamble, G. M.; Sparks, D. L. *J. Colloid Interf. Sci.* **1997**, *186*, 118-128.
17. Scheidegger, A. M.; Lamble, G. M.; Sparks, D. L. *J. Phys. IV France* **1997**, *7*, C2-773-775.
18. Johnston, C. T.; Sposito, G.; Earl, W. L. In *Environmental Particles*; Buffle, J.; vanLeeuwen, H. P., Eds.; Lewis Publ: Boca Raton, FL, 1993, pp. 1-36.
19. Koppelmann, M. H.; Emerson, A. B.; Dillard, J. G. *Clays Clay Miner.* **1980**, *28*, 119-124.
20. Dillard, J. G.; Koppelmann, M. H. *J. Colloid Interf. Sci.* **1982**, *95*, 298-309.
21. Schenk, C. V.; Dillard, J. G. *J. Colloid Interf. Sci.* **1983**, *95*, 398-409.
22. Hochella, M. F., Jr.; Carim, A. H. *Surf. Sci.* **1988**, *197*, 260-268.

23. Davison, N.; Whinnie, W. R. *Clays Clay Miner*. **1991**, *39*, 22-27.
24. Stipp, S. L., Hochella, M. F., Jr. *Geochim. Cosmochim. Acta*. **1994**, *58*, 3023-3033.
25. Scheidegger, A. M.; Borkovec, M.; Sticher, H. *Geoderma* **1993**, *58*, 43-65.
26. Junta, J. L.; Hochella, M. F., Jr. *Geochim. Cosmochim. Acta*. **1994**, *58*, 4985-4999.
27. Wersin, P.; Hochella, M. F., Jr.; Persson, P.; Redden, G.; Leckie, J. O.; Harris, D. W. *Geochim. Cosmochim. Acta*. **1994**, *58*, 2829-2843.
28. Papelis, C. *Environ. Sci. Technol*. **1995**, *29*, 1526-1533.
29. Perry, D. L.; Taylor, A.; Wagner, C. D. In *Instrumental Surface Analysis of Geologic Materials*; Perry, D. L., Ed.; VCH Publishers: New York, NY, 1990, pp. 45-86.
30. Fendorf, S. E.; Sparks, D. L.; Lamble, G. M.; Kelley, M. J. *Soil Sci. Soc. Am. J.* **1994**, *58*, 1583-1595.
31. Hochella, M. F., Jr. In *Spectroscopic Methods in Mineralogy and Geology*; Hawthorne, F. C., Ed.; Reviews in Mineralogy 18; Mineralogical Society of America: Washington, DC, 1988, pp. 573-630.
32. Hochella, M. F., Jr. In *Mineral-Water Interface Geochemistry*; Hochella, M. F., Jr.; White, A. F., Eds.; Reviews in Mineralogy 23; Mineralogical Society of America: Washington, DC, 1990, pp. 87-132.
33. Brown, G. E., Jr. In *Mineral-Water Interface Geochemistry*; Hochella, M. F., Jr.; White, A. F., Eds.; Reviews in Mineralogy 23; Mineralogical Society of America: Washington, DC, 1990, pp. 309-353.
34. Brown, G. E., Jr.; Parks, G. A.; Chisholm-Brause, C. J. *Chimia* **1989**, *43*, 248-256.
35. Charlet, L.; Manceau, A. In *Environmental Particles*; Buffle, J.; vanLeeuwen, H. P., Eds., Lewis Publ: Boca Raton, FL, 1993, pp. 117-164.
36. Schulze, D. G.; Bertsch, P. M. In *Advances in Agronomy*; Sparks, D. L., Ed.; Academic Press: San Diego, CA, 1995, *Vol.* 55; pp. 1-66.
37. Scheidegger, A. M.; Sparks, D. L. *Soil Sci. Soc. Am. J.* **1996**, *60*, 1763-1772.
38. Dove, P. M.; Chermak, J. A. In *Scanning Probe Microscopy of Clay Minerals*; Blum, A. E.; Nagy, K., Eds.; Clay Minerals Soc.: Boulder, CO, 1994, pp. 149-169.
39. Hochella, M. F., Jr. In *Mineral Surfaces*; Vaughn, D. J.; Pattrick, R. A. D., Eds.; Chapman and Hall: New York, NY, 1995, pp. 17-60.
40. Maurice, P. A. In *Environmental Particles*; Huang, P. M.; Senesi, N.; Buffle, J., Eds.; Wiley: New York, NY, 1998, Vol. 4; *In Press*.
41. Maurice, P. A. In *Advances in Agronomy*; Sparks, D. L., Ed.; Academic Press: San Diego, CA, 1997, *Vol.* 62; pp. 1-43.
42. Sparks, D. L. *Kinetics of Soil Chemical Processes*; Academic Press: San Diego, CA, 1989.
43. Jardine, P. M.; Sparks, D. L. *Soil Sci. Soc. Am. J.* **1984**, *48*, 39-45.
44. Comans, R. N. J.; Hockley, D. E. *Geochim. Cosmochim. Acta*. **1992**, *56*, 1157-1164.

45. Hachiya, K.; Sasaki, M.; Ikeda, J.; Mikami, N.; Yasunaga, T. *J. Phys. Chem.* **1984**, *88*, 27-31.
46. Hayes, K. F.; Leckie, J. O. In *Geochemical Processes at Mineral Surfaces*; Davis, J. A.; Hayes, K. F., Eds.; Am. Chem. Soc. Symp. Ser. 323; Am. Chem. Soc.: Washington, DC, 1986, pp. 114-141.
47. Grossl, P. R.; Sparks, D. L.; Ainsworth, C. C. *Environ. Sci. Technol.* **1994**, *28*, 1422-1429.
48. Bunzl, K.; Schmidt, W.; Sansoni, B. *J. Soil Sci.* **1976**, *27*, 32-41.
49. Sparks, D. L. In *Environmental Particles*; Huang, P. M.; Senesi, N.; Buffle, J., Eds.; Wiley: New York, NY, *Vol.* 4; *In Press*.
50. Strawn, D. G.; Sparks, D. L. In *Fate and Transport of Metals in the Vadose Zone*; Selim, H. M.; Iskandar, A., Eds.; Lewis Publishers: Boca Raton, FL, 1998, *In Press*.
51. McBride, M. B. *Environmental Chemistry of Soils*; Oxford Univ. Press: New York, NY, 1994.
52. Ainsworth, C. C.; Pilou, J. L.; Gassman, P. L.; Van Der Sluys, W. G. *Soil Sci. Soc. Am. J.* **1994**, *58*, 1615-1623.
53. Pignatello, J. J.; Xing, B. *Environ. Sci. Technol.* **1996**, *30*, 1-11.
54. Burgos, W. D.; Novak, J. T.; Berry, D. F. *Environ. Sci. Technol.* **1996**, *30*, 1205-1211.
55. Bruemmer, G. W.; Gerth, J.; Tiller, K. G. *J. Soil Sci.* **1988**, *39*, 37-52.
56. Coughlin, B. R.; Stone, A. T. *Environ. Sci. Technol.* **1995**, *29*, 2445-2455.
57. Axe, L.; Anderson, P. R. *J. Colloid Interf. Sci.* **1997**, *185*, 436-448.
58. Sposito, G. *The Chemistry of Soils*; Oxford Univ. Press: New York, NY, 1989.
59. McBride, M. B. *Clays Clay Miner.* **1982**, *30*, 21-28.
60. Lehmann, R. G.; Harter, R. D. *Soil Sci. Soc. Am. J.* **1984**, *48*, 769-772.
61. O'Day, P. A.; Chisholm-Brause, C. J.; Towle, S. N.; Parks, G. A.; Brown, G. E., Jr. *Geochim. Cosmochim. Acta.* **1996**, *60*, 2515-2532.
62. Chisholm-Brause, C. J.; O'Day, P. A.; Brown, G. E., Jr.; Parks, G. A.; Leckie, J. O. *Nature* **1990**, *348*, 528-530.
63. Chisholm-Brause, C. J.; Hayes, K. F.; Roe, A. L.; Brown, G. E., Jr.; Parks, G. A.; Leckie, J. O. *Geochim. Cosmochim. Acta.* **1990**, *54*, 1897-1909.
64. Roe, A. L.; Hayes, K. F.; Chisholm-Brause, C. J.; Brown, G. E., Jr.; Parks, G. A.; Hodgson, K. O.; Leckie, J. O. *Langmuir* **1991**, *7*, 367-373.
65. Charlet, L.; Manceau, A. *J. Colloid Interf. Sci.* **1992**, *148*, 443-458.
66. Bargar, J. R.; Brown, G. E., Jr.; Parks, G. A. 209th Am. Chem. Soc. National Meeting, extended abstracts, Vol. 35(1).
67. Towle, S. N.; Bargar, J. R.; Brown, G. E., Jr.; Parks, G. A. *J. Colloid Interf. Sci.* **1997**, *187*, 62-82.
68. Xia, K.; Mehadi, A.; Taylor, R. W.; Bleam, W. F. *J. Colloid Interf. Sci.* **1997**, *185*, 252-257.
69. Sposito, G. In *Geochemical Processes at Mineral Surfaces*; Davis, J. A.; Hayes, K. F., Eds.; Am. Chem. Soc. Symp. Ser. 323; Am. Chem. Soc.: Washington, DC, 1986, pp. 217-228.

70. Magini, M.; Licheri, G.; Paschina, G.; Piccaluga, G.; Pirna, G. *X-ray Diffraction of Ions in Aqueous Solutions: Hydration and Complex Formation*; CRC Press: Boca Raton, FL. 1988.
71. d'Espinose de la Caillerie, J. B.; Kermarec, M.; Clause, O. *J. Am. Chem. Soc.* **1995**, *117*, 11471-11481.
72. Allmann, R. *Chimia* **1970**, *24*, 99-108.
73. Taylor, R. M. *Clay Miner.* **1984**, *19*, 591-603.
74. Brindley, G. W.; Kikkawa, S. *Am. Mineral.* **1979**, *64*, 836-843.
75. Hashi, K.; Kikkawa, S.; Koizumi, M. *Clays Clay Miner.* **1983**, *31*, 152-154.
76. Lothenbach, B.; Furrer, G.; Schulin, R. *Environ. Sci. Technol.* **1997**, *31*, 1452-1462.
77. Shannon, R. D. *Acta Cryst.* **1976**, *A32*, 751-767.
78. Hayes, K. F. Ph.D. Dissertation, Stanford University. **1987**.

Chapter 8

Evaluation of Oxyanion Adsorption Mechanisms on Oxides Using FTIR Spectroscopy and Electrophoretic Mobility

D. L. Suarez[1], S. Goldberg[1], and C. Su[2]

[1]**U.S. Salinity Laboratory, ARS, U.S. Department of Agriculture, Riverside, CA 92507**
[2]**National Risk Management Laboratory, U.S. Environmental Protection Agency, Ada, OK 74820**

Use of surface speciation models for prediction of adsorption and transport requires specification of the mode of bonding and speciation of oxyanions on oxide surfaces. FTIR spectroscopy (especially ATR and DRIFT) offers the potential to establish symmetry of surface species, protonation, and determination of monodentate or bidentate bonding. Determination of surface speciation is greatly enhanced when the spectroscopic information is combined with measurements of electrophoretic mobility (EM), calculation of point of zero charge and proton balance measurements before and after adsorption. We review adsorption of phosphate, carbonate, boron, selenate and selenite on Fe and Al oxides. New preliminary spectra and EM and proton balance information for arsenate and arsenite adsorption on amorphous Fe and Al oxide suggest that $HAsO_4$ and H_2AsO_3 are the dominant surface species.

Identification of the specific species of the adsorbed oxyanion as well as mode of bonding to the oxide surface is often possible using a combination of Fourier Transform Infrared (FTIR) spectroscopy, electrophoretic mobility (EM) and sorption-proton balance data. This information is required for selection of realistic surface species when using surface complexation models and prediction of oxyanion transport. Earlier, limited IR research on surface speciation was conducted under dry conditions, thus results may not correspond to those for natural systems where surface species may be hydrated. In this study we review adsorbed phosphate, carbonate, borate, selenate, selenite, and molybdate species on aluminum and iron oxides using FTIR spectroscopy in both Attenuated Total Reflectance (ATR) and Diffuse Reflectance Infrared Fourier Transform (DRIFT) modes. We present new FTIR, EM, and titration information on adsorbed arsenate and arsenite. Using these techniques we distinguish inner- from outer-sphere adsorption mechanisms and monodentate, bidentate, and binuclear attachments. We also utilize EM measurements of oxide suspensions to determine shifts in point of zero charge (PZC) upon oxyanion adsorption. Inner-sphere adsorption can be inferred from a shift in zero point of charge but lack of a shift in PZC is a possible outcome for both inner- and outer-sphere complexes. Based on these experiments, we deduce inner-sphere adsorption mechanisms for all oxyanions studied. Hydroxyl release was also consistent with ligand exchange. Reaction rates for both adsorption and desorption provide the opportunity to obtain additional information regarding bonding mechanisms. Limited information available suggests that these reactions are not readily reversible.

Recent developments in spectroscopic techniques offer the opportunity to increase our understanding of oxyanion surface speciation and binding. This understanding is essential to properly use mechanistic sorption models, such as the constant capacitance model and the triple layer model. A recent criticisms of these models is that selection of the surface species and reaction from the sorption data alone results in an empirical model which could be replaced with the traditional Langmuir model (*1*). However determination of the surface species and reactions will constrain the parameterization and allow for mechanistic evaluation of the sorption models. Knowledge of the actual species and reaction should also enable more generalized prediction of sorption behavior outside the range of the actual experiment, which is not possible at present.

Among the current spectroscopic methods extended x-ray absorption fine structure spectroscopy (EXAFS) has received the most attention, however other methods such as FTIR, may be equally as promising. EXAFS is considered to provide definitive information on inner- vs. outer-sphere bonding and is suitable for determining mode of attachment to the surface (monodentate, bidentate, binuclear) but does not resolve questions of surface speciation since it is not sensitive to H atoms. In addition, examination of the same system by different researchers has in some instances resulted in different conclusions.

The emphasis in examining sorption with spectroscopic techniques has been mostly related to reaction with oxide surfaces, primarily Fe and Al. In addition to the simplicity of the systems, these are regarded as the primary adsorbent minerals for oxyanions in natural systems.

The specific advantage of FTIR spectrometers over the older dispersion IR instruments is the increased throughput, since all the radiation passes through the sample and the entire spectrum of wavenumbers is detected at once, resulting in dramatic increases in signal to noise ratio (*2*). Among the new methods possible with FTIR are DRIFT and ATR spectroscopy. In reflectance techniques the IR beam is bounced off the sample, rather than passed through it as is done in transmission methods (such as in the use of KBr pellets). Since the DRIFT method results in a strong signal contribution from the sample surface (typical penetration depths are 1 to 10 μm (*2*)), it is well suited for the detection of surface adsorbed species. When using the DRIFT method, a ground sample is diluted with KBr such that the sample is 1 to 10% by weight of the mixture (*2*).

A limitation in the use of DRIFT for adsorption studies is that the spectrum is obtained from a relatively dry sample. Use of high water contents is not possible due to the strong aqueous adsorption. Drying of an oxide suspension may or may not favor the conversion of a monodentate to a bidentate complex, depending on the reaction. For example both reactions

$$\begin{aligned} SHPO_4^- + SOH_2^+ &\rightarrow S_2HPO_4 + H_2O \\ SHPO_4^- + SOH &\rightarrow S_2HPO_4 + OH^- \end{aligned} \tag{1}$$

where S represents a metal ion in the oxide mineral, are possible. The first reaction in equation 1 provides a possible explanation for the observation by Parfitt and Atkinson (*3*) that drying shifted the P-O bands from 1080 and 1040 cm^{-1} to 1192 and 1000 cm^{-1}, respectively in spectra of phosphate sorbed onto goethite. In addition, Parfitt and Russell (*4*) considered that air-drying was sufficient to cause the conversion from outer-sphere to inner-sphere complexes for nitrate and chloride ions adsorbed on goethite. Outer-sphere refers to complexes in which at least one water molecule is located between the sorbate anion and the surface functional group. Inner-sphere complexes are complexes without water between the surface functional group and the sorbing anion.

The use of ATR provides the potential to examine adsorption in wet systems. A solution or "paste" (mixture of finely ground sample and solution) is placed in contact with a crystal of material having a high refractive index and optimally transparent to IR radiation in the wavelength range to be analyzed. In this method the radiation is reflected off the crystal surface and into the crystal multiple times. The relatively short pathlength through the sample provides the capability to obtain radiation through aqueous solutions or suspensions. The depth of penetration is on the order of 0.1 to 5 μm into the sample (*2*); thus ATR is well suited to measurement of adsorbed species in the presence of water.

Electrophoretic mobility measurements, including PZC, and potentiometric titration data are very useful for distinguishing inner- from outer-sphere complexes. Ions that retain their hydration sphere are adsorbed only at the outer Helmholtz plane (*5*). The presence of outer-sphere complexes should not alter the EM of the suspended solid relative to the solid without the sorbed species because the sorbed species is considered to reside outside of the shear plane of the solid. The shear plane is considered to lie very close to the outer Helmholtz plane (*5*). In contrast, inner-sphere complexes (specific adsorption) will reside inside the shear plane and depending on the reaction may result in a change in the particle charge inside the shear plane. Shifts in PZC and reversal of EM with increasing ion concentration are considered characteristic of specific ion adsorption (*5*). It does not follow that all specific or inner-sphere adsorption results in shifts in EM. As discussed below, specific adsorption reactions need not result in changes in surface charge. It follows that all adsorption which results in shifts in EM is inner-sphere but adsorption which does not result in a shift in EM may be inner- or outer-sphere.

The combination of shifts in EM with surface charge measurements and titration data can be interpreted to provide important information on the surface speciation, which is complimentary to spectroscopic data. Titration data provide information on the net proton or hydroxyl release from the surface. Among the techniques that can be utilized is suspension of the oxide in the background electrolyte, adjustment of the pH to the value of interest, adjustment of the background electrolyte with the adsorbing ion, then mixing the two and measuring the proton or hydroxyl mass necessary to return to the specified pH, in conjunction with determination of the concentration of anion adsorbed. These data are then reported in mol H^+ or OH^- per mmol anion adsorbed.

The EM and titration data are also essential to evaluate the possibility of mixed reaction mechanisms. For example, both FTIR and EXAFS can be used to determine the presence of specific bonds or complexes. Pure inner- and outer-sphere complexes may in fact be endmembers in some systems, while a mixture of hydrated and nonhydrated bonds may result in average hydration values intermediate between zero and one. Indifferent ions such as Cl^-, ClO_4^-, NO_3^-, Na^+ and K^+ do not shift the PZC of oxides (*6*). In contrast, SO_4^{2-} has been found to adsorb specifically and produce a shift in the PZC of hematite (*7*).

Reaction kinetics and bonding energies are additional tools useful for examination of ligand bonding to oxide surfaces. Non-specific bonds should result in rapid desorption reaction kinetics, while monodentate and bidentate and binuclear bonds should be correspondingly more difficult to break.

Surface Configuration of Oxyanions

Phosphate Speciation. Differences in position and number of absorption bands between ATR-FTIR spectra of phosphate in aqueous solution and phosphate at the goethite/solution interface indicate phosphate coordination with surface iron atoms (*8*). By analogy with ferric phosphate solution complexes, this was interpreted as indicative of formation of protonated and nonprotonated bridging bidentate and nonprotonated monodentate phosphate surface complexes on goethite. Points of zero charge of goethite obtained from EM measurements decrease with increasing phosphate addition, indicating an inner-sphere adsorption mechanism for phosphate on goethite (*8, 9*).

Persson et al. (*10*) investigated phosphate adsorption on goethite using DRIFT-FTIR and concluded that phosphate forms strictly monodentate surface complexes of different degrees of protonation. These surface configurations agree with those postulated in the macroscopic surface complexation models. These authors argue that their spectra are incompatible with a bidentate bridging structure of adsorbed phosphate since such a complex would have a surface site symmetry other than C_{3v}. However Persson et al. (*10*) indicate that in the intermediate pH range a bidentate complex cannot be ruled out. Based on their modeled surface species distribution and their FTIR spectra, they estimate surface pK_{a2} and pK_{a3} values of 5.6 and 7.6, respectively. The neutral species H_3PO_4 was not detected on the surface but the lowest pH examined was 3. The aqueous pK_a values for phosphoric acid are 2.2, 7.2, and 12.3 for pK_{a1}, pK_{a2}, and pK_{a3}, respectively. The surface thus has a strong preference for the negatively charged species, consistent with the positive surface charge of goethite.

Earlier, Tejedor-Tejedor and Anderson (*8*) interpreted C_{2v} and C_{3v} symmetries of adsorbed phosphate as bidentate bridging complexes. These authors indicted that some of their stretching modes are also compatible with a monodentate complex. Controversy in interpretation of FTIR spectra of adsorbed phosphate, as with interpretation of Se XAFS results (discussed below) suggests that spectroscopic results by themselves, are not always definitive.

Magic angle spinning Nuclear Magnetic Residence (NMR) experiments have been used to distinguish protonated surface complexes from fully deprotonated complexes (*11*). These authors found that phosphate adsorbed on the Al oxide surface became fully deprotonated between pH 9 to 11. The pK_{a3} of phosphoric acid is 12.3, thus the Al oxide surface favors the deprotonated PO_4^{3-} species relative to the HPO_4^{2-} species in the solution (by a factor of 10^2).

Carbonate. Schulthess and McCarthy (*12*) determined carbonate and acetate adsorption by δ-Al_2O_3 and concluded that competitive adsorption of carbonate must be considered for prediction of organic ion adsorption under field conditions. Competitive adsorption of dissolved CO_2 species has also been established for chromate on am-Fe hydroxide (*13*) and goethite (*14*). These titration/adsorption studies demonstrate and quantify the importance of adsorption of dissolved CO_2 species but do not provide information as to the specific species and mode of bonding with the oxide surface.

Su and Suarez (*15*) examined carbonate adsorption on Al and Fe oxides using ATR and DRIFT-FTIR in combination with EM and titration data. The presence of carbonate lowered the EM and decreased the PZC to lower pH for all oxides examined (am-$Al(OH)_3$ and am-$Fe(OH)_3$, gibbsite, and goethite). Since shifts in PZC are caused by changes in the charge within the shear plane and within the outer Helmholtz plane (*5*), these data indicate the presence of inner-sphere complexation of a carbonate species with the oxide surface of these minerals.

The spectra shown in Figure 1, from Su and Suarez (*15*), are for Na carbonate solutions at varied pH. These spectra demonstrate that carbonate species in solution can be readily distinguished by ATR-FTIR spectroscopy, since the pK_a values for carbonic acid in solution are 6.35 and 10.33. The IR spectrum of HCO_3^- can be interpreted as a planar tetratomic group $CO_2(OH)^-$ with C_{2v} symmetry (*16*). The aqueous HCO_3^- spectrum, (Figure 1a) has been characterized by ν_1 C-OH stretching at 1010 cm^{-1} ν_2 symmetric stretching of CO_2 at 1310 and 1360 cm^{-1}, ν_4 asymmetric stretching of CO_2 at 1605 and 1668 cm^{-1} and ν_6 out of plane bending of CO_2 at 843 cm^{-1} (*15*). As shown in Table

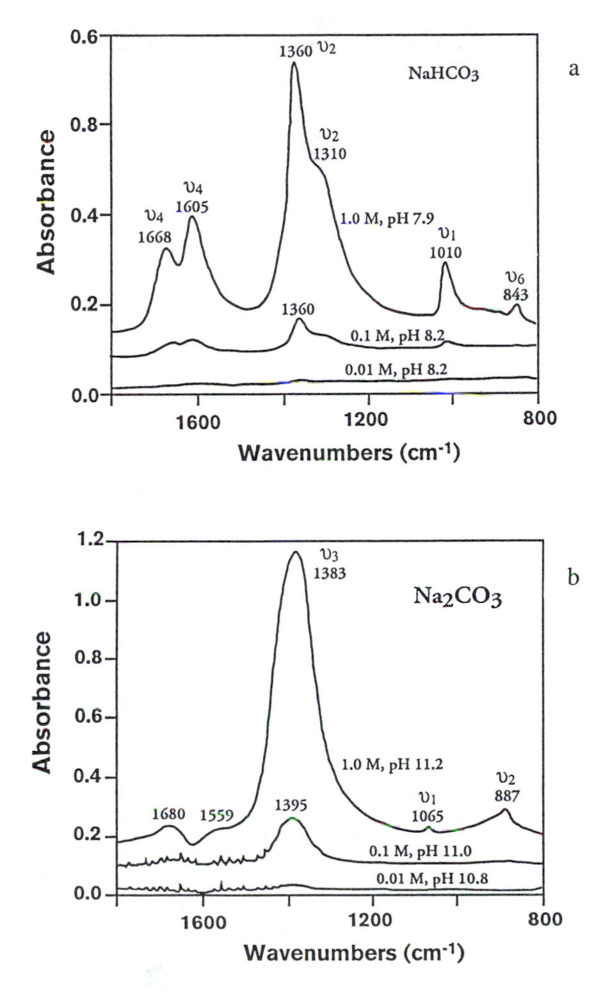

Figure 1. ATR-FTIR difference spectra of a) aqueous bicarbonate and b) carbonate ions from solutions of 1.0 M Na_2CO_3 at pH 11.2 and 7.9, respectively. Subtracted spectra were of NaCl at the same concentration and pH. Spectra taken from Su and Suarez (*15*).

I, the assignments are in good agreement with those obtained for spectra of solid $NaHCO_3$ and $KHCO_3$ (*16*). The aqueous CO_3^{2-} spectrum has ν_1 symmetric stretching at 1065 cm^{-1}, ν_2 out-of-plane bending at 887 cm^{-1}, and ν_3 asymmetric stretching at 1383 cm^{-1} (Figure 1b). These are matched to the CO_3 spectra from $CaCO_3$ minerals (calcite and aragonite) in Table I.

Adsorbed carbonate and bicarbonate can be readily distinguished both from each other as well as from the aqueous species. The ATR-FTIR difference spectrum shown in Figure 2a, from Su and Suarez (*15*) represents am-$Al(OH)_3$ in the presence of bicarbonate in solution subtracted from the spectrum of am-$Al(OH)_3$ in the absence of carbonate species. The pH region examined was 4.1 to 7.8. The bands at 1490 and 1420 cm^{-1} correspond to a doublet splitting of the ν_3 vibration for CO_3^{2-}, while the band at 1028 cm^{-1} is assigned to ν_2 vibration of CO_3^{2-}. The lack of the ν_4 doublet for HCO_3^- (even at pH 4.3) indicates that only carbonate sorbs onto the am-$Al(OH)_3$ surface, despite the fact that HCO_3^- is the dominant solution species (*15*). The authors conclude that carbonate forms a monodentate complex, based on a splitting of 70 cm^{-1} for the ν_3 vibration doublet. This value is consistent with a splitting of 137, 79, and 84 cm^{-1} for the monodentate carbonato complexes of $[Co(NH_3)_5CO_3]Cl$, $[Co(NH_3)_5CO_3]Br$, and $[Co(NH_3)_5CO_3]I$, respectively and much less than the splitting of 308, 313, and 311 cm^{-1} reported for the bidentate carbonate complexes $[Co(NH_3)_4CO_3]Cl$, $[Co(NH_3)_4CO_3]Br$, and $[Co(NH_3)_4CO_3]I$ (*16*). The following reaction was proposed

$$AlOH + CO_3^{2-} = AlCO_3^- + OH^- \quad (2)$$

In contrast to the Al oxides, Su and Suarez (*15*) detected the presence of both bicarbonate and carbonate complexes sorbed on the Fe oxide surfaces. Figure 2b, taken from Su and Suarez (*15*) shows the difference spectrum for am-$Fe(OH)_3$ in the presence of carbonate species in the pH range of 4.5 to 10.5. The characteristic ν_4 asymmetric stretching indicative of HCO_3^- is seen in the region of 1667-1674 and 1586 cm^{-1} only for difference spectra at or below pH 6.2. The lack of peaks for HCO_3^- above pH 6.2 is particularly significant since the total carbon content was orders of magnitude higher for the higher pH spectra. The ν_3 splitting for the surface complexed carbonate species was observed at all pH and concentration values examined; the splitting was 75 cm^{-1} for carbonate on goethite, suggesting monodentate surface complexation. For carbonate on the am-$Fe(OH)_3$, the ν_3 splitting was dependent on concentration, with splitting of 118 cm^{-1} at 0.1 M, 136 cm^{-1} at 0.1 M, and 149 cm^{-1} at 1.0 M carbonate. These results suggest that the bonding on am-$Fe(OH)_3$ is predominantly monodentate, but with increasing bidentate complexation with increasing surface loading. A change from monodentate to bidentate complexation with increasing surface coverage has also been observed for other ligands, as discussed below.

At pH 7.0 and above only the carbonate surface complex was observed (*15*). Based on these results it appears that surface complexation modeling of carbonate should consider monodentate carbonate and bicarbonate bonding with

Table I. Vibrational modes for oxyanions

Species	ν_1	ν_2	ν_3	ν_4	ν_5	ν_6	Reference
$NaHCO_3$	1046, 1035	1395, 1300, 1340	650	1659, 1616	700, 690	837	*(16)*
$KHCO_3$	1005	1400, 1370	657	1650, 1625	698	831	*(16)*
$HCO_3^-{}_{(aq)}$	1010	1360, 1310		1668, 1605		843	*(15)*
$CaCO_3$, calcite	1087	874, 855	1432	712			*(16)*
$CaCO_3$, aragonite	1083	857	1489, 1411	713, 700			*(16)*
$CO_3^{2-}{}_{(aq)}$	1065	887	1383				*(15)*
monodentate CO_3	1062	877	1448, 1354				*(16)*
bidentate CO_3	1043	865	1590, 1282				*(16)*
carbonate on $Al(OH)_3$		1028	1490, 1420				*(15)*
carbonate on $Fe(OH)_3$ 1.0 M, pH 10.5	1070	944	1485, 1336				*(15)*
carbonate on $Fe(OH)_3$ 0.01 M, pH 4.5	1020		1474, 1358	1674, 1586			*(15)*
$B(OH)_3$ solid	1060	668, 648	1490-1428	545			*(24)*
$LaBO_3$	930	708	1310, 1244	606, 587			*(24)*
$[B(OH)_4]^-$ solid	754	379	945	533			*(24)*

Continued on next page.

Table I. continued

Species	ν_1	ν_2	ν_3	ν_4	ν_5	ν_6	Reference
$B(OH)_{3(aq)}$	1148		1410				*(18)*
$B(OH)_{4\ (aq)}^{-}$	1170		955				*(18)*
B on $Al(OH)_3$ pH 6.8	1280		1420				*(18)*
B on $Al(OH)_3$ pH 10.2	1266		1412				*(18)*
$(SeO_4)^{2-}$ solid	833	335	875	432			*(24)*
$SeO_{4\ (aq)}^{2-}$ [a]	837	345	873	314			*(16)*
$BaSeO_4$	840	357, 344	926, 884, 862	452, 416			*(16)*
$SeO_{4\ (aq)}^{2-}$			872				*(23)*
selenate on $Fe(OH)_3$	820		916, 890				*(23)*
selenate on goethite	815		911, 885				*(23)*
$SeO_{3\ (aq)}^{2-}$	807	432	737	374			*(24)*
$SeO_{3\ (aq)}^{2-}$ 1.0 M	851, 822		731				*(23)*
selenite on $Fe(OH)_3$	842	757	478	594			*(23)*
bidentate $Co(en)_2O_2SeO^+$	830	762	515	405	678	360	*(26)*

monodentate $Co(en)OSeO_2H^{2+}$	830	770	530	400	580	350	*(26)*
$MoO_4^{2-}{}_{(aq)}$	894	407	833	320			*(16)*
$NaMoO_4$	898, 903	855, 820, 805					*(16)*
$(MoO_4)^{2-}$, solid	897	317	837	317			*(24)*
molybdate on $Fe(OH)_3$	928, 878		828				*(27)*
$AsO_4^{3-}{}_{(aq)}$	810	342	810	398			*(16)*
$(AsO_4)^{3-}$, solid	837	349	878	463			*(24)*
Na_2HAs_4	725	350	857	396			*(45)*
KH_2AsO_4	745		870, 860				*(46)*
$H_2AsO_4^{-}{}_{(aq)}$			909, 872				this study
$H_2AsO_4^{2-}{}_{(aq)}$			873				this study
arsenate on $Fe(OH)_3$, pH 5			832				this study
arsenate on $Fe(OH)_3$, pH 8			872, 820				this study
$H_3AsO_{3(aq)}$ [a]	710		655				*(32)*
$H_2AsO_3^{-}{}_{(aq)}$ [a]	790	570	370, 320		610		*(32)*
$HAsO_3^{2-}{}_{(aq)}$ [a]	770[b]						*(32)*

Continued on next page.

Table I. continued

Species	ν_1	ν_2	ν_3	ν_4	ν_5	ν_6	Reference
$AsO_3^{3-}{}_{(aq)}$ [a]	752	340	680				(*32*)
H_3AsO_3 [a] pH > 8	700		655				(*33*)
$H_2AsO_3^{-}{}_{(aq)}$ [a]	790				600		(*33*)
arsenite on $Fe(OH)_3$, pH 5	794				630		this study
arsenite on $Fe(OH)_3$, pH 8	789				633		this study
arsenite on $Al(OH)_3$, pH 5	828				600		this study
arsenite on $Al(OH)_3$, pH 8	823						this study

All results are from IR experiments unless otherwise indicated.
[a] Results from Raman experiments.
[b] Estimated by Loehr and Plane (1968).

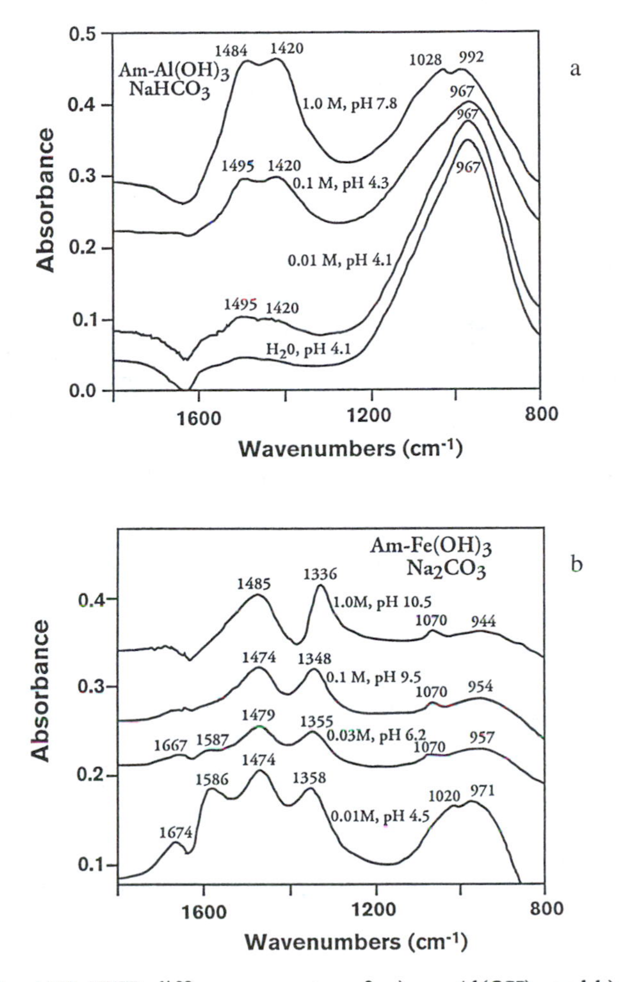

Figure 2. ATR-FTIR difference spectra of a) am-$Al(OH)_3$ and b) am-$Fe(OH)_3$ with and without the presence of 1.0 M Na_2CO_3 at various pH values. Spectra taken from Su and Suarez (*15*).

the am-$Fe(OH)_3$ surface and a pK_a between pH 4 and 6 for the surface carbonate species. In contrast, the solution pK_a is 10.33, indicating that the am-$Fe(OH)_3$ surface has a preference for carbonate over bicarbonate of about 10^5, similar to that observed for PO_4^{3-} over HPO_4^{2-}. Su and Suarez (*15*) proposed the following reactions for bicarbonate and carbonate reactions with the am-$Fe(OH)_3$ surface

$$FeOH_2^+ + HCO_3^- = FeHCO_3 + H_2O \tag{3}$$

$$FeOH + CO_3^{2-} = FeCO_2^- + OH^- \tag{4}$$

Boron. There is relatively little information on the coordination and speciation of B on mineral surfaces. Palmer et al. (*17*) using B isotopic fractionation in the pH range of 6.65 to 8.45, suggested that B is tetrahedrally coordinated with clay surfaces. Su and Suarez (*18*) examined coordination of B adsorbed on Al and Fe oxides, allophane, and kaolinite using ATR-FTIR. The dominant B species in solution are boric acid, $B(OH)_3$ and borate anion, $B(OH)_4^-$, with a pK_a of 9.23. These two species can be readily distinguished since the B atom in boric acid is trigonally coordinated while B in the borate anion is tetrahedrally coordinated.

The solid boric acid spectrum, as well as those of solid phase BO_3 and BO_4 anion are presented in Table I. The solution boric acid spectrum shown in Figure 3a at pH 7 (*18*) shows major peaks at 1410 and 1148 cm^{-1}, corresponding to ν_3 and ν_1 modes for trigonal B. At pH 11 (Figure 3b) the 955 cm^{-1} band is assigned to ν_3 tetrahedral B-OH bending, and the broad band at 1170 cm^{-1} to ν_1 asymmetric stretching of tetrahedral B. At pH 9 the 1410 and 955 cm^{-1} peaks were approximately equal (*18*), suggesting that despite the fact that absorbance rather than Kubelka-Monk units were used, peak size could be used to approximate relative amounts of the two B species. Since at pH 9.23 the activities of $B(OH)_3$ and $B(OH)_4^-$ are equal, the pH at which the concentrations of these two species are equal is shifted to lower pH, closer to pH 9.0, after accounting for activity coefficients. In these spectra the bands in the region 1148-1170 cm^{-1} can be due either to trigonal or to tetrahedral B.

The ATR difference spectrum presented by Su and Suarez (*18*), shown in Figure 4a for am-$Al(OH)_3$ and adsorbed B at pH 6.8 shows large bands at 1420 and 1280 cm^{-1}. The ν_1 modes were shifted to higher frequencies for adsorbed B relative to the solution B (Table I), an indication of increased bonding strength. Similarly, the spectrum in Figure 4b for am-$Al(OH)_3$ and B at pH 10.2 shows only the two large bands at 1412 and 1266 cm^{-1}. The 1412-1420 cm^{-1} ν_3 mode is indicative of trigonal B bound to the am-$Al(OH)_3$ surface. The lack of a decrease in the peak size of this band at pH 10.2 relative to that at pH 6.8 suggests that the surface has a high affinity for $B(OH)_3$. Affinity of the $B(OH)_3$ species on the surface at pH 10.2 may be explained in part by considering that the surface is negatively charged at this pH and $B(OH)_3$ still

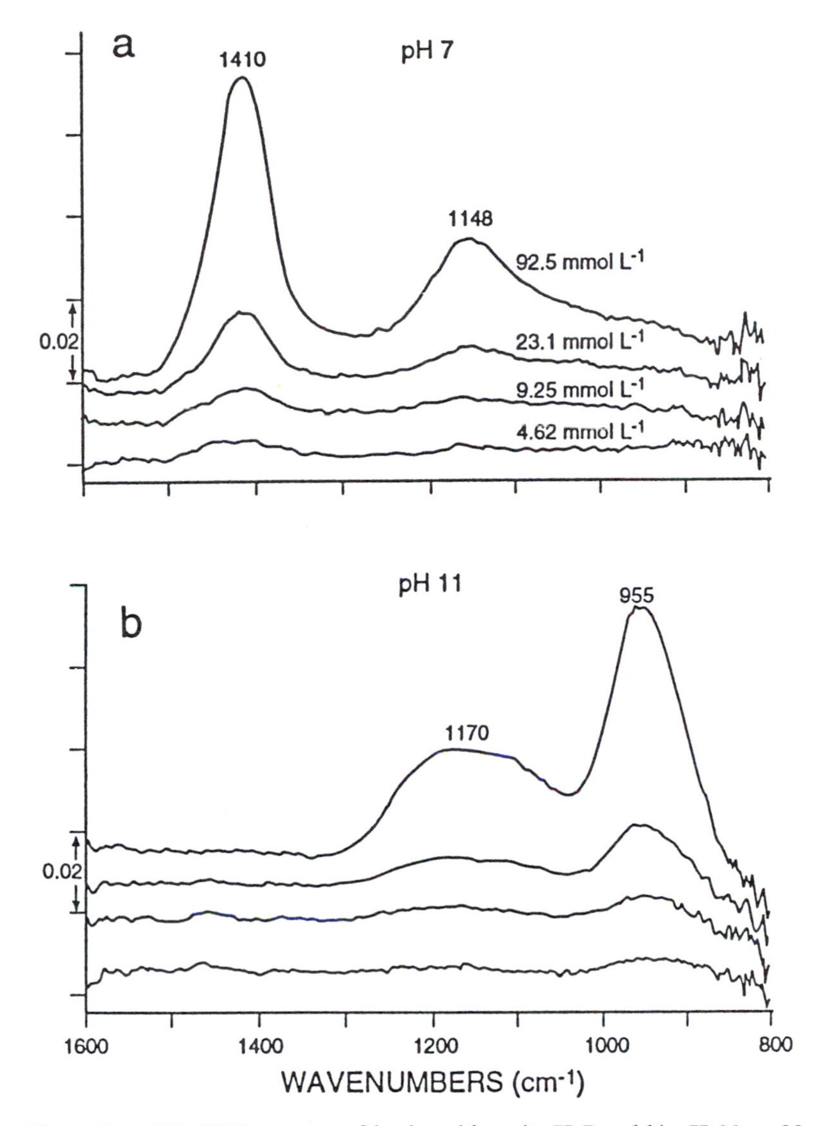

Figure 3. ATR-FTIR spectra of boric acid at a) pH 7 and b) pH 11 at 92 mmol L^{-1} B. Each spectrum is the difference between sample in 0.1 M NaCl and 0.1 M NaCl at the same pH. Spectra taken from Su and Suarez (*18*).

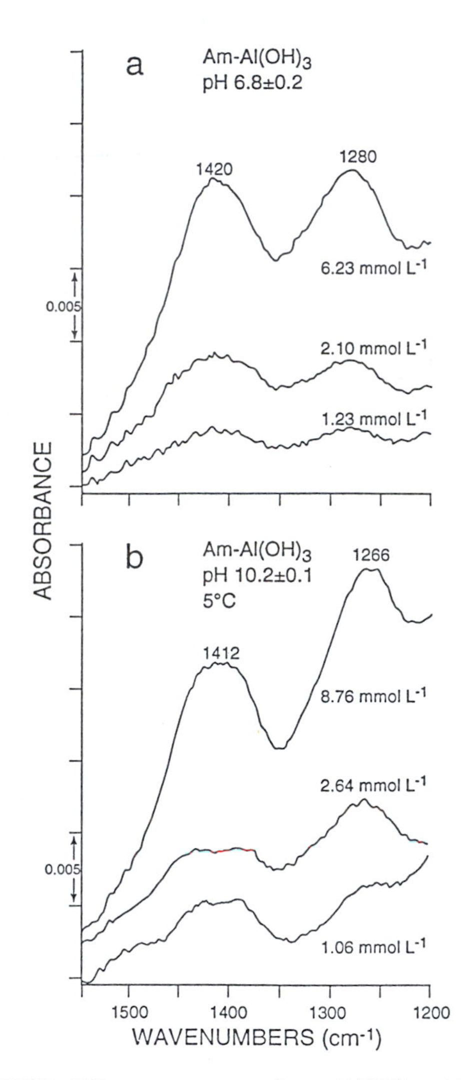

Figure 4. ATR-FTIR difference spectra of am-$Al(OH)_3$ with and without the presence of B at a) pH 6.8 and b) pH 10.2. Spectra taken from Su and Suarez (*18*).

constitutes more than 10% of the total soluble B. Tetrahedral B could not be evaluated due to interference from the strong Al-O band at 969 cm^{-1}. However, if borate were effectively competing with boric acid on the surface, there would be a decrease in the trigonal 1412-1420 cm^{-1} band with increasing pH, and this was not observed.

In contrast to the Al oxide, Fe oxide absorbs both trigonal and tetrahedral B. The ATR difference spectrum presented by Su and Suarez (*18*) and shown in Figure 5a for am-$Fe(OH)_3$ and B at pH 7 showed a large peak at 1400 cm^{-1}, a broad band in the range of 1293-1257 cm^{-1}, and a small peak at 985 cm^{-1}. As discussed above, the 1400 cm^{-1} ν_3 peak is diagnostic of the asymmetric stretching of trigonal B and the 985 cm^{-1} ν_1 peak of the asymmetric stretching mode of tetrahedral B. The ν_1 mode for adsorbed tetrahedral B is shifted upward in wavenumber relative to the same tetrahedral band for aqueous $B(OH)_4^-$ (where ν_1 is at 955 cm^{-1}). From this spectrum the authors concluded that at pH 7 the predominant surface species on am-$Fe(OH)_3$ is trigonal, boric acid, with minor sorption of $B(OH)_4^-$.

At pH 10.2 the difference spectrum for am-$Fe(OH)_3$ with and without B showed a small peak at 1394 cm^{-1} and a large peak at 962 cm^{-1} (Figure 5b). The authors interpreted this spectrum as indicative of the dominance of the $B(OH)_4^-$ species on the am-$Fe(OH)_3$ surface at pH 10.2. Based on these spectra and evaluating the changes in the size of the 1400 and 960 cm^{-1} peaks as a function of pH, it is estimated that the surface pK_a for boric acid on am-$Fe(OH)_3$ is in the pH range 8 to 9.

Measurements of EM for am-$Al(OH)_3$ and am-$Fe(OH)_3$ show a decrease in the PZC when B is adsorbed (*18*). Therefore, only reactions which show a decrease in surface charge can be consistent with the EM results. The following reaction appears to be the only reaction consistent with both the FTIR and EM data for B adsorption on am-$Al(OH)_3$ and am-$Fe(OH)_3$ at acid pH conditions

$$SOH_2^+ + B(OH)_3^0 \rightleftharpoons SBO(OH)_2 + H_2O + H^+ \tag{5}$$

Equation 5 considers adsorption on positively charged surface sites, proton release, and a downward shift in the PZC, with adsorption of monodentate trigonal B. The following equation

$$2SOH + B(OH)_3^0 \rightleftharpoons S_2BO_2(OH)_2^- + H_2O + H^+ \tag{6}$$

represents adsorption on a neutral surface site with proton release and a downward shift in the PZC, with adsorption of bidentate tetrahedral B. This reaction is applicable to B adsorption on Fe oxides at alkaline pH. We note that the adsorption maxima for Al oxides are lower than those for Fe oxides (*19*), despite the fact that the PZC is greater for Al oxides. This is consistent with the above reactions since at high pH the concentration of $B(OH)_3$ adsorbed by the Al oxide decreases, while the Fe oxide adsorbs the tetrahedral B species

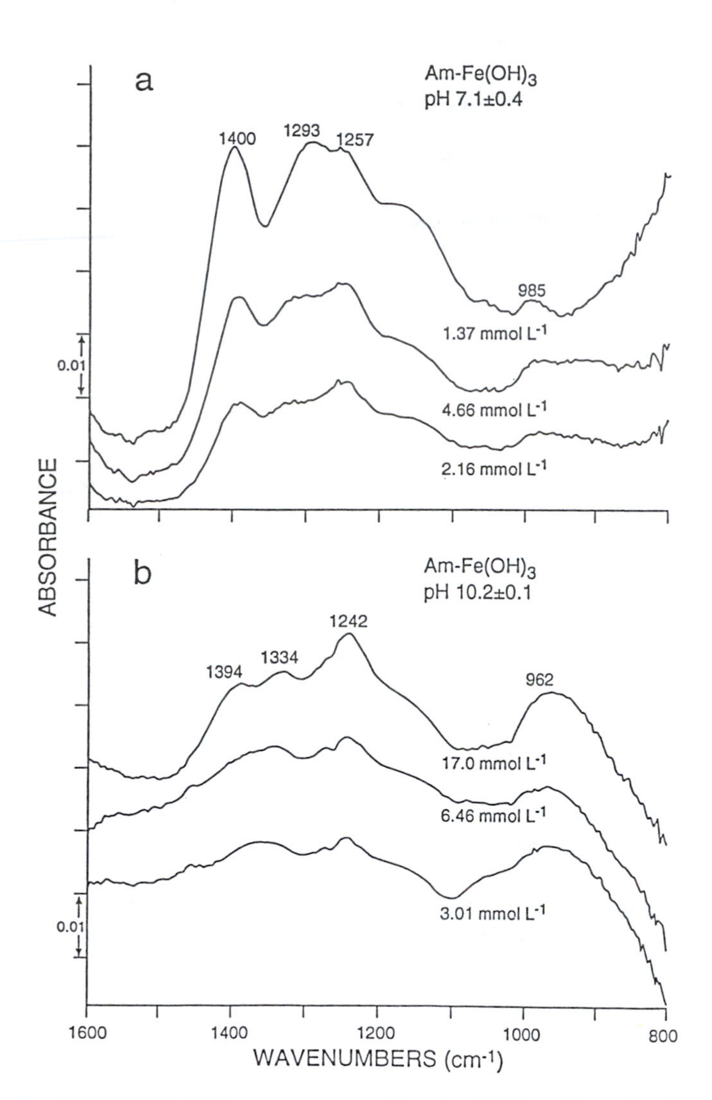

Figure 5. ATR-FTIR difference spectra of am-$Fe(OH)_3$ with and without the presence of B at a) pH 7.1 and b) pH 10.2. Spectra taken from Su and Suarez (*18*).

as well as the trigonal, and thus would be expected to have greater adsorption at higher pH.

Futher evaluation of B surface reactions on oxides is possible using measurements of net OH^- or H^+ release after reacting oxide suspensions and adsorbing ligand solutions which were previously equilibrated to the same specified pH. At pH 5 and 7 the reaction released 0.55 and 0.66 moles, respectively of H^+ per mole of B adsorbed; this corresponds to OH/B ratios of -0.55 and -0.66 respectively. The reaction represented in equation 5 predicts OH/B ratios of -1.0 for both pH 5 and 7. These results suggest that another surface reaction

$$SOH + B(OH)_3^0 \rightleftharpoons SBO(OH)_2 + H_2O \tag{7}$$

is equally important. Note that the surface species are the same for equations 5 and 7 but that equation 7 results in no shift in pzc and no net proton release. At pH 5-7 adsorption of trigonal B occurs by reaction with both neutral and positive surface sites, resulting in a shift downward in PZC and a net release of protons to solution (OH/B ratio of -0.5).

The following equation

$$2SOH + B(OH)_4^- \rightleftharpoons S_2BO_2(OH)_2^- + 2H_2O \tag{8}$$

represents a bidentate tetrahetdral surface complex consistent with the FTIR results and downward shift in EM and surface charge at high pH. This cannot be the only reaction since it does not predict a net release of OH^-. At pH 9 and 10 we measured a net release of 0.79 and 1.17 moles of OH^-, respectively per mole of B adsorbed. The reaction

$$SO^- + B(OH)_4^- \rightleftharpoons (SO)B(OH)_3^- + OH^- \tag{9}$$

represents a monodentate tetrahedral surface complex with net release of OH^-. The species $B(OH)_4^-$ is dominant in solution above pH 9 in 0.1 M NaCl. This reaction is consistent with the FTIR results and OH release data (but not the EM data). At high pH a combination of equations 8 and 9 are consistent with the FTIR, EM, and OH^- release data.

Selenate. Harrison and Berkheiser (*20*) examined selenate sorption onto freshly precipitated hydrous ferric oxide using dispersive IR spectroscopy on air dried samples. They concluded that selenate forms a bidentate bridging complex, replacing both protonated and unprotonated hydroxyls.

Hayes et al. (*21*) and Manceau and Charlet (*22*) examined selenate sorption onto goethite, and goethite and hydrous ferric oxide, respectively using EXAFS. Hayes et al. (*21*) concluded that the surface complex was outer-sphere while Manceau and Charlet (*22*) concluded that the complex was inner-sphere. This discrepancy demonstrates how spectroscopic studies are not always conclusive. The reasons for these differences are speculative but could be due to various factors including the use of reagents of one oxidation state contaminated with material of the other oxidation state, possible oxidation-reduction reactions occurring during the EXAFS experiments, differences in EXAFS data analysis software, or differences in surface coverage by the ligand.

Points of zero charge of both goethite and am-$Fe(OH)_3$ were shifted to lower pH values in the presence of selenate, suggesting inner-sphere complexation (*23*). Increase in pH during selenate sorption is indicative of hydroxyl release. Su and Suarez (*23*) also examined selenate sorption on am-$Fe(OH)_3$ and goethite at pH 5 and 8 using DRIFT and ATR-FTIR. The Raman spectra for aqueous SeO_4^{2-} from Ross (*16*) and the vibrational spectra for aqueous SeO_4^{2-} and solid SeO_4 (from $BaSeO_4$) from Nakamoto (*24*) are presented in Table I. The ATR spectra for aqueous SeO_4^{2-} determined by Su and Suarez (*23*) showed only a single peak at 870 cm^{-1}. The ν_1 and ν_2 modes are not IR active for T_d symmetry (*24*) and the ν_4 mode is below the detection range of the ATR cell. The DRIFT spectrum of selenate on am-$Fe(OH)_3$ (Figure 6, from (*23*)) shows splitting of the ν_3 vibration of aqueous SeO_4^{2-} into bands at 916 and 890 cm^{-1}, with the 820 cm^{-1} band assigned to ν_1. The ν_3 mode is also shifted to higher frequency for adsorbed as compared to aqueous selenate (Table I), consistent with increased bond strength. Splitting of the ν_3 mode for XO_4 ligands indicates reduced symmetry (to C_{3v}) and is consistent with formation of a bidentate bridging complex (*16*). This designation is not definitive since reductions in symmetry are also possible due to crystal structure (*16*), as evidenced by the ν_3 splitting seen in $BaSeO_4$ solid (Table I). In either case this splitting is indicative of adsorbed selenate. Consistent with this reasoning, the ν_1 mode is IR active for C_{3v} symmetry and is observed for adsorbed selenate. The proposed reaction is

$$FeOH + FeOH_2^+ + SeO_4^{2-} \rightarrow Fe_2SeO_4 + H_2O + OH^- \qquad (10)$$

The dominant selenate species above pH 2 is SeO_4^{2-}, thus this reaction should result in release of one hydroxyl ion for each SeO_4^{2-} ion sorbed, as well as a decrease in the PZC of the oxide. The measured OH/Se ratio ranged from 0.56 to 0.94 and the PZC decreased in the presence of adsorbed selenate (*23*). These measurements, similar to those discussed earlier, are based on reactions in which the oxide suspension at a fixed ionic strength and pH is reacted with a solution containing the adsorbing ligand at the same ionic strength and pH. The reaction is quantified by adding acid or base to adjust the pH back to the initial value. This avoids difficulties associated with titration where the initial and reacted pH are different.

Since we can write no surface reactions with SeO_4^{2-} (using neutral or positively charged surface sites) which release less than one hydroxyl, these results suggest that some selenate sorption may be non-specific or outer-sphere. Alternative surface complexes can be written using the sites defined by the MUSIC model (*25*), $FeOH^{½-}$, Fe_2O^-, Fe_2OH, and $Fe_3O^{½-}$. Consideration of alternative sites does not change the interpretation of the titration and EM data.

We conclude that selenate sorption is mostly inner-sphere and possibly bidentate.

Selenite. Su and Suarez (*25*) examined selenite sorption on goethite and am-$Fe(OH)_3$ at pH 5 and pH 8 using FTIR in ATR and DRIFT mode, as well as measurements of EM in the presence and absence of selenite. Selenious acid, H_2SeO_3, dissociates to form $HSeO_3^-$ and SeO_3^{2-} with pK_a values of 2.5 and 8.0, respectively. At pH 5, the dominant species in solution is thus $HSeO_3^-$, while at pH 8.0 both $HSeO_3^-$ and SeO_3^{2-} are present. The SeO_3^{2-} anion has C_{3v} symmetry with four IR active frequencies, as shown in Table I. Coordination through either one or two oxygen atoms or presence in a crystalline state lowers the symmetry and the ν_3 and ν_4 modes may be split (*24*). Su and Suarez (*23*) did not see any difference in the liquid spectra of selenite at pH 5 and 8, indicating that ATR could not be used to distinguish $HSeO_3^-$ and SeO_3^{2-}. They observed peaks at 851, 822 cm^{-1} and the ν_3 mode at 731 cm^{-1}.

The ATR difference spectrum of adsorbed selenite on am-$Fe(OH)_3$ showed two distinct peaks at 844 and 750 cm^{-1}, corresponding to the ν_1 and ν_3 modes of selenite, with the ν_3 mode shifted upward as observed for all solid phase selenites, consistent with the presence of adsorbed selenite. The DRIFT spectrum, as shown in Figure 7, confirmed the presence of these peaks at 842 and 757 cm^{-1} as well as additional bands at 594 and 478 cm^{-1}. In accordance with reduced symmetry (from C_{3v} to C_s) ν_3 is split into ν_2 and ν_5 and ν_2 is shifted to ν_3 (*24*). The band assignments are listed in Table I, with the splitting of the 737 band into bands at 757 and 594 cm^{-1}. The previously IR inactive ν_2 mode is now assigned to ν_3 at 478 cm^{-1}. This additional IR band in addition to the peak splitting is evidence of the reduced symmetry associated with formation of a selenito surface complex.

Monodentate selenito complexes can be distinguished from bidentate complexes by examining the ν_3 splitting discussed above. Monodentate complexes split into a doublet while bidentate complexes split into two distinct bands (*26*). Inspection of the peaks in Figure 7 suggests that they are separate peaks consistent with bidentate complexation. This is not definitive since the separation of the peaks (161 cm^{-1}) is closer to the separation seen for the monodentate bands than for that reported for the bidentate selenito complex (Table I).

The EM data showed a downward shift in the PZC of both am-$Fe(OH)_3$ and goethite, consistent with ligand exchange and inner-sphere complexation. The proposed reaction

$$FeOH + FeOH_2^+ + SeO_3^{2-} \rightarrow Fe_2SeO_3 + H_2O + OH^- \qquad (11)$$

must also consider that the predominant solution species is $HSeO_3^-$. Thus the above reaction must be combined with the deprotonation reaction in solution

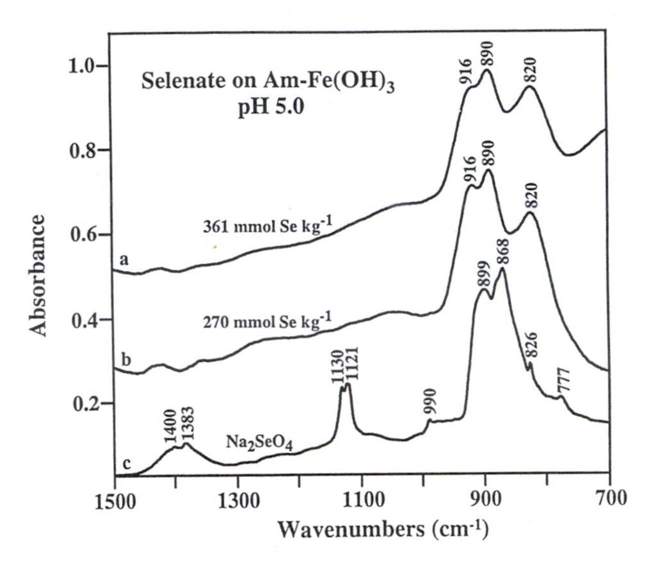

Figure 6. DRIFT difference spectrum of am-$Fe(OH)_3$ with and without the presence of selenate at pH 5.0. Spectra taken from Su and Suarez (*23*).

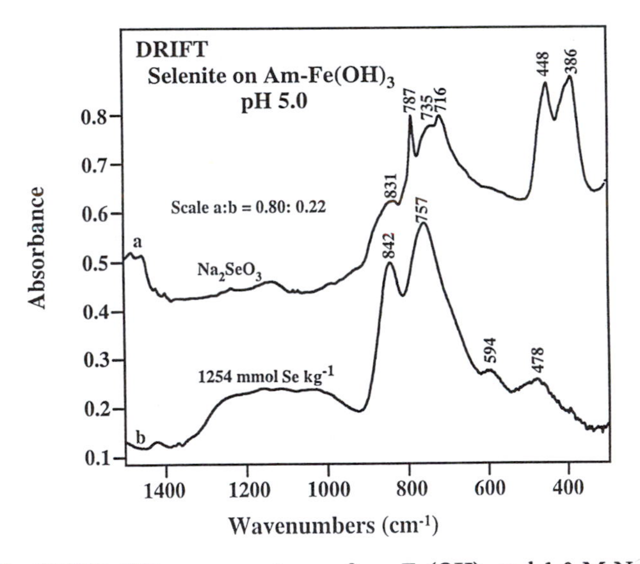

Figure 7. DRIFT difference spectrum of am-$Fe(OH)_3$ and 1.0 M NaCl with and without reaction with 0.05 M Na_2SeO_3. Also shown is the DRIFT spectrum for analytical grade Na_2SeO_3. Spectra taken from Su and Suarez (*23*).

$$HSeO_3^- \rightarrow SeO_3^{2-} + H^+ \tag{12}$$

suggesting that there is no net hydroxyl release during selenite surface complexation. Since the observed OH/Se ratio ranged from 0.51 to 0.81 (*23*) it is necessary to consider that a significant part of the overall reaction includes

$$2FeOH + SeO_3^{2-} \rightarrow Fe_2SeO_3 + 2OH^- \tag{13}$$

Equation 13 in combination with equation 12 results in net hydroxyl release (OH/Se ratio equal to one) and no shift in PZC. Overall the FTIR, PZC shift, and OH/Se release data are consistent with a combination of equation 11 and equation 13. From the above, it is clear that although a shift in PZC is indicative of inner-sphere complexation, lack of a shift in PZC cannot be used as evidence for an outer-sphere complex.

Molybdenum. Sorption of Mo onto am-$Fe(OH)_3$ was examined by Goldberg et al. (*27*) using FTIR. The MoO_4^{2-} aqueous species is dominant in most natural systems (pK_{a2} for molybdic acid = 4.2). The ATR-FTIR spectrum for MoO_4^{2-} reported by them showed peaks at 933, 885, and 835 cm^{-1}, in reasonable agreement with the earlier published vibrational spectra (Table I). The Mo, am-$Fe(OH)_3$ ATR difference spectrum showed peaks at 928 and 880 cm^{-1}, similar to aqueous MoO_4^{2-} thus adsorbed Mo was not detected.

The DRIFT spectrum (Figure 8) showed bands at 924 and 874 cm^{-1}, and a broad band at 830-780 cm^{-1}. The spectra for monodentate and bidentate complexes are distinguished by the degree of ν_3 splitting. The $[Co(NH_3)_5MoO_4]Cl$ monodentate complex has a ν_1 mode at 910 cm^{-1} and a splitting of the ν_3 mode into two peaks at 877, and 833 cm^{-1}. The $[Co(NH_3)_4MoO_4]NO_3$ bidentate complex, has the ν_1 mode at 920 cm^{-1} and a triplet splitting of the ν_3 mode with peaks at 868, 845, and 795 cm^{-1} (*16*). The broad band seen in the DRIFT spectrum at 828 cm^{-1} cannot be resolved, thus distinction of mono or bidentate bonding was not possible based on IR.

The EM data reported by Goldberg et al. (*28*) indicated a large downward shift in the PZC to lower pH, indicating ligand exchange. The OH/Mo (release/sorption ratio) exchange data of Goldberg et al. (*27*) increased from 1.36 to 1.81 with increasing Mo solution concentration. The reaction for the monodentate surface species

$$SOH + MoO_4^{2-} \rightarrow SMoO_4^- + OH^- \tag{14}$$

requires a OH/MoO_4^{2-} exchange ratio of 1.0 while the reaction for the bidentate surface species

$$2SOH + MoO_4^{2-} \rightarrow S_2MoO_4 + 2OH^- \tag{15}$$

requires a OH/MoO_4^{2-} ratio of 2.0. Together these data suggest a mix of inner-

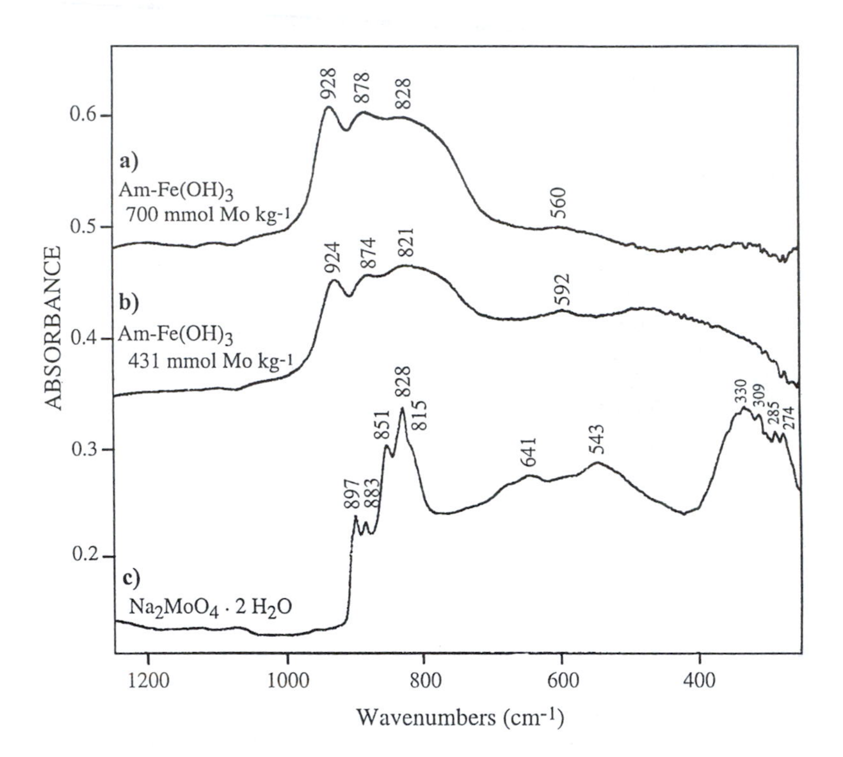

Figure 8. DRIFT difference spectra of am-$Fe(OH)_3$ with and without reaction with solutions of Mo. Suspensions were reacted twice with deionized water. Also shown is the DRIFT spectrum for analytical grade $Na_2MoO_4{\cdot}2H_2O$. Spectra taken from Goldberg et al. (*27*).

sphere, mono and bidentate surface sorption of MoO_4^{2-} with increasing bidentate relative to monodentate surface species with increasing concentration.

Arsenate. Waychunas et al. (*29*) examined arsenate sorption onto goethite, lepidocrocite, akaganeite and ferrihydrite using EXAFS. They concluded that the As tetrahedral complex underwent little distortion upon sorption, retaining essentially identical As-O distances as As(V) in solution. Arsenate sorption was primarily as inner-sphere bidentate complexes with monodentate complexes occurring only at low surface coverage on goethite. The bidentate arsenate complexes were attached to adjacent edge-sharing Fe octahedra while the monodentate arsenate complexes were attached to Fe octahedra apices.

Arsenate sorption onto goethite has also been examined by Fendorf et al. (*30*) using EXAFS. They concluded that the mode of arsenate bonding to the surface was related to the extent of surface coverage. Surface complexation of arsenate was transformed from monodentate to bidentate mononuclear and bidentate binuclear with increasing surface concentration. Fendorf et al. (*30*) increased surface concentration by decreasing pH from 9 to 6, which will also decrease the number of negatively charged sites (SO^-), and affect the aqueous phase speciation. The pK_a values for arsenic acid are 2.3, 6.8 and 11.6, thus the dominant aqueous arsenic species in natural systems is $H_2AsO_4^-$ at low pH and $HAsO_4^{2-}$ under alkaline conditions.

Sun and Doner (*31*) examined the OH stretching modes of deuterated goethite in the presence of arsenate using FTIR. Based on transmission and ATR modes in dry systems, the authors concluded that arsenate oxyanions replaced singly coordinated surface hydroxyl groups to form bidentate binuclear complexes. Reactions presented by Sun and Doner (*31*) indicate $HAsO_4^{2-}$ as the surface species, bound to metal centers via two O bonds.

In Figure 9 we present the EM of am-$Fe(OH)_3$ in comparison to the EM of am-$Fe(OH)_3$ in the presence of 0.01 and 1.0 mM As(V) all as a function of solution pH. These data were collected using a Zeta-Meter 3.0 system (Zeta Meter, Long Island City, NY; mention of trade names is for the benefit of the reader only and does not indicate endorsement by the USDA or USEPA). Suspensions containing 0.2 g L^{-1} solid in 0.01 M NaCl (with and without As) were acidified with HCl to pH 3.0, equilibrated for 1 h and then titrated with 0.01 M NaCl to various pH values for EM measurements. The PZC was determined by interpolation of the data to zero mobility.

As shown in Figure 9, the sorption of As decreased the measured PZC from pH 8.7 in the absence of As to pH 6.9 at 0.01 mM As and to below pH 3 at 1.0 mM As. These data indicate a large negative shift in surface charge. The hydroxyl release data were obtained by reacting 20 mL of 0.2 M As solution adjusted to pH 5 or 8 with 4 g of oxide suspension also adjusted to pH 5 or 8, or solutions of 20 mL of 0.2 M As(V) reacted with 4 g of am-$Fe(OH)_3$. Upon mixing the pH was adjusted back with NaOH or HCl, and the supernatant analyzed for As. The hydroxyl release, expressed as the ratio of OH released /As adsorbed, was 0.19 and 1.03 for pH 5 and pH 8 respectively.

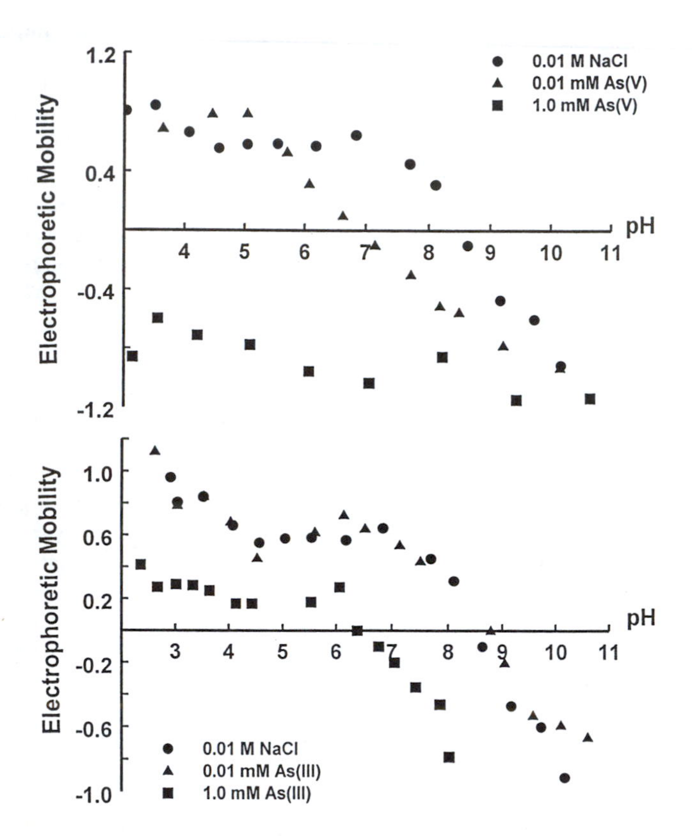

Figure 9. EM of am-$Fe(OH)_3$ suspensions reacted with 0, 0.01 and 1.0 mM As(V) concentrations as a function of pH and b) 0, 0.01 and 1.0 mM As(III) concentrations as a function of pH. As(V) suspensions underwent a large shift in PZC as compared to the As(III) suspensions.

The EM data for am-$Al(OH)_3$ showed a large shift in the PZC in the presence of 1 mM As(V) with a shift in the PZC to 5.4 (data not shown). In contrast, the sorption of As(III) (1.0 mM) on am-$Al(OH)_3$ showed a very small shift in PZC to 9.2.

The ATR-FTIR spectrum of a solutions of Na_2HAsO_4 at pH 5 and 8 were collected using a Bio-Rad FTS-7 spectrometer with a DTGS detector and a CsI beamsplitter (Bio-Rad Digilab Division, Cambridge, MA). Single beam spectra were collected from 2000 co-added interferograms with a 4 cm^{-1} resolution. Spectra were obtained by subtracting the spectra of the solution of interest from the spectra of the reference NaCl solution. Both spectra were ratioed against the empty cell. A subtraction factor of 1.0 was used for all subtractions.

The vibrational spectra for AsO_4 (solids) are listed in Table I, along with the Raman spectra for AsO_4^{3-}. The ATR spectrum of a solution of Na_2HAsO_4 at pH 8, where $HAsO_4^{2-}$ is the dominant aqueous As species, has a large broad peak at 862 cm^{-1}, as shown in Figure 10a. This band is assigned to the IR active ν_1 and ν_3 modes and suggests that $HAsO_4^{2-}$ has a similar spectrum to AsO_4^{3-}. In contrast, the spectrum shown in Figure 10b, at pH 5, where $H_2AsO_4^-$ is dominant, has two peaks at 909 and 878 cm^{-1}, corresponding to the splitting of the ν_3 mode. By analogy with the PO_4 system, AsO_4^{3-} has tetrahedral symmetry. Protonation to $HAsO_4^{2-}$ should lower the symmetry to C_{3v}, with further protonation to $H_2AsO_4^-$ lowering the symmetry to C_{2v}. Consistent with this we see a splitting of the ν_3 mode for $H_2AsO_4^-$. The ν_2 and ν_4 modes are found below 500 cm^{-1} and thus are not seen in our IR spectra.

The vibrational data for As solids, shown in Table I indicates that for $H_2AsO_4^-$ there is a splitting of the ν_3 mode, as compared to the single ν_3 mode reported for $HAsO_4^{2-}$ solids, consistent with our ATR data described above. DRIFT difference spectra, were collected using both the Bio-Rad FTS-7 and a Bio-Rad FTS 175 with a MCT detector and a KBr beamsplitter. The DRIFT difference spectrum of As(V) sorbed onto am-$Fe(OH)_3$ at pH 5, collected using the Bio-Rad FTS 175 is shown in Figure 11a. The large broad peak at 834 cm^{-1}, with a shoulder at 880 cm^{-1} is interpreted as the ν_3 mode suggesting that $HAsO_4^{2-}$ is the surface species.

At pH 9 the DRIFT difference spectrum for As(V) on am-$Fe(OH)_3$ appears to have two peaks, at 872 and 820 cm^{-1}, as shown in Figure 11b. Similar results were obtained with the FTS-7 and the CsI beamsplitter in a completely replicated experiment, except that additional peaks were identified around 700, 605, and 480 cm^{-1}. The two peaks at 605 and 480 cm^{-1} are tentatively identified as the ν_1 and ν_4 modes, which are below the detection of our FTS-175 system. Interpretation of the spectra obtained at pH 9 is not clear as the two peaks at 872 and 820 cm^{-1} appear split much more than expected for the ν_3 split assigned to $H_2AsO_4^-$. It also does not appear reasonable to consider $HAsO_4^{2-}$ adsorption at pH 5 and $H_2AsO_4^-$ adsorption at pH 9. The following reaction

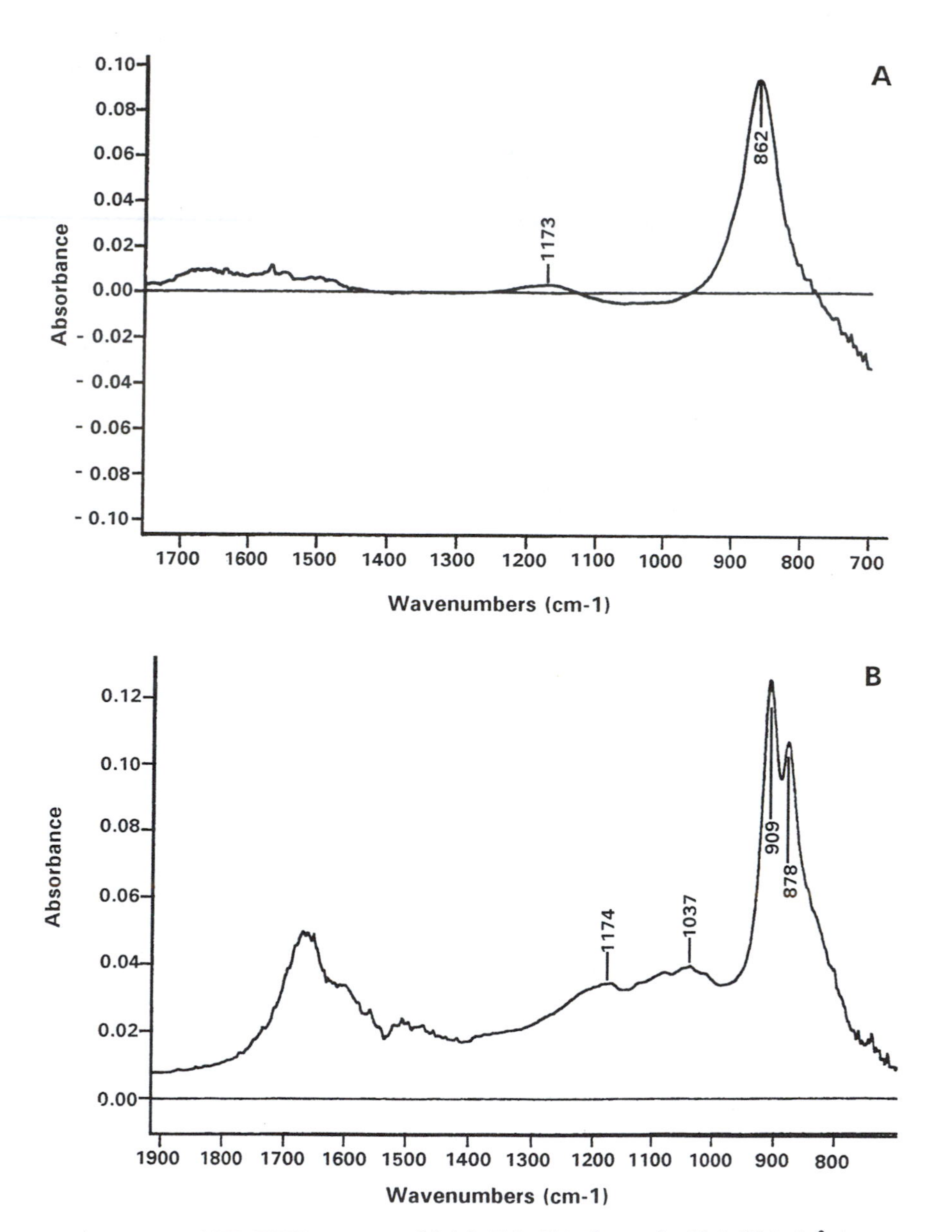

Figure 10. ATR-FTIR spectra of 0.1 M Na_2HAsO_4 at a) pH 8 ($HAsO_4^{2-}$ is the dominant aqueous As(V) species) and b) pH 5 ($H_2AsO_4^-$ is the dominant aqueous As(V) species).

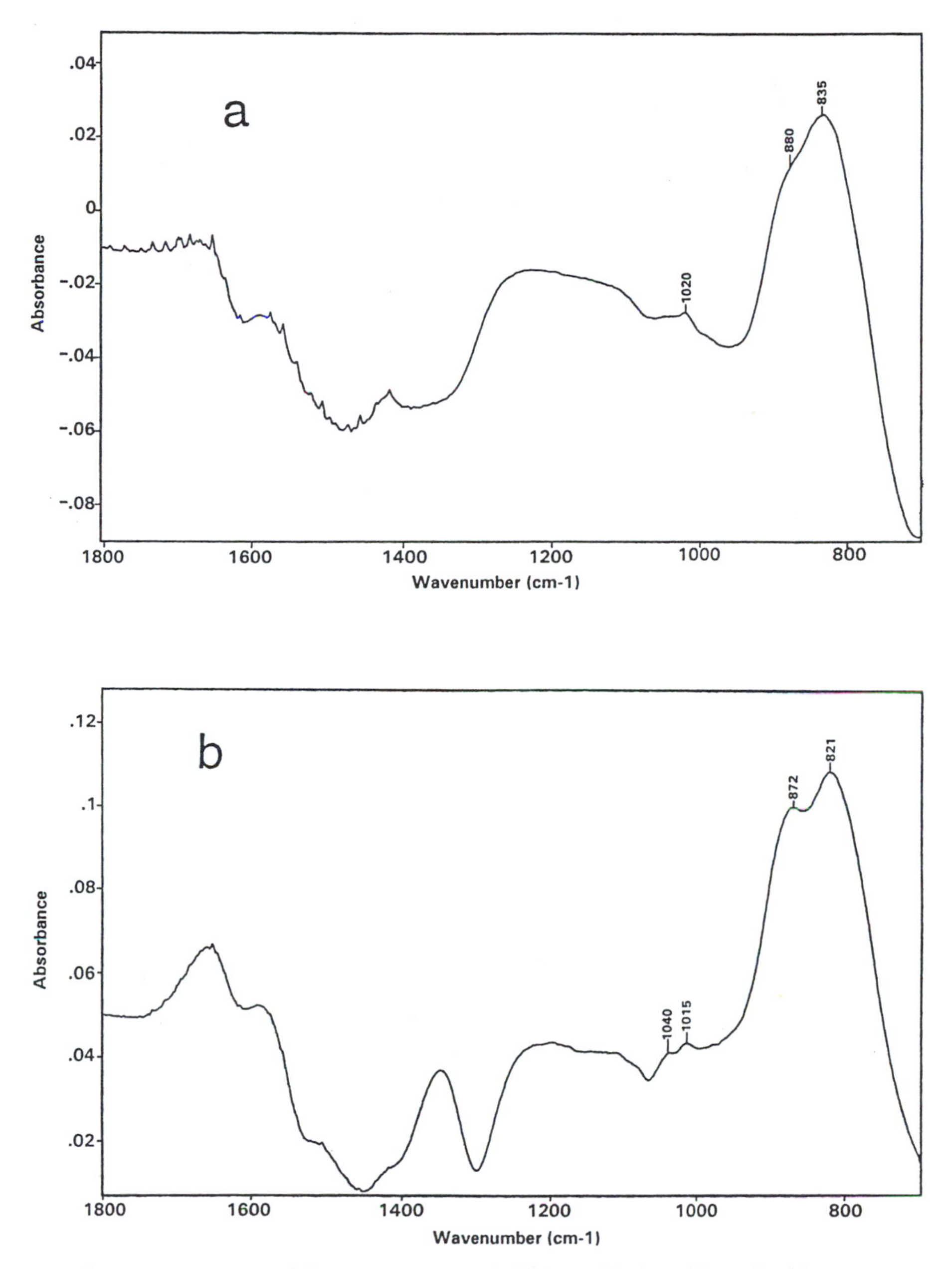

Figure 11. DRIFT difference spectra of a) am-$Fe(OH)_3$ with and without 0.1 M As(V) and 1.0 M NaCl at pH 8 and b) am-$Fe(OH)_3$ with and without 0.1 M As(V) and 1.0 M NaCl at pH 5.

$$SOH + H_2AsO_4^- \rightleftharpoons SHAsO_4^- + H_2O \tag{16}$$

is consistent with the spectroscopy at pH 5, large shift in PZC, and OH/As data. At pH 5 this reaction results in no net release of hydroxyl (vs. 0.2 measured) since $H_2AsO_4^-$ is the dominant solution species. At pH 8 this reaction results in a net release of 1.0 hydroxyl per As adsorbed (vs. 1.03 measured) since $HAsO_4^{2-}$ is the dominant solution species. The reaction

$$2SOH + H_2AsO_4^- \rightleftharpoons S_2HAsO_4^0 + H_2O + OH^- \tag{17}$$

is not possible since it does not account for the shift in PZC nor titration data. The reaction

$$2SOH_2^+ + H_2AsO_4^- \rightleftharpoons S_2HAsO_4^0 + 2H_2O + H^+ \tag{18}$$

accounts for the shift in PZC but results in a net H^+ release at pH 5 and no *net* hydroxyl release pH 8, which is clearly inconsistent with the titration data.

Formation of a bidentate $HAsO_4^{2-}$ complex is inconsistent with the PZC and titration data. Only formation of the bidentate AsO_4^{3-} complex

$$2SOH + H_2AsO_4^- \rightleftharpoons S_2AsO_4^- + 2H_2O \tag{19}$$

appears to be consistent with the EXAFS, PZC, and sorption/titration data (H_3AsO_4 and $H_2AsO_4^-$ presumably cannot form bidentate complexes). In contrast only formation of the monodentate $HAsO_4^{2-}$ complex (equation 16) appears consistent with the FTIR, PZC, and titration data. Some insight into the relative stability of the various complexes may be obtained by examination of the stability of the aqueous Fe-AsO_4 complexes. The reaction

$$FeH_2AsO_4^{2+} \rightleftharpoons FeHAsO_4^+ + H^+ \tag{20}$$

has a pK of 0.24. The AsO_4^{3-} anion forms a stable complex

$$FeHAsO_4^+ \rightleftharpoons FeAsO_4^0 + H^+ \tag{21}$$

with a pK of 0.92, suggesting that AsO_4^{3-} is a likely surface species. Additional FTIR study is needed, including spectra of various Fe-AsO_4 aqueous complexes.

Arsenite. Sun and Doner (*31*) examined the OH stretching modes of deuterated goethite in the presence of arsenite using FTIR. Based on transmission and ATR modes in dry systems, as done for arsenate, they concluded that arsenite oxyanions replaced singly coordinated surface hydroxyl groups to form bidentate binuclear complexes. The assumed surface species was $HAsO_3^{2-}$.

The pK_{a1} of arsenious acid is 9.2 and the pK_{a2} is 12.7, thus $H_3AsO_3^0$ is the dominant aqueous species.

EM measurements were made using a Zeta-Meter model 3.0 system on suspensions of Fe oxide consisting of 0.2 g of solid and 50 mL of solution. Solutions consisted of 0.01 M NaCl and 0, 0.01, and 1.0 mM arsenite. The EM data presented in Figure 9 show a downward shift in the PZC, but considerably less than observed for As(V) at comparable levels of As sorption. The sorption reaction also caused a net decrease in pH. The amount of OH consumed (H^+ released) relative to As sorbed was 0.103 and 0.021 at pH 5 and 8, respectively.

Loehr and Plane (*32*) reported on the Raman spectra of arsenious acid. The spectra for H_3AsO_3, $H_2AsO_2^-$, and AsO_3^{3-} are given in Table I, along with a tentative assignment of $HAsO_3^{2-}$. The ν_3 asymmetric stretching mode and ν_1 symmetric stretching mode reported for H_3AsO_3 in Raman spectra are at 655 and 710 cm^{-1}, respectively. For C_{3v} symmetries all Raman modes are IR active. The $H_2AsO_3^-$ species (which is predominant in solution at pH above 9.2), has a ν_1 mode at 790 cm^{-1}, a ν_2 mode at 570 cm^{-1}, ν_3 and ν_4 modes at 370 and 320 cm^{-1}, and a ν_5 mode at 610 cm^{-1} (*32*). Loehr and Plane (*32*) noted a decrease in intensity of the band at 700 cm^{-1} as they went from H_3AsO_3 to $H_2AsO_3^-$, as well as an increase in intensity at 575 and 775 cm^{-1}. The $HAsO_3^{2-}$ species were characterized by a very weak band at 810 cm^{-1} (As-O asymmetric stretching), a strong band at 770 cm^{-1} (As-O symmetric stretching), and bands at 670, 520 cm^{-1} (As-OH bending modes).

The aqueous spectrum of H_3AsO_3 ($As(OH)_3$) at pH 5 confirmed the presence of a large peak extending below the 700 cm^{-1} lower limit of our ATR cell, corresponding to the ν_1 stretching mode. The aqueous spectrum of $H_2AsO_3^-$ at pH 10.4 showed a large peak at 800 cm^{-1}, corresponding to the ν_1 As-O stretching mode at 790 cm^{-1} reported by Loehr and Plane (*32*) and Gout et al. (*33*), thus indicative of the $H_2AsO_3^-$ ($AsO(OH)_2^-$) species.

The DRIFT difference spectrum of As(III) sorbed onto am-$Fe(OH)_3$ at pH 5 is shown in Figure 12a. Two large peaks were observed at 794 and 630 cm^{-1} with a minor peak at 453 cm^{-1}. This spectrum resembles that of $H_2AsO_3^-$, with 794 cm^{-1} being the ν_1 As-O stretching mode and 630 cm^{-1} being the ν_5 As-OH asymmetric stretching mode. We do not see evidence for $HAsO_3^{2-}$ on the surface since the ν_1 As-O stretching mode for this species is expected at 770 cm^{-1} (*32*). Also, the As-OH symmetric stretching modes at 670 and 520 cm^{-1}, and the As-O asymmetric mode at 810 cm^{-1} were not detected.

Based on the FTIR DRIFT results we consider that the adsorbed As(III) species on am-$Fe(OH)_3$ is $H_2AsO_3^-$. The following reaction is considered dominant

$$SOH + H_2AsO_3^- \rightleftharpoons SH_2AsO_3 + OH^- \quad (22)$$

This reaction results in an overall net OH/As ratio of 0 (when we consider that H_3AsO_3 is the dominant solution species), in close agreement with the

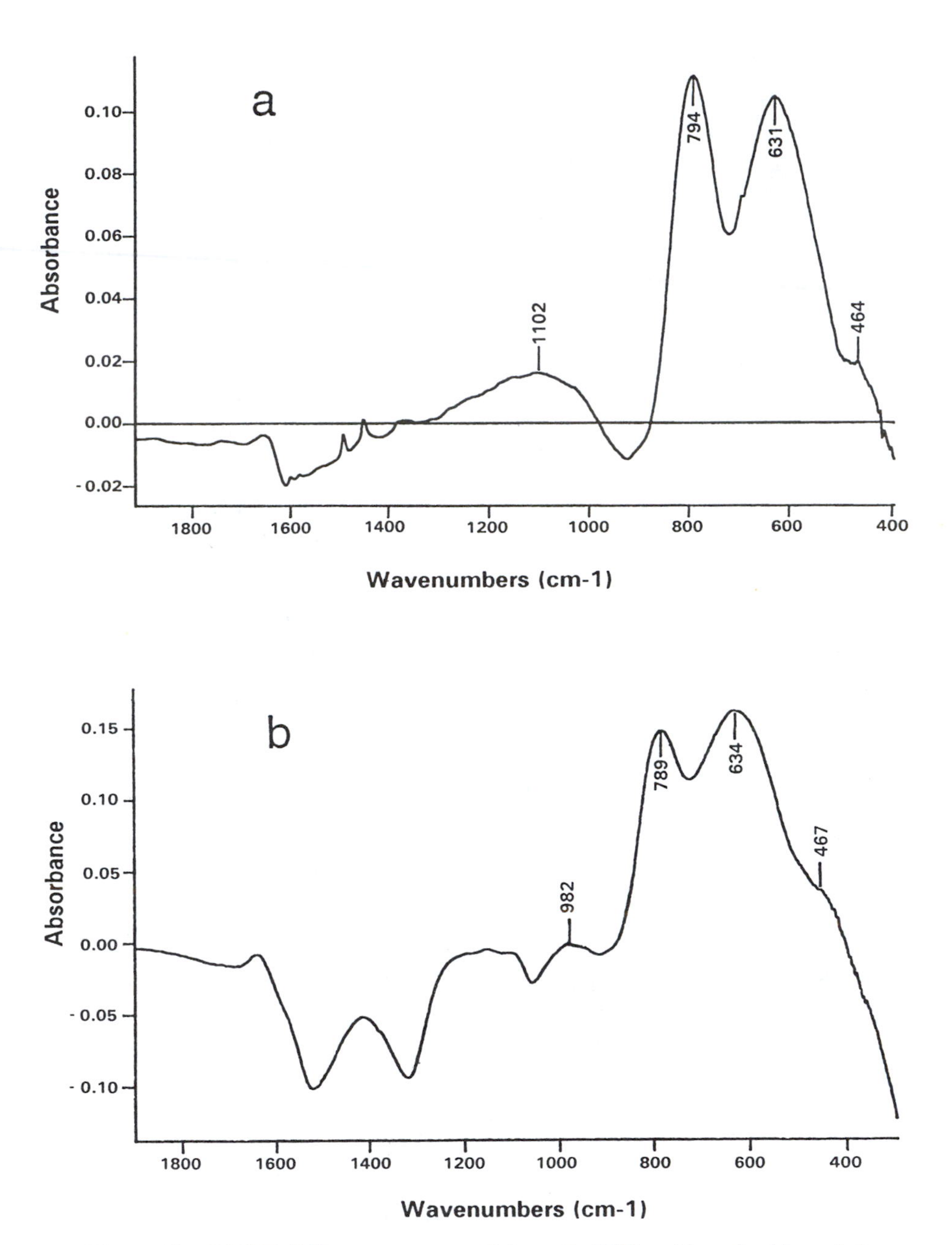

Figure 12. DRIFT difference spectra of a) am-$Fe(OH)_3$ with and without 0.5 M As(III) and 0.1 M NaCl at pH 5 and b) DRIFT difference spectrum of am-$Fe(OH)_3$ with and without 0.5 M As(III) and 0.1 M NaCl at pH 8.

measured values of 0.1 and 0.02 reported above. This reaction predicts no shift in the PZC, although a slight shift was observed.

Alternatively, considering the adsorption of $HAsO_3^{2-}$, the following two reactions are possible for bidentate and monodentate complexation

$$2SOH + HAsO_3^{2-} \rightleftharpoons S_2HAsO_3 + 2OH^- \quad (23)$$

$$SOH + HAsO_3^{2-} \rightleftharpoons SHAsO_3^- + OH^- \quad (24)$$

Equation 23 in combination with the dissociation of the arsenious acid, H_3AsO_3, results in no *net* hydroxyl release (similar to equation 22) while equation 24 (in combination with the dissociation of arsenious acid), results in net H^+ release (OH/As ratio of -1.0). Equation 23 predicts no shift in PZC (similar to equation 22), while equation 24 predicts a downward shift in PZC. The following reaction

$$2SOH + H_2AsO_3^- \rightleftharpoons S_2AsO_3^- + 2H_2O \quad (25)$$

in combination with the dissociation of H_3AsO_3 to $H_2AsO_3^-$ result in a net OH/As ratio of -1.0 and a downward shift in the PZC. If we assume that approximately 10 % of the As partakes in reaction 24, or 25 and 90% in reaction 22 or 23, we satisfy the constraints of the OH/As data at pH 5 and the slight shift in the PZC. At pH 8, reaction 22 or 23 would have to constitute 98% of the As adsorption to be consistent with the titration data. In summary only reaction 22 is consistent with our FTIR data, and we need only consider a small amount of $HAsO_3^{2-}$ (reaction 24) to explain the FTIR, hydroxyl release, and EM results. Reaction 23 is consistent with the bidentate EXAFS results, and if we consider a small amount of monodentate $HAsO_3^{2-}$ (reaction 24) consistent with hydroxyl release and EM results. Additional characterization of the aqueous and solid phase arsenite ligands is needed to better interpret the As(III) on am-$Fe(OH)_3$ DRIFT data.

DRIFT difference spectra of As(III) sorbed on am-$Al(OH)_3$ showed a large peak at 828 cm^{-1} at pH 5 with a minor peak around 600 cm^{-1}, as shown in Figure 13a. At pH 8 (Figure 13b) there is a large peak at 823 cm^{-1} with additional peaks at 669, 610, and 397 cm^{-1}. The peak around 825 cm^{-1} is considered too strong to be the ν_1 mode of the surface $HAsO_3^{2-}$ complex (since the ν_1 mode is shifted to higher frequency), and is tentatively assigned as the ν_3 mode of $H_2AsO_3^-$ (at 790 cm^{-1} for the aqueous species) .

Chromate. Increasing chromate adsorption shifted the PZC of amorphous iron oxide to increasingly lower pH value indicating an inner-sphere adsorption mechanism (*34*). FTIR analyses of freeze-dried samples containing adsorbed chromate mixed with KBr showed shifts in absorption bands from those

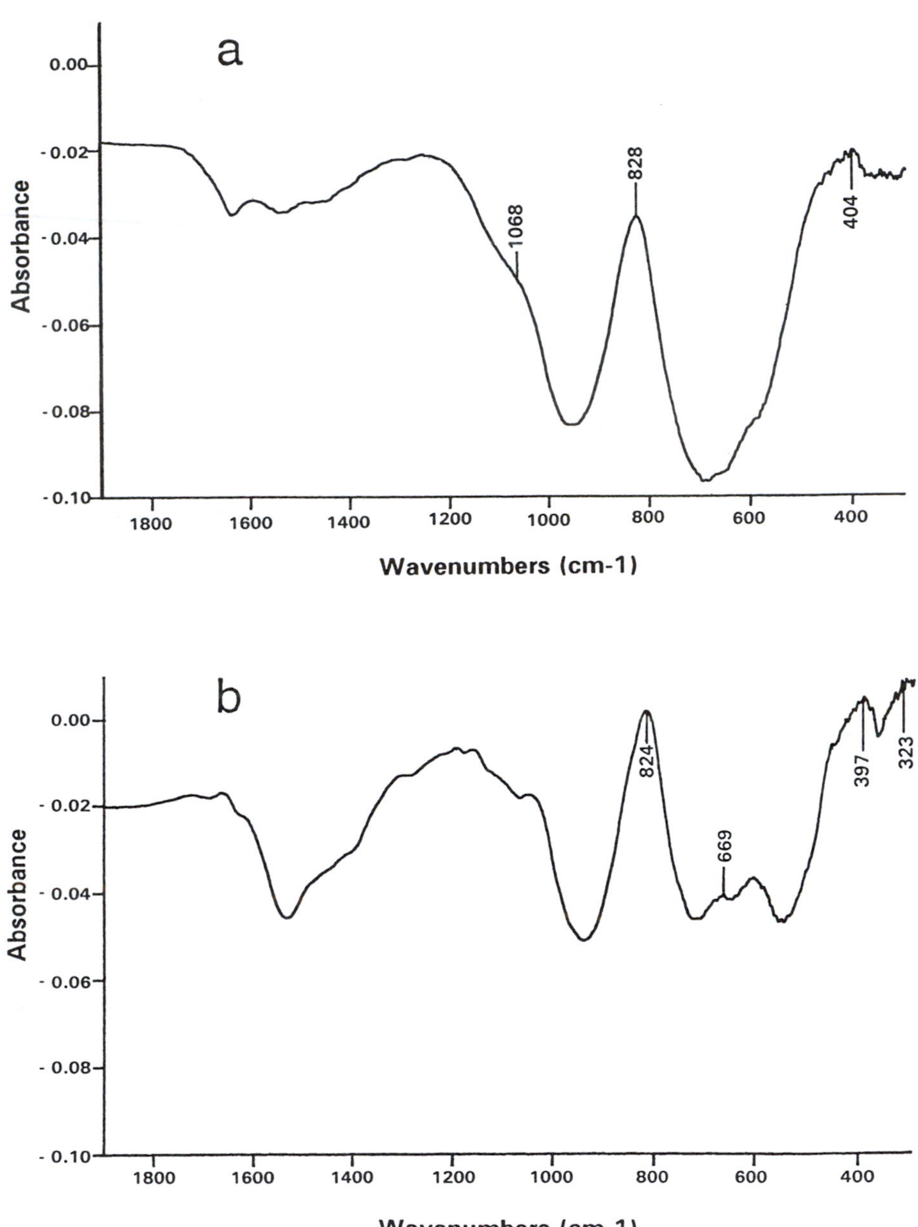

Figure 13. DRIFT difference spectra of a) am-$Al(OH)_3$ with and without 0.5 M As(III) and 0.1 M NaCl at pH 5 and b) DRIFT difference spectrum of am-$Al(OH)_3$ with and without 0.5 M As(III) and 0.1 M NaCl at pH 8.

observed for solid sodium chromate indicating specific adsorption of chromate on amorphous iron oxide.

Kinetics of Oxyanion Adsorption

Reaction rates of anion adsorption have been related to degree of crystallinity. For example, Strauss et al. (*35*) observed that desorbability of phosphate from goethite increased with increasing crystallinity. These rates are related to the long term diffusion and migration of surface species into the particles, likely into surface sites inside pores (internal surface sites). In contrast, surface adsorption is typically very rapid, occurring on the order of milliseconds (*36*). There is relatively little information on the rapid reactions between the oxyanions and the oxide surface, since traditional batch and flow experiments are primarily influenced by the mass transfer processes rather than the surface reaction.

Pressure jump is one of the few techniques suitable for measuring rapid surface reactions involving oxyanions, for which data is available. The data from these experiments have been utilized to infer reaction mechanisms, however most of these studies have utilized assumptions about specific surface configurations. Comparison of the relative reaction rates for various anions may provide additional information on the bonding mechanism.

Kinetics of Phosphate Adsorption. A coagulation model has been postulated for phosphate adsorption on goethite consisting of initial adsorption, followed by particle aggregation and rearrangement (*37*). This model can explain various experimental observations of phosphate adsorption behavior, such as the reduction in isotopically exchangeable phosphate with increasing phosphate uptake. Electron micrographs showed increased particle aggregation, increased particle size, and increased particle order of goethite with the addition of phosphate. BET surface areas decreased following phosphate adsorption implying an interparticle bridging mechanism of phosphate adsorption whereby phosphate becomes buried and inactivated with respect to exchange.

Similar to the adsorption of phosphate on goethite, the slow reaction of phosphate adsorption on ferrihydrite was attributed to the migration of phosphate to surface sorption sites of decreasing accessibility within aggregates (*38*). These authors obtained rapid equilibrium of phosphate adsorption on a goethite consisting of large crystals with few aggregates.

Investigation of the effect of crystal morphology on phosphate adsorption on various hematites indicated that the more platy crystals had a lower proportion of high affinity sites (*39*). Desorbability was positively correlated with relative affinity of adsorbing sites, suggesting that slow adsorption and desorption take place preferentially at low affinity sites. Phosphate is adsorbed initially onto the charged external surfaces of goethite particles and subsequently diffuses into poorly crystalline goethite particles via meso- and micro-pores between goethite crystal domains and adsorbs on internal surface sites (*35*). Internally sorbed phosphate is not readily desorbed.

Differences in phosphate adsorption sites result mainly from location on either external or internal surfaces.

Kinetics of Selenium Adsorption. Zhang and Sparks (*40*) examined selenate and selenite adsorption and desorption on goethite using pressure jump relaxation techniques. Selenate produced a single relaxation, that was interpreted as outer-sphere complexation with surface protonation based on fitting to the triple layer model. The forward rate constant was $10^{8.55}$ $L^2 \cdot mol^{-2} \cdot s^{-1}$. Selenite adsorption was proposed to occur via two steps, an initial outer-sphere complex and subsequent replacement of a water molecule by formation of inner-sphere complexes of both $HSeO_3^-$ and SeO_3^{2-}, based on optimized fits using the triple layer model. The model optimized fit for the pK_a of the surface species was approximately 8.7. Forward rate constants for the first step were on the order of 10^{14} $L^3 \cdot mol^{-3} \cdot s^{-1}$ for $HSeO_3^-$ and 10^{13} $L^3 \cdot mol^{-2} \cdot s^{-1}$ for SeO_3^{2-}. Forward rate constants for the formation of the inner-sphere complexes were 100 and 13 s^{-1}, respectively for $HSeO_3^-$ and SeO_3^{2-}. Agreement between the equilibrium constant obtained from batch and kinetic studies was taken as confirmation of the proposed reactions.

Backward rate constants obtained from the pressure jump experiments suggest that the desorption reaction should be essentially instantaneous (relative to the time frame of batch reactions). In contrast, actual measurements of desorption indicate irreversibility. Su and Suarez (*23*) examined the desorption of selenate and selenite on am-$Fe(OH)_3$. Shown in Figure. 14 is a sorption isotherm and a series of desorption isotherms. Each desorption isotherm was initiated from a point on the sorption isotherm. Desorption was performed by removing the supernatant from the suspension, adding deionized water, reacting for 24 h in a shaker, reanalyzing the supernatant, and calculating the quantity of Se still sorbed on the solid. The series of parallel desorption isotherms indicated that, in all instances, regardless of the initial Se sorbed, only low amounts of the Se were desorbed. Similar data were obtained by Su and Suarez (*23*) on goethite, suggesting that irreversibility might not be simply related to crystallinity and migration into the structure. These data also suggest that sorption isotherms may not be suitable for describing oxyanion desorption and transport under field conditions.

Kinetics of Molybdenum Adsorption. Zhang and Sparks (*41*) examined molybdate adsorption on goethite using pressure jump relaxation experiments. Molybdate adsorption was proposed to occur via two steps, an initial outer-sphere complex and subsequent replacement of a water molecule by formation of an inner-sphere complex of MoO_4^{2-}, based on optimized fits using the triple layer model. Forward rate constants were on the order of $4x10^3$ $L \cdot mol^{-1} \cdot s^{-1}$ and 40 s^{-1} for the first and second reaction steps.

Kinetics of Arsenic Adsorption. Arsenite adsorption on hematite is rapid, attaining equilibrium within 50 minutes and follows a first-order adsorption rate expression (*42*). The removal of arsenite is partially diffusion controlled and

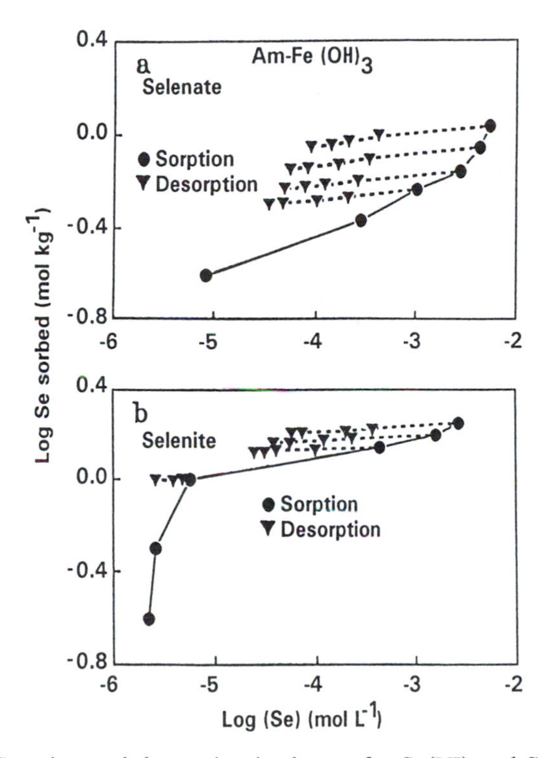

Figure 14. Sorption and desorption isotherms for Se(VI) and Se(IV) reacted with am-$Fe(OH)_3$ at pH 5. Adsorption and desorption points were determined after reaction for 24 h. From Su and Suarez (23).

the velocity of mass transfer is sufficiently rapid to recommend hematite as an adsorbent for treatment of waters elevated in arsenite.

Arsenate adsorption on ferrihydrite consisted of a period of rapid uptake followed by slow adsorption for at least 8 days (*43*). The rate of the slow adsorption reaction is considered to be limited by diffusion into the ferrihydrite aggregates. Slow adsorption kinetics similar to those for phosphate are expected for arsenate because of the similar chemistry of these two anions. Arsenate adsorption data adhere to the Elovich kinetic model indicating a diffusion limited reaction. Arsenate desorption rates were much slower than arsenate adsorption rates, also consistent with a diffusion limited process. A model was developed that assumes that 63% of adsorbing sites are located at the exteriors of aggregates and reach arsenate equilibrium rapidly, while 37% of adsorbing sites are located in the interiors of aggregates with access being diffusion limited.

Pressure-jump relaxation kinetics have been examined by Grossl et al. (*44*). They start with the assumption that arsenate forms an inner-sphere bidentate complex with the goethite surface, as evidenced by EXAFS studies (*29*). These authors obtained a good fit to the data with the proposed two-step mechanism of arsenate adsorption on the goethite surface. The first, fast step is considered a ligand exchange reaction between aqueous $H_2AsO_4^-$ and hydroxyl groups at the goethite surface forming an inner-sphere monodentate complex. The forward rate constant was $10^{6.3}$ $L{\cdot}mol^{-1}{\cdot}s^{-1}$. The second, slow step is an additional ligand exchange reaction forming an inner-sphere bidentate surface complex (forward rate constant of 8 s^{-1}) . The overall equilibrium constant for the formation of the bidentate complex was 1.8, suggesting a predominance of the bidentate species over the monodentate.

Kinetics of Chromate Adsorption. Pressure-jump relaxation kinetic results suggest a two-step mechanism of chromate adsorption on goethite (*44*). The first, fast step is a ligand exchange reaction between aqueous $HCrO_4^-$ and hydroxyl groups at the goethite surface forming an inner-sphere monodentate complex (forward rate constant was $10^{5.8}$ $L{\cdot}mol^{-1}{\cdot}s^{-1}$). The second, slow step is an additional ligand exchange reaction forming an inner-sphere bidentate surface complex (forward rate constant of 16 s^{-1}). The overall equilibrium constant for the formation of the bidentate complex was 0.4 s^{-1}, suggesting a preference of the monodentate complex over the bidentate complex.

Conclusion

Detailed information on oxyanion bonding mechanism and adsorbed oxide surface species can be obtained by utilizing a combination of spectroscopic techniques and macroscopic measurements, particularly, EM and measurements of net OH change/anion adsorption. ATR-FTIR spectroscopy in suspensions is complemented by DRIFT spectroscopy under conditions where small quantities of adsorbed water are still present. Neither FTIR nor EXAFS provide complete information on surface speciation and bonding. FTIR may not always be sufficient to distinguish monodentate from bidentate bonding and EXAFS does not provide information on the protonation of the adsorbed surface complex. The combination of FTIR spectroscopy, EM measurements, measurements of oxyanion sorption, and titrations for determination of proton balance shows special promise to provide the information required for modeling studies. Application of these methods to various oxyanions has provided information on the surface speciation which constrains the number of adjustable parameters in the surface complexation models and provides chemical significance to the reactions modeled.

New FTIR DRIFT data on As(V) reacted with am-$Fe(OH)_3$ suggest that at pH 8 and below, $HAsO_4^-$ is the dominant surface species. The titration data and large shift in PZC are consistent with this conclusion. New FTIR DRIFT data suggest that $H_2AsO_3^-$ is the dominant As(III) species on am-$Fe(OH)_3$ and am-$Al(OH)_3$ surfaces. We observed little shift in the PZC and essentially no change in the proton balance at pH 5.0 and 8.0 when As(III) was adsorbed onto am-$Fe(OH)_3$. A summary of the present knowledge regarding adsorbed oxyanion species and mode of bonding is presented in Table II.

Acknowledgment
Gratitude is expressed to Mr. H.S. Forster for technical assistance.

Literature Cited

(*1*) Westall, J.C.; Jones, J.D., Turner, G.D., Zachara, J.M. *Environ. Sci. Technol.* **1995**, *29*, 951-959.
(*2*) Smith, B.C. *Fundamentals of fourier transform infrared spectroscopy.* CRC Press:Boca Raton, 1996.
(*3*) Parfitt, R.L.; Atkinson, R.J. *Nature* **1976**, *264*, 740-742.
(*4*) Parfitt, R.L.; Russell, J.D. *J. Soil Sci.* **1977**, *28*, 297-305.
(*5*) Hunter, R.J. *Zeta potential in colloid science*, Academic Press: London, 1981.
(*6*) Lewis-Russ, A. *Adv. Agron.* **1991**, 46, 199-243.
(*7*) Breeuwsma, A.; Lyklema, J. *J. Colloid Interface Sci.* **1973**, 43, 437-448.
(*8*) Tejedor-Tejedor, M.I.; Anderson, M.A. *Langmuir* **1990**, *6*, 602-611.
(*9*) Hansmann, D.D.; Anderson, M.A. *Environ. Sci. Technol.* **1985**, *19*, 544-551.
(*10*) Persson, P.; Nilsson, N., Sjöberg, S. *J. Colloid Interface Sci.* **1996**, *177*, 263-275.
(*11*) Bleam, W.F.; Pfeffer, P.E., Goldberg, S., Taylor, R.W., Dudley, R. *Langmuir*, **1991**, *7*, 1702-1712.
(*12*) Schulthess, C.P.; McCarthy, J.F. *Soil Sci. Soc. Am. J.* **1990**, *54*, 688-694.
(*13*) Zachara, J.M.; Girvin, D.C., Schmidt, R.L., Resch, C.T. *Envron. Sci. Technol.* **1987**, *21*, 589-594.
(*14*) van Geen, A.; Robertson, A.P. *Geochim. Cosmochim. Acta* **1994**, *58*, 2073-2086.

Table II. Oxyanion binding mechanisms and surface speciation

Species	pH	Oxide	Bonding Mechanism	Dominant Surface Species	Method	Reference
B	7 11	$Fe(OH)_3$	inner-sphere inner-sphere	$B(OH)_3$ $B(OH)_4^-$	ATR-FTIR DRIFT-FTIR EM	(*18*)
B	7 10	$Al(OH)_3$	inner-sphere	$B(OH)_3$	ATR-FTIR DRIFT-FTIR EM	(*18*)
P	3-12.8	goethite	inner-sphere monodentate	$H_2PO_4^-$ HPO_4^{2-} PO_4^{3-}	DRIFT-FTIR	(*10*)
P	4-8	goethite	inner-sphere bidentate inner-sphere monodentate	HPO_4^{2-} $H_2PO_4^-$	ATR-FTIR EM	(*8*)
P	4-11	boehmite	inner-sphere	deprotonated above pH 9-11	MAS-NMR EM	(*11*)
CO_2	4 9.5		inner-sphere	HCO_3^- CO_3^{2-}	ATR-FTIR EM	(*15*)
Se(VI)		$Fe(OH)_3$	bidentate inner-sphere		dispersion IR	(*20*)
Se(VI)	4	goethite	outer-sphere		EXAFS	(*21*)
Se(VI)	3	goethite $Fe(OH)_3$	inner-sphere		EXAFS	(22)

Se(VI)	5 8	$Fe(OH)_3$	inner-sphere bidentate	SeO_4^{2-}	ATR-FTIR DRIFT-FTIR EM titration sorption	(*23*)
Se(IV)	5 8	$Fe(OH)_3$ goethite	inner-sphere bidentate	SeO_3^{2-}	ATR-FTIR DRIFT-FTIR EM titration sorption	(*23*)
Se(IV)	3	goethite $Fe(OH)_3$	inner-sphere bidentate		EXAFS	(*22*)
Mo	6	$Fe(OH)_3$	inner-sphere bidentate, monodentate	MoO_4^{2-}	ATR-FTIR DRIFT-FTIR EM titration sorption	(*27*)
Cr(VI)		$Fe(OH)_3$	inner-sphere		FTIR (dry) EM	(*34*)
Cr(VI)	5 6	goethite	inner-sphere monodentate to bidentate		EXAFS	(*30*)

Continued on next page.

Table II. continued

Species	pH	Oxide	Bonding Mechanism	Dominant Surface Species	Method	Reference
As(V)	5 8	$Fe(OH)_3$	inner-sphere	$HAsO_4^{2-}$ $HAsO_4^{2-}$	ATR-FTIR DRIFT-FTIR EM titration sorption	this study
As(V)	6 8 9	goethite	inner-sphere monodentate to bidentate		EXAFS	*(30)*
As(V)	8	goethite $Fe(OH)_3$	inner-sphere monodentate to bidentate		EXAFS	*(29)*
As(V)	3-8.5	goethite	inner-sphere	$HAsO_4^{2-}$	ATR-FTIR (dry)	*(31)*
As(III)	5 8	$Fe(OH)_3$	inner-sphere	$H_2AsO_3^-$	ATR-FTIR DRIFT-FTIR EM titration sorption	this study
As(III)	3-8.5	goethite	inner-sphere	$HAsO_3^{2-}$	ATR-FTIR (dry)	*(31)*

(*15*) Su, C.; Suarez, D.L. *Clays Clay Miner.* **1997**, *45*, 814-825.
(*16*) Ross, D. *Inorganic infrared and Raman spectra;* McGraw Hill: London, 1972.
(*17*) Palmer, M.R.; Spivack, A.J., Edmond, J.M. *Geochim. Cosmochim. Acta* **1987**, *51*, 2319-2323.
(*18*) Su, C.; Suarez, D.L. *Environ. Sci. Technol.* **1995**, *29*, 302-311.
(*19*) Goldberg, S.; Glaubig, R. *Soil Sci. Soc. Am. J.* **1985**, *49*, 1374-1379.
(*20*) Harrison, J.B.; Berkheiser, V.E. *Clays Clay Miner.* **1982**, *30*, 97-102.
(*21*) Hayes, K.F.; Roe, A.L., Brown, G.E., Hodgson, K.O., Leckie, J.O., Parks, G.A. *Nature* **1987**, *238*, 783-786.
(*22*) Manceau, A.; Charlet, C. *J. Colloid Interface Sci.* **1994**, *168*, 87-93.
(*23*) Su, C.; Suarez, D.L. *Soil Sci. Soc. Am. J.* **1998**, *In review*.
(*24*) Nakamoto, K., *Infrared and Raman spectra of inorganic and coordination compounds*. 4th ed. John Wiley & Sons. New York. 1986.
(*25*) Hiemstra, T.; de Wit, J.C.M; van Riemsdijk, W.H. *J. Colloid Interface Sci.*, **1989**, 133, 105-117.
(*26*) Fowless, A.D.; Stranks, D.R. *Inorganic Chem.* **1977**, *16*, 1271-1276.
(*27*) Goldberg, S.; Su,C., Forster, H.S. In *Adsorption of metals by geomedia: Variables, mechanisms, and model applications*; Jenne, E.A., Ed.; Proc. Am. Chem. Soc. Symp. Ser., Academic Press: San Diego, CA, 1997 (in press).
(*28*) Goldberg, S.; Forster, H.S., Godfrey, C.L. *Soil Sci. Soc. Am. J.* **1996**, *60*, 425-432.
(*29*) Waychunas, G.A.; Rea, B.A., Fuller, C.C., Davis, J.A. *Geochim. Cosmochim. Acta.* **1993**, *57*, 2251-2269.
(*30*) Fendorf, S.; Eick, M.J., Grossl, P., Sparks, D.L. *Environ. Sci. Technol.* **1997**, *31*, 315-319.
(*31*) Sun, X.; Doner, H.E. *Soil Sci.* **1996**, *161*, 865-872.
(*32*) Loehr, T.M.; Plane, R.A. *Inorganic Chem.* **1968**, *7*, 1708-1714.
(*33*) Gout, R.; Pokrovski, G. Schott, J. Zwick, A. *J. Raman Spectrosc.* **1997**, *28*, 725-730.
(*34*) Hsia, T.H.; Lo, S.L., Lin, C.F., Lee, D.Y. *Chemosphere* **1993**, *26*, 1897-1904.
(*35*) Strauss, R.; Brümmer, G.W., Barrow, N.J. *Europ. J. Soil Sci.* **1997**, *48*, 101-114.
(*36*) Sparks, D.L. *Kinetics of soil chemical processes.* Academic Press: New York, NY, 1989.
(*37*) Anderson, M.A.; Tejedor-Tejedor, M.I., Stanforth, R.R. *Environ. Sci. Technol.* **1985**, *19*, 632-637.
(*38*) Willett, I.R.; Chartres, C.J., Nguyen, T.T. *J. Soil Sci.* **1988**, *39*, 275-282.
(*39*) Colombo, C.; Barrón, V., Torrrent, J. *Geochim. Cosmochim. Acta* **1994**, *58*, 1261-1269.
(*40*) Zhang, P.; Sparks, D.L. *Environ. Sci. Technol.* **1990**, *24*, 1848-1856.
(*41*) Zhang, P.C.; Sparks, D.L. *Soil Sci. Soc. Am. J.* **1989**, *53*, 1028-1034.
(*42*) Singh, D.B.; Prasad, G., Rupainwar, D.C., Singh, V.N. *Water, Air, Soil Pollut.* **1988**, *42*, 373-386.

(*43*) Fuller, C.C.; Davis, J.A., Waychunas, G.A. *Geochim. Cosmochim. Acta* **1993**, *57*, 2271-2282.
(*44*) Grossl, P.R.; Eick, M., Sparks, D.L., Goldberg, S., Ainsworth, C.C. *Environ. Sci. Technol.* **1997**, *31*, 321-326.
(*45*) Miller, F.A.; Carlson, G.L, Bentley, F.F, Jones, W.H. *Spectrochim. Acta* **1960**, *20*, 799-808.
(*46*) Murphy, G.M.; Weiner, G; Oberley, J.J. *J. Chem. Phys.* **1954**, *22*, 1322

Chapter 9

Kinetics and Reversibility of Radiocesium Sorption on Illite and Sediments Containing Illite

Rob N. J. Comans

Netherlands Energy Research Foundation (ECN), P.O. Box 1, NL-1755 ZG Petten, Netherlands (Telephone: (+31) 224 564218; fax: (+31) 224 563163; e-mail: comans@ecn.nl)

The kinetics and reversibility of radiocesium sorption on illite and natural sediments are reviewed and interpreted in terms of a mechanistic framework. This framework is based on the premise that radiocesium is almost exclusively and highly-selectively bound to the frayed particle edges of illitic clay minerals in the sediments. *In-situ* solid/liquid distribution coefficients of radiocesium in sediments are shown to be consistent with this ion-exchange process on illite. Several processes with distinctly different rates can be distinguished in radiocesium sorption. 2- and 3-box kinetic models can describe both the overall adsorption/desorption, as well as the reversible (exchangeable) and irreversible (non-exchangeable or "fixed") fractions of radiocesium on illite and natural sediments, over time scales relevant for natural aquatic systems. The obtained rate parameters indicate that reversible partitioning of radiocesium dominates over the first few days following a contamination event, whereas irreversible sorption kinetics becomes important over time scales of weeks to months. The slow process, which reduces the exchangeability of sediment-bound radiocesium over time, is believed to result from a migration of radiocesium from exchangeable sites on the frayed edges of illite towards less-exchangeable interlayer sites. Long-term extraction of radiocesium from historically contaminated sediments has given evidence for a reverse (remobilization) process with a half-life of the order of years to a few tens of years. These findings suggest that the long-term exchangeability of radiocesium in sediments may be higher than the few % which is generally assumed.

The relevance of the kinetics and reversibility of radiocesium sorption is immediately obvious from observations following the introduction of the anthropogenic radioisotopes of cesium in the environment, e.g. by nuclear weapons tests and nuclear accidents such as Chernobyl. For example, Mundschenk *(1)* measured one order of magnitude higher *in-situ* sediment/water distribution coefficients (K_Ds) for native ^{133}Cs, relative to the more recently introduced ^{137}Cs isotope, in suspended and bottom sediments from the river Rhine. Ten-day laboratory experiments with the same solids and ^{134}Cs gave even lower values *(2)*. Evans and co-workers *(3)* have shown that increasing the contact time with sediments reduces the availability of radiocesium isotopes for exchange by the competing ammonium ion. The issue of radiocesium sorption reversibility is particularly relevant because both Evans and co-workers *(3)* and Comans and co-workers *(4)* have shown that sediment-bound ^{137}Cs can be remobilized in anoxic pore waters by ion-exchange with high concentrations of NH_4^+. This remobilization process may re-introduce radiocesium in the aquatic food chain. Therefore, a thorough understanding of the interlink between the kinetics and reversibility of radiocesium binding to environmental particles is a prerequisite for a reliable assessment of the long-term risks of the release of radiocesium into aquatic environments. This paper reviews earlier, and presents some new, data and models for the kinetics and reversibility of radiocesium sorption, which are interpreted in the framework of the highly selective ion-exchange interaction of radiocesium with illitic clay minerals.

Radiocesium binding in sediments and ion-exchange theory

The traditional way to express mobility, e.g. in contaminant/radionuclide transport models, is by means of the solid/liquid distribution coefficient or K_D (L/kg):

$$^{137}\mathrm{Cs}^+_{\mathrm{aq}} \xleftrightarrow{K_D} {}^{137}\mathrm{Cs}^+_{\mathrm{ads}} \quad (1)$$

or

$$K_D = \frac{[^{137}\mathrm{Cs}^+]_{\mathrm{ads}}}{[^{137}\mathrm{Cs}^+]_{\mathrm{aq}}} \quad (2)$$

where $[^{137}\mathrm{Cs}^+]_{\mathrm{aq}}$ and $[^{137}\mathrm{Cs}^+]_{\mathrm{ads}}$ represent the activity of radiocesium in the aqueous (Bq/L) and particulate phase (Bq/kg), respectively. The K_D-model cannot, however, account for differences in binding properties of different sediments, which has resulted in published K_D-values for radiocesium which vary over more than 3 orders of magnitude *(5)*. These large differences hamper the general use of K_D-based sediment transport models.

Studies since the 1960s have shown that Cs is sorbed very selectively by micaceous clay minerals, particularly illite, a process which is believed to be analogous to the fixation of potassium in the clay interlayer structures *(e.g. 6,7)*. The more recent work by Cremers and co-workers *(8)* has enabled a quantitative characterisation of the Cs-selective sorption sites in soils and sediments, which are believed to be located on the frayed edges of illite (the "frayed edge sites" or FES). These authors have shown that the interaction of cesium with sediment and soil particles can be described more mechanistically in terms of an ion exchange process:

$$\mathrm{FES-M} + \mathrm{Cs}^+ \overset{K_c^{Cs/M}}{\leftrightarrow} \mathrm{FES-Cs} + \mathrm{M}^+ \quad (3)$$

The relative affinity of the illite surface for Cs^+, relative to the competing metal M^+, can then be expressed in terms of a selectivity coefficient, $K_c^{Cs/M}$:

$$K_c^{Cs/M} = \frac{([Cs^+_{FES}]/[FES]) \cdot [M^+]}{([M^+_{FES}]/[FES]) \cdot [Cs^+]} \quad \textbf{(4)}$$

where $[Cs^+_{FES}]$ and $[M^+_{FES}]$ are the concentrations of cesium and the competing metal M^+ adsorbed on the frayed edge sites (eq/kg), $[Cs^+]$ and $[M^+]$ represent the aqueous concentrations (eq/L), and [FES] is the frayed edge site capacity of the sediment (eq/kg). At the negligible stable cesium concentrations in the natural aquatic environment $[Cs^+_{FES}]/[FES] \rightarrow 0$ and $[M^+_{FES}]/[FES] \rightarrow 1$, hence:

$$K_c^{Cs/M} = \frac{[M^+] \cdot ([Cs^+_{FES}]/[FES])}{[Cs^+]} \quad \textbf{(5)}$$

substituting for K_D (L/kg) which, for radiocesium, can be calculated from its activity in the particulate (Bq/kg) and aqueous (Bq/L) phase:

$$K_c^{Cs/M} = \frac{[M^+] \cdot K_D}{[FES]} \quad \textbf{(6)}$$

hence, K_D is given by:

$$K_D = \frac{K_c^{Cs/M} \cdot [FES]}{[M^+]} \quad \textbf{(7)}$$

or

$$\log[K_D] = \log K_c^{Cs/M} \cdot [FES] - \log[M^+] \quad \textbf{(8)}$$

The major metal ions (M^+) competing with radiocesium on the frayed edge sites on illitic clays in natural freshwater environments, are generally potassium (K^+; in oxic environments, e.g. in unsaturated soils and surface waters) and ammonium (NH_4^+, under anoxic conditions in sediments). These ions, like Cs^+, can easily dehydrate and enter the edge-interlayer structure of illite. Divalent ions such as calcium, which are surrounded by a large and stable hydration shell, do not fit easily in this structure and are much less competitive *(7)*.

Figure 1 shows a compilation of *in-situ* K_D-values from a number of different European sediments *(9)* and reveals a single relationship between the (total, see below) radiocesium K_D and the pore-water ammonium concentration. This observation is

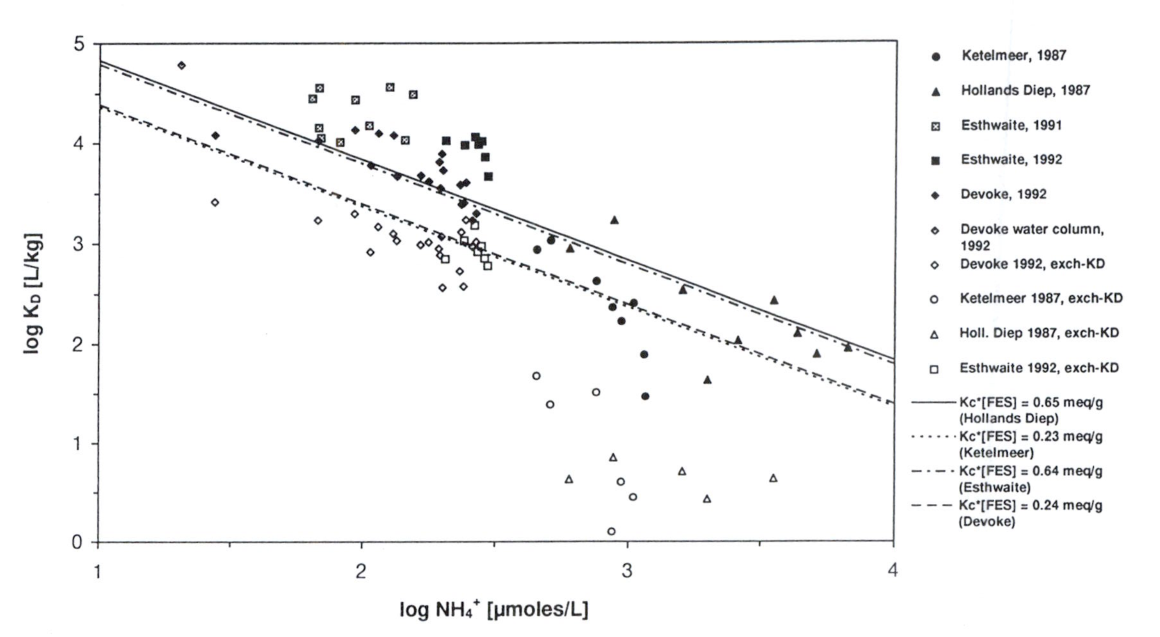

Figure 1. Total (closed symbols), exchangeable (open symbols), and predicted (lines) in-situ K_D-values for different w-European lake sediments. (Adapted from ref. 9).

consistent with the above ion-exchange theory in that, among the major ions that compete with cesium for binding sites on illite clays, NH_4^+ outcompetes K^+ in anoxic sediments as it reaches higher concentrations and is about 5-6 times more selectively bound than K^+ *(10)*. Figure 1 clearly indicates that the data follow a straight line with a slope close to -1.

The intercept $K_c^{Cs/NH4}$·[FES] can be independently measured when the FES are "isolated" by blocking all other exchange sites in sediments with AgTU *(8,11)*. Using this procedure, Wauters *(12)* has measured $K_c^{Cs/NH4}$·[FES] for the sediments in Figure 1. Based on these values, Figure 1 includes the theoretical relationship between K_D and [NH_4^+] according to equation (8). The interpretation of the radiocesium K_Ds in Figure 1 clearly shows that ^{137}Cs in the aquatic environment obeys ion-exchange theory. The ion-exchange model allows the K_D of this radionuclide in sediment transport models to be predicted from environmental variables (i.e. the quantity of FES in the sediments and the pore-water NH_4^+ concentration), rather than to be erroneously treated as a constant.

Modelling kinetics and reversibility of radiocesium partitioning

Another important feature of radiocesium partitioning in sediments is its apparently irreversible behaviour. It has generally been observed that K_D increases, and the exchangeability of radiocesium bound to aquatic particles decreases, with increasing contact time between the radionuclide and the particles *(e.g. 1,2,13)*. This observation has been interpreted as a slow migration of radiocesium from the frayed edge sites towards the deeper interlayer spaces between the illite layers, from where it cannot easily be released *(3,14)*. It is, therefore, important to obtain rate parameters for this process in order to enable prediction of the long-term availability of radiocesium to the aqueous phase, and thus for further transport and uptake in the aquatic food chain.

Laboratory sorption measurements of up to 1 month reaction time have shown that the overall kinetics of the sorption process of radiocesium on illite cannot be described by a single rate, but involves at least a fast and a slower process *(14)*. Figure 2 clearly shows these features for a Ca- and a K-saturated illite, and includes predictions of 2- and 3-box kinetic models developed by Comans & Hockley *(15)* which are described in detail below. The large variation in the fraction of radiocesium in solution (C_L/C_{L0}) at different initial (total) concentrations of cesium in the experiments shows that the sorption process on illite is strongly non-linear. Figure 2 shows also that the uptake process is strongly influenced by the nature of the saturating/competing cation: both the extent, as reflected in the fraction of radiocesium in solution, and the rate of uptake are higher when there is less competition, i.e. in the Ca- relative to the K-environment.

In order to study the kinetics and reversibility of radiocesium sorption on natural sediments, the experimental setup described in *(14)* was applied to sediments sampled from the lakes Ketelmeer *(4)* and Hollands Diep *(16)* in the Netherlands. The top 2 (Ketelmeer) and 4 cm (Hollands Diep) from sediment cores were dialysed for 2 weeks

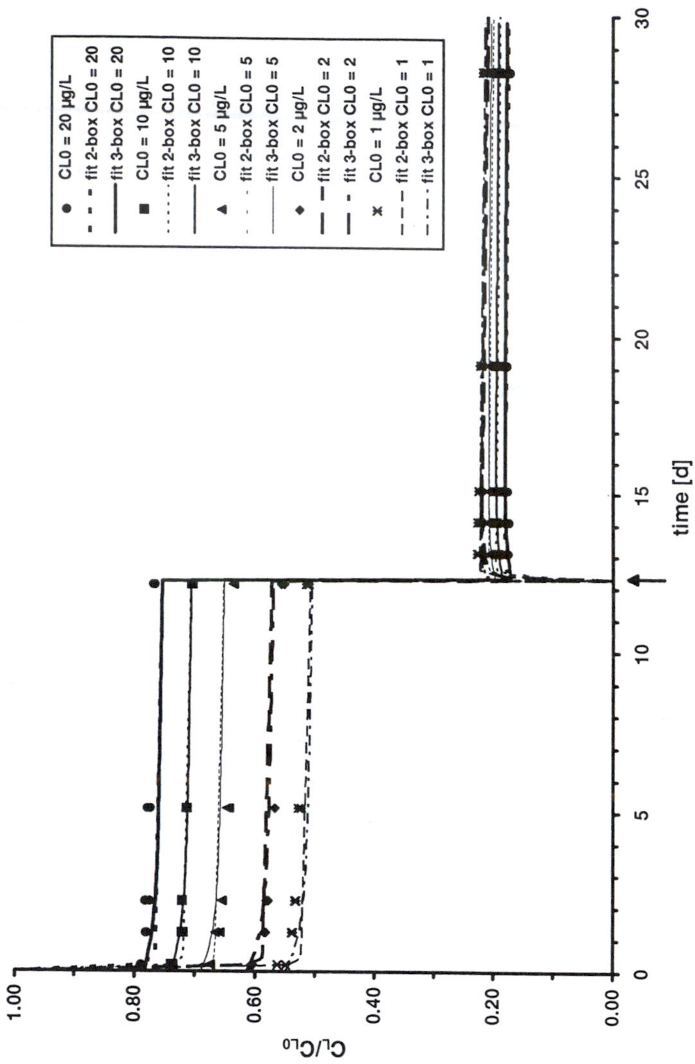
CL0 = 20 µg/L
fit 2-box CL0 = 20
fit 3-box CL0 = 20
CL0 = 10 µg/L
fit 2-box CL0 = 10
fit 3-box CL0 = 10
CL0 = 5 µg/L
fit 2-box CL0 = 5
fit 3-box CL0 = 5
CL0 = 2 µg/L
fit 2-box CL0 = 2
fit 3-box CL0 = 2
CL0 = 1 µg/L
fit 2-box CL0 = 1
fit 3-box CL0 = 1
CL/CL0
1.00
0.80
0.60
0.40
0.20
0.00
0
5
10
15
20
25
30
time [d]

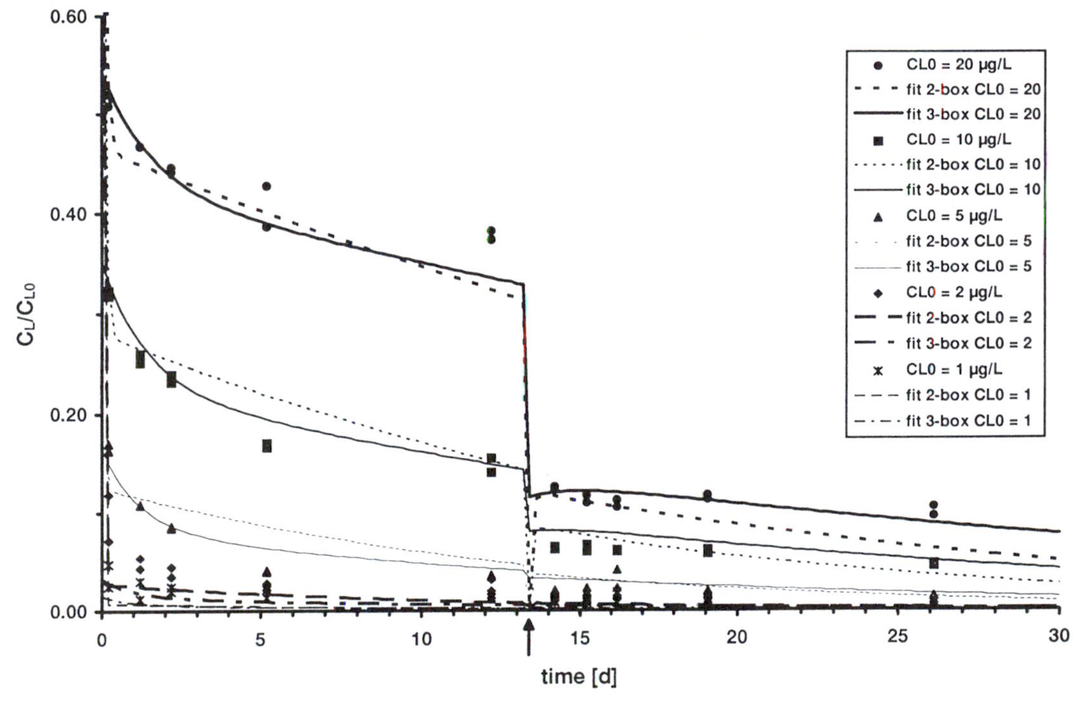

Figure 2. Measurements and 2- and 3-box model predictions of Cs adsorption and desorption on K-saturated illite (top) and Ca-saturated illite (bottom). Solutions were 10^{-3}M in either K^+ or Ca^{2+}. C_{L0} = concentration of Cs in solution at t = 0; C_L = concentration of Cs in solution at time t (i.e. C_L/C_{L0} = fraction of total Cs in solution); particle concentration = 100 mg/L. The arrow indicates the start of desorption. (Adapted from ref. 15)

in a regularly refreshed 10^{-3} M $Ca/HCO_3/NO_3$ solution at pH ≈ 7.9. The sediments were then used, at a particle concentration of 100 mg/L, in the same solution for the sorption experiments. Figure 3 shows the results and illustrates that radiocesium adsorption/desorption on the sediments, at different total-Cs concentrations, shows features very similar to those observed for illite in Figure 2. Again, at least a fast and a slower process are observed, as well as non-linear sorption. Therefore, multiple-process kinetic models are needed to interpret the interaction of radiocesium with illite and natural sediments and to obtain reliable rate parameters.

Linearization of kinetic data. Jannasch and co-workers *(17)* have developed a very useful approach to identify the number of distinct processes that contribute to the observed overall kinetics and, hence, need to be considered in the development of a suitable kinetic model to evaluate the data. These authors have derived a linearized equation for kinetic sorption controlled by a number of separate processes. Their analysis assumes that only one process controls overall sorption at any time and that each process can be described by a first order reversible reaction:

$$-\ln\left[\frac{[M_d]-[M_d]_e}{[M_t]-[M_d]_e}\right]=\left(r_f^i+r_b^i\right)\cdot t \qquad (9)$$

where:

$[M_d]$ = aqueous metal concentration
$[M_d]_e$ = aqueous metal concentration at equilibrium
$[M_t]$ = total metal concentration
r_f^i = forwards rate constant for i^{th} first order process
r_b^i = backwards rate constant for i^{th} first order process

Plotting the left hand side of equation (9) versus time, using arbitrary approximations of $[M_d]_e$, allows for the distinct first order processes to be identified. Of course, the slopes and intercepts of the linearizations change for different choices of $[M_d]_e$, but the number of distinct processes and the time scales over which they dominate remain the same.

Figure 4 shows the linearization applied to the Cs sorption data on illite *(14)* and Ketelmeer and Hollands Diep sediments. Both the illite and the sediment data show an apparently instantaneous reaction and two other distinct processes. As noted by Jannasch and co-workers *(17)*, the sampling schedule affects the number of processes revealed by the linearization procedure. In particular, the initial reaction may, with the appropriate experimental technique and sampling schedule, be subdivided into processes operating on minute or sub-minute time scales. The purpose here, however, is to model Cs sorption kinetics on the longer time scales which are more relevant for natural systems.

The above "Jannasch" linearization implicitly assumes a linear K_D relationship between dissolved and sorbed concentrations. Earlier kinetic models, such as that of Nyffeler and co-workers *(18)*, also adopt that assumption. However, the illite and sediment data in Figures 2 and 3 cover a range of initial Cs concentrations for which sorption isotherms are non-linear. This non-linearity implies that K_D-models cannot be used to model sorption kinetics over the whole concentration range. Kinetic models that include non-linear isotherm equations have been used successfully to describe metal mobilities in soil systems (e.g. *19,20* and references therein). It should be noted that isotherm non-linearity does not invalidate the above conclusion that a number of distinct processes are active. Figure 4 clearly shows that the individual processes operate within similar time frames over a wide concentration range.

2- and 3-box models for cesium sorption. Comans & Hockley *(15)* have developed 2- and 3-box (four and five parameter) models to describe the multiple-process and non-linear kinetics and reversibility of cesium sorption on illite. The equations defining these models are shown in Figure 5. The formulations in Figure 5 have been used earlier to describe reversible non-linear kinetics, see e.g. *(21)* and references therein.

The first "two-box" model includes a reversible Freundlich reaction followed by an irreversible first order process. There are four independent parameters: the Freundlich parameters, K and n, a reversible rate constant, r, and an irreversible rate constant, k. Initial estimates for the Freundlich parameters were obtained from independent isotherm fits of the data after approximately 2 weeks of adsorption. Fitting the kinetic data by eye provided the initial estimates for r and k.

The second model provides an additional reaction by splitting the initial process of the two-box model into an equilibrium reaction and a parallel reversible reaction. Both these reactions tend towards the same Freundlich equilibrium. This model can be conceptualised as dividing a single set of Freundlich sorption sites into a portion which is rapidly accessible and a portion which is kinetically controlled. The Freundlich processes are followed in series by a third, irreversible process. In addition to the four independent parameters of the first model, the three-box model includes f, the fraction of the Freundlich sorption sites reaching equilibrium instantaneously.

Figure 2 shows the fits of these models superposed on the data of cesium sorption on K- and Ca-saturated illite. Although the 2-box model describes the K-illite data well, Comans & Hockley *(15)* concluded that the 3-box model is preferred for a good prediction of cesium sorption on illite in both the K- and Ca-environment. Table I shows the parameter values obtained after fitting the models to the data, using a finite difference method to solve the differential equations defining the models, and a non-linear least squares regression package for the parameter fitting.

Both models have also been fitted to the sediment data of Ketelmeer and Hollands Diep. No three-parameter model could adequately describe either the illite or the sediment data. The model fits to the sediment data are included in Figure 3. The parameter values are listed in Table I. The 2-box model provides an adequate description of the data, but tends to overpredict radiocesium sorption during the first 24 to 48 hours, particularly at low total-Cs concentrations. As was needed for Cs

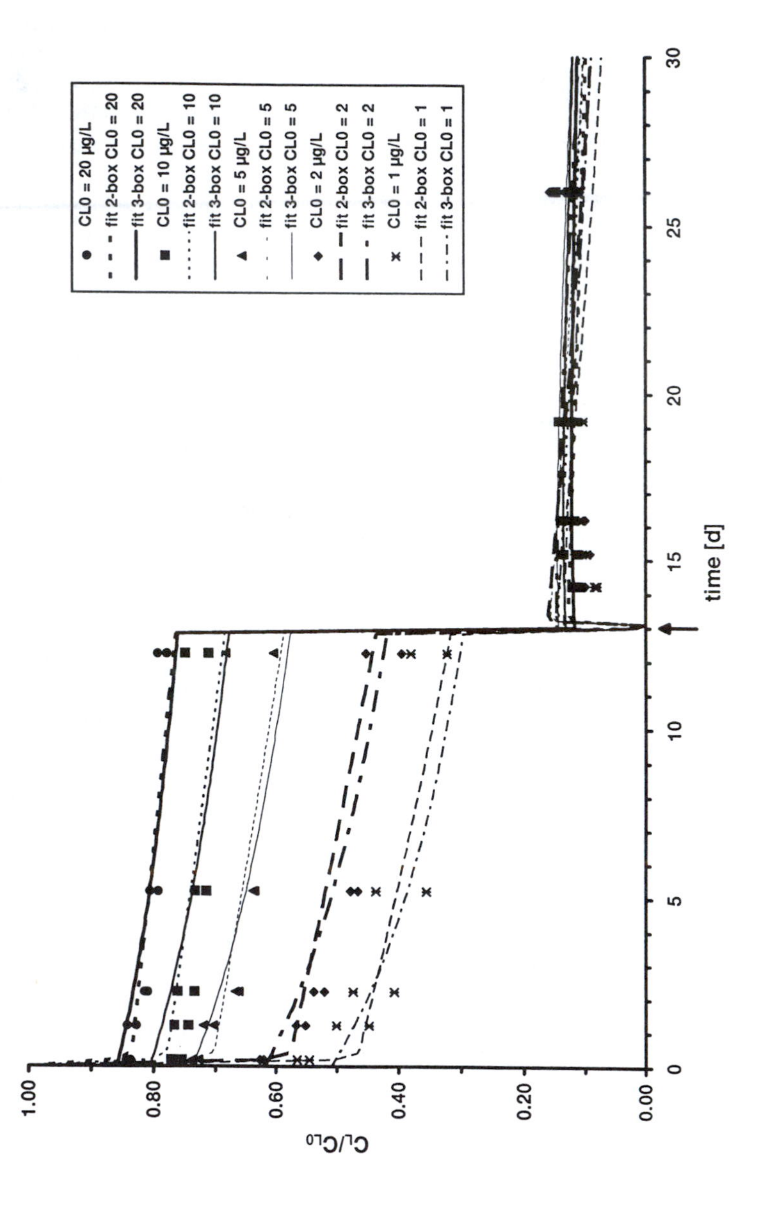

CL0 = 20 µg/L
fit 2-box CL0 = 20
fit 3-box CL0 = 20
CL0 = 10 µg/L
fit 2-box CL0 = 10
fit 3-box CL0 = 10
CL0 = 5 µg/L
fit 2-box CL0 = 5
fit 3-box CL0 = 5
CL0 = 2 µg/L
fit 2-box CL0 = 2
fit 3-box CL0 = 2
CL0 = 1 µg/L
fit 2-box CL0 = 1
fit 3-box CL0 = 1
C_L/C_L0
1.00
0.80
0.60
0.40
0.20
0.00
0
5
10
15
20
25
30
time [d]

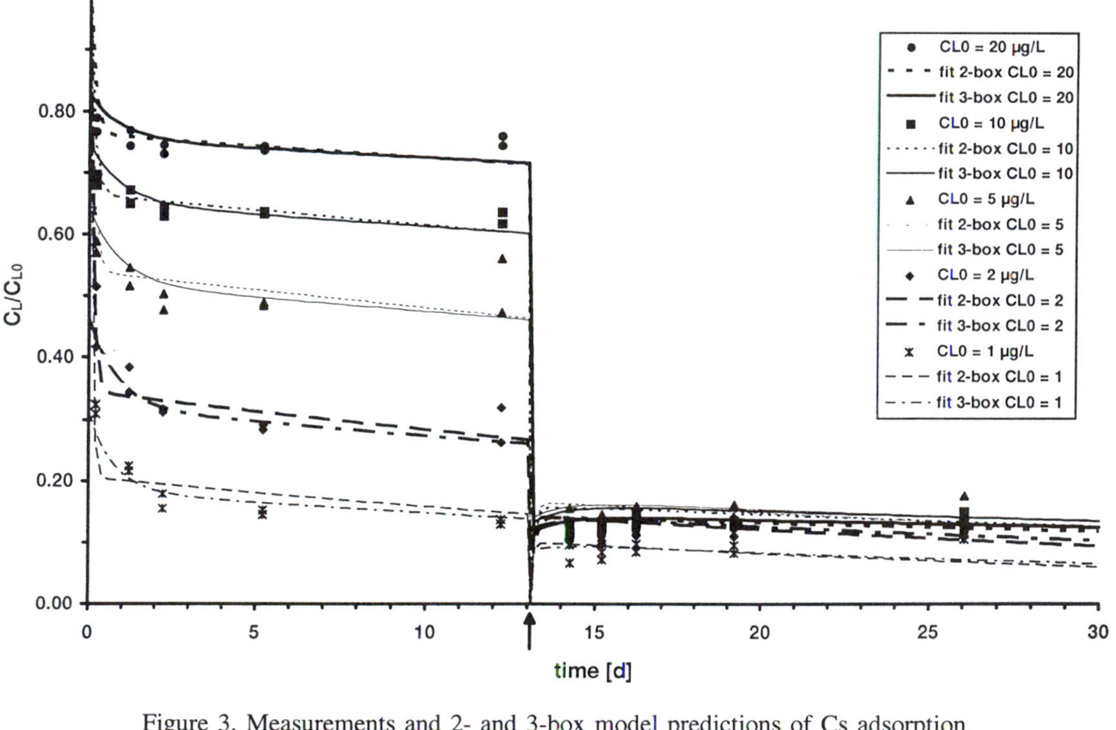

Figure 3. Measurements and 2- and 3-box model predictions of Cs adsorption and desorption on sediments from Ketelmeer (top) and Hollands Diep (bottom) in 10^{-3}M $Ca/NO_3/HCO_3$. C_{L0} = concentration of Cs in solution at t = 0; C_L = concentration of Cs in solution at time t (i.e. C_L/C_{L0} = fraction of total Cs in solution); particle concentration = 100 mg/L. The arrow indicates the start of desorption.

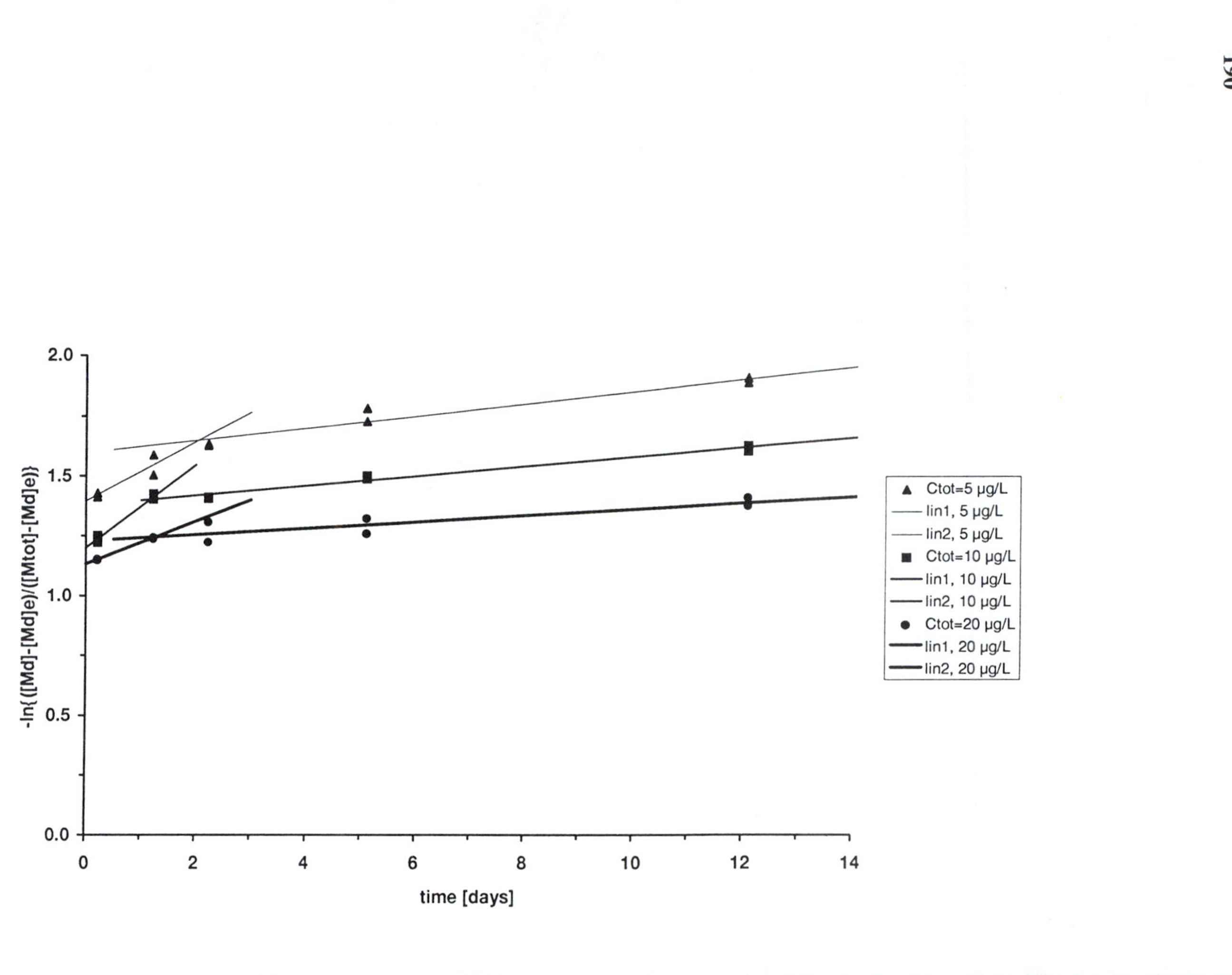

2.0
1.5
1.0
0.5
0.0
-ln{([Md]-[Md]e)/([Mtot]-[Md]e)}
0
2
4
6
8
10
12
14
time [days]
Ctot=5 µg/L
lin1, 5 µg/L
lin2, 5 µg/L
Ctot=10 µg/L
lin1, 10 µg/L
lin2, 10 µg/L
Ctot=20 µg/L
lin1, 20 µg/L
lin2, 20 µg/L

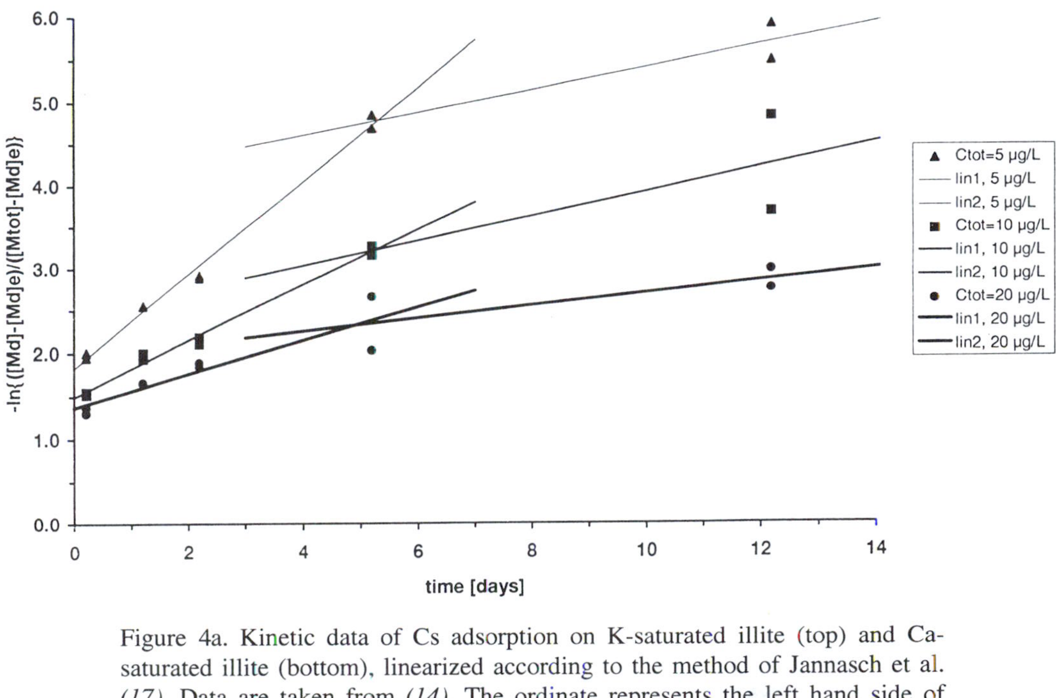

Figure 4a. Kinetic data of Cs adsorption on K-saturated illite (top) and Ca-saturated illite (bottom), linearized according to the method of Jannasch et al. *(17)*. Data are taken from *(14)*. The ordinate represents the left hand side of equation (9); C_{tot} = total Cs concentration of the suspension; lines (lin1, lin2) indicate distinct first order processes identified by the linearization method.

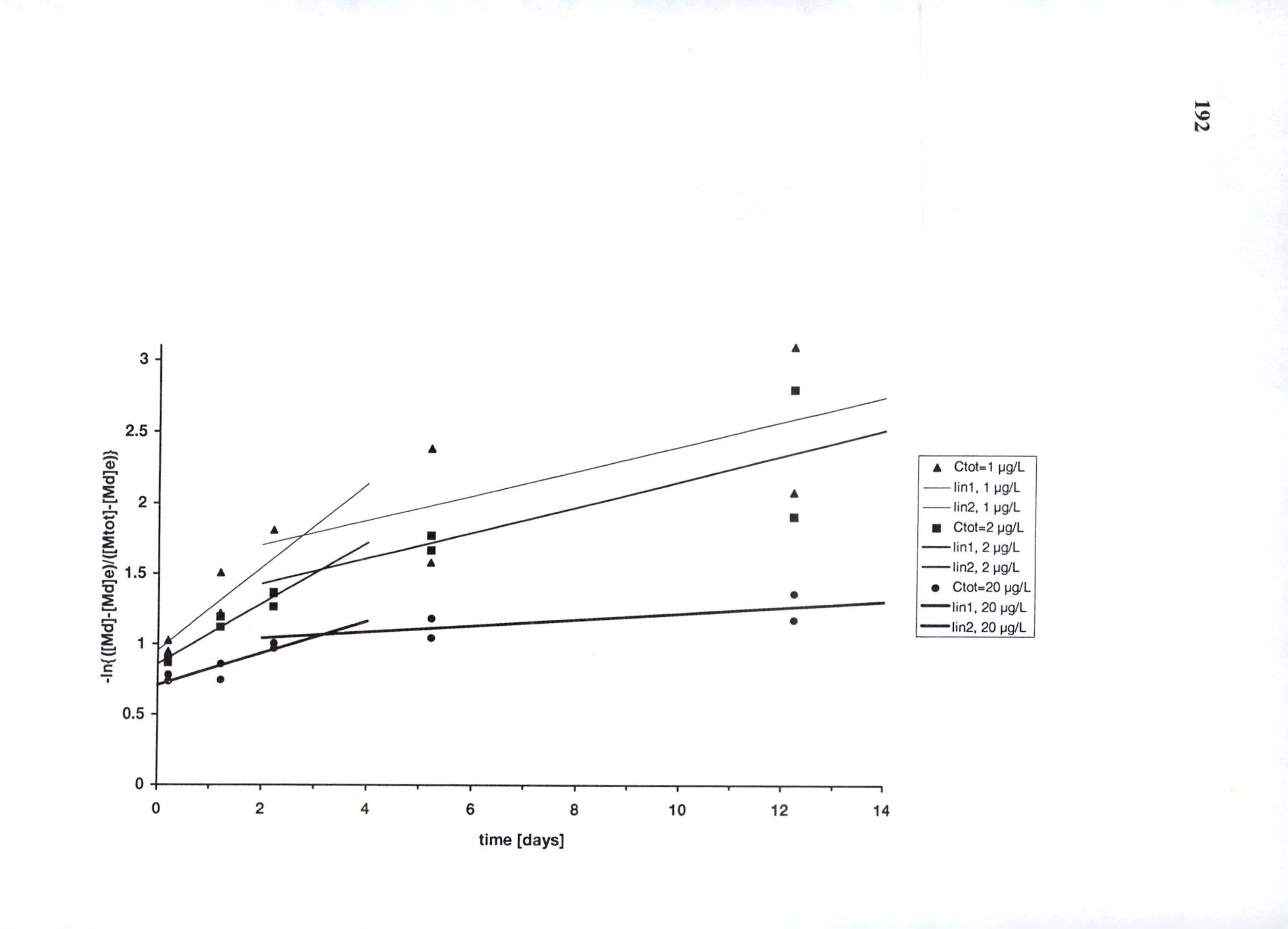

Ctot=1 µg/L
lin1, 1 µg/L
lin2, 1 µg/L
Ctot=2 µg/L
lin1, 2 µg/L
lin2, 2 µg/L
Ctot=20 µg/L
lin1, 20 µg/L
lin2, 20 µg/L
-ln{([Md]-[Md]e)/([Mtot]-[Md]e)}
0
0.5
1
1.5
2
2.5
3
0
2
4
6
8
10
12
14
time [days]

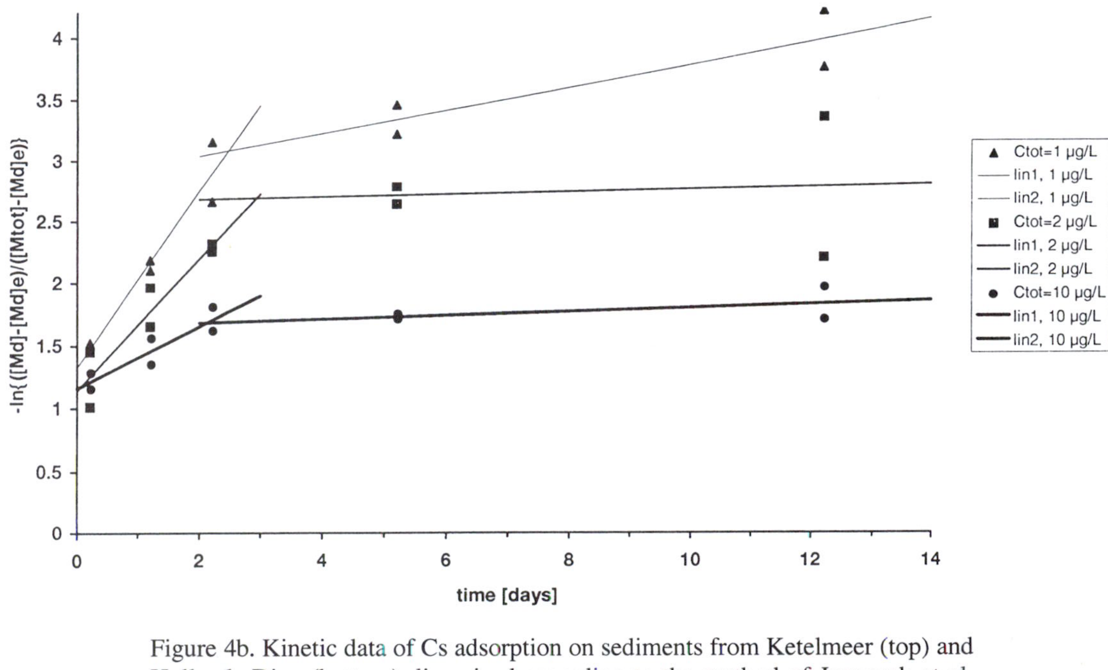

Figure 4b. Kinetic data of Cs adsorption on sediments from Ketelmeer (top) and Hollands Diep (bottom), linearized according to the method of Jannasch et al. *(17)*. The ordinate represents the left hand side of equation (9); C_{tot} = total Cs concentration of the suspension; lines (lin1, lin2) indicate distinct first order processes identified by the linearization method.

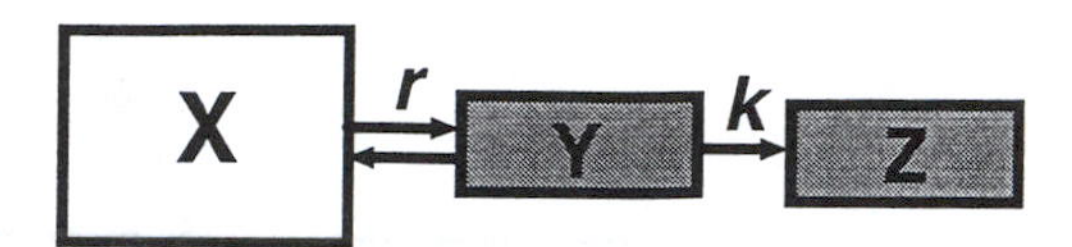

$$\frac{d[Y]}{dt} = r(K[X]^n - [Y]) - k[Y]$$

$$\frac{d[Z]}{dt} = k[Y]$$

Mass Balance:

$$[X] + ([Y] + [Z])C_p = [X]_0 + ([Y]_0 + [Z]_0)C_p$$

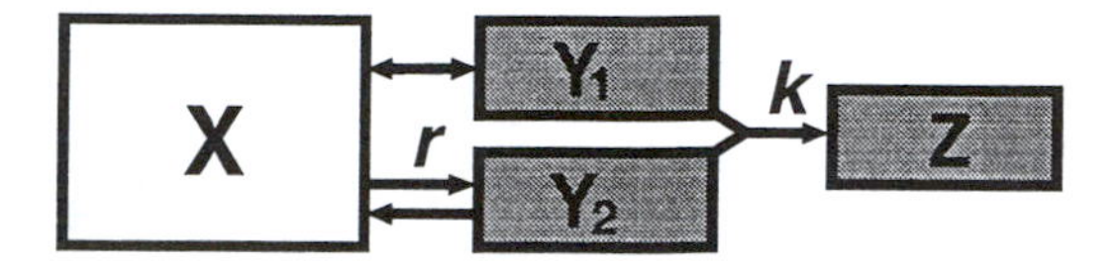

$$[Y_1] = fK[X]^n$$

$$\frac{d[Y_2]}{dt} = r\{(1-f)K[X]^n - [Y_2]) - k[Y_2]\}$$

$$\frac{d[Z]}{dt} = k([Y_1] + [Y_2])$$

Mass Balance:

$$[X] + ([Y_1] + [Y_2] + [Z])C_p = [X]_0 + ([Y_1]_0 + [Y_2]_0 + [Z]_0)C_p$$

Figure 5. Schematic representation and equations defining the 2-box (top) and 3-box (bottom) kinetic models. X = dissolved metal; Y, Y_1 and Y_2 = reversibly sorbed metal on Freundlich sorption sites; f = fraction of Freundlich sorption sites reaching equilibrium instantaneously; K and n are the Freundlich-isotherm constants; r and k are the reversible and irreversible rate constant, respectively; Z = irreversibly sorbed metal. The subscript "0" in the mass balance equations denotes concentrations at time zero, and C_p = particle concentration. (Adapted from ref. 15)

sorption on illite (Figure 2), an additional (fast) reaction may be considered for a better overall prediction of the measurements. The apparently irreversible reaction in the model, which predicts readsorption during the desorption phase (period between 13-26 days in Figure 3), and which is clearly noticeable in the illite data (Figure 2), is not reflected in the sediment data. Possibly, some ammonium was produced by the breakdown of organic matter in the sediments at the later times during the experiments (sediments were not spiked with bacteriocides). Competition by ammonium for frayed edge sites could have prevented the readsorption of radiocesium.

Table I. Parameter values of the 2-box and 3-box kinetic models (Fig. 5) obtained after fitting sorption data of Cs on K- and Ca-saturated illite *(15)* and on sediments from Ketelmeer and Hollands Diep (this study).

Parameter*	K-illite#	Ca-illite#	Ketelmeer	Hollands Diep
K (2-box)	7.4 ± 0.1	51 ± 2	7.82 ± 0.40	15.34 ± 0.64
n	0.68 ± 0.01	0.33 ± 0.02	0.51 ± 0.026	0.42 ± 0.019
r (d^{-1})	9.8 ± 1.0	7.4 ± 1.4	8.55 ± 2.25	5.81 ± 0.95
k (d^{-1})	0.0049 ± 0.0015	0.032 ± 0.005	0.040 ± 0.0071	0.017 ± 0.0045
RSS	6.77E-03	4.55E-02	8.92E-02	8.84E-02
K (3-box)	7.5 ± 0.2	61 ± 5	10.25 ± 4.22	16.36 ± 1.00
n	0.67 ± 0.01	0.30 ± 0.02	0.50 ± 0.026	0.41 ± 0.017
f	0.89 ± 0.03	0.73 ± 0.06	0.67 ± 0.26	0.68 ± 0.070
r (d^{-1})	0.81 ± 0.50	0.58 ± 0.36	0.16 ± 0.25	1.00 ± 0.62
k (d^{-1})	0.0038 ± 0.0013	0.020 ± 0.006	0.025 ± 0.020	0.013 ± 0.0051
RSS	6.16E-03	3.21E-02	8.69E-02	6.93E-02

*K and n are the Freundlich-isotherm constants, using µg/L and µg/g for aqueous and sorbed concentrations, respectively; r and k are the reversible and irreversible rate constant, respectively; RSS is the residual sum of squares of the model fits to the data.
#data from Comans & Hockley *(15)*

The fit of the 3-box model to the sediment data is slightly better, particularly for the first 48 hours. Again, readsorption is predicted for the desorption period, albeit slightly less than the extent predicted by the 2-box model. However, one might argue whether the addition of a fifth fitting parameter in the model is statistically justified for the data of cesium sorption to sediments.

The sorption experiments with Ketelmeer and Hollands Diep sediments (Figure 3) were carried out in a 10^{-3} M Ca-solution. Table I shows that the rate parameters are indeed very similar, for both models, to those for Ca-saturated illite (in the same aqueous solution). If we use the 3-box model for the best fit to the early data points, the fast reversible process *(r)* has a half life of a few (1-4) days. The slower, irreversible, process *(k)* has a half life of approximately 20-50 days.

Mechanistic interpretation. Although kinetic models such as those used above and outlined in Figure 5 are empirical in nature and origin, we may compare their predictions and parameter values with the theory of highly-selective interaction of radiocesium with the frayed edges of illite clays (in the sediments). The non-linear sorption of cesium, which is apparent in Figures 2 and 3, has been interpreted by Comans and co-workers *(14)* in terms of an earlier equilibrium ion-exchange model for illite *(22)*. That model considers three types of sites on illite, two of which are highly selective for cesium and believed to be located at the frayed edges of the clay mineral. The least selective sites were attributed to the planar surfaces of the clay and comprised >95% of the exchange capacity. According to this model, >75% of the Cs sorbed on K-illite, and >99% on Ca-illite in Figure 2, are located on the frayed edge sites. Therefore, the higher Freundlich K-values (Table I) in the Ca-, relative to the K-environment, are representative of the higher selectivity of Cs relative to Ca. The lower Freundlich n-value reflects the greater saturation of the frayed edge sites, and the larger differences in the Cs-selectivity between the three sites on the Ca-clay *(14)*.

The 3-box model considers a fraction of the Freundlich sites to be in instantaneous equilibrium with dissolved Cs. The larger fraction for K-illite (Table I) is consistent with the greater contribution of the planar sites to Cs sorption on this clay. However, the instantaneous-equilibrium fractions predicted by the model are much greater than can be explained by planar sites alone, and suggest that a substantial portion of the frayed edge sites may also reach very rapid (instantaneous) equilibrium with the solution. The reaction rates for the remaining fraction of reversible sites are similar for both the Ca- and K-illite.

The initial reversible sorption of Cs, is followed by a much slower process which is considered in the models to be irreversible. This process is believed to represent a slow Cs-migration towards the interlayers of illite *(14)*, similar to the ^{40}K-uptake by illite from solution as observed by De Haan and co-workers *(23)*. The removal of the latter radiotracer from solution was interpreted as isotopic exchange with interlayer-potassium, i.e. as a slow diffusion of ^{40}K into the illite lattice. This process is often referred to as (potassium) "fixation". The faster rates (and lower reversibility) of Cs sorption on Ca- relative to K-saturated illite have been ascribed to the hydrated and expanded state of the frayed edge sites in the case of Ca-saturation, and the collapsed/dehydrated state in the case of K-saturation *(14)*.

It is important to realise that the above models should not be used to predict Cs-sorption reversibility over time frames of more than a few months. Observations that radiocesium, which has been in contact with contaminated sediments for many years, can still be mobilised by enhanced ammonium concentrations *(3,4)* give evidence for a backwards reaction. This reaction, which is interpreted as a remobilization of radiocesium from "fixed" edge-interlayer sites on illite (i.e. box "Z" in Figure 5), is apparently too slow to be observed in laboratory experiments of many weeks. Therefore, extractions of radiocesium from historically contaminated sediments have been used, and are described below, to obtain a first estimate of this slow remobilization rate.

Exchangeability and slow remobilization of radiocesium in sediments

Exchangeability of sediment-bound radiocesium. The exchangeability of sediment-bound radiocesium has been investigated by extracting sediments with 0.1 M NH_4-acetate *(9)*. These extractions have been performed on all sediments shown in Figure 1. Sediment samples from different depths were extracted three times sequentially, each step for 24 hours, with 0.1 M NH_4-acetate at a liquid/solid ratio of 10 L/kg. Further details on the extraction procedure and radiocesium measurements are given in *(9)*.

The exchangeable amounts of ^{137}Cs are given in Table II. The amount of ^{137}Cs that can be released by the three sequential NH_4-extractions from the more mineral sediments of Hollands Diep, Ketelmeer and Esthwaite is very low; on average 2, 3, and 7%, respectively. The more organic-rich sediments of Devoke show a much higher exchangeability of 16%. The values for the three mineral sediments are low when compared, for instance, with those of Evans and co-workers *(3)*, who have measured ^{137}Cs exchangeabilities (also in 0.1 M NH_4^+) of 10-20% in the sediments of the Par Pond reservoir, 15-20 years after contamination. These authors attribute their relatively high exchangeabilities to the high kaolinite content of the Par Pond sediments. Western European sediments generally contain illite as the major clay mineral, which likely causes the strong fixation of radiocesium that has been observed in the Hollands Diep, Ketelmeer and Esthwaite sediments. Devoke apparently behaves more like the Par Pond sediments in that similar amounts of ^{137}Cs are exchangeable.

Table II: Average fraction of exchangeable-^{137}Cs in sediments after 3 sequential 24-hr NH_4-extractions and the additional fraction mobilized after a 4th 400-842 d extraction. A reverse rate constant and half-life for the slow remobilization of ^{137}Cs from the sediments has been calculated on the basis of the 4th extraction, assuming a first order process *(9)*.

Sediment	exch. ^{137}Cs after 3x 24-hr extraction [fraction of total]	additional exch. ^{137}Cs after 400-842 d extraction* [fraction of total]	reverse rate const. [y^{-1}]	$t_{1/2}$ [y]
Hollands Diep	0.022 ± 0.038	0.0089 ± 0.0050	0.0082	85
Ketelmeer	0.030 ± 0.030	0.0095 ± 0.0030	0.0087	79
Esthwaite	0.070 ± 0.012	0.0156 ± 0.0053	0.0068	102
Devoke	0.164 ± 0.050	0.0380 ± 0.0355	0.0220	31

*equilibration time 4th extraction = 400 days for Hollands Diep & Ketelmeer; 560 days for Devoke; 842 days for Esthwaite.

If we accept that radiocesium in all four of the above sediments is bound solely to frayed edge sites on illitic clays in the sediments, the up to one order of magnitude differences between the exchangeable fractions of radiocesium in these sediments is unexpected. The differences are clearly not related to the different contact times of radiocesium with the sediments, as the lowest values are found for the sediments sampled only 1.5 years after the Chernobyl accident (Hollands Diep and Ketelmeer).

"Exchangeable" K_D-values for radiocesium in the sediments, calculated from the exchangeable rather than total amount of ^{137}Cs in the sediments, are included in Figure 1. We would expect "exchangeable" K_Ds to correspond better with values predicted on the basis of (short-term) laboratory measurements of $K_c^{Cs/NH4}$·[FES] than the total K_Ds. Although this may be the case for the Devoke sediments, total and exchangeable K_D-values for Esthwaite correspond about equally with the predictions, whereas the "exchangeable" K_Ds for Hollands Diep and Ketelmeer deviate much more from the predicted values than the total K_Ds. These observations strongly suggest that short-term exchangeability measurements of radiocesium, especially in the more mineral sediments, underestimate the amount of the radionuclide that is actually taking part in ion-exchange with the pore waters and, hence, is available for remobilization by high concentrations of ammonium.

Slow (reverse) migration of radiocesium from clay-mineral interlayers into solution. In the kinetic models discussed above, a reverse process of radiocesium remobilization from "fixed" edge-interlayer sites was not considered. The equilibration times of up to 4-weeks were too short for the reverse process to become apparent. Nevertheless, the fact that radiocesium in sediments is still exchangeable to a certain extent after more than 20 years of contact with sediments *(3)*, indicates that such a reverse process must exist.

Because of its relevance for the long-term availability of sediment-bound radiocesium, the long-term release of particle-bound radiocesium has been studied in more detail *(9)*. The primary objective was to estimate the existence and magnitude of a slow remobilization (reverse rate) of sediment-bound radiocesium during contact with a high concentration of competing ions. This process is interpreted as a slow migration of radiocesium from the "fixed" edge-interlayer positions on illite back into solution. After the three sequential 0.1 M NH_4-acetate extractions, the sediments were resuspended for a fourth time in a fresh 0.1 M NH_4-acetate solution and have been allowed to equilibrate for more than one year (400-842 days). After this period, the additional amount of extracted radiocesium was measured.

Table II shows the average fraction of exchangeable-^{137}Cs in sediments after 3 sequential 24-hr NH_4-extractions and the additional fraction mobilized after the fourth, long-term (400-842 days) extraction. Assuming (1) that all (rapidly) "exchangeable" radiocesium had been removed by the three prior extractions, and (2) a first order remobilization process, we can roughly calculate the reverse rate constant that describes the slow remobilization of ^{137}Cs from the sediments. Table II indicates that the half-life of this reaction, which is interpreted as slow release from the edge-interlayer sites, is of the order of 30-100 y^{-1}. These findings suggest that radiocesium on these interlayer sites, which is generally referred to as being "fixed", is not truly immobilised but can, at least partly, be very slowly remobilized.

Comparison of rate parameters with other estimates

Slow uptake rates. As was mentioned above, the slow uptake of radiocesium by illite, and sediments containing illite, is interpreted as a migration towards interlayer sites on this clay mineral. This process has been recognised for many years to cause the so-called "potassium fixation" in (agricultural) soils. De Haan and co-workers *(23)* have studied the kinetics of that process by following the ^{40}K-uptake from solution by illite clay over a period of 16 months. Re-evaluated in terms of a first order uptake process, their data indicate a rate constant for the interlayer migration of the order of 10^{-3} d^{-1}. This value is very similar to the values for Cs-sorption on illite in a K-dominated environment (Table I) , which lends further support to the hypothesis of the similar binding mechanism for Cs and K. The slow Cs-uptake rate was found to be approximately an order of magnitude faster for a Ca-illite, which is supported by the values found for the illite-containing sediments of Ketelmeer and Hollands Diep in a Ca-environment (Table I). As was discussed above, the slow interlayer migration of Cs proceeds faster when the edges of illite are expanded by reaction with calcium, relative to the collapsed structure of a K-saturated illite.

Evans and co-workers *(3)* have studied the "fixation" of ^{134}Cs by freshwater sediments over a period of 180 days. The increasingly incomplete extractability of radiocesium, using 0.1 M NH_4, with Cs-sediment contact time was interpreted as a slow movement of radiocesium to clay interlayer sites in the sediment. If the data of *(3)* are interpreted in terms of a first order process, a rate of 10^{-2}-10^{-3} d^{-1} is obtained, which is again of the same order as the values found for illite and the Hollands Diep and Ketelmeer sediments (Table I).

Finally, Nyffeler and co-workers *(18)* have modelled radiocesium uptake by marine sediments, using a multi-reaction model with an irreversible process similar to that in the 2- and 3-box models of Fig. 5. These authors have also obtained a similar rate of 10^{-2} d^{-1}.

Reverse (remobilization) rates. Smith & Comans *(16)* have developed a transport model that includes radiocesium sorption kinetics to simulate radiocesium in each of three phases in the sediment profiles of Ketelmeer and Hollands Diep: aqueous, exchangeably-bound and slowly reversible ("fixed"). The model gave evidence for a reverse reaction from less-exchangeable to exchangeable sites with a half-life of order 10 years, which is close to the independent estimates by the long-term extractions in Table II.

Using the exchangeability measurements of *(3)*, 85% fixed/15% exchangeable (after 180 days contact of ^{134}Cs with the sediment), and a forward rate constant of half-life 100 days, a reverse rate of half-life 2 years is estimated. The same estimate can be made from the short-term exchangeability measurements shown in Table II (84-98% fixed/2-15% exchangeable). It is reasonable to assume that the long-term extractions discussed above have led to an upper estimate of the remobilization rate, as a new equilibrium between NH_4 and remobilized Cs may have been reached during the single long-term extraction step. Given the limited available data, the remobilization half-life of sediment-bound radiocesium may, therefore, be expected to be of the order of a few years to a few tens of years.

Conclusions

The kinetics and reversibility of radiocesium sorption on illite and natural sediments have been reviewed and interpreted in terms of a mechanistic framework. This framework is based on the premise that radiocesium is almost exclusively and highly-selectively bound to the frayed particle edges of illitic clay minerals. It is shown that *in-situ* K_Ds of radiocesium in sediments are consistent with this ion-exchange process on illite.

Several processes with distinctly different rates can be distinguished in radiocesium sorption. 2- and 3-box models can describe both the overall sorption of cesium on illite and sediments and the reversible (exchangeable) and irreversible (non-exchangeable or "fixed") fractions of radiocesium over time scales relevant for natural aquatic systems. The models are consistent with knowledge of radiocesium sorption mechanisms. The obtained rate parameters indicate that reversible sorption of radiocesium dominates over the first few days following a contamination event, whereas apparently irreversible sorption kinetics becomes important over time scales of weeks to months. The slow process, which reduces the exchangeability of sorbed radiocesium over time, is believed to result from a migration from exchangeable sites on the frayed edges of illite clays towards less-exchangeable interlayer sites. This hypothesis is supported by comparison of model results with previous findings.

Observations that radiocesium, which has been in contact with contaminated sediments for many years, can still be mobilised have given evidence for a backwards reaction. This reaction is interpreted as a remobilization of "fixed" radiocesium from interlayer sites on illite, and is apparently too slow to be observed in laboratory experiments of many weeks. Long-term extraction of radiocesium from historically contaminated sediments has confirmed the existence of a reverse reaction and has provided a first estimate of the half-life of this slow remobilization rate of the order of years to a few tens of years. These findings suggest that the long-term exchangeability of radiocesium in sediments may be higher than the few % which is generally assumed. Interpretation of *in-situ* distribution coefficients in sediments in terms of radiocesium ion-exchange on illite lends support to a relatively high exchangeability of radiocesium on longer time scales.

References

1. Mundschenk H. *Deutsche Gewässerkundl. Mitt.* 1983, 27, 12-20.
2. Mundschenk H. *Deutsche Gewässerkundl. Mitt.* 1983, 27, 62-68.

3. Evans D.W., Alberts J.J. & Clark R.A. *Geochim. Cosmochim. Acta* 1983, 47, 1041-1049.
4. Comans, R.N.J., Middelburg, J.J., Zonderhuis, J., Woittiez, J.R.W., De Lange, G.J., Das, H.A. & Van Der Weijden, C.H. *Nature* 1989, 339, 367-369.
5. Smith, J.T. *Mathematical modeling of ^{137}Cs and ^{210}Pb transport in lakes, their sediments, and the surrounding catchment;* Ph.D Thesis, University of Liverpool, UK, 1993.
6. Bolt, G.H., Sumner, M.E. & Kamphorst, A. *Soil Sci. Soc. Am. Proc.* 1963, 27, 294-299.
7. Sawhney, B.L. *Clays Clay Miner.* 1972, 20, 93-100.
8. Cremers A., Elsen A., De Preter P. & Maes A. *Nature* 1988, 335, 247-249.
9. Comans, R.N.J., Hilton, J., Cremers, A., Bonouvrie, P.A. & Smith, J.T. *Predicting radiocesium ion-exchange behaviour in freshwater sediments (submitted for publication).*
10. Wauters, J., Elsen, A., Cremers, A., Konoplev, A.V., Bulgakov, A.A. & Comans, R.N.J. *Applied Geochem.* 1996, 11, 589-594.
11. De Preter P. *Radiocesium retention in the aquatic, terrestrial and urban environment: a quantitative and unifying analysis;* Ph.D Thesis. Faculty of Agronomy, Katholieke Universiteit Leuven, Belgium, 1990.
12. Wauters, J. *Radiocesium in aquatic sediments: sorption, remobilization and fixation;* Ph.D Thesis. Faculty of Agronomy, Katholieke Universiteit Leuven, Belgium, 1994.
13. Konoplev, A.V., Bulgakov, A.A., Hilton, J., Comans, R.N.J. & Popov, V.E. *Radiation Protection Dosimetry* 1996, 64, 15-18.
14. Comans, R.N.J., Haller, M. & De Preter, P. *Geochim. Cosmochim. Acta* 1991, 55, 433-440.
15. Comans, R.N.J. & Hockley, D.E. *Geochim. Cosmochim. Acta* 1992, 56, 1157-1164.
16. Smith, J.T. & Comans, R.N.J. *Geochim. Cosmochim. Acta* 1996, 60, 995-1004.
17. Jannasch, H.W., Honeyman, B.D., Balistrieri, L.S. & Murray, J.W. *Geochim. Cosmochim. Acta* 1988, 52, 567-577.
18. Nyffeler, U.P., Li, Y.-H., & Santschi, P.H. *Geochim. Cosmochim. Acta* 1984, 48, 1513-1522.
19. Amacher M.C., Kotuby-Amacher J., Selim H.M. & Iskandar I.K. *Geoderma* 1986, 38, 131-154.
20. Amacher M.C., Selim H.M. & Iskandar I.K. *Soil Sci. Soc. Am. J.* 1988, 52, 398-408.
21. Sparks, D.L. *Kinetics of soil chemical processes;* Academic Press: San Diego, CA, 1989.
22. Brouwer E., Baeyens B., Maes A. & Cremers A. *J. Phys. Chem.* 1983, 87, 1213-1219.
23. De Haan, F.A.M., Bolt, G.H. & Pieters, B.G.M. *Soil Sci. Soc. Am. Proc.* 1965, 29, 528-530.

Sorption of Organic Species

Chapter 10

A Revised Physical Concept of Natural Organic Matter as a Sorbent of Organic Compounds

J. J. Pignatello

Department of Soil and Water, The Connecticut Agricultural Experiment Station, 123 Huntington Street, P.O. Box 1106, New Haven, CT 06504–1106

Natural organic matter (NOM) is a penetrable, polymer-like phase that varies continuously from rubbery (flexible, expanded) to glassy (rigid, dense) in character. The nature of sorption of nonionic organic compounds changes along this continuum, from one in which solid-phase dissolution (partitioning) predominates, to one in which a site-specific hole-filling (i.e., void-filling) process becomes increasingly important. This is referred to as dual-mode sorption. The hole-filling component, which is not conditional on the presence of a polar group on the sorbing molecule, gives rise to nonlinear and competitive sorption behavior. The fraction of hole-filling sorption can be estimated from competitive effects or by analysis of the Freundlich isotherm. The hole-filling fraction is an inverse function of sorbed concentration, and may reach as high as two-thirds of total sorption at infinitely small concentration. Removal of the humic acid fraction of NOM appears to uncover hole sites; whereas, addition of natural aromatic acids which are structural analogs of humic aromatic subunits leads to their blockage. Hole-filling increases with equilibration time due to slower penetration of the more glassy NOM phases, or steric hindrance at hole openings, or both. Soot particles, which have been suggested to affect the magnitude and kinetics of polycyclic aromatic hydrocarbon sorption in sediments, were found to play no significant role in the nonideal behaviors of chlorinated benzenes in one particular soil and its derivatives.

Sparingly soluble (hydrophobic) organic compounds have a high affinity for the NOM component of geosolids. Recent studies of sorption to metal oxides of weakly polar compounds, such as chlorinated benzenes and polycyclic aromatic hydrocarbons (PAHs), suggests that sorption to mineral surfaces in soil is insignificant when the fraction of organic carbon is above ~0.01% of total solids (*1*). Hence, an understanding of organic molecular interactions with NOM is important both for predicting the fate of organic compounds in the environment and for achieving remediation objectives. The prevailing concept of sorption since at least the late 1970's is that of a linear *partitioning*, or solid-phase dissolution process (*2-4*). Recent evidence, however, indicates that this concept must be modified to include a second, adsorption-like process that takes place in the interior of NOM. This paper summarizes that evidence and describes the process.

The Dual-Mode Model

Many researchers believe that solid NOM is closer in physical structure to a penetrable organic polymer phase than it is to an impenetrable inorganic material with a fixed pore network. Based on our knowledge of humic substance structure (*5*), the NOM polymer phase can be visualized as a water-swollen, three-dimensional mesh of macromolecules, in which small sorbing molecules may commingle. However, NOM is not homogeneous but rather exists in a continuum of states, from relatively expanded, flexible, and solvated— 'rubbery', using the polymer analogy (*6-9*)—to relatively condensed, rigid, and less solvated— 'glassy' in the polymer analogy. It is reasonable to expect that the nature of the sorption process changes along this continuum.

Using polymer theory, we have proposed that sorption to NOM occurs by dual *dissolution* and *hole-filling* mechanisms (*10-12*). A schematic of the dual-mode mechanism is depicted in Figure 1. Dissolution is depicted as the white and gray background areas in Figure 1. Dissolution occurs in regions of the polymer that are thermally flexible owing to a combination of weak interactions among polymer segments, a low degree of cross-linking between segments, and a high degree of solvation. In these relatively expanded regions, sorption 'sites' have only momentary existence and sorbed molecules experience an average environment, analogous to that experienced by dissolved molecules in a viscous liquid. Consequently the dissolution process is expected to be concentration-independent (linear) and noncompetitive, at least when the sorbate concentration is dilute.

Holes are represented by the small geometric shapes in Figure 1. Hole-filling occurs in regions of the polymer that are thermally rigid owing to stronger interactions between polymer segments, a higher degree of cross-linking, and a lower degree of solvation. In these relatively more condensed regions, the inflexibility of the macromolecules results in the creation of relatively long-lived sorption sites. As we shall see, the presence of specific sites may result in nonlinear and competitive behavior. In glassy polymers, the sites are believed to be unrelaxed free volume in the form of nanometer-size voids within the polymer (hence the term "hole-filling"). Penetrant molecule(s) filling the voids undergo an adsorption-like interaction with the polymer strands making up the void walls. We postulate that a similar mechanism occurs in NOM. In NOM, the holes are regarded as nanovoids within the macromolecular matrix. These voids are distributed in energy owing to their varying steric and electronic properties. It is not necessary that the sorbing molecule have a functional group that can orient in specific manner, like a hydrogen bond donor or acceptor group, in order to undergo this adsorption-like interaction. In glassy polymers, dual-mode sorption takes place even when oriented interactions are not possible—such as between vinyl chloride (*9*) or chlorinated benzenes (*10,11*) and glassy poly(vinyl chloride) (PVC). The same point can be made for NOM (see below).

Total sorption is the sum of dissolution S(D) and a hole-filling S(H) components (eq 1). Sorption in the dissolution domain is given by a linear term where C is the fluid-phase concentration and K_D is a lumped coefficient representing all available dissolution regions j (i.e., $K_D = \sum_j K_{D,j}$). Sorption in the hole-filling domain is given by a sum of Langmuir terms, where b_i and S^o_i are the affinity and capacity constants, respectively, for each unique hole site.

$$S = S(\mathrm{D}) + S(\mathrm{H}) = K_D C + \sum_{i=1}^{n} \frac{S_i^o b_i C}{1 + b_i C} \tag{1}$$

Due to the contribution of the hole-filling mechanism, the dual-mode model predicts that the sorption potential of NOM is concentration-dependent (i.e., nonlinear) and that sorption is subject to competitive effects.

Weber and co-workers have independently arrived at a depiction of NOM that is similar to the one described above for the dual-mode model (see W.J. Weber, Jr. in an accompanying chapter in this book). In their 'dual domain model,' sorption to "amorphous" NOM is considered to be linear, non-competitive, and rapidly-reversible, while sorption to "crystalline" NOM results in the opposite behaviors.

A mainstay of the classical partition model is the experimental observation that the major thermodynamic driving force for sorption is the *hydrophobic effect*. The hydrophobic effect results from gain in free energy when non- or weakly-polar molecular surface is transferred out of the polar medium of water (*2-4*). The hydrophobic effect is manifested by a linear free energy relationship (LFER) between the NOM-normalized partition coefficient (K_{NOM}) and the n-octanol-water partition coefficient (K_{ow}) [i.e., $\ln K_{NOM} = a \ln K_{ow} + b$ where a and b are regression constants], or the inverse of the compound's liquid (or theoretical subcooled liquid) saturated water solubility (C_{sat}) [i.e., $\ln K_{NOM} = -c \ln C_{sat} + d$].

The demonstration that such LFERs exist, however, does not necessarily conflict with a hole-filling contributing to sorption for the following reasons. First, the hydrophobic effect plays an important role in all physisorption processes from aqueous solution, whether site-specific or not. Examples include cation exchange of tetraalkylammonium ions on charged clays (*3*) and sorption of hydrocarbons on metal oxides (*1*). This is because the hydrophobic effect originates from solution-phase rather than sorbent-phase interactions. Hydrophobic effects undoubtedly play a role in hole-filling. Thus, the demonstration of one of the above LFERs does not constitute proof that partition occurs exclusively. Second, sorbate-sorbent intermolecular forces at holes will be the same as those in the dissolution domain except that steric effects are expected to play a role. This role is complex because there are multiple kinds of sites and because steric effects at a given site vary with molecular size in relation to the available space. Some data suggest size exclusion of polyaromatic hydrocarbons (PAHs) in humic acid colloids (*13*). The identification of factors other than hydrophobic effects that influence hole filling requires a separation of hole filling and dissolution components of sorption. As will be discussed below, the hole filling contribution depends on the source of organic matter, equilibration time, and solute concentration.

Evidence in Support of the Dual-Mode Model

Nonlinear Isotherms. For multiple sites, eq 1 is approximated by the Freundlich isotherm,

$$S = K_F C^N \tag{2}$$

where K_F and N are constants, and N <1. (An N equal to 1 means the isotherm is linear.) When constructed with data of sufficient precision and range, the isotherms of typical pollutants in soils where sorption is dominated by the NOM fraction are often found to be at least slightly nonlinear. A nonlinear isotherm indicates that the sorption potential of the solid is a function of solute concentration.

Figure 2a shows a plot of the log-transformed Freundlich isotherm of 1,3-dichlorobenzene in a suspension of a mineral soil (1.4% organic carbon). The slope (N value) is 0.858 or 0.801, depending on the equilibration time, 1 or 30 days, respectively. Sorption increases with equilibration time due to slow kinetics (*14*). In Figure 2b, the low-, middle-, and high-concentration regions of each isotherm in Figure 2a are replotted on a linear scale and separately fit by linear regression to obtain the "partition coefficient" K_p of the assumed linear isotherm, $S = K_p C$. The range of solute concentration in each region is 1.0-1.5 orders of magnitude. The curves appear linear and the r^2 are all above 0.98; on this basis, most investigators would reasonably conclude that sorption fits a linear model. The purpose of this exercise is to show that it is easy to mistake a clearly nonlinear isotherm for a linear isotherm when too narrow a concentration range is used in its construction. Notice that K_p increases as we go from the high- to the low-concentration regions, in accord with the concave-down curvature of the isotherm as a whole. A further point can be made that increasing scatter in the data due to experimental uncertainties makes it more and more difficult to distinguish between linear and nonlinear curves.

Table I gives Freundlich parameters for selected additional systems where the isotherm was established with many data taken over a wide concentration range. Most of these isotherms are nonlinear. Inspection of Figure 2a and Table I reveal the following. *i*) Nonlinearity applies to both polar and apolar compounds. *ii*) Nonlinearity applies to soils both high and low in NOM. *iii*) Nonlinearity is a characteristic of sorption to humic materials because it occurs even in soils with very little mineral content. Many of our studies have been carried out on Pahokee peat soil, a high-organic (~7% ash), well-humified reference material. This material consists of jagged amber-like translucent particles containing opaque inclusions, and is absent of undecomposed plant fibers. Furthermore, nonlinearity persists in cases where the mineral component of the peat has been stripped by HF/HCl treatment (*cf.*, entry 1 *vs* entry 15, Table I). *iv*) Nonlinearity is not an artifact of insufficient equilibration time, since isotherms typically become increasingly nonlinear with time.

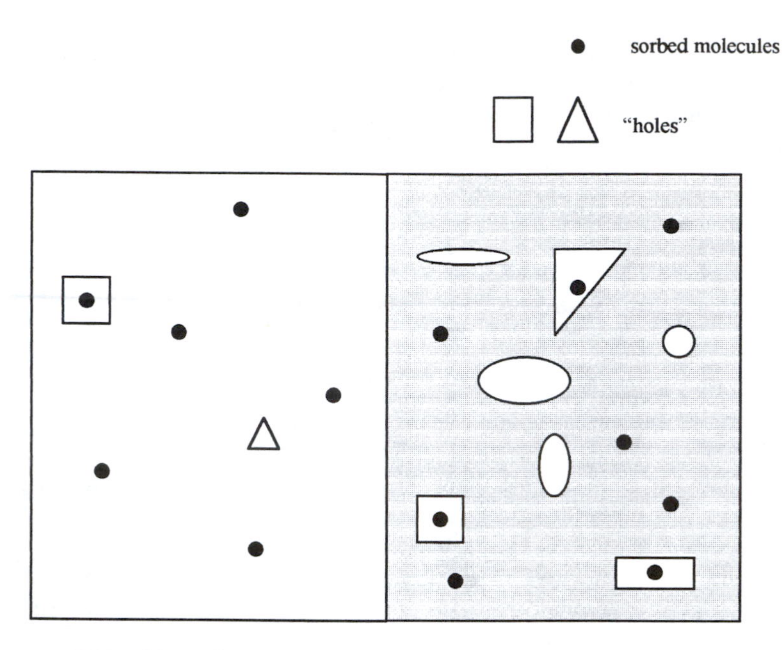

Figure 1. Schematic depiction of dual-mode sorption in NOM solid showing a more rubbery phase with few holes (sites) and a more glassy phase with abundant holes. Dissolution occurs in the background white and gray shaded areas. The stereo-electronic variability of the holes is represented by different shapes and sizes.

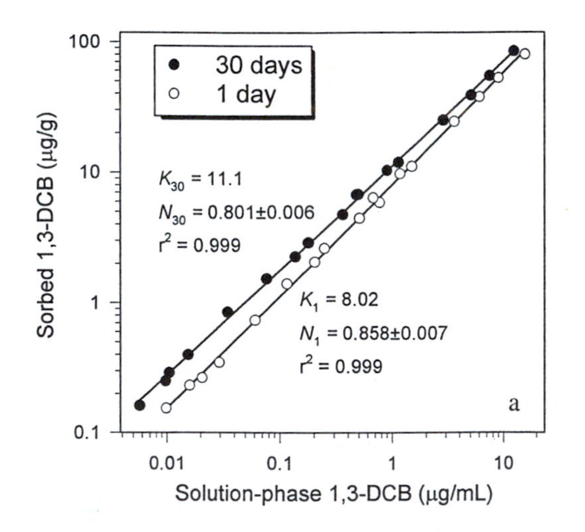

Figure 2. (a) Isotherm of 1,3-dichlorobenzene in Cheshire fine sandy loam soil suspension as a function of contact time. Adapted from data in ref. 10. (b) Same data but in linear form for the high, medium, and low concentration regions, with linear regression lines and the K_p values shown.

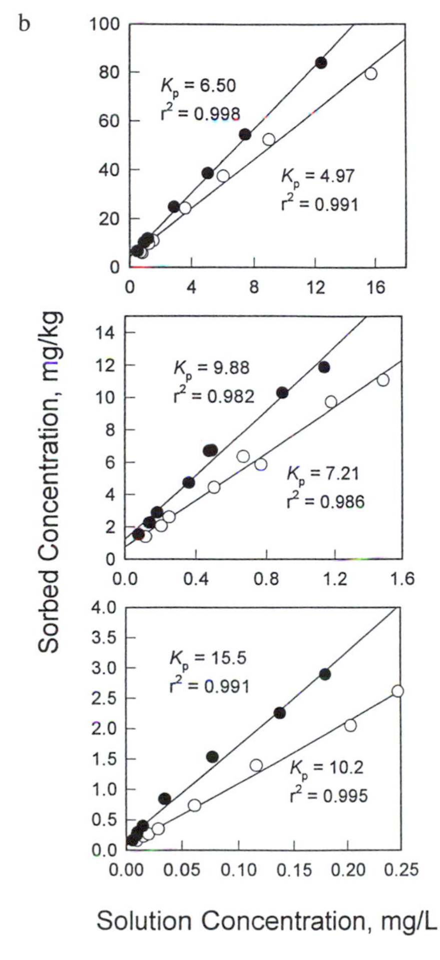

Figure 2. *Continued.*

Table I. Freundlich Isotherm Parameters in Soils and Soil Organic Matter Materials for Various Compounds[a].

	Soil (% organic matter)	Compound	conditions[b]	Equil. Time, d	K_F, (μg/g)·(μg/mL)^N	N^c	Ref.
1.	Pahokee Peat Soil (93%)	1,3-dichlorobenzene		1	253	0.839±0.008	*10*
2.	"	1,3-dichlorobenzene		30	321	0.799±0.004	*10*
3.	"	1,3-dichlorobenzene	at 6 °C	1	244	0.809 ±0.006	*11*
4.	"	1,3-dichlorobenzene	at 90 °C	1	103	0.887 ±0.015	*11*
5.	"	1,3-dichlorobenzene	in 20% CH_3OH	1	89.7	0.881 ± 0.021	*11*
6.	"	1,3-dichlorobenzene	in 20% DMSO	1	83.2	0.918 ± 0.020	*11*
7.	"	1,2-dichlorobenzene		2	331	0.833 ± 0.007	*11*
8.	"	2,4-dichlorophenol		1	319	0.777±0.005	*10*
9.	"	2,4-dichlorophenol		180	494	0.727±0.002	*10*
10.	"	metolachlor		30	107	0.875±0.006	*10*
11.	"	atrazine		2	74	0.916 ± 0.007	*12*
12.	"	prometon		2	123	0.799 ± 0.018	*12*
13.	"	trichloroethene		2	35.5	0.933 ±0.023	*12*
14.	"	phenanthrene		1	4950	0.83	new
15.	P.P. Soil, HF/HCl-treated (98%)	1,3-dichlorobenzene		1	402	0.856 ± 0.004	*11*
16.	P.P. Soil humic acid (94.4%)	1,2-dichlorobenzene		2	155	0.896 ± 0.008	*11*
17.	"	1,3-dichlorobenzene		2	161	0.936 ± 0.008	*11*
18.	Windsor humic acid (91.3%)	1,2-dichlorobenzene		2	225	0.974 ± 0.005	*11*
19.	P.P. Soil humin (74.4%)	1,2-dichlorobenzene		2	437	0.780 ± 0.012	*11*
20.	Cheshire fsl (2.8%)[d]	2,4-dichlorophenol		30	21.1	0.720±0.011	*10*
21.	"	metolachlor		30	2.46	0.841±0.005	*10*
22.	"	atrazine		2	2.15	0.923 ± 0.004	*12*
23.	"	prometon		2	2.6	0.930 ± 0.035	*12*
24.	"	cyanazine		2	1.82	0.963 ± 0.014	*12*
25.	"	trichloroethene		2	1.90	1.02 ± 0.02	*12*

26.	PVC microspheres (100%)	1,3-dichlorobenzene		1	65.9	0.879 ± 0.009	*10*
27.	"	1,3-dichlorobenzene		15	671	0.879 ± 0.018	*10*
28.	"	1,3-dichlorobenzene	at 90 °C	1	236	0.997 ± 0.014	*11*
29.	EPA-23[e] (4.7%)[d]	phenanthrene		14	258	0.727	*31*
30.	EPA-22[e] (3.4%)[d]	phenanthrene		14	341	0.890	*31*
31.	EPA-20[e] (2.6%)[d]	phenanthrene		14	118	0.756	*31*
32.	EPA-15[e] (2.0%)[d]	phenanthrene		14	161	0.726	*31*
33.	Sycamore Soil 1 (1.3%)	fenuron (<469 mg/L)		3.33	0.664	0.781	*32*
34.	Sycamore Soil 2 (6.8%)	monuron		3.33	5.24	0.824	*32*

[a] Metolachlor: 2-Chloro-N-[2-ethyl-6-methylphenyl]-N-[2-methoxy-1-methylethyl] acetamide; atrazine: 2-chloro-4-ethylamino-6-isopropylamino-*s*-triazine. Prometon: 2,4-bis(isopropylamino)-6-methoxy-*s*-triazine. Cyanazine: 2-[[4-chloro-6-(ethylamino)-*s*-triazin-2-yl]amino]-2-methylpropionitrile. Fenuron: 1,1-dimethyl-3-phenylurea. Monuron: 1,1-dimethyl-3-(4-chlorophenyl)urea. [b] Isotherms in water suspensions at room temperature unless otherwise indicated. [c] Standard error of the slope. [d] Assuming organic matter is twice the organic carbon. [e] U.S. Environmental Protection Agency reference sediments.

Parallel studies have also been carried out with synthetic polymers as models for NOM. Under dilute conditions, sorption of polar and nonpolar compounds from either aqueous solution or the vapor state is *linear in rubbery polymers* and *nonlinear in glassy polymers* (*6,8,10-12*). This is true regardless of the match or mismatch in polarity between sorbate and sorbent, confirming that oriented functional group interactions are irrelevant. Heating polymers or exposing them to swelling solvents relaxes the glassy state and eliminates holes, causing the isotherm to become linear. NOM responds similarly: the Freundlich N for chlorinated benzenes approaches unity with increasing temperature (Table I, entries 3, 1, and 4) and increasing concentration of swelling co-solvents such as methanol or dimethyl sulfoxide (entries 5 and 6).

Competitive Effects. Competitive sorption is another indication of the non-uniform sorption potential of NOM. Competitive effects can be established by measuring the influence of increasing co-solute concentration on the solid-liquid distribution ratio (S/C) of the principal solute held at constant total concentration (Method 1), or by comparing the isotherm of the principal solute with and without co-solute (Method 2). Using these methods, we have observed competitive effects in both mineral and high-organic soils, and between polar compounds, such as triazine herbicides, as well as between apolar compounds, such as chlorinated benzenes (*11,12*) or halogenated aliphatic hydrocarbons (*15*). In experiments using Method 1, the rate of decline in S/C of the principal solute with co-solute concentration is sharpest at low co-solute concentration and then levels off as the co-solute concentration increases. This is consistent with competitive sorption and inconsistent with indirect effects of the co-solute such as induced changes in the structure of NOM or the activity of the principal solute in solution.

Competitive sorption is commonly observed in glassy polymers (*8,11,12*), but is absent in rubbery polymers (*12*). Furthermore, competitive sorption is possible between compounds that interact with the sorbent only by non-directional van der Waals forces; for example, 1,2-dichlorobenzene competes with chlorobenzene in PVC microspheres (*11*) and CO_2 competes with C_2H_4 in glassy poly(methyl methacrylate) (*8*). This shows that specificity of sorption is not conditional on the presence of strongly-interacting functional groups.

Figure 3 shows an example of competitive sorption in a soil using Method 2. The isotherm of atrazine [2-chloro-4-ethylamino-6-isopropylamino-*s*-triazine] in Cheshire fine sandy loam is displayed in the presence and absence of a structural analog, prometon [2,4-bis(isopropylamino)-6-methoxy-*s*-triazine], whose concentration was chosen on the basis of Method 1 experiments to be maximally competitive. Atrazine sorption is clearly suppressed—though not eliminated—by prometon, and its isotherm becomes more linear. This behavior is consistent with the dual-mode mechanism in that prometon blocks atrazine holes, but does not compete with atrazine in the dissolution domain. Analogous results were obtained for sorption of 1,2-dibromoethane to Pahokee peat soil in the absence and presence of trichloroethene (Pignatello, J.J., In, *Advances in Colloid and Interface Science,* Elsevier, in press). A thermodynamic competitive model, Ideal Adsorbed Solution Theory (IAST), simulated the competition between atrazine and prometon accurately (Figure 3), indicating that they share an "adsorptive" surface.

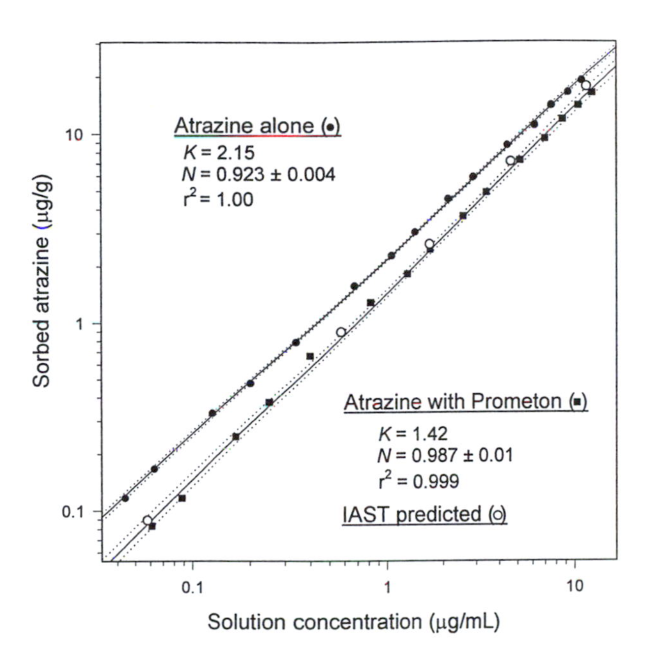

Figure 3. Isotherm of atrazine in Cheshire fsl with and without prometon. Reproduced with permission from ref. 14. Copyright 1996 American Chemical Society.

Additional studies show that the degree of competition in NOM may be related to the structural similarity of the competing molecules, indicating a degree of selectivity in the hole-filling domain. For example, atrazine is strongly competitive with prometon, but only slightly competitive with the much smaller and less polar TCE molecule—even though all three compounds had about the same affinity for sorbent when present separately (*12*). The IAST model simulated the effect of prometon but not of TCE. This means that atrazine and prometon share much of their respective hole-filling domains, but atrazine and TCE share little. In another case (*15*), the degree of competitive sorption between pairs of halogenated hydrocarbons varied inversely with the difference in their van der Waal's molecular free surface areas, although the scatter in the data was large. Verification of this relationship would provide support for a steric effect.

Evidence of Nanoporosity in NOM. Nitrogen adsorption at 77 K is a standard method for determining surface area and pore size distribution. This method gives apparent specific surface areas of NOM of only $\sim 10^0$-10^1 m^2/g (*11,16-18*), suggesting that NOM is a non-microporous material. The low surface area, yet high affinity for organic molecules has been used in support of a partition mechanism and in opposition to an adsorption mechanism for NOM. However, diffusion of molecules the size of N_2 through pores $\leq \sim 1$ nm and across the organic surface is kinetically restricted at 77 K, and thus too slow to measure conveniently. This is well-known for other carbonaceous materials like coal, coke and activated carbon (numerous refs. cited in *11,18*).

Carbon dioxide has been used as an alternative probe for characterizing carbonaceous materials (*18*) and homogeneous polymers (*6,9*). Carbon dioxide is about the same size as N_2, but has a lower vapor pressure, allowing its isotherms to be determined at higher temperatures where the activation energy for diffusion in nanopores is overcome. Specific interactions of CO_2 due to its quadrupole moment appear to play an unimportant role in its adsorption on carbonaceous and mineral surfaces (*18* and references therein). In support of this, we have shown that adsorption of N_2 and CO_2 on a mesoporous silica gel are of similar magnitude (unpublished results).

The CO_2 isotherms at 273 K for a various high-organic soils and humic acid particles (*11,18*) are Type I, characteristic of microporous solids (*19*). Furthermore, CO_2 adsorption is two or more orders of magnitude greater than N_2 adsorption at comparable relative pressures. Using a model that assumes liquid condensation of CO_2 in the pores, porosities of up to several percent of total solid volume are indicated. Given that this porosity is not revealed in the N_2 isotherms, it is reasonable to infer that the pores are small ($<\sim 1$ nm in aperture), internal, and accessible only by diffusion through the solid state.

The gas adsorption results are consistent with the dual-mode model upon assumption that some of the nanopores available to CO_2 within the organic matrix are potential adsorption sites for organic compounds. Consistent with this assumption, Xing and Pignatello (*11*) found an inverse correlation between CO_2-determined porosity and the Freundlich *N* parameter, and a direct correlation between porosity and the magnitude of the competitive effect for chlorinated benzenes in NOM particle suspensions. Obviously, the correlation needs to be tested with other systems.

Supporting Evidence in the Literature. A number of other studies support the dual-mode concept directly or indirectly. X-ray diffraction patterns of humic and fulvic acid particles typically show a broad band near 0.35 nm attributable to tightly condensed regions enriched in aromatic groups, and also a broad band near 0.41 nm attributable to less condensed, more aliphatic regions (*20*). The rubber-glassy character of NOM is supported by LeBouef and Weber (*21*) who observed a transition temperature between the glassy and rubbery states (T_g) of Aldrich humic acid of 43 °C wet and 62 °C dry using differential scanning calorimetry. Unaltered soil NOM did not give a distinct T_g up to 110 °C (*11*); presumably, this is because it is more glassy than humic acid (hence, a higher T_g) and less homogeneous than humic acid (hence, a broader T_g). Young and Weber (*22*) found that sorption of phenanthrene to a shale and a topsoil with moderate NOM content was, in each case, nonlinear and that the heat of sorption decreased with increased sorption. These observations are consistent with the dual-mode mechanism if sorption can be attributed entirely to NOM, which is reasonable for phenanthrene.

Studies of purified humic substances are also supportive of the dual-mode concept. Sorption of a PAH apparently depends not only on its hydrophobicity but also on its steric ability to fit into cavities in NOM (*13*). The abundance and "microviscosity" of the cavities (*23,24*) seems to increase with $[H^+]$, ionic strength, and divalent metal ion concentration. Schulten (*25,26*) has performed molecular mechanics calculations for a hypothetical humic macromolecule in the presence or absence of water. He found that small molecules—atrazine, pentachlorophenol, a peptide, and a trisaccharide— "bind" in nanometer-size voids in the humic structure. With continued energy minimization, the voids shrink and become more hindered.

Importance of Hole-Filling

Hole-filling is concentration dependent. One way to estimate hole-filling is by eliminating it through the effect of a competing solute (*11,12*). The Langmuir isotherm for the principal solute at a given site where competition with a co-solute occurs is given by:

$$S_i = \frac{S_i^o b_i C}{1 + b_i C + b_{co} C_{co}} \tag{3}$$

where b_{co} and C_{co} are the affinity constant and aqueous concentration, respectively, of the co-solute, and the other terms are as defined in eq 1. At sufficiently high C_{co}, S_i is reduced to zero. If competition occurs at all possible sites in the hole-filling domain occupied by the principal solute, the Langmuir summation term in eq 1 will drop out at high C_{co}, so that $S = S(\mathrm{D}) = K_\mathrm{D} C$ and K_D is therefore obtained straightforwardly. The single-solute isotherm is expected to approach linearity, which it does—see Figure 3. The calculated K_D is an upper limit estimate because we cannot expect perfect competition for any pair of non-identical solutes.

If competitive data are lacking, K_D may be estimated from the single-solute isotherm by realizing that the solute undergoes self-competition; that is, a point on the isotherm is reached where the holes become completely filled but where dissolution continues linearly with the same K_D. Locating the precise transition to linearity requires the isotherm to be highly defined in that region. Since such definition is usually not attainable, a conservative approach is to assume that hole-filling continues up to C_{sat}. By definition, sorption ceases at C_{sat} because all sites of equal or greater energy than the energy of liquid condensation of the solute have been filled. If we assume that the Freundlich parameters established on the main part of the isotherm are applicable up to C_{sat}, then K_D is the tangent of the Freundlich curve at C_{sat}:

$$K_D = \frac{d}{dC} K_F C^N \bigg|_{C_{sat}} = K_F N C_{sat}^{N-1} \tag{4}$$

This approach yields a lower limit estimate of K_D due to our assumption that hole filling sorption continues all the way up to C_{sat}.

Having obtained K_D from the competitive effect or by eq 4, the fractions of sorbate in the hole-filling and dissolution domains at any concentration are given by:

$$f_{\mathrm{H}} = \frac{S - K_D C}{S}; \qquad f_{\mathrm{D}} = 1 - f_{\mathrm{H}} \tag{5ab}$$

In order to obtain $f_{\mathrm{H}}(C)$ as a smooth function we first fit the isotherm data by nonlinear regression to the dual-mode equation, allowing a single composite site (eq 1, $n = 1$) and plugging in the value of K_D. This yields the composite site parameters, S_c^0 and b_c. At infinite dilution of solute, f_{H} is simply $S_c^0 b_c/(K_D + S_c^0 b_c)$. At any other concentration, f_{H} is given by eq 5a, where S is calculated from the dual-mode parameters.

A comparison of the f_{H} estimated by the competitive and Freundlich methods is given elsewhere (Pignatello, J.J., In, *Advances in Colloid and Interface Science*, Elsevier, in press). Figure 4 shows how f_{H} calculated by the Frendlich method varies with solution-phase 1,3-dichlorobenzene concentration in selected systems. It can be seen that hole-filling is relatively small at C_{sat} (123 mg/L), but plays an increasingly greater role as $C \rightarrow 0$, ultimately accounting for as much as two-thirds of total sorption. It is also important to note that f_{H} is non-trivial even though the isotherm may be only 'slightly' nonlinear. For example, the N for 1,3-dichlorobenzene in Cheshire fine sandy loam after 24-h equilibration is 0.858, yet hole-filling contributes about 60% to sorption at low concentration (Fig. 4a).

Role of Structural Components of NOM in the Dual-Mode Mechanism

Insight into the nature of NOM as a sorbent of organic compounds can be obtained by examining sorption on derivative materials. We have studied sorption on humic acid particles, as well as on the resulting insoluble *humin* fraction left after extraction (*11*).

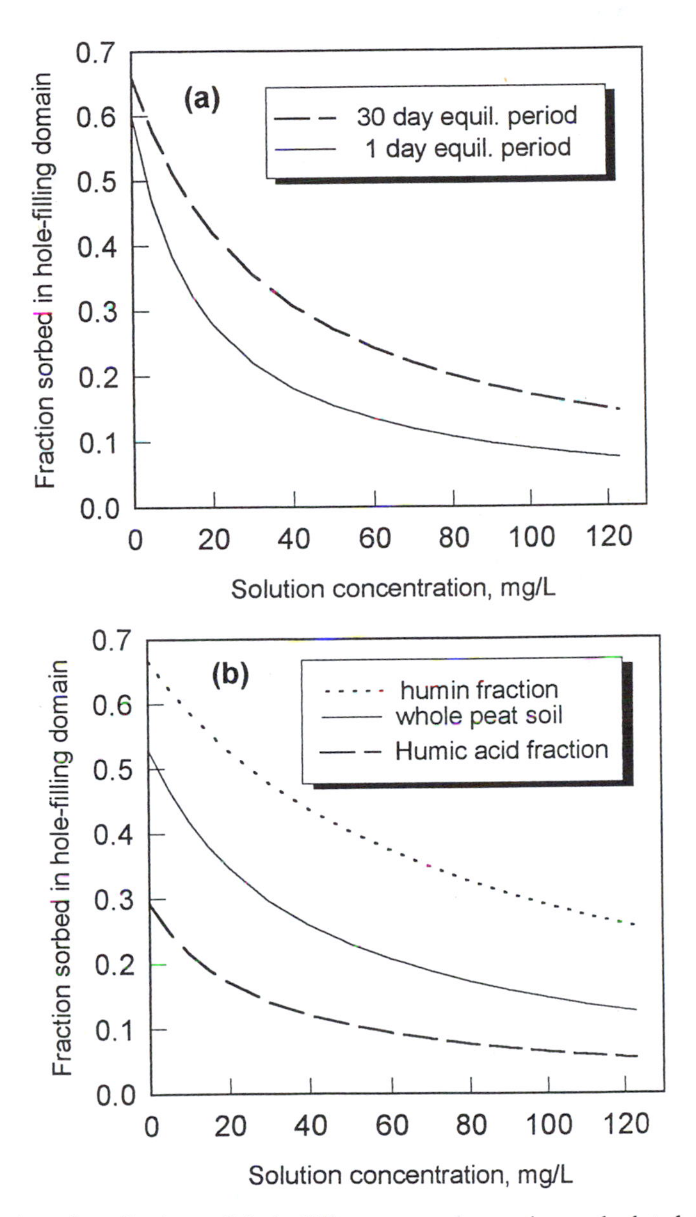

Figure 4. Contribution of hole-filling to total sorption calculated by the Freundlich slope method for 1,3-dichlorobenzene in: (a) Cheshire fsl.; and (b) Pahokee peat soil and its humin and humic acid fractions for a 48-h equilibration time. Data taken from isotherms in Figures 2 and 5.

The extraction was performed with a weakly basic metal-complexing agent ($Na_2P_4O_7$). The Freundlich *N* for 1,2- and 1,3-dichlorobenzenes follows the order:

humin < whole soil < humic acid

The isotherms for the 1,3-isomer are shown in Figure 5.

The degree of competition between the two isomers follows the reverse order:

humin > whole soil > humic acid

These results are consistent with dual-mode polymer theory. Humin is more condensed and less hydrated than humic acid (*5*), while the whole soil is expected to be intermediate. Thus, nonlinearity and competition increase with increasing glassy character (i.e., greater hole population) of the organic matter. Figure 4b shows that the hole-filling fraction increases from humic acid to whole soil to humin.

Further inspection of the 48-h isotherms of 1,3-dichlobenzene in the whole peat, peat humic acid, and peat humin (Figure 5) reveals another interesting trend: although the three sorbents have about the same affinity for 1,3-DCB at concentrations approaching its solubility, their affinities diverge as the concentration declines. This trend suggests that removal of the looser, easily extractable humic acid fraction from NOM *uncovers* sorption sites in NOM. Furthermore, we found that structural analogs of the aromatic subunits of humic macromolecules— *t*-cinnamic, vanillic, coumeric, *p*-hydroxybenzoic, and others—are competitive with anthropogenic compounds in soil NOM; that is, they plug up sorption sites (Xing, B.; Pignatello, J.J. *Environ. Sci. Technol.*, in press). Taken together, these results suggest that NOM can be visualized as a rigid, nanoporous framework composed primarily of humin that is partially filled with smaller, more flexible humic acid macromolecules. The inclusion of humic acid adds dissolution capacity to the NOM.

In addition, since the competitive aromatic acids above are naturally occurring, it is possible that other simple natural molecules in the environment are competetive with pollutants for sorption sites.

Recent papers have raised the question of whether soot particles may be responsible for enhanced sorption and slow desorption rates of PAHs in sediments (*27,28*). As a product of incomplete combustion, soot contains highly condensed carbonaceous residue that may occlude PAHs during its creation in forms that are highly immobile. In addition, soot may act as a microporous adsorbent similar to activated carbon that has the potential to cause nonlinearity and competitive effects of contaminants in soil containing such particles. The "soot carbon" content of the Pahokee peat and the peat humin (after $Na_2P_4O_7$ extraction) was measured courtesy of Phillip Gschwend and his group at MIT by a procedure (*28*) in which the inorganic carbon is removed by HF treatment and the "natural organic matter" is selectively removed by low-temperature combustion, leaving the soot carbon behind. The ratio of soot carbon to total organic carbon in the peat is 0.015 ± 0.002 and in the humin is 0.0031 ± 0.0003 g soot/ g TOC. (By comparison, the soot content of a humic acid reported in ref. 28 is 0.0015 g soot/ g TOC). The fact that the soot carbon is a factor of

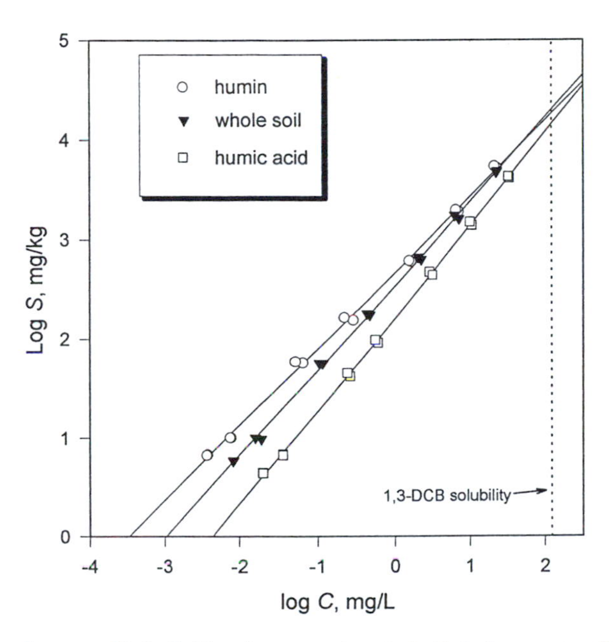

Figure 5. Isotherms of 1,3-dichlorobenzene in whole Pahokee peat soil (K_F = 340, N =0.8501) and its derivatives, humin (K_F = 464, N =0.7662) and humic acid (K_F = 161, N =0.936). Equilibration period, 48 h. The soil was extracted with sodium pyrophosphate. Adapted from data in ref. 11.

five less in the material that gives greater nonlinearity and competitive effects (i.e., the humin) argues against soot carbon playing a major role in those behaviors.

Influence of Mechanism on the Rates of Sorption and Desorption

Recently there has been a great deal of interest in sorption kinetics and how it may influence the physical and biological availability of contaminants. It is known that intraparticle sorption processes reach equilibrium only slowly—in many cases on weeks- or months-long timescales (*14*). It is logical to look to NOM as the matrix in which sorption rates are retarded, since most organic compounds sorb preferentially to NOM. Based on the results of experiments carried out on humic acid disks (*29*), diffusion through rubbery humic phases appears to be too rapid to account for the very slow sorption/desorption rates that have been observed in soils by many investigators. However, it is well-known that the glassy state of polymers resists molecular diffusion more than the rubbery state by—depending on diffusant size—many orders of magnitude due to its higher viscosity and the extra time the diffusant spends trapped in holes (*6,9,30*).

Experimentally in soil systems, we find a tendency toward increasing nonlinearity (several examples, Table I) and stronger competitive effect with increasing contact time between chemical and soil (*10,11,31*). In fact, isotherms tend to be fairly linear at very short times (*31*). Figure 4a shows a general increase in the hole-filling fraction f_H with equilibration time at all concentrations. Notice in Table I, entries 26 vs 27, that the same trend of increasing nonlinearity of 1,3-dichlorobenzene sorption with increasing contact time is not followed in glassy PVC where the holes are homogeneously distributed throughout the polymer. The implications of this for NOM are: a) that the holes are inhomogeneously distributed; b) that there is faster penetration of hole-poor rubbery phases relative to hole-rich glassy phases; and c) that the preponderance of sites responsible for nonlinear behavior are not located on the available (external) surfaces of the solid. Finally, it is conceivable that the slowest sorption/desorption rates in NOM are due to an activated hole-filling process—i.e., steric hindrance at the entrance/exit of holes.

Acknowledgments. I thank my collaborators, Baoshan Xing and Alexander Neimark, and the U.S. Department of Agriculture, National Research Initiative (Agreement Nos. 93-37102-8975 and 97-35102-4201).

References

1. Mader, B.T.; Goss, K.; Eisenreich, S.J. *Environ. Sci. Technol.* **1997**, *31*, 1079.
2. Chiou, C.T. In *Reactions and Movement of Organic Chemicals in Soil*; Sawhney, B.L. and K. Brown, K., Eds.; Soil Science Society of America: Madison, WI, 1989; Chap. 1.; pp 1-30.
3. Schwarzenbach, R.P.; Gschwend, P.M.; Imboden, D.M. *Environmental Organic Chemistry;* J. Wiley: New York, 1993.
4. Karickhoff, S.W. *J. Hydraul Eng.* **1984**, *110*, 707.
5. Hayes, M.H.B.; MacCarthy, P.; Malcolm, R.L.; Swift, R.S., Eds., *Humic Substances II* ; J. Wiley & Sons: London, UK, 1989.

6. Vieth, W.R. *Diffusion In and Through Polymers*; Oxford Univ. Press, NY, 1991.
7. Berens, A.R. *J. Membrane Sci.* **1978**, *3*, 247.
8. Sanders, E.S.; Koros, W.J. *J. polym. Sci. Polym. Phys.* **1986**, *24*, 175.
9. Berens, A.R. *Macromol. Chem., Macromol. Symp.* **1989**, *29*, 95.
10. Xing, B.; Pignatello, J.J. *Environ. Toxicol. Chem.* **1996**, *15*, 1282.
11. Xing, B.; Pignatello, J.J., *Environ. Sci. Technol.* **1997**, *31*, 792.
12. Schlautman, M.A.; Morgan, J.J. *Environ. Sci. Technol.* **1993**, *27*, 961-969.
13. Pignatello, J.J.; Xing, B. *Environ. Sci. Technol.* **1996**, *30*, 1-11.
14. Xing, B.; Pignatello, J.J.; Gigliotti, B. *Environ. Sci. Technol.* 30 (1996) 2432.
15. Pignatello, J.J. In *Organic Substances and Sediments in Water, vol. 1*; Baker, R.A., Ed.; Lewis Publ.: Chelsea, MI, 1991; 291-307.
16. Chiou, C.T.; Rutherford, D.W.; Manes, M. *Environ. Sci. Technol.* **1993**, *27*, 1587-1594.
17. Pennell, K.D.; Boyd, S.A.; Abriola, L.M. *J. Environ. Qual.* **1995**, *59*, 1012.
18. de Jonge, H.; Mittelmeijer-Hazeleger, M.C. *Environ. Sci. Technol.* **1996**, *30*, 408-413.
19. Gregg, S.J.; Sing, K.S.W. *Adsorption, Surface Area, and Porosity,* Second Edition, Academic Press: London, 1982.
20. Schnitzer, M.; Kodama, H.; Ripmeester, J.A. *Soil Sci. Soc. Am. J.* **1991**, *55*, 745-750.
21. LeBoeuf, E.J.; Weber, W.J., Jr. *Environ. Sci. Technol.* **1997**, *31*, 1697-1702.
22. Young, T.M.; Weber, W.J., Jr. *Environ. Sci. Technol.* **1995**, *29*, 92.
23. Engebretson, R.R.; von Wandruszka, R. *Environ. Sci. Technol.* **1994**, *28*, 1934-1941.
24. Engebretson, R.R.; von Wandruszka, R. *Environ. Sci. Technol.* **1994**, *28*, 1934-1941.
25. Schulten, H.-R. *Intern. J. Environ. Anal. Chem.* **1996**, *64*, 147-162.
26. Schulten, H.-R. *Fresenius J. Environ. Anal. Chem.* **1995**, *351*, 62-73.
27. McGroddy, S.E.; Farrington, J.W.; Gschwend, P.M. *Environ. Sci. Technol.* **1996**, *30*, 172-177.
28. Gustafsson, Ö; Haghseta, F.; Chan, C. MacFarlane, J.; Gschwend, P.M. *Environ. Sci. Technol.* **1997**, *31*, 203-209.
29. Chang, M.-L.; Wu, S.-C.; Chen, C.Y. *Environ. Sci. Technol.* **1997**, *31*, 2307-2312.
30. Bandis, A.; Cauley, B.J.; Inglefield, C.E.; Wen, W.-Y.; Inglefield, P.T.; Jones, A.A.; Melc'uk, A. *J. Polymer Sci. Part B.* **1993**, *31*, 447.
31. Weber, W.J., Jr.; Huang, W. *Environ.. Sci. Technol.* **1996**, *30*, 881.
32. Spurlock, F.C.; Huang, K.; van Genuchten, M. Th. *Environ. Sci. Technol.* **1995**, *29*, 1000.

Chapter 11

A Three-Domain Model for Sorption and Desorption of Organic Contaminants by Soils and Sediments

Walter J. Weber, Jr., Weilin Huang, and Eugene J. Leboeuf

Environmental and Water Resources Engineering Program, Department of Civil and Environmental Engineering, EWRE Building, 1351 Beal Avenue, The University of Michigan, Ann Arbor, MI 48109–2125

A model predicated on a hypothesis that the sorbing components of soils and sediments comprise three principally different types of domains is proposed. Two of these are soil organic matter (SOM) domains, one highly amorphous and the other relatively condensed. The third consists of mineral surfaces. The existence of two phenomenologically different SOM domains is evidenced by observations of a glass transition for soil-derived humic acids and of the similarity of sorption behaviors between synthetic organic polymers and SOM. Extensive examinations of phenanthrene sorption and desorption for a broad range of soils, sediments, shales, kerogens, and model inorganic sorbents reveals that: i) sorption by condensed SOM matrices is nonlinear and apparently hysteretic; ii) sorption by amorphous SOM domains and mineral surfaces is linear and completely reversible; iii) mineral domains contribute little to the overall sorption and desorption of hydrophobic organic compounds by soils and sediments. The results of this study indicate that the three-domain model not only quantitatively describes non-ideal partitioning sorption phenomena in subsurface systems, but also provides fundamental insight to, and understanding of, the binding and sequestration of organic contaminants by soils and sediments.

Sorption by soils and sediments in aquatic surface and subsurface systems functions to retard the transport, regulate the toxicity, and constrain the bioavailability of organic contaminants (*1–3*). The process is frequently observed to be nonlinear, slow, and apparently hysteretic (*4–23*). Such behavior is often invoked to explain the ineffectiveness of various technologies to fully cleanup and remediate contaminated subsurface systems (*2*). Until recently, no mechanistically rigorous model for explaining such behaviors or for predicting their occurrence has been forthcoming.

This paper describes our pursuit of a scientific rationale for interpreting and quantitatively describing non-ideal sorption phenomena in subsurface systems, and of a better resulting understanding of the binding and sequestration of organic contaminants by soils and sediments. Characterizations of the organic matter fractions of a broad range of geosorbents (soils, sediments, shales and kerogens) and synthetic organic polymers by various techniques are described, and measured equilibrium data for the sorption and desorption of a representative hydrophobic organic solute, phenanthrene, are presented. The observed sorption and desorption behaviors are correlated with the physical and chemical properties of the sorbents tested, and the results are used to establish and verify a three-domain model that reconciles various non-ideal sorption phenomena.

Sorption Models

The Linear Partitioning Model. Several conceptual and empirical models have been proposed to quantify sorption of hydrophobic organic contaminants (HOCs) by soils and sediments. The simplest of these is the linear partitioning model, in which the solid-phase solute concentration, q_e, is assumed to be directly proportional to the aqueous-phase solute concentration, C_e; i.e.,

$$q_e = K_D C_e \quad (1)$$

where K_D is the distribution coefficient. The simplicity of this model with respect to fate and transport model predictions is appealing. Several investigators, assuming that soil organic matter (SOM) is homogeneous and acts as a gel- or liquid-like phase, have suggested that sorption from aqueous solution into SOM is analogous to a solute partitioning into an organic liquid, such as octanol (*24–28*). Such reasoning has led to a hypothesis that the process is strictly dependent upon the hydrophobicity of the solute in question and the amount of the organic phase present, suggesting the concept of an organic carbon-normalized sorption distribution coefficient, K_{OC}, defined as

$$K_{OC} = K_D / f_{OC} \quad (2)$$

where f_{OC} is the weight fraction of organic carbon associated with a soil or sediment. If K_{OC} values are assumed to be approximately the same for all soils and sediments they should then be correlatable to the octanol-water partitioning coefficient (K_{OW}) or to the aqueous-phase solubility limit (C_S) of an HOC solute (*24–31*). This is clearly inconsistent with the reality that soils and sediments are comprised by organic materials having different physicochemical properties (*11, 12, 14–17, 32*).

The linear partition model has been challenged with respect to its implied mechanistic interpretation of HOC sorption by soils and sediments (*5, 6, 11, 14–17, 32*), and with respect to its validity for the prediction of such processes (*11, 14, 17, 19, 21*).

The slow, nonlinear and apparently hysteretic behavior commonly observed for soils and sediments is phenomenologically and mechanistically incompatible with the linear partition model. The clear dependence of sorption properties on chemical structures and elemental compositions of SOM found by several investigators (*11, 14, 19, 21–23, 32–36*) confirms that SOM is in most cases neither homogeneous nor completely gel-like in character, as assumed implicitly by the linear partitioning model.

Distributed Reactivity Model. Noting that soils and sediments are highly heterogeneous at both microscopic and particle scales, Weber et al. (*11*) invoked the concept of multiple discrete reactive mineral and SOM domains to explain the non-linear and nonideal behavior of HOC sorption, and defined a composite distributed reactivity model (DRM) to quantify such behavior. They also introduced the notions of "hard carbon" (oxidation resistant) and "soft carbon" (easily oxidized) SOM to distinguish operationally between diagenetically-altered kerogen-like and geologically-younger humic SOMs. They demonstrated that hard SOMs exhibit more nonlinear, thermodynamically more favorable, and most likely pore-surface-related sorption for organic contaminants, while soft SOMs exhibit less nonlinear, thermodynamically less favorable, and possibly partition-related sorption. The logic underlying the DRM is that soils and sediments can be treated as heterogeneous combinations of active organic and inorganic components with respect to sorption, each component having its own sorption energy and sorptive properties and exhibiting either a non-linear or linear sorption with respect to a particular solute. The overall sorption isotherm for a natural solid is therefore the sum of the sorption isotherms of its M linear and N nonlinear active components (*11*). That is,

$$q_e = \sum_{i=1}^{M+N} q_{e,i} \tag{3}$$

which can be expressed in terms of conventional sorption equilibrium relationships as

$$q_e = \sum_{i=1}^{M} X_{L,\,i} K_{D,\,i} C_e + \sum_{j=1}^{N} \frac{X_{NL,\,j} Q^{o}_{a,j} b_j C_e}{1 + b_j C_e} \tag{4}$$

where: $X_{L,\,i}$ and $X_{NL,\,j}$ are the mass fractions of solid phase exhibiting linear (L) and nonlinear (NL) sorption behavior; $K_{D,\,i}$ is the mass-averaged sorption coefficient for linear component i; $Q^{o}_{a,j}$ and b_j are Langmuir sorption capacity and energy parameters respectively for a discrete reactive sorption domain j; and, M and N are the numbers of discrete linear and nonlinear reactive sorption domains, respectively.

Because linear contributions are directly additive, the first term on the right side of Equation 4 can be simplified to

$$q_{e,L} = \sum_{i=1}^{M} X_{L,\,i} K_{D,\,i} C_e = X_L K_D C_e \tag{5}$$

In the context of the DRM, sorption by soft carbon is described as a partitioning process, and is represented by a linear model. Conversely, sorption by hard carbon is treated as a nonlinear adsorption process and described by the Langmuir sorption model. Because the sorption sites and associated energies of hard carbon matrices are typically heterogeneous (i.e., N is large), the second right-hand term in Equation 4 can be approximated well by the Freundlich equation (*37*)

$$q_{\mathrm{e,\,NL}} = \sum_{\mathrm{j=1}}^{\mathrm{N}} X_{\mathrm{NL}}^{\mathrm{j}} \frac{Q_{\mathrm{a,j}}^{\mathrm{o}} b_{\mathrm{j}} C_{\mathrm{e}}}{1 + b_{\mathrm{j}} C_{\mathrm{e}}} \approx K_{\mathrm{F}} C_{\mathrm{e}}^{\,n} \tag{6}$$

where K_F and n are the nonlinear (Freundlich) sorption capacity and intensity parameters for lumped nonlinear reactive sorption domains. By assuming a certain distribution pattern for sorption-site energies for heterogeneous soils, Equation 6 can be derived from fundamental adsorption theory using a patchwise approach (*37, 38*).

Three Domain Model. Weber and Huang (*16*) devised a temporal phase-distribution relationship (PDR) approach for measuring sorption under non-equilibrium conditions. They observed that sorption rate data obtained at a given time, *t*, for systems initiated at different solution-phase concentrations can be fitted using a Freundlich-like equation having the form

$$q(t) = K_{\mathrm{F}}(t) C(t)^{n(t)} \tag{7}$$

where $q(t)$ and $C(t)$ are time-dependent solid-phase and aqueous-phase solute concentrations, respectively, and $K_{\mathrm{F}}(t)$ and $n(t)$ are time-dependent parameters expressive of utilized sorption capacity and sorption energy/heterogeneity, respectively. As shown in Figures 1 and 2, these investigators observed that the parameter $n(t)$ for all of the systems they tested decreased from ~1.0 initially to significantly smaller values as a function of time, whereas the parameter $K_{\mathrm{F}}(t)$ increased continuously, but not necessarily regularly. A three-domain model was invoked to explain the characteristic changes in the PDR parameters they observed as resulting from the sequential accessing of different types of sorption "sites" by sorbing molecules. They hypothesized that the multiple discrete sorption reactions suggested in the DRM are actually grouped within three principal domains; i.e., an exposed mineral surface Domain I, a highly amorphous SOM Domain II, and a relatively condensed SOM Domain III, as illustrated in Figure 3. They further suggested that sorptions attributable to Domains I and II are nearly linear, fast and completely reversible, whereas sorption occurring in Domain III is non-linear, slow and only partially reversible (i.e., hysteretic) (*16, 19, 21*). They concluded that Domain III is responsible for the non-ideal partitioning behavior commonly observed for soils and sediments, and that the overall sorption behavior for a specific soil or sediment is a function of its particular domain composition and distribution.

Analogy between SOM and Synthetic Organic Polymers

Natural organic matter in the environment varies widely in composition. It may be comprised of: newly deposited bits and pieces of biopolymers; moderately aged humins,

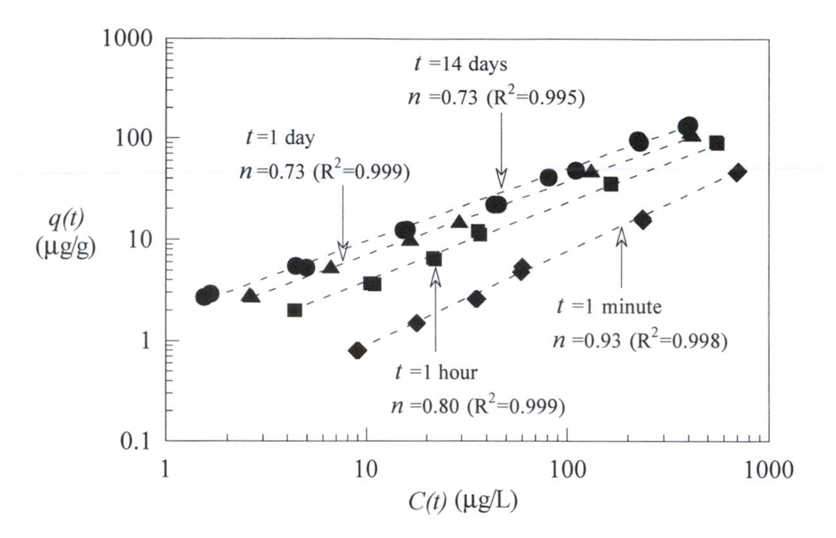

Figure 1. Time-dependent PDRs for sorption of phenanthrene by EPA-23 sediment (Adapted from ref *16*).

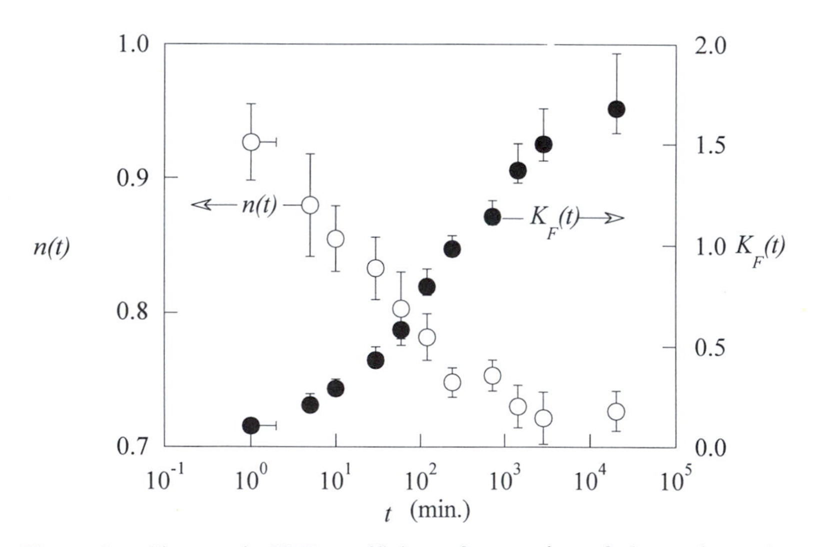

Figure 2. Changes in PDR coefficients for sorption of phenanthrene by EPA-23 sediment as a function of log time. Error bars represent 95% confidence level (Adapted from ref *16*).

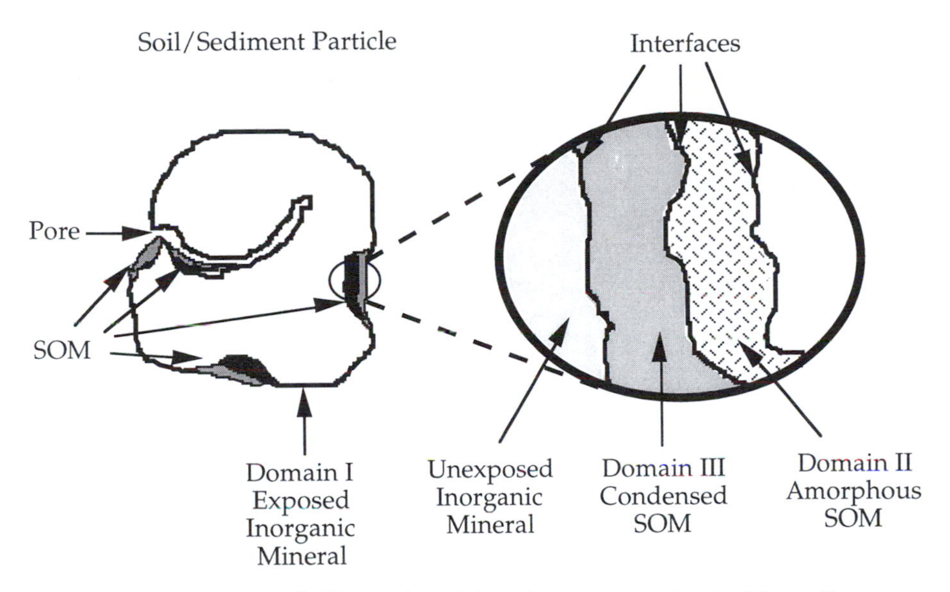

Figure 3. Schematic illustration of domain types associated with a soil or sediment particle (Adapted from ref *16*).

humic acids, and fulvic acids; well-aged thin coatings of kerogen on mineral surfaces constituting particles of shale; or chunks of coal formed over millions of years (*21, 35, 39–41*). Nonetheless, it is likely to maintain some semblance of macromolecular structure regardless of age or origin. It is therefore reasonable to hypothesize that soil and sediment organic matter, which is comprised of portions of one or all of the aforementioned constituents, exhibits macromolecular properties analogous to those of synthetic polymers (*12, 14, 15, 17, 32*).

Synthetic polymers are known to exhibit two distinct forms of mechanical behavior, depending on whether they exist in so-called glassy or rubbery thermodynamic states. Glassy polymers are differentiated from rubbery polymers by their hard, rigid, glass-like structure (e.g., poly(methyl methacrylate (PMMA); i.e., Lucite®, Plexiglas®) at room temperature. Rubbery polymers (e.g., polyethylene, rubber) are soft, and flexible at room temperature. This flexible nature of rubbery polymers can be attributed to long-range (multi-molecular) motions of polymer chains not fully present in glassy polymers. For example, heating a glassy polymer such as PMMA significantly above room temperature (to approximately 120°C) will result in its transition to rubber-like behavior. This is a result of the additional thermal energy available at elevated temperatures that allows increasing numbers and magnitudes of molecular motions, with resulting increases in volume (larger molecular motions create greater free volume) and heat capacity (larger molecular motions are better able to dissipate energy) of the polymer. In similar fashion, cooling a rubbery polymer results in less long-range molecular motion, and consequent polymer stiffness and rigidity. The temperature at which increased molecular motions lead to rubbery behavior is referred to as the glass transition temperature, T_g.

The glass transition temperature of a polymer marks a second-order phase transition in which there is continuity of the free energy function and its first partial derivatives with respect to state variables such as temperature or pressure, but there is a discontinuity in the second partial derivatives of free energy. There is, therefore, continuity in enthalpy, entropy, or volume at the transition temperature, but not in the constant-pressure heat capacity, Q_H^o (*42*). Hence, measurements of changes in Q_H^o with increasing temperature yield information about T_g, as well as about the magnitude of change in Q_H^o that occurs in the transition from glassy state to rubbery state. Because the rubbery state allows greater molecular motion, it exhibits a greater ability to disperse heat, and thus manifests a correspondingly higher Q_H^o. Such measurements can be made using differential scanning calorimetry (DSC) techniques, as described, for example, by LeBoeuf and Weber (*32*) and references given therein.

Because T_g is a function of macromolecular mobility, any changes to the macromolecular structure that increase or decrease this mobility will have similar effects on T_g. For example, increased cross-linking restricts chain mobility of larger macromolecular segments, while increased attractive forces between molecules (as measured by the solubility parameter, σ_p) require more thermal energy to produce molecular motion. Thus, T_g will generally increase with increased cross-linking and increased σ_p. In addition, an increase in the free volume of the macromolecule (i.e., that volume not occupied by the component molecules themselves) allows more room for molecular movement, and thus yields an accompanying reduction in T_g. Swelling of a macromolecular sorbent by thermodynamically compatible solutes (i.e., those possessing similar σ_p values) will therefore tend to increase the free volume and lower T_g (*43*).

Figure 4 presents results from a calorimetric investigation by LeBoeuf and Weber (*32*) of glass transitions for dry and water-wet samples of a purified, soil-derived humic acid. The humic acid shows a less sharp phase transition than that of a typical synthetic polymer. LeBoeuf and Weber (*32*) hypothesized that the spreading of the glass transition over a greater temperature range is the result of the heterogeneous nature of the humic acid relative to the homogeneous synthetic polymer, with respect to both chemical structure and molecular weight. As illustrated by comparing Figures 4a and 4b, equilibration of the humic acid with water brought about a reduction in T_g, from approximately 62°C to 43°C, as well as an almost complete elimination of an endothermic peak immediately following the phase transition. The significant (19°C) drop in T_g for the water-wet humic acid can be attributed to increased micromolecular mobility of the humic acid matrix. Water, with a solubility parameter of 23.4 $(cal/cm^3)^{0.5}$ (*44*), interacts favorably with humic acid, which has a σ_p value of approximately 11.5 $(cal/cm^3)^{0.5}$ (*45*). This interaction results in sorption of water within the humic acid matrix, causing swelling (and thus an increase in the macromolecular mobility), and resulting reduction in the T_g.

In explaining the polymer-like behavior of the natural humic acid with respect to temperature, LeBoeuf and Weber (*32*) described how the process of diagenesis can be viewed as the action of converting relatively young, expanded, lightly crosslinked, relatively rubbery organic matter into more condensed, highly crosslinked, more aromatic (more "stiff"), glassy structures having increased glass transition temperatures. This perspective suggests that "younger" soil organic matter such as humic and fulvic acids, although polar in nature and thus likely having large σ_p values, may manifest lower glass transition temperatures than more diagenetically advanced, or "aged" SOMs such as kerogens. Coal, a highly aromatic form of kerogen, has been shown to have glass transition temperatures ranging from 307°C to 359°C (*46*), with increasing values shown to be a function of increasing presence of carbon in the coal structure. Biopolymers, such as cellulose, also exhibit glass transition temperatures, with a remarkable reduction in their T_g values in the presence of water, a "good" swelling solvent for cellulose (*47*). Similarly, reductions in the glass transition temperatures of other polar polymers such as humic and fulvic acids in aqueous solutions are also expected. Further, the organic matter found in most natural systems is composed of several different organic macromolecules or components of varying crosslink densities and molecular weight, and should therefore be expected to exhibit a composite range of glass transition temperatures.

Similarities of Sorption and Desorption Behavior between SOM and Synthetic Organic Polymers

The Dual Reactive Domain Model. The existence of a macromolecule in either the rubbery or glassy state can result in widely varying sorption behavior. Nonlinear

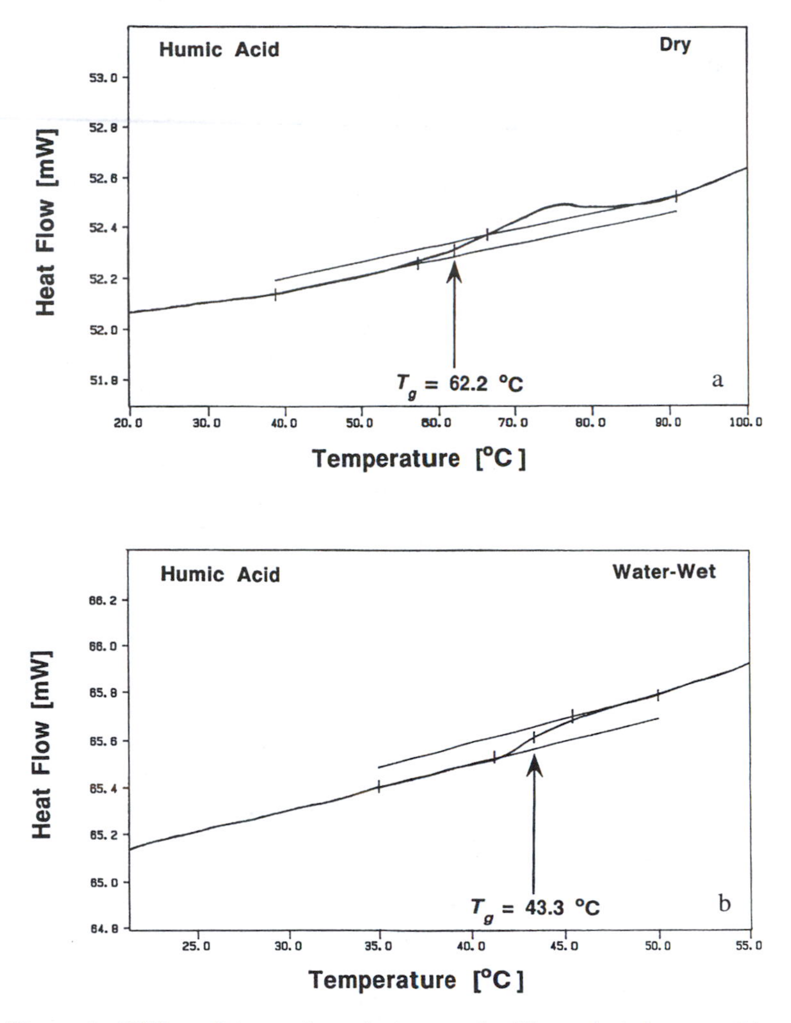

Figure 4. Differential scanning calorimetry plot illustrating glass transition for soil-derived humic acid under dessicator-dry (a) and water-wet (b) conditions (Adapted from ref *32*).

sorption, slow non-Fickian diffusion, and competitive multi-solute sorption has been attributed to glassy regions, whereas linear, partition-like sorption, relatively fast Fickian-type diffusion, and no multi-solute competitive sorption has been attributed to rubbery regions (*12, 14, 15, 17, 19, 21, 32*). LeBoeuf and Weber (*32*) proposed a dual organic matrix model to explain their observations. That model, a limiting case of the DRM proposed earlier by Weber et al. (*11*) and termed the Dual Reactive Domain Model (DRDM), includes a linear phase partitioning component and a single, limited-site Langmuir-type isotherm component; i.e.,

$$q_e = q_{e,L} + q_{e,NL} = K_{D,C}C_e + \frac{Q_a^o bC_e}{1 + bC_e} \tag{8}$$

where all variables and parameters are as defined previously.

The simplifications of the DRM that lead to the polymer-based expression given in Equation 8 are two-fold. The first simplification assumes that a single partitioning reaction into a homogeneous rubbery phase accounts for all of the linear component, whereas the more robust DRM addresses linear sorption reactions onto mineral surfaces as well as partitioning into highly amorphous organic matter. Most significantly, the second simplification treats sorption by a glassy polymer phase as a singular site-limited and relatively constant energy process; i.e., a Langmuir-type adsorption. The DRM treats the non-linear component of adsorption as a set of multiple reactions involving different sites of different energy, thus manifesting Freundlich-type behavior, i.e., the summation of several Langmuir-type adsorptions (*37*).

Figures 5 and 6 illustrate the utility of the DRDM for describing sorption in rubbery and glassy matrices. At 5°C, both the humic acid (T_g = 43°C, water-wet) and a synthetic poly(isobutyl methacrylate (PIMA) polymer (T_g = 50°C, water-wet) display significant nonlinear contributions to the overall sorption isotherm, while at 45°C (i.e., at or near their water-wet T_g values) these non-linear contributions disappear for the PIMA, and are several orders of magnitude less than the linear contributions for the humic acid, giving rise to almost complete partitioning behavior, consistent with the existence of a predominantly rubbery state.

Desorption. Apparent hysteresis in the sorption/desorption process can originate from a number of sources, including: i) experimental artifacts such as solute losses to the experimental system (e.g., solids effect, sorption to reactor walls, volatile losses through caps, and sampling errors) (*22*); ii) abiotic reactions such as polymerization of the sorbing solute to the organic matter (*23*); iii) nonequilibrium conditions (*22, 48*); iv) changes in equilibrium macromolecular configurations from the sorption process to the desorption process (*49, 50*); and, v) immobilization of solute within fixed microvoids present in glassy matrices (*21, 49*). Experimental artifacts and abiotic reactions can generally be eliminated by the careful design of experimental systems, leaving the last three circumstances listed above as process related sources of observed apparent hysteresis.

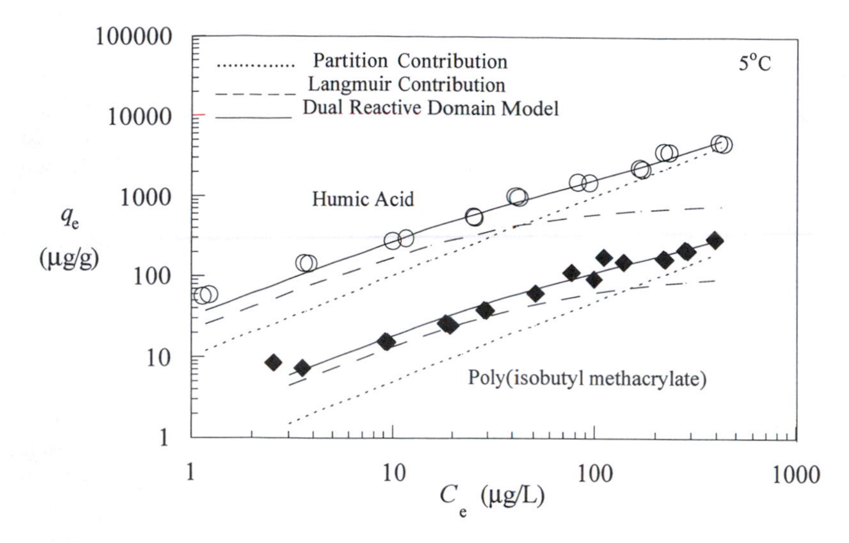

Figure 5. Dual Reactive Domain Model characterization of the sorption behavior of humic acid and poly(isobutyl methacrylate) at 5°C (Adapted from ref *32*).

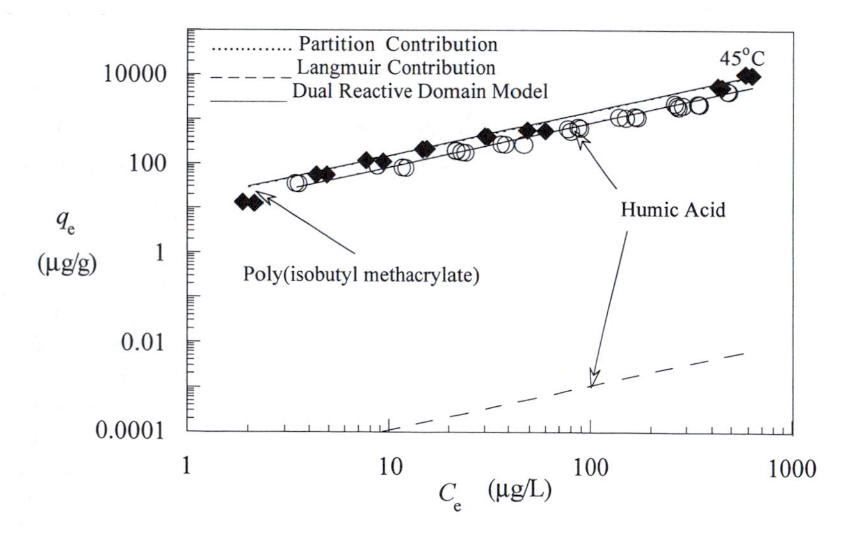

Figure 6. Dual Reactive Domain Model characterization of the sorption behavior of humic acid and poly(isobutyl methacrylate) at 45°C (Adapted from ref *32*).

A recent study reported by LeBoeuf and Weber (*49*) illustrates the impact of rubbery and glassy matrices on long term sorption/desorption behavior of phenanthrene. In that study, both sorption and desorption isotherm data for these systems were fit to the Freundlich equation, as a simple means for characterizing relative linearity

$$q_e = K_F C_e^{\,n} \tag{9}$$

where all parameters are as defined previously. To quantify hysteresis phenomena, we have defined a residual-concentration-specific relative apparent sorption-desorption Hysteresis Index (HI) (*21-23*):

$$\text{Hysteresis Index} = \left.\frac{q_e^d - q_e^s}{q_e^s}\right|_{T,\,C_e} \tag{10}$$

where q_e^s and q_e^d are solid phase solute concentrations for single-cycle sorption and desorption, respectively, and the subscripts T and C_e specify constant conditions of temperature and residual solution phase concentration, respectively.

Figure 7 shows there is little to no apparent desorption hysteresis (HI = 0.07 at $C_e = 100$ μg/L) for the rubbery sorbent cellulose. Figure 8 displays prominent sorption-desorption hysteresis (HI = 1.01 at $C_e = 100$ μg/L) for glassy poly(phenyl methacrylate). This significant finding suggests that resistant sorption processes observed for organic compounds in natural particles are likely the results of sorption and desorption processes associated with the more condensed, glass-like organic matter domain characteristic of more diagenetically altered soil and sediment organic matter.

Sorption and Desorption by Natural Geosorbents

Ends of Spectrum Examples. Two geosorbents, Canadian peat and Lachine shale, are selected here to demonstrate distinctly different sorption and desorption behavior by amorphous and condensed SOM matrices, respectively. Canadian peat is a brownish-colored fiber-dominated younger horticultural soil having total organic carbon (TOC) content of 47.5 wt% and ash content of 4 wt%. Its SOM characterized by a high oxygen to carbon (O/C) atomic ratio of ~0.7, is representative of the early humification products of cellulose-like biopolymers (*21*). The Canadian peat is selected as an example of a highly amorphous and rubbery SOM. Lachine shale, collected from the Paxton Quarry west of Alpena, Michigan, is a diagenetically-altered geosorbent having a TOC content of 8.3 wt% and a geological age of 365 to 375 million years (*21*). The organic matter (i.e., kerogen) isolated from this solid has a low O/C atomic ratio of 0.13, representative of the diagenetic products of biopolymers of lower-level living organisms such as phytoplankton and zooplankton. The shale sample is selected as an example of a condensed and glassy SOM.

Figures 9 and 10 show phenanthrene sorption and desorption isotherms measured for the peat and the shale. The values of organic carbon-normalized K_F are 21.7 and

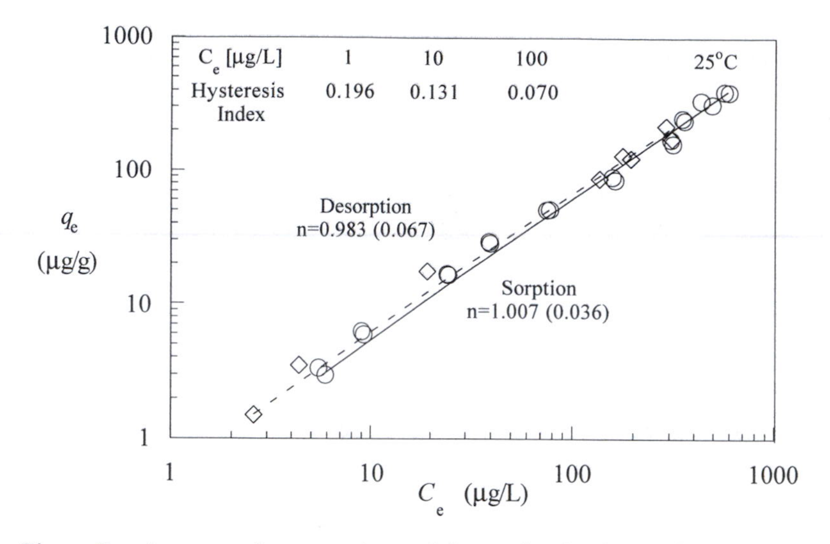

Figure 7. Aqueous-phase sorption and desorption isotherms for rubbery cellulose (T_g = −45°C) (Adapted from ref *49*).

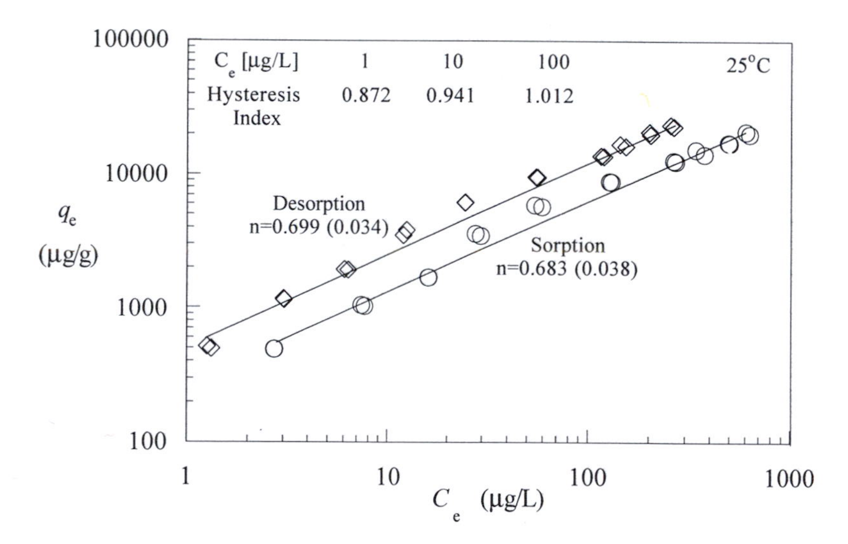

Figure 8. Aqueous-phase sorption and desorption isotherms for glassy poly(phenyl methacrylate) (T_g = 110°C) (Adapted from ref *49*).

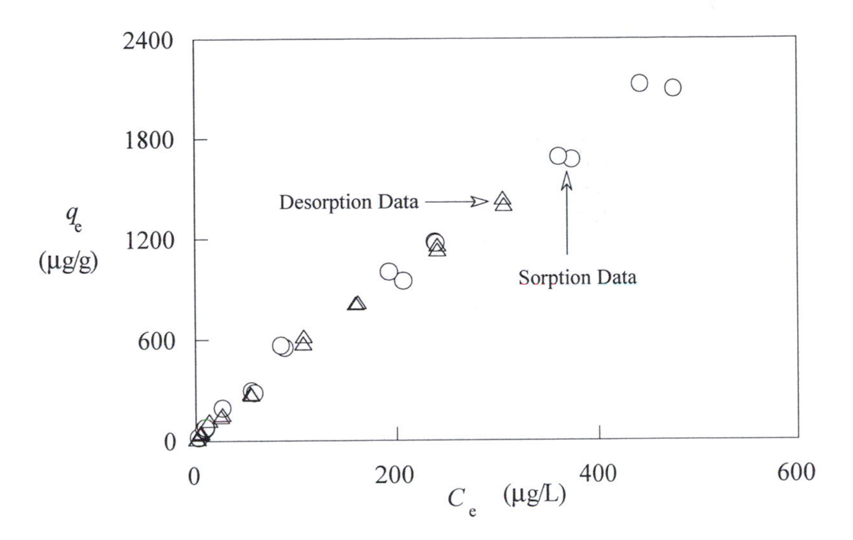

Figure 9. Phenanthrene sorption and desorption data for Canadian peat (sorption time = 21 days; desorption time = 14 days) (Adapted from ref *21*).

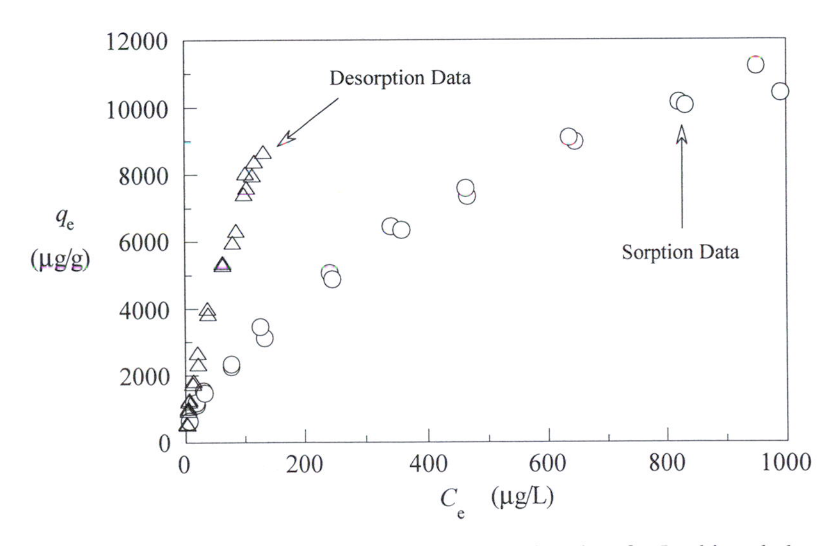

Figure 10. Phenanthrene sorption and desorption data for Lachine shale (sorption time = 28 days; desorption time = 14 days) (Adapted from ref *21*).

2801.7 μg/g-OC/(μg/L)^n, the values of isotherm linearity, n, are about 0.88 and 0.53, and the HI values calculated at C_e = 100 μg/L are 0.0 and 0.92 for the peat and the shale, respectively. These results indicate that the condensed SOM has much higher sorptive affinity for phenanthrene than does the amorphous SOM, and that sorption by the amorphous SOM is nearly linear and completely reversible whereas sorption by the condensed SOM is highly nonlinear and apparently hysteretic.

It is apparent that the distinctly different sorption-desorption behaviors observed for the two solids relate to differences in the physical and chemical characteristics of the associated SOMs. The SOM associated with the peat comprises primarily oxygen-containing functional groups such as carbohydrates, as revealed by ^{13}C-NMR analysis (*21*). These functional groups can interact with water molecules so strongly that the SOM matrices become swollen and amorphous under water-saturated conditions. Sorption into these SOM matrices follows a linear partitioning process having no history dependent behavior (i.e., no apparent hysteresis). Conversely, the SOM associated with the shale comprises primarily aromatic and aliphatic oxygen-free functional groups, again as shown by ^{13}C-NMR analysis (*21*). These functional groups favor interactions with HOC molecules rather than with water. Under water-saturated conditions the SOM matrices remain rigid structures and exhibit strong affinities for HOCs because of their hydrophobic nature. Sorption into such matrices follows simultaneous external/internal surface adsorption and matrix penetration processes. The adsorption process is limited by the total number of "sites" available and the overall sorption is thus nonlinear (Langmuir-like). The apparent sorption-desorption hysteresis observed for the shale sample is likely attributable to both slow rates of sorption and desorption and long-term (essentially irreversible) entrapment of sorbing molecules within condensed SOM matrices; i.e., mechanisms (iii) to (v) described earlier.

General Behavior of a Broad Range of Geosorbents. We have measured phenanthrene sorption and desorption isotherms for a total of 33 different soils, sediments, shales and kerogens (*19, 21–23*). The results show that: i) sorption isotherms for all sorbents tested are nonlinear, with n values ranging from 0.43 to 0.91; ii) apparent sorption-desorption hysteresis indices vary from 0.0 for the geosorbents having the least diagenetically-altered SOM to > 1.0 for the geosorbents having the most diagenetically-altered SOM; and, iii) organic carbon normalized-sorption capacities vary from sample to sample. Single-point K_{OC} values calculated for all 33 geosorbents from best-fit sorption isotherm parameters using Equations 2 and 9 range from 6.1 to 210 L/g-OC at C_e = 1000 μg/L and from 17.8 to 5123 L/g-OC at C_e = 1 μg/L. Average K_{OC} values calculated for 20 soils and sediments are 13 ± 4 and 59 ± 29 L/g-OC at C_e = 1000 and 1 μg/L, respectively. It should be noted that the predicted K_{OC} value for phenanthrene based on log K_{OC} – log K_{OW} and log K_{OC} – log C_S correlations (*24-31*) ranges from 6 to 18 L/g-OC. While some of the correlation-predicted K_{OC} values compare favorably with the average K_{OC} value (13 (± 4) L/g-OC at C_e = 1000 μg/L) for soils and sediments, all are approximately an order of magnitude smaller than the K_{OC} values calculated for shales and kerogens over a

C_e range of 1–1000 μg/L, and several times lower than the average K_{OC} value for the soil and sediment samples at C_e = 1 μg/L.

An important finding of this study is that sorption and desorption behavior can be correlated with the oxygen/carbon (O/C) atomic ratio of the associated SOM. Statistical regressions of K_{OC} values calculated at C_e = 1 and 1000 μg/L, *n,* and HI values against O/C atomic ratios for seven organic-rich geosorbents (three peats, three kerogens and one humic acid) yield the following four correlations (*21*)

$$\log K_{OC} = 4.39 - 4.63\,(O/C) \quad \text{at } C_e = 1\ \mu g/L; \qquad R^2 = 0.918 \qquad (11)$$

$$\log K_{OC} = 2.62 - 2.54\,(O/C) \quad \text{at } C_e = 1000\ \mu g/L; \qquad R^2 = 0.826 \qquad (12)$$

$$n = 0.409 + 0.704\,(O/C); \qquad R^2 = 0.911 \qquad (13)$$

$$HI = 0.774 - 1.09\,(O/C) \quad \text{at } C_e = 100\ \mu g/L; \qquad R^2 = 0.979 \qquad (14)$$

The above correlations define sorption/desorption behavior in the context of the chemical and physical characteristics of soil organic matter evolved under different physical, chemical, and biological conditions in different stages of diagenesis occurring over geological time. It should be noted that correlations between linear model-based K_{OC} and chemical characteristics of SOM similar to those given in Equations 11 and 12 have been reported by others (*33-36*). The correlations given above, however, indicate that K_{OC} values not only are functions of the O/C ratio of given associated SOM, but also that they are dependent upon aqueous-phase solute concentration. Correlations 13 and 14 suggest further that sorption isotherm nonlinearity and apparent sorption-desorption hysteresis are phenomenologically interrelated; i.e., the more nonlinear the sorption isotherm, the greater the apparent sorption-desorption hysteresis.

Sorption by Exposed Inorganic Mineral Domains

Inorganic mineral matrices are often the major bulk constituent (>90 wt%) of soils and sediments, and these may in some cases play important roles in equilibria and rates of HOC sorption. Schwarzenbach and Westall (*30*) have shown that exposed external and internal pore surfaces of inorganic mineral domains contribute significantly to equilibrium sorption of HOCs if a soil or sediment has a low f_{OC} value (e.g., < 0.001). It is widely believed that the diffusion of HOC molecules into meso- (pore diameter = 20–500 Å) and micropores (pore diameter < 20 Å) is very slow, and thus responsible for the overall slow rates of sorption typically exhibited by soils and sediments.

To investigate the capacity and rate roles of the mineral domain, we examined sorption and desorption of phenanthrene using seven pure inorganic model solids as sorbents (*51*). The solids tested include nonporous amorphous SiO_2, quartz, α-Al_2O_3, kaolinite, and three porous silica gels (silica gel-40, -100, and -150) having average pore diameters of 40, 100, 119 Å, and their N_2-gas BET specific surface areas (SSA) are 8.2, 6.5, 13.3, 12.2, 521. 314, 314 m^2/g, respectively.

The results showed that for all the systems examined: i) sorption is fast and reversible; ii) sorption and desorption isotherms are nearly linear, with n values ranging from 0.93 to 1.10; iii) surface-area-based apparent K_D values for the four non-porous sorbents increase in order from the amorphous SiO_2 (5.6×10^{-5} L/m^2), to kaolinite (9.4×10^{-5} L/m^2), quartz (1.8×10^{-4} L/m^2), and α-Al_2O_3 (2.6×10^{-4} L/m^2); and, iv) iii) despite their remarkably different pore sizes and pore size distributions, the three porous silica gels exhibit coincident sorption isotherms having apparent K_D values of 0.86 to 1.1×10^{-5} L/m^2. The mass-based K_D value (1.2×10^{-3} L/g) for quartz is equivalent to that of a soil organic carbon fraction (f_{OC}) of < 0.01 wt% assuming K_{OC} >12 L/g-OC. This accords with the results of a study by Ball and Roberts (*52*), suggesting that non-expandable mineral surfaces may not be as important as commonly believed for sorption of organic contaminants on low-organic sandy aquifer materials.

A typical pattern of change in the time-dependent SSA-normalized distribution coefficient, $K_D(t)$, as a function of t is shown in Figure 11 for α-Al_2O_3, silica gel-40 and -100. It is clear from this Figure that sorption is completed within a few minutes, regardless of that facts that: i) the two gel samples have enormous internal surface areas and ii) the particles of these solids have diameters of 63–150 μm. Complete reversible sorption is shown in Figure 12 for silica gel-40, in which both sorption and desorption isotherms coincide, indicating no hysteresis.

The low SSA-based K_D values and fast rates of sorption exhibited by the three mesoporous silica gels suggest that the majority of the surface areas internal to gel particles are not accessible by phenanthrene under water-wet conditions because water is preferentially adsorbed on hydrophilic surfaces, forming a highly structured water phase within pores. Phenanthrene molecules can not enter such pores due to entropic effects.

Conclusions

The study summarized in this paper supports a hypothesis that the sorbing components of soils and sediments comprise three principally different types of domains; a highly amorphous SOM domain, a relatively condensed SOM domain, and an exposed mineral surface domain. Discovery of a glass transition for a soil-derived humic acid provides the necessary bridge to link observed sorption behaviors of natural and synthetic macromolecules to their rubbery and glassy thermodynamic states. Extensive examinations of phenanthrene sorption and desorption for a broad range of soils, sediments, shales, kerogens, model inorganic sorbents, and synthetic organic polymers reveals that: i) sorption by condensed SOM matrices is nonlinear and apparently hysteretic; ii) sorption by amorphous SOM domains and mineral surfaces is linear and completely reversible; iii) mineral domains contribute little to the overall sorption and desorption of hydrophobic organic compounds by soils and sediments. The three-domain model has been found to quantitatively describe non-ideal partitioning sorption phenomena in subsurface systems. The model additionally provides insight to, and better understanding of, the binding and sequestration of organic contaminants by soils and sediments.

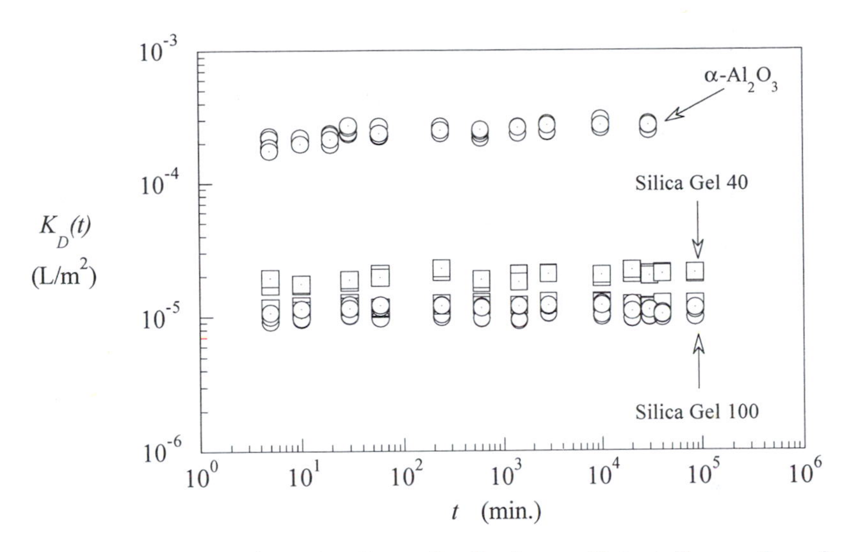

Figure 11. Time-dependent linear distribution coefficients for sorption of by α-Al_2O_3, silica gel 40, and silica gel 100. Symbols with a dot at the center represent CMBR systems initiated at an aqueous-phase phenanthrene concentration of 50 µg/L; open symbols represent CMBR systems initiated at 900 µg/L (Adapted from ref *51*).

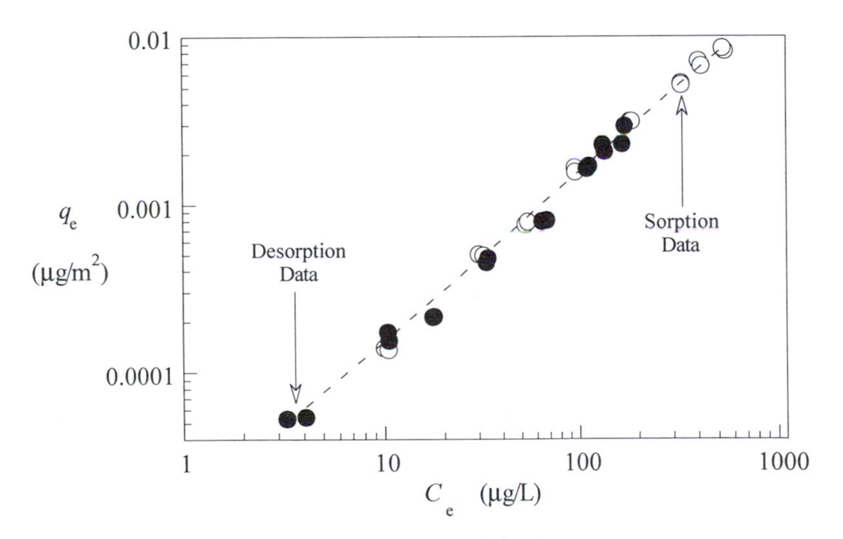

Figure 12. Phenanthrene sorption and desorption data for silica gel-40 (sorption time = 14 days; desorption time = 14 days).

Acknowledgments

This research was funded in part by the U. S. Environmental Protection Agency, Office of Research and Development and Grant R-819605 to the Great Lakes and Mid-Atlantic Center (GLMAC) for Hazardous Substance Research. Partial funding of the research activities of GLMAC is also provided by the State of Michigan Department of Environmental Quality and by a fellowship grant to E.J.L. from the Regents of The University of Michigan.

Literature Cited

(1) Weber, W.J., Jr.; McGinley, P.M.; Katz, L.E. *Water Res.* **1991**, *25*, 499-528.
(2) NRC, *Alternative for Ground Water Cleanup*, National Academy Press, Washington DC, 1994.
(3) Alexander, M. *Environ. Sci. Technol.* **1995**, *29*, 2713-2717.
(4) Di Toro, D. M.; Horzempa, L. M. *Environ. Sci. Technol.* **1982**, *16*, 594-602.
(5) Mingelgrin, U; Gerstl, Z. *J. Environ. Qual.* **1983**, *12*, 1-11.
(6) Miller, C. T.; Weber, W. J., Jr. *J. Contam. Hydrol.* **1986**, *1*, 243-261.
(7) Steinberg, S. M.; Pignatello, J. J.; Sawhney, B. L. *Environ. Sci. Technol.* **1987**, *21*, 1201-1208.
(8) Pignatello, J. J. *Environ. Toxicol. Chem.* **1990**, *9*, 1107-1115.
(9) Ball, W. P.; Roberts, P. V. *Environ. Sci. Technol.* **1991**, *25*, 1237-1249.
(10) Pavlostathis, S. G.; Mathavan, G. N. *Environ. Sci. Technol.* **1992**, *26*, 532-538.
(11) Weber, W. J., Jr.; McGinley, P. M.; Katz, L. E. *Environ. Sci. Technol.* **1992**, *26*, 1955-1962.
(12) Carroll, K. M.; Harkness, M. R.; Bracco, A. A.; Balcarcel, R. R. *Environ. Sci. Technol.* **1994**, *28*, 253-258.
(13) Kan, A. T.; Fu, G.; Tomson, M. B. *Environ. Sci. Technol.* **1994**, *28*, 859-867.
(14) Young, T. M.; Weber, W. J., Jr. *Environ. Sci. Technol.* **1995**, *29*, 92-97.
(15) Pignatello, J. J.; Xing, B. *Environ. Sci. Technol.* **1996**, *30*, 1-11.
(16) Weber, W. J., Jr.; Huang, H. *Environ. Sci. Technol.* **1996**, *30*, 881-888.
(17) Xing, B.; Pignatello, J. J.; Gigliotti, B. *Environ. Sci. Technol.* **1996**, *30*, 2432-2440.
(18) Allen-King, R. M.; Groenevelt, H.; Warren, C. J.; Mackay, D. M. *J. Contam. Hydrol.* **1996**, *22*, 203-221.
(19) Huang, W.; Young, T. M.; Schlautman, M. A.; Weber, W. J., Jr. *Environ. Sci. Technol.* **1997**, *31*, 1703-1710.
(20) Jafvert, C. T.; Vogt, B. K.; Fabrega, J. R. *J. Environ. Engin.* **1997**, *123*, 225-233.
(21) Huang, W.; Weber, W. J., Jr. *Environ. Sci. Technol.* **1997**, *31*, 2562-2569.
(22) Huang, W.; Yu, H.; Weber, W. J., Jr. *J. Contaminant Hydrol.* 1997, (in press).
(23) Weber, W. J., Jr.; Huang, W.; Yu, H. *J. Contaminant Hydrol.* 1997, (in press).
(24) Chiou, C. T.; Peters, L. J.; Freed, V. H. *Science* **1979**, *206*, 831-832.
(25) Karickhoff, S. W.; Brown, D. S.; Scott, T. A. *Water Res.* **1979**, *13*, 241-248.
(26) Karickhoff, S. W. *Chemosphere* **1981**, *10*, 833-846.
(27) Chiou, C. T.; Porter, P. E.; Schmedding, D. W. *Environ. Sci. Technol.* **1983**, *17*, 227-231.
(28) Karickhoff, S. W. *J. Hydraul. Eng.* **1984**, *110*, 707-735.

(29) Means, J. C.; Wood, S. G.; Hassett, J. J.; Banwart, W. L. *Environ. Sci. Technol.* **1980**, *14*, 1524-1528.

(30) Schwarzenbach, R. P.; Westall, J. *Environ. Sci. Technol.* **1981**, *15*, 1360-1367.

(31) Means, J. C.; Wood, S. G.; Hassett, J. J.; Banwart, W. L. *Environ. Sci. Technol.* **1982**, *16*, 93-98.

(32) LeBoeuf, E. J.; Weber, W. J., Jr. *Environ. Sci. Technol.*, **1997**, *31*, 1696-1702.

(33) Garbarini, D. R.; Lion, L. W. *Environ. Sci. Technol.* **1986**, *20*, 1263-1269.

(34) Gauthier, T. D.; Seitz, W. R.; Grant, C. L. *Environ. Sci. Technol.* **1987**, *21*, 243-248.

(35) Grathwohl, P. *Environ. Sci. Technol.* **1990**, *24*, 1687-1693.

(36) Rutherford, D. W.; Chiou, C. T.; Kile, D. E. *Environ. Sci. Technol.* **1992**, *26*, 336-340.

(37) Weber, W. J., Jr.; DiGiano, F. A. *Process Dynamics in Environmental Systems*, John Wiley & Sons: New York, 1996; Chapters 6 and 11.

(38) Carter, M. C.; Kilduff, J. E.; Weber, W. J., Jr. *Environ. Sci. Technol.* **1995**, *29*, 1773-1780.

(39) Durand, B. *Kerogen: Insoluble Organic Matter from Sedimentary Rocks;* Technip: Paris, 1980.

(40) Tissot, B. P.; Welte, D. H. *Petroleum Formation and Occurrence;* Springer-Verlag: New York, 1984.

(41) Engel, M. H.; Macko, S. A. *Organic Geochemistry: Principles and Applications;* Plenum Press: New York, 1993.

(42) McKenna, G. B. *Comprehensive Polymer Science: Vol. 2, Polymer Properties*, Booth, C.; Price, C., Eds.; Pergamon: Oxford, 1989; Chap. 10.

(43) Barton, A. F. M. *Handbook of Solubility Parameters and Other Cohesion Parameters*; CRC Press: Boca Raton, FL, 1983.

(44) Grulke, E. A. *Polymer Handbook*, Brandrup, J.; Immerfut, E. H., Eds.; Wiley-Interscience: New York, 1989.

(45) Curtis, G. P.; Reinhard, M.; Roberts, P. V. *Geochemical Processes at Mineral Surfaces*, Davis, J. A.; Hayes, K. F., Eds.; ACS Symposium Series 323, American Chemical Society, Washington, D.C., 1986.

(46) Lucht, L. M.; Larson, J. M.; Pepas, N. A. Energy Fuels **1987**, *1*, 56.

(47) Akim, E. L. *Chemtech* **1978**, *8*, 676.

(48) Miller, C. T.; Pedit, J. A. *Environ. Sci. Technol.* **1992,** *26*: 1417-1427.

(49) LeBoeuf, E. J.; Weber, W. J., Jr., 214th ACS National Meeting, Vol. 37, No. 2, 1997.

(50) Pignatello, J. J. (1989) In *Reactions and Movement of Organic Chemicals in Soils* (Edited by Sawhney, B. L.; Brown, K.), Soil Science Society of America, Special Publication No. 22, Madison, WI.

(51) Huang, W.; Schlautman, M. A.; Weber, W. J., Jr. *Environ. Sci. Technol.* **1996**, *30*, 2993-3000.

(52) Ball, W. P.; Roberts, P. V. *Environ. Sci. Technol.* **1991**, *25*, 1223-1237.

Heterogeneous Electron Transfer

Chapter 12

Interfacial Kinetics Through the Lens of Solution Chemistry: Hydrolytic Processes at Oxide Mineral Surfaces

William H. Casey[1], Brian L. Phillips[2], and Jan Nordin[1]

[1]**Department of Land, Air, and Water Resources, Department of Geology, and** [2]**Material Science and Chemical Engineering, University of California, Davis, CA 95616**

Mineral dissolution and growth can be understood, in part, by drawing analogies with reactions where a dissolved metal exchanges ligands in solution. The rates of dissolution are generally controlled by slow hydrolytic dissociation of surface complexes at monomolecular steps on the mineral surfaces. Rates of oxygen exchange around metals in dissolved metal-ligand complexes also involve hydrolysis. In both environments these rates depend on association to protons, or certain ligands, in the inner-coordination sphere of the metal. An enormous literature of useful data describes microscopic acid-base chemistry of oxide oligomers and the molecular controls on their condensation and dissociation. This literature may be extraordinarily useful in understanding oxide interfacial chemistry. It is presently unknown whether rate coefficients from the solution state can be directly transferred to surface reactions, but useful comparisons can be made to identify reactivity trends.

Geochemistry is much simpler if one ignores the shallow Earth. At elevated temperatures and pressures, such as those in the deeper crust of the Earth, equilibrium thermodynamics gives meaningful predictions of solution chemistries and mineral assemblages. Many reactions in the shallow Earth, however, do not proceed to equilibrium and time is a critical variable. Fluids exist in pores that are interstitial to minerals and rates of cycling of organic matter, degradation of pesticides, metal mobilities, and the diagenesis of minerals are influenced by interactions between solutes and surfaces. Reactions at surfaces modulate the flow of toxicants and nutrients to, and from, the biosphere.

Most reactions at mineral surfaces either add or remove mass from the mineral and can therefore be thought of as forms of dissolution or growth. Surface complexes form when ligands in solution combine with surface metals or where metal solutes adsorb and interact with ligand functional groups at the mineral surface. No techniques exist for determining elementary reaction rates in these surface complexes and the bulk of our current understanding arises from measurements of rates of net mass transfer. We can, however, rank the rates of the reactions by examining other better-constrained processes, including elementary reactions in dissolved complexes that are models for mineral surface complexes.

In this paper we discuss the hypothesis that depolymerization of metal-ligand adsorbates, and the dissolution of minerals in general, can be described using a ligand-exchange framework and existing data on metal-(hydr)oxide oligomers. A wide range of possible oligomers have been prepared and characterized. Furthermore, it is possible to determine microscopic equilibrium constants for individual Brønsted acid-base reactions and rate coefficients for elementary reactions in oligomers in ways that are, so far, impossible for extended surface structures on minerals [Table 1]. One can construct Linear-Free Energy Relations (LFER) to predict rate coefficients for more complicated processes, like the dissolution of an oxide mineral or the adsorption of a metal to the oxide surface, using either the elementary reaction rate coefficients or equilibrium thermodynamic parameters.

One particularly important class of reactions is the exchange of water between the bulk solvent and inner-coordination sphere of a dissolved metal complex. Not only is water elimination a direct, rate-controlling step in some key geochemical processes, but the dissociation rate of structurally similar bonds is a useful measure of reactivity for other more complicated paths that may not involve elimination of water. The water-exchange reactions are also *elementary*--that is, they proceed as written on the molecular scale and the rates range over 10^{13} from diffusion-controlled rates for alkali metals to rates of exchange that take many months for inert transition metals, like Cr(III). Exchange of oxygens in some oxyanions and oligomers are slow even on a geologic time scale. Knowledge of the rates of water exchange is at the most fundamental level of reaction kinetics.

Structural Considerations

Formalism. Surface complexes have features in common with both solutes and solids and the mineralogic description of metals and oxygens is limiting. We here employ the formalism of structural inorganic chemistry that is used to describe oligomers, where μ_i sites bridge '*i*' metals and η^i sites are nonbridging sites with '*i*' ligand atoms bonding to the metal (see 2, p. 38). Hydration waters in $Mn(H_2O)_6^{2+}(aq)$, for example, are each denoted η^1-OH_2 and the hydroxyl bridge in the dimer $(H_2O)(en)_2Cr(OH)Cr(en)_2(H_2O)$ is denoted a μ_2-OH site. We use the symbol '<' and '>' to denote surface sites. We apply this formalism to sites at monomolecular steps because scanning-probe microscopy (Figure 1) shows us that these steps are the most reactive sites on a mineral surface (1).

Table 1: Variation in the rates of hydrolytic processes with the OH/Cr(III) ratios in monomers and [a]oligomers at 298 K and I=1.0 M (from 33).

reactant	OH/Cr(III)	10^5·rate (s^{-1})	Comment
	1. Exchange of Water from Inner-coordination sphere to bulk		
$Cr(H_2O)_6^{3+}$	0	0.24	Fully-hydrated monomer
$Cr(OH)^{2+}$	1.0	18	Hydrolyzed monomer
$[Cr(\mu\text{-}OH)_2Cr]^{4+}$			
trans to μ-OH	1	36	Doubly-bridged dimer, fully hydrated
cis to μ-OH	1	6.6	Doubly-bridged dimer, fully hydrated
$[Cr(\mu\text{-}OH)_2CrOH]^{3+}$			
trans to μ-OH	1.5	1260	Doubly-bridged dimer, hydrolyzed
cis to μ-OH	1.5	490	Doubly-bridged dimer, hydrolyzed
	2. Intramolecular Bridge Formation		
$[Cr(\mu\text{-}OH)Cr]^{5+} + H_2O$			
→ DBD + H^+	0.5	10	SBD→DBD; fully hydrated reactant
$[Cr(\mu\text{-}OH)CrOH]^{4+} + H_2O$			
→ DBD + H^+	1.0	40	SBD→DBD; singly hydrolyzed reactant
$[HOCr(\mu\text{-}OH)CrOH]^{3+} + H_2O$			
→ DBD-H	1.5	1140	SBD→DBD; doubly hydrolyzed reactant
$[Cr_4(OH)_6]^{6+}$→			
closed tetramer	1.5	8700	
$[Cr_4(OH)_7]^{5+}$→			
closed tetramer	1.75	24000	
	3. [b,c]Sulfate anation		
$Cr(H_2O)_6^{3+}$	0	1.1	Hydrated monomer
$Cr(OH)^{2+}$	1.0	61	Hydrolyzed monomer
$[Cr(\mu\text{-}OH)_2Cr]^{4+}$	1	35	Hydrated DBD
$[Cr(\mu\text{-}OH)_2CrOH]^{3+}$	1.5	1700	Hydrolyzed DBD

[a]The abbreviation SBD and DBD correspond to a singly-bridged dimer and doubly bridged dimer, respectively; e.g., $[Cr(\mu\text{-}OH)Cr]^{5+}$ and $[Cr(\mu\text{-}OH)_2Cr]^{4+}$.

[b]Anation refers to exchange of a water molecule in the inner-coordination sphere by an anion, usually monodentate, such as sulfate.

[c]The rate coefficients are pseudo-first order for these reactions because the precursor complex is an outer-sphere ion pair including both the hydrated ion pair and the anion; this ion pair exists at equilibrium concentrations.

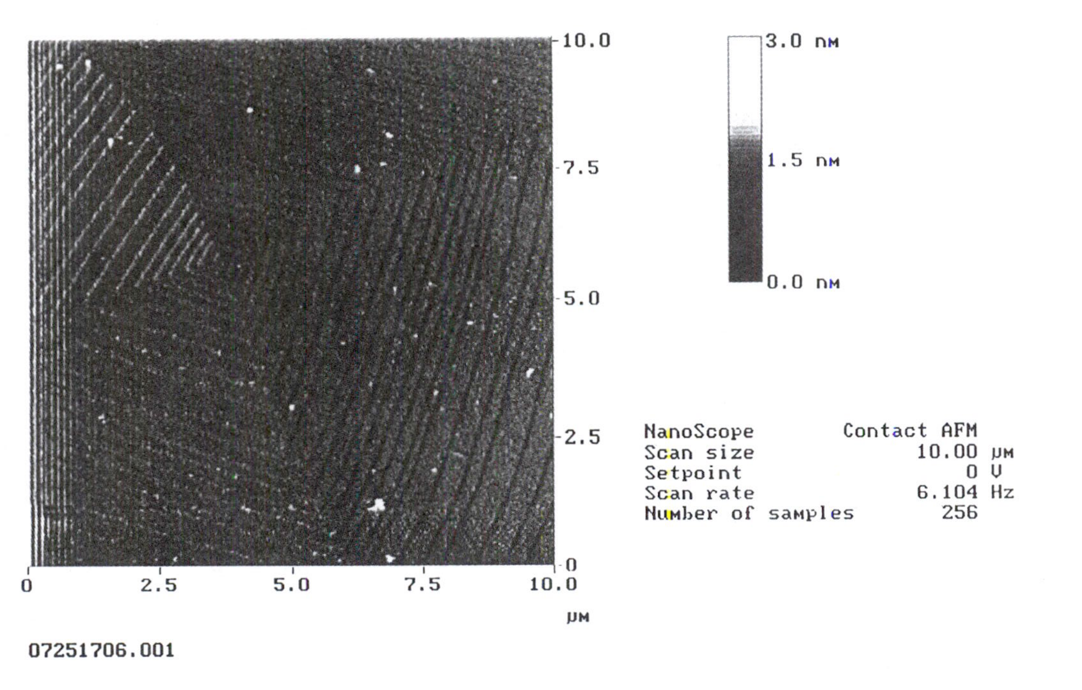

Figure 1. An atomic-force microscope image showing monomolecular steps being generated at a dislocation on the surface of calcite (image courtesy of R. Shiraki, T. Land and P. Rock). The steps grow and retreat as mass is accumulated or lost at these steps.

Structural Similarities and Dissimilarities Between Minerals and Dissolved Oligomers. There are a couple of general points to remember when comparing solutes and surfaces:

1) The number and distance to nearest neighbors from the metal in surface complexes is commonly similar to those in the solute.

2) Mineral surfaces contain oxygens with higher coordination numbers to metals than in dissolved complexes; this is where attention should be focused.

For example, the Mn(II) metal center is hexacoordinated to oxygens in the $Mn(H_2O)_6^{2+}(aq)$ complex, at the MnO(s) surface, and even at the surface of a molecular solid like $Mn_2SiO_4(s)$. Therefore, the <Mn-O> bond lengths are roughly 2.20Å in each environment. While it is possible to create undercoordinated metals in high vacuum, [such as a five- or four-coordinated Mn(II)], in an aqueous solution these defects are instantaneously reacted away to give metals with full coordination spheres. A few metals, such as Zn(II) and Al(III), can change coordination numbers to oxygen as they pass from a solid to a solute, but most metals have the same number of coordinated oxygens both at the surface, in the solid, and in solution.

In contrast, oxygens in $Mn(H_2O)_6^{2+}(aq)$ are coordinated to a single metal (excluding protons) while those as monomolecular steps on even simple surface steps (Figure 2) range from μ_6-O to η^1-OH_2. In Figure 2 a monomolecular step on the [100] surface of MnO(s) is shown with the coordination numbers of oxygens to metals labeled. Only one site, the dangling η^1-OH_2 site labeled '1' in Figure 2, is analogous to oxygens in the dissolved $Mn(H_2O)_6^{2+}(aq)$ complex. In only one case has a dissolved complex been observed with a μ_5-O site (2) and no μ_6-O sites have been reported, yet these sites are not uncommon on surfaces (Figure 2)'.

The coordination of these oxygens at the monomolecular step will, of course, change as the mineral reacts with the aqueous solution because the MnO_6 octahedra shown with the wireframe (Figure 2) must detach as a surface complex as the step migrates. This detachment causes newly uncovered metals to reestablish their inner-coordination spheres by movement and deprotonation of water molecules. If the reaction proceeds at steady state, these water molecules dissociate to maintain a fixed charge density and a fixed numbers of different metal-ligand coordination numbers (7, 8).

Low-symmetry solids expose a more narrow range of coordination numbers than high-symmetry solids, and the coordination chemistry varies with the structural orientation of the exposed surface. The low-symmetry Bayerite surface, for example, has only terminal η^1-OH_2 sites and μ_2-OH sites (Figure 3-middle) and not the rich array of coordination chemistries illustrated in Figure 2, where μ_2- through μ_6-oxo sites surround each metal. Assigning microscopic Brønsted acidity constants to these oxygens is difficult and is the subject of much current research (e.g., 9,10).

A Useful Comparison. There are striking similarities in the local bonding configuration of <u>some</u> surface sites and <u>some</u> dissolved complexes. Compare, for example, the structure of the $Al_{13}O_4(OH)_{24}(H_2O)_{12}^{+7}(aq)$ oligomer (Figure 3-top) with the [010] surface of Bayerite [β-$Al(OH)_3(s)$] (Figure 3-middle) and the [001] surface of corundum [α-Al_2O_3] (Figure 3-bottom). Both the dissolved oligomer and the surfaces are dominated by six-membered rings of shared AlO_6 octahedra. The oligomer has only edge-shared octahedra that are each bound to an AlO_4 tetrahedron, while the α-Al_2O_3 structure includes shared octahedral faces perpendicular to [001] as well as edge-shared octahedra.

The most striking similarities between the oligomer and surface structures lie with the configuration of surface oxygen sites. At monomolecular steps on α-$Al_2O_3(s)$ and $AlOOH(s)$, there exist terminal waters (η^1-OH_2) on the kink octahedra as well as protonable oxo- or hydroxo-bridges linking octahedra together. These sites are nearly identical to those found on the $Al_{13}O_4(OH)_{24}(H_2O)_{12}^{+7}(aq)$ oligomer (Figure 3-top). On the apices of the AlO_6 octahedra in the $Al_{13}O_4(OH)_{24}(H_2O)_{12}^{+7}(aq)$ oligomer are twelve -OH_2 sites (η^1-OH_2) that can undergo Brønsted acid-base reactions with the aqueous phase. There are also two types of μ_2-OH sites and one type of μ_4-O site on the $Al_{13}O_4(OH)_{24}(H_2O)_{12}^{+7}(aq)$ oligomer (Figure 3-top).

How could this be useful? By hypothesis, knowledge of the rates of exchange of η^1-OH_2 sites on the $Al_{13}O_4(OH)_{24}(H_2O)_{12}^{7+}(aq)$ tridecamer can be applied to estimate the rates of other surface reactions. For example, consider metal adsorption where the η^1-OH_2 groups on the oxide minerals, or the conjugate hydroxyl bases (η^1-OH sites), are loci for adsorption of metals as inner-sphere complexes. The key step controlling the adsorption rate is dehydration of a precursor, outer-sphere complex to form a strong oxo- or hydroxo-bridge between the surface and adsorbate metals (e.g., 5, 20, 26). Dehydration can occur by elimination of a water from either the inner-coordination sphere of the adsorbing metal or the -OH_2 site coordinated to a metal at the mineral surface. In this example we arbitrarily assign $k = 10^3\ s^{-1}$ for the rate of exchange of water from the η^1-OH_2 sites on an oxide mineral, such as the sites shown in Figure 3, and bulk solution:

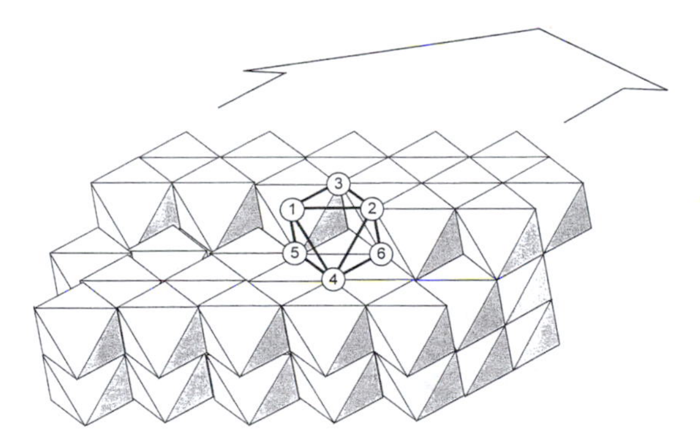

Figure 2. A retreating monomolecular step on the [100] surface of an oxide mineral with the rocksalt structure, such as MnO(s). Numbers identified in the wireframe diagram indicate the coordination numbers of oxygens to metal; i.e., the number 4 identifies a μ_4-oxo site.

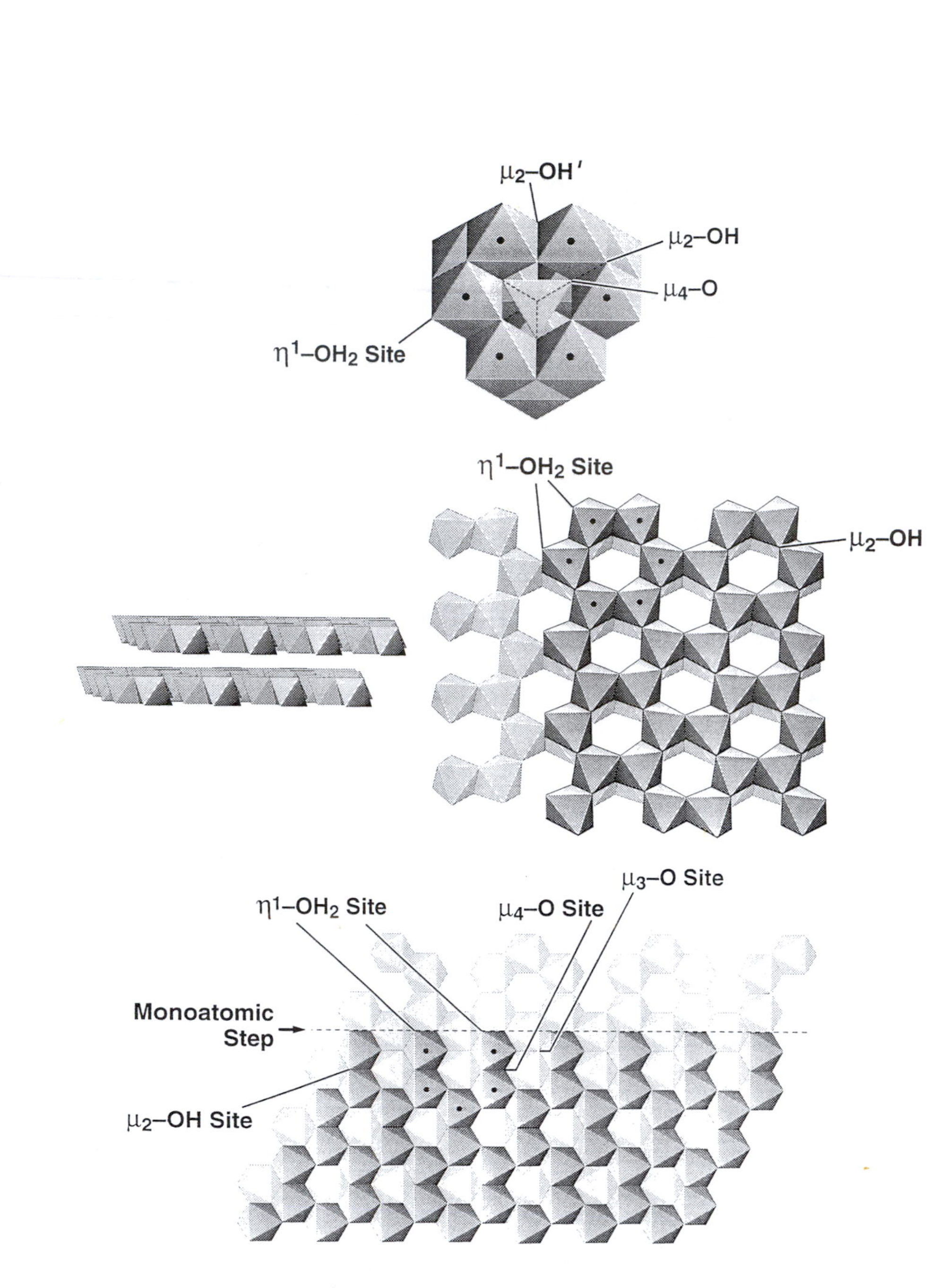
μ2–OH′
μ2–OH
μ4–O
η1–OH2 Site
η1–OH2 Site
μ2–OH
μ3–O Site
η1–OH2 Site
μ4–O Site
Monoatomic Step
μ2–OH Site

Figure 3 **Top:** The $Al_{13}O_4(OH)_{24}(H_2O)_{12}^{7+}(aq)$ tridecamer consists of AlO_6 octahedra that share edges and a single, central, AlO_4 tetrahedron in a Keggin structure (3). At the apices of the AlO_6 octahedra are twelve $-OH_2$ sites (η^1-OH_2) that can undergo Brønsted acid-base reactions with the aqueous phase. There are two types of μ_2-OH sites and one type of μ_4-O site.
Middle: A monomolecular step on Bayerite [β-$Al(OH)_3$], which consists of AlO_6 octahedra that share edges. Truncation of the surface exposes μ_2-OH groups and η^1-OH_2 groups to the aqueous phase. Solid dots are added to six-membered rings on the uppermost layer of AlO_6 octahedra to make them easier to identify. Also shown is a side view of the structure to emphasize that the layers are linked via Van der Waals forces.
Bottom: A monomolecular step on the [001] surface of α-Al_2O_3(s). Solid circles are placed on the face-shared AlO_6 octahedra that make up six-membered rings that surround an unoccupied octahedral site. The octahedra are linked to one another via faces and edges, and expose μ_2-OH and η^1-OH_2 sites that can change protonation state through reaction with the aqueous solution.

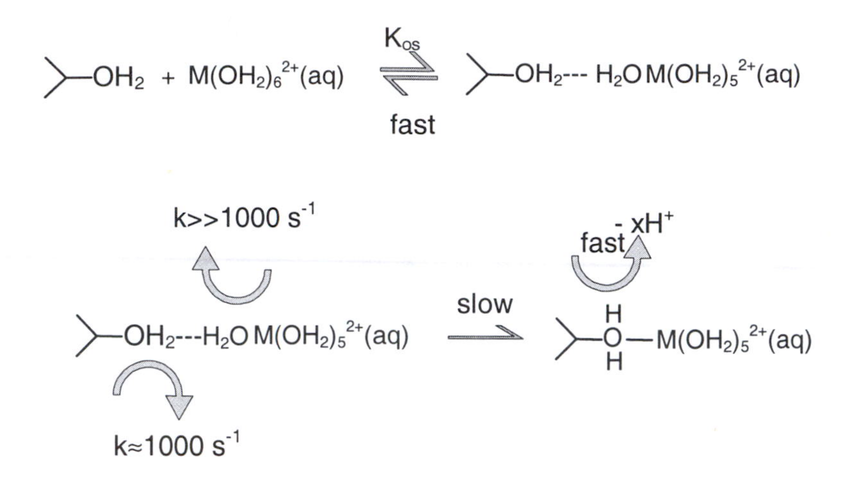

If water exchange from the adsorbate is relatively rapid ($k>>10^3\ s^{-1}$), adsorption rates will correlate with the rates of exchange of waters around the corresponding hydrated metal. In other words, Pb(II) will adsorb more rapidly than Mn(II) because the rates of exchange of inner-sphere water in $Pb(H_2O)_6^{2+}(aq)$ is much more rapid than for $Mn(H_2O)_6^{2+}(aq)$ (see 4, 5). Conversely, if rates of elimination of the η^1-OH_2 site at the surface are more rapid than rates of elimination of a water from the inner-coordination sphere of the hydrated metal adsorbate, then the overall rate of reaction will be constant.

Reaction Kinetics

Although this subject is immense, there are some general rules for understanding reaction rates.

Rate Expressions. Far from equilibrium, the overall dissociation of an oligomer or hydrolytic complex on a mineral surface is resolvable into pathways corresponding to different sets of elementary reactions. Usually, the pathways are assumed to be independent, and therefore additive, and the rate-controlling step reaction is the disruption of bonds at, or near, the metal-ligand complex. The rate-controlling step is not movement of the reactive solutes through the bulk aqueous phase to the site(s) of reaction.

For dissolved oligomers, the overall rate of dissociation has contributions from all stoichiometrically distinct species and it is usually possible to analytically distinguish the various stoichiometries. For example, the dissociation of the μ_2-hydroxo sites in the $(en)(H_2O)_2Cr(\mu\text{-}OH)_2Cr(H_2O)_2(en)^{4+}(aq)$ complex can be studied separately from dissociation in $(H_2O)_4Cr(\mu\text{-}OH)_2Cr(OH_2)_4]^{4+}(aq)$. The rate laws are expressable with complexes with defined, real stoichiometries that can be analytically determined, and the implicit concentration of water. The assumption of independent reaction pathways at surfaces will fail as more information becomes available about the actual molecular motions, but the assumption is adequate for more cases and has proven to be enormously useful.

In contrast, the stoichiometries and concentrations of different reactive complexes at the mineral surface are rarely known. The rate laws for dissolution are therefore usually expressed in terms of total adsorbate concentrations, and we know that these expressions are approximate. If hydration or hydrolysis by a water molecule of a detaching surface complex at steady state controls the rate of reaction, then rate laws such as those proposed by Furrer and Stumm (7) result. These rate laws are characterized by rates that are proportional to single adsorbate concentrations (see 8). Implicit in these rate laws is the idea that water is present in large and constant concentration and that only a single complex stoichiometry affects the reaction at the surface.

To derive true mechanistic information from dissolution experiments, the rate laws must be expressed in terms of the concentration of all distinct metal-ligand surface complexes that vary in reactivity and stoichiometry. Distinction must be made, for example, between protons that adsorb onto the ligand and those that adsorb onto the mineral surface in addition to the ligand. These adsorbed ligands all potentially change conformation with solution chemistry and these changes would be manifested in the rate laws.

Because of our ignorance about the stoichiometries of surface complexes, comparisons between dissolved complexes and surfaces work only at the scale of reactivity trends and then only in tightly constrained conditions. One could use the relative dissociation rates of $(H_2O)_4Cr(\mu\text{-}OH)_2Cr(OH_2)_4]^{4+}(aq)$ and $(en)(H_2O)_2Cr(\mu\text{-}OH)_2Cr(H_2O)_2(en)^{4+}(aq)$ complexes, for example, to gauge the effect that coordination of two $-NH_2$ ligands to a metal site has on the dissolution rates of a $Cr(OH)_3(s)$ mineral if amine ligands could be induced to adsorb in a structurally simple manner (11,12).

Reactivity Trends for Dissociation of Oligomers. Rules useful for assigning reactivities to oxygens in dissolved multimers that may apply to surfaces include:

1) The important forces are local. Substitutions of other ligands in the complex are, to a first approximation, only important if they are in the inner-coordination sphere.

2) Most metals at an exposed aqueous surface are coordinatively saturated meaning that they have all the ligands allowed in the inner-coordination sphere by geometric arguments. Mg(II) at the surface of a mineral is virtually always hexacoordinated to ligand atoms; the vacant coordination site in an underbonded five-coordinated Mg(II) surface metal would disappear instantaneously when exposed to water molecules. In cases of coordinative saturation, the rate-controlling step in ligand exchange is usually slow dissociation of a ligand from the metal, which opens up a site for a new ligand to flop into place. The most important variable correlating with dissociation rate is the metal-ligand (generally metal-oxygen) bond strength,

which can be estimated from electronic structures and from positions in the Periodic Chart of the Elements. The reactivities of hard Lewis acids scale with the ionic potential of the metal so, for example, Ca-O bonds are more reactive than Mg-O bonds, which are themselves more reactive than Be-O bonds. Reactivities for the first-row transition metals are controlled by the d-electron structure (e.g., 13).

3) Electron exchange changes the metal-oxygen bond strengths and introduces *reductive* and *oxidative* pathways for dissolution and ligand-exchange. Co(II) has the electronic structure $t_{2g}^6 e_g^1$ and exchanges oxygens in the inner-coordination sphere of the hydrated ion much more rapidly than Co(III) with t_{2g}^6 electronic structure; Co(III) hydrolysates are very unreactive relative to the Co(II) complexes.

4) The rates of exchange of oligomeric oxygens decreases with increased coordination to metals, so μ_3-oxo groups are less reactive than μ_2-oxo sites. Within an oligomer, η^1-OH_2 sites are most reactive relative to the μ-hydroxo or μ-oxo sites, and μ_3-oxo bridges are less reactive than μ_2-oxo sites. The differences are profound: Richens et al. (14) found no evidence for exchange of the μ_2- or μ_3-oxo bridges in the $[Mo_3O_4(OH_2)_9]^{4+}$(aq) complex over two years. In contrast, although all the terminal waters of this complex exchange readily, rates of exchange of those *cis*- or *trans*- to the μ-oxo bridges differ by a factor of 10^5! Furthermore, there is correlation between the reactivity of an oxygen in an oligomer and its Brønsted acidity.

5) Equilibrium coordination of protons to the oxygens dramatically changes their reactivities, as measured by the rates of dissociation or rate of exchange with solution oxygens. The *proton-assisted* path arises from adsorption of a proton to a bridging hydroxyl, which converts it to a poorly bridging water molecule. These protonation reactions are rapid (orders of magnitude more rapid than ligand exchange or dissociation of a metal-oxygen bond) and introduce a strong pH-dependence to an otherwise slow and pH-independent reaction.

6) The lability and Brønsted acidity of sites in an oligomer reflects the structural position; terminal waters *trans* to a μ-hydroxo or μ-oxo site are commonly more reactive and more acidic than waters *cis* to the bridges. The Brønsted acidity of waters in oligomers is affected also by strong intermolecular hydrogen bonding, commonly between terminal waters and hydroxyls (i.e., between η^1-OH_2 sites and η^1-OH sites) or with a hydroxyl bridge (e.g., between η^1-OH_2 and μ_2-OH sites). For example, the pK_{a1}, and pK_{a2} values for terminal water molecules in $(H_2O)(en)_2Cr(OH)Cr(en)_2(H_2O)$ differ by 10^7, a much larger value than expected from the monomers (15). Such hydrogen-bonding occurs at mineral surfaces, but is not now separable from other effects.

7) Substitution of a stable ligand into the inner-sphere of a coordination complex can enhance or retard the depolymerization rates at sites away from the substitution. Because this class of reactions is so important, it is discussed in greater detail in the next section.

In summary, the depolymerization reactions of metal oxo- and hydroxo-oligomers have *proton-assisted*, *base-assisted*, *ligand-assisted,* and *pH-independent* pathways for bridge cleavage. These pathways depend upon whether protons or ligands are configured to the reactive site at equilibrium and whether this configuration increases the rate.

Stable Ligands Can Change Reactivities of Other Sites in the Inner-Coordination Sphere. Some ligands affect the reactivity of other metal-oxygen bonds in the inner-coordination sphere, and particularly those *trans* to the substitution. These are *ligand-directed labilizations.* Experiments on dissolved complexes can help identify the important ligands and to elucidate the mechanisms of rate enhancement.

For example, measured and estimated rates of water exchange between the inner-coordination sphere of $AlF_x(H_2O)_{6-x}^{3-x}(aq)$ complexes and bulk are shown in Figure 4. The rate of exchange of water between the inner-coordination sphere of $Al(H_2O)_6^{3+}(aq)$ and bulk solution is approximately 1-3 s^{-1} at 298 K (16). Dynamically stable substitution of one and two fluorides for hydration waters in the $Al(H_2O)_6^{3+}(aq)$ complex [forming $AlF_x(H_2O)_{6-x}^{3-x}(aq)$ complexes] causes rates of exchange of the remaining water molecules to increase by factors of 10^2 and 10^4, respectively (16). The effect is progressive so that the reactivity of the water molecule in the $AlF_5(H_2O)^{2-}(aq)$ would be many, many times greater than in the monofluoro complex, if it could be measured.

The simplest such *ligand-directed labilization* is deprotonation of a hydration water (η^1-OH_2) to form a hydroxyl (η^1-OH sites) in a dissolved monomer. This deprotonation is effectively a ligand substitution because an equilibrium concentration of such hydroxyl ligands is maintained dynamically by rapid proton exchange with the solvent. It is distinct from other substitutions, however, because no metal-oxygen bond is broken. The rates of exchange of protons are so fast that the waters convert to hydroxyls virtually instantaneously.

Rates of water exchange around dissolved complexes typically increase by a factor of $\approx 10^2$ as one proton is removed from a hydration water. Rates of water exchange from the inner-coordination sphere of $Cr(H_2O)_6^{3+}(aq)$, for example, are a factor of 75 times slower than around the deprotonated $Cr(H_2O)_5OH^{2+}(aq)$ complex [Table 1]. Other reactions are accelerated as well. Rates of dimerization of Cr(III) [Table 1] increase by factors of 60 for each extra hydroxyl group that is involved in the reaction and the rates of dimerization of Fe(III) monomers increase by a factor of 10^{10} as the hydration waters deprotonate to become hydroxyls (18). Stable hydroxyls are enormously labilizing to other inner-sphere oxygens.

The labilities of metal-oxygen bonds to hydration waters are similarly enhanced when aminocarboxylate ligands are configured to the metal in a chelate complex (e.g., 18). The effect is progressive: rates increase dramatically with the number of amino- or carboxylate- atoms coordinated to the metal (see 15) so that rates of exchange of waters between the inner-coordination sphere of Ni(II) and bulk solution increase by factors of 100-200 for each ammonia or amine substitution in the inner-coordination sphere. In Figure 4, data are presented for $Ni(NH_3)_x(H_2O)_{6-x}^{2+}(aq)$ complexes.

The important point of these observations is that similar increases in the reactivity of surface M-O bonds can be expected with replacement of terminal waters. In other words, one expects increases the dissolution rates of an aluminum oxide mineral if the surface η^1-OH_2 sites are replaced with either hydroxyl or fluoride, forming η^1-OH and η^1-F sites respectively (see 34, 35).

Proton Catalysis and Proton Induction. The term 'catalysis' is used loosely by the chemists who study multimer dissociation to describe any proton-assisted dissociation, even in cases where it is inappropriate. Understanding a distinction between *catalysis* by and *induction* (see 36, p. 20-16) by adsorbed protons is important for mineral dissolution since the concentration of stable adsorbates is used to interpret the rate order (e.g., 8). For catalysis (see 37)**:** (i) the catalyst must be involved in the reaction in an actual molecular configuration that increases the rate by lowering the activation energy; and (ii) the catalyst must be recycled and not consumed in the reaction. A catalyst cannot affect the equilibrium state of reactants and products and it must not appear in the overall reaction stoichiometry.

As an example of a proton-catalyzed reaction, consider the hypothetical dissociation of a μ_2-oxo site at the surface of silica [SiO_2(s)]. The first step is equilibrium protonation of the neutral siloxane (the μ_2-O site) to form a hydroxyl bridge (the μ_2-OH site):

$$[\rangle Si\,(\mu_2 - O)\,Si\langle]^o \;+\; H^+(aq) \xleftrightarrow{equilibrium} [\rangle Si\,(\mu_2 - OH)\,Si\langle]^+$$

followed by coordination to an incoming water molecule, which we portray here to form an unstable bridge:

$$[\rangle Si\,(\mu_2 - OH)\,Si\langle]^+ \;+\; H_2O \longleftrightarrow [\rangle Si\,(\mu_2 - OH)(\mu_2 - H_2O)\,Si\langle]^+$$

The unstable complex slowly hydrolyzes to release the proton:

$$[\rangle Si\,(\mu_2 - OH)(\mu_2 - H_2O)\,Si\langle]^+ \xrightarrow{slow} 2\,\rangle\,Si\,(\eta_1 - OH) \;+\; H^+(aq)$$

and the proton rate order would correspond to a single proton adsorbing to bridging bonds and repeatedly catalyzing dissociation.

It is not hard to imagine that repeated reaction would release a silicic acid molecule [$Si(OH)_4^o(aq)$] to solution. Note also that a proton is released because we

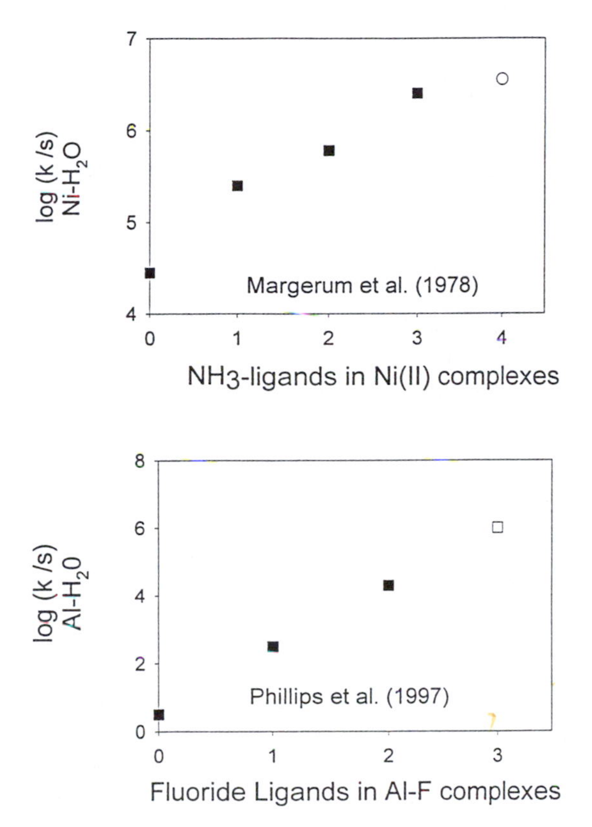

Figure 4. Rate coefficients at 298 K for exchange of water molecules from the inner-coordination sphere and bulk solution in $AlF_x(H_2O)_{6-x}^{3-x}(aq)$ (top) and $Ni(NH_3)_x(H_2O)_{6-x}^{2+}(aq)$ (bottom) complexes. Filled symbols indicate measured values and unfilled symbols indicate estimates. In the case of $AlF_x(H_2O)_{6-x}^{3-x}(aq)$ complexes, the estimates are a simple extension of the reactivity trend for less-fluoridated complexes; for the case of $Ni(NH_3)_x(H_2O)_{6-x}^{2+}(aq)$, the estimate is from exchange of ammonia (see 19).

defined the hyperprotonated η^1-OH_2 site on Si(IV) to be a much stronger acid than the reactant: $[\rangle Si\,(\mu_2-OH)(\mu_2-H_2O)\,Si\langle]^+$. Had the Si($\eta^1$-$OH_2$) site retained the proton after dissociation from the surface site, the proton could not recycle and the reaction would not be catalytic. It is not difficult to understand how this reaction could be catalytic as there are no known monomers of dissolved Si(IV) with η^1-OH_2 sites (3); the Si(η^1-OH_2) sites on the dissolved monomers deprotonate immediately to form sets of Si(η^1-OH) sites, as in silicic acid.

Now consider an *induced* reaction for proton-assisted dissociation of a singly bridged dimer (see 15, 17) where the reaction products retain the proton. The first step, as before, is equilibrium association of a proton to the μ_2-OH site between the metals, thereby forming a weakly bridging water molecule between two Fe(III) atoms:

$$(H_2O)_5Fe(OH)Fe(H_2O)_5{}^{5+} + H^+ \xleftrightarrow{equilibrium} (H_2O)_5Fe(HOH)Fe(H_2O)_5{}^{6+}$$

(In this example we eliminate the η^i, μ_i notation in order to save space.) As in the previous example, this protonation is followed by slow hydration and dissociation, perhaps assisted by association of the hydrating water molecule to the bridge. This hydrolysis releases two monomeric $Fe(H_2O)_6^{3+}(aq)$ species:

$$(H_2O)_5Fe(HOH)Fe(H_2O)_5{}^{6+}(aq) + H_2O(l) \xrightarrow{slow} 2\,Fe(H_2O)_6{}^{3+}(aq)$$

In this example, the adsorbing proton converts the hydroxyl bridge to a hydration water in the stable monomer, a $Fe(H_2O)_6^{3+}(aq)$. The reaction is not catalyzed because one stable ligand (a hydroxyl) converts rapidly to another (a water molecule) via protonation. This ligand remains in the inner-coordination sphere of the reaction product as the monomer is released.

One can immediately see the inductive or catalytic role that protons play in dissociation depends upon the relative Brønsted acidities of the dissociating dimer and the stable monomer. If the hydrated, dissociating dimer is a stronger acid than the reaction products, protons are retained and the reaction is induced. If the dissociating dimer is a weaker acid than the monomeric reaction products (as in the example of silica, above) the protons are catalytic. Because protonation reactions are so much more rapid that depolymerization, the actual protonation state of the immediate reaction product is difficult to ascertain. How would one know if the dissociation step released fully hydrated metals or partly hydrolyzed metals and protons if the hydrolysis species equilibrate in 10^{-12} s?

This point is important because a dissolving mineral surface can be thought of as a polymer that detaches monomers from monomolecular steps. In the limiting case of pure induction by adsorbed protons, the rate order for dissolution comes from a charge balance at the surface (8) and, in acid solutions, commonly equals the formal metal valence in the oxide. In these cases, for example, rates of dissolution of BeO(s) increase with the square of adsorbed proton concentrations; likewise the dissolution rates of Al_2O_3(s) increase with the cube of adsorbed proton concentration (see 7,8). In the case of catalysis, the rate order bears no necessary similarity to the formal charge of the metal in the oxide.

Solute Adsorption is Usually Faster than Metal Detachment In the above example, protonation of the bridging hydroxyl groups in a dissolved dimer introduces a distinct pH dependence to a slow, and otherwise pH-independent dissociation reaction. Hydration is the slow step and the released monomers are coordinatively saturated. The rates of protonation-deprotonation reactions have been measured for simple oxide solids in electrolyte solutions via relaxation spectroscopy (20). In general, the adsorption of protons and hydroxyl ions to form surface charge commonly proceeds at time scales of 0.001 to 10 s (e.g., 21). These time scales are much more rapid than the characteristic times for detachment of a metal from a mineral surface. The adsorption rates of potential rate-enhancing adsorbates other than protons have not been widely studied, although some data exist for oxyanion adsorption (22-24), acetate (25), and metals (4, 26). These studies show that rates of adsorption of the rate-enhancing ligand are (usually) much more rapid than the subsequent bond cleavage.

This rule, that adsorption of the ligand is more rapid than the subsequent metal detachment, is not inviolate. Recent work by Nowack and Sigg (27) shows that the ligand-enhanced dissolution of ferric hydroxide by EDTA complexes in nature can be quite complicated. This ligand-enhanced dissolution proceeds by a free EDTA ligand adsorbing to an Fe(III)-oxide mineral and inducing detachment of an Fe(III)-EDTA complex. There is, however, very little free EDTA in most natural solutions because the metal concentrations are relatively high and free metal coordinates rapidly to a free ligand. The EDTA in these solutions exists as metal-EDTA complexes that must dissociate before the ligand is free to attach to a mineral surface. In the dissolution of very reactive solids (e.g., ferrihydrite), dissociation of a metal-EDTA complex near the surface is so slow that it actually influences the overall rate of dissolution. Adsorption of free EDTA to the surface, and subsequent detachment of an Fe(III)-EDTA complex from the surface, are relatively rapid in these rare cases.

The Reactivities of Different Oxide Minerals. Metal releases from a mineral surface, when they can be isolated from other side reactions, proceed with reactivity trends that resemble the exchange of solvation waters around the corresponding dissolved, hydrated metal (Figure 5). In other words, rates of dissolution of isostructural, molecular solids, such as the orthosilicate minerals shown in Figure 5, scale like the bond strengths. Rates for minerals containing alkaline-earth cations (Be_2SiO_4: phenakite; Mg_2SiO_4: forsterite; Ca_2SiO_4: calcium olivine) scale inversely with ionic radius of the cation. Rates for minerals containing first-row transition metals (e.g., Ni_2SiO_4: liebenbergite) vary according to the d-electron structure and are independent of ionic radius. These familiar reactivity trends are not accidental, but result from the similarity in coordination chemistry between metals at the mineral surfaces and the hydrated metals in solution.

The data in Figure 5 indicate that detachment of the metal from the surface proceeds via a largely dissociative pathway with at least some mechanistic similarities to a ligand-exchange reaction in solution. Dissociation of similar metal-oxygen bonds, to either water molecules or to structural oxygens at the mineral surface, controls the rates of reactions in both environments. The orthosilicate minerals do not have a polymerized silicate anion (28) and the solids have similar proton-adsorption affinities so that experiments at constant pH are also at approximately equal proton adsorbate concentrations.

The results can be much more complicated for rock-forming minerals that have an extensively polymerized silicate anion because the near-surface region of the mineral commonly has a different bulk composition than the interior as more-reactive sites are selectively leached away. The leached zone can extend to hundreds of nanometers and the extent of reaction depends sensitively on the solution pH; that is, the extent of protonation of surface sites (see 29).

Reactivity Trends with Adsorbed Ligands. We showed above that the reactivity of a particular oxygen in dissolved metal complex depends upon: (i) the coordination number to metals; (ii) other ligands also present in the inner-coordination sphere of the metal and (iii) on the ligand position *cis* or *trans* to a bridging group. There are many reasons why adsorbed ligands should not induce similar chemical changes as ligands in dissolved complexes. Nevertheless, for some simple ligands and metals, there is apparent consistency between the reactivities observed in dissolved metal-ligand complexes and the dissolution of oxide minerals.

We saw earlier that the substitution of ammonia (or amino groups) for water molecules in the inner-coordination sphere of Ni(II), to form $Ni(NH_3)_x(H_2O)_{6-x}^{2+}(aq)$ complexes, increases the rates of exchange of remaining water molecules in the inner-coordination sphere (Figure 4). The accelerating effect of these groups is also progressive; rates of water exchange in bis-amino complexes are more rapid than in mono-amino complexes. Furthermore, the same effect is observed when the amino groups are linked to a carbon backbone to make an amino- or aminocarboxylate chelate. Ethylenediamine (*en*), for example, forms a bidentate complex with Ni(II) and diethylenediamine (*dien*) forms a tridentate complex. Both of these ligands cause the rates of exchange of the distal Ni(II)-oxygen bonds to water to become more labile; by distal we mean other sites also in the inner-coordination sphere. The ligand dien is more effective than *en*, as one expects from the extra $-NH_2$ group coordinated to the metal center. Likewise, the rates of water exchange between the inner-coordination sphere of a Ni(II)-H_2O-ligand complex and bulk water increases when aminocarboxylate ligands are introduced into the complex. The rates increase with the number of ligand functional groups (-COOH and $-NH_x$) coordinated to the Ni(II) metal.

Ludwig et al. (11,12) measured the effect that these ligands have on the dissolution rate of bunsenite [NiO(s)] and some of the results are shown in Figure 6. The flux of Ni(II) from dissolving NiO(s) increases with the number of carboxyl- and amino-functional groups which are presumed to coordinate to a metal center, in a similar fashion as the rates of exchange of waters around a hydrated Ni(II) metal in solution increase (11) (c.f., Figure 4).

This reactivity trend is identical to the enhancement of rates of water exchange from the inner-coordination sphere of dissolved Ni-H_2O-ligand complexes to the bulk solution. Some ligands that form inner-sphere complexes with the metal

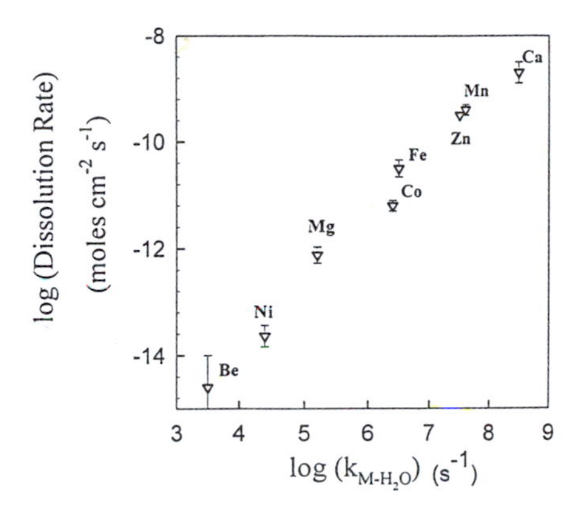

Figure 5. The dissolution rates at pH=2 of orthosilicate minerals plotted against the rate of exchange of water from the bulk aqueous phase into the inner-coordination sphere of the corresponding hydrated metal in solution. The symbol Mg, for example, stands both for the dissolution rate of forsterite [Mg_2SiO_4(s)] at pH=2 and the rate of exchange of waters from the hydration sphere of $Mg(H_2O)_6^{2+}$(aq) and the bulk solution.

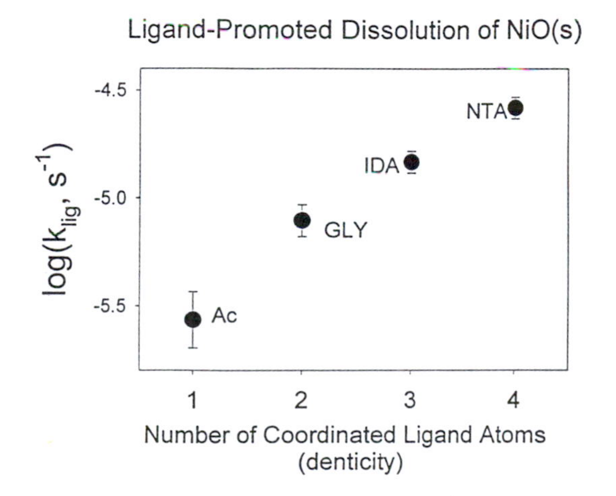

Figure 6. The rate coefficients (k_L) for ligand-induced dissolution of bunsenite [NiO(s)] for a set of ligands that differ in the number of ligand atoms coordinated to the metal center at the mineral surface. Acetate (Ac) is unidentate; glycine (GLY), iminodiacetic acid (IDA), and nitrilotriacetic acid (NTA) are bidentate, tridentate, and tetradentate, respectively.

decrease the effective charge of that metal center, and thereby reduce the metal-oxygen bond strength. In addition, ligands such as fluoride, ammonia and water molecules form strong hydrogen bonds. Substitution of these ligands into the symmetric hydration sphere must disrupt the existing hydrogen bonds, further labilizing inner-sphere metal-oxygen bonds.

One important result of Ludwig et al. (11,12) is that the rate coefficients correlate with the equilibrium constant in solution for these carefully chosen metal-ligand complexes. This LFER relates the rate coefficient for ligand-promoted dissolution of NiO(s), which is difficult to measure, to the equilibrium constant for forming the Ni(II)-ligand species in solution, which is already known. This correlation works only because there is some structural similarity between the metal-ligand complexes in solution and the surface complexes that are important to dissolution. It will not be observed for all adsorbates.

Mechanisms of Dissolution. Little is really known about the molecular mechanisms of oxide-mineral dissolution because the surface-speciation information is so difficult to acquire. Speculations about the mechanism generally fall into two groups. The first model is noncatalytic and treats adsorbate effects as due to *ligand-directed labilization* in the case of adsorbed ligands and *induction* in the case of protons.

The stoichiometry of the surface complex formed by proton induction is thought to resemble the stable metal hydrolysate in solution. Many of these experiments are conducted in acidic solutions, so the detaching surface complex is a monomer generally thought to be similar in stoichiometry to the fully hydrated monomer. A reaction with water molecules is the rate-controlling step, as we saw in the above examples of oligomer dissociation. Within this model, the proton rate order for dissolution comes from a charge balance at the surface (e.g., 8) and provides very little information about the stoichiometry of the activated complex. It is unique to oxide solids that the reaction proceeds at a steady state surface where new sites must constantly be formed by movement and dissolution of water molecules (see 5, 7, 30). Data to evaluate this inductive model of dissolution are few and are solely limited to simple oxide minerals [e.g., NiO(s), BeO(s), Al_2O_3(s)].

The second model requires catalysis by adsorbed protons and is applied to silicate minerals with covalent, polymerized structures, like quartz. In this model, protons react quickly with oxide bonds at the surface, accelerate cleavage, and return to solution. As bonds are progressively cleaved, a monomer or small oligomer is released from the surface. The weakness of this model is the enormous difficulty in simultaneously determining rates of dissolution and proton adsorption densities on complicated multi-oxide mineral structures. Protons taken up by leaching alkaline-metal cations from the mineral must be separated from those involved in protonation of bonds in order to assign a value to the rate order.

Conclusions.

The future increase of understanding of mineral surface chemistry will involve more molecular-level interpretations and fewer interpretations based upon macroscopic features of the mineral structure. Using a molecular approach, which requires the establishment of model complexes for study, catalysis at surface sites, oligomerization of adsorbates, and other reactions of geochemical importance, can be related to well-understood chemical principles.

The analogy between complexes at the surface of a mineral and oligomers is strong, but imperfect. While the coordination of metals in surface complexes commonly resembles that in hydrolyzed solutes, the coordination chemistry of coordinated oxygens differs considerably. Only by studying oligomers that have oxygens in high coordination to metals, and which can be studied spectroscopically, can we establish properties that control reactions at oxide surfaces. Reactions at this scale are also more appropriate for computer simulation of reaction processes (e.g., 32) than full surfaces.

Acknowledgments

The authors thank two anonymous referees and Prof. D. Sparks for perceptive comments on the manuscript. Support for this research was from the U.S. NSF grants #96-25663; 94-14103 and U.S. DOE DE-FG03-96ER 14629.

References

1 Burton, W. K.; Cabrera, N.; Frank, F. C. *Phil. Trans. Roy. Soc. London* **1951,** *A243*, 299.

2 Cotton, F. A; Wilkinson, G. *Advanced Inorganic Chemistry*, Wiley-Interscience. New York, 1988, 1455 p.

3 Baes, C. M.; Mesmer, R. E. *The Hydrolysis of Cations*. Wiley. New York, 1976.

4 Hachiya, K.; Sasaki, M.; Ikeda, T.; Mikami, N.; Yasunaga, T. *J. Phys Chem* **1984,** *88,* 27.

5 Stumm, W. (1992) *The Chemistry of the Solid-Water Interface*. John Wiley. New York, 1992, 428 p.

6 Nordin, J. unpubl. Ph.D thesis, Dept. Inorganic Chemistry, University of Umeå, Umeå, Sweden. 1997.

7 Furrer, G.; Stumm, W. *Geochim Cosmochim Acta* **1986,** *50*, 1847.

8 Casey, W.H.; Ludwig, Chr. *Nature* **1996,** *381,* 506.

9 Rustad, J.R.; Felmy, A. R.; Hay, B. P. *Geochim. Cosmochim. Acta* **1996,** *60*, 1563.

10 Hiemstra, T.; Venema, P.; and Van Riemsdijk, W. H. *J. Coll. Interf. Sci.* **1996,** *184*, 680.

11 Ludwig, C.; Casey, W.H.; Rock, P.A. *Nature* **1995**, *375*, 44.

12 Ludwig, Chr.; Casey, W.H.; Devidal, J-L. *Geochim. Cosmochim Acta* **1996,** *60*, 213.

13 Basolo, F.; Pearson, R. G. *Mechanisms of Inorganic Reactions*, 2nd edition, Wiley-Interscience, New York, 1967, 157 p.

14 Richens, D. T.; Helm, L.; Pittet, P.-A.; Merbach, A. E.; Nicolo, F.; Chapuis, G. *Inorg. Chem.* **1989,** *28*, 1394.

15 Springborg, J. *Adv. Inorg. Chem.* **1988,** *32,* 55.

16 Phillips, B.L.; Casey, W. H.; Neugebauer-Crawford, S. *Geochim. Cosmochim. Acta.* **1997,** *15,* 3041.

17 Johnston, R. F.; Holwerda, R. A. *Inorg. Chem.* **1983**, *22*, 2942.

18 Wilkens, R.G. (1991) *Kinetics and Mechanisms of Reactions of Transition Metal Complexes*, VCH., New York, 1991, 465 p

19 Margerum, D.W.; Cayley, G.R.; Weatherburn, D.C.; Pagenkopf, G.K. (1978) Ch. 1 *in*: *Coordination Chemistry* v. 2 (A. Martell ed.), ACS Monograph **174**, American Chemical Society, Washington, D.C. 1-220 p.

20 Sparks DL *Kinetics of Soil Chemical Processes*. Academic Press, New York, 1989, 210 p

21 Astumian, R. D.; Sasaki, M.; Yasunaga, T.; Schelly, Z.A. *J. Phys. Chem.* **1981,** *85*, 3832.

22 Zhang P.C.; Sparks, D.L. *Soil Sci. Soc. Am. J.* **1989,** *53*,1028.

23 Zhang P.C.; Sparks D.L. *Soil Sci. Soc. Am. J.* **1989,** *54*,1266.

24 Zhang P.C.; Sparks, D.L. *Env Sci Tech.* **1990**, *24*, 1848.

25 Ikeda, T,; Sasaki, M.; Hachiya, K.; Astumian, R.D.; Yasunaga, T.; Schelly, Z.A. *J Phys Chem* 1982, *86*, 3861.

26 Hachiya, K,; Ashida, M.; Sasakı, M.; Kan, H.; Inoue, T.; Yasunaga, T. *J Phys Chem* **1979,** *83*, 1866.

27 Nowack, B.; Sigg, L. *Geochim. Cosmochim. Acta* **1997**, *61*, 951.

28 Casey, W. H.; Westrich, H. R. *Nature* **1992,** *355*, 147.

29 Casey, W. H.; Bunker, B. (1988) in: *Mineral-Water Interface Geochemistry*, Reviews in Mineralogy (M. F. Hochella and A. F. White, eds), Washington, D.C., 1988, Vol. 23; Chapter 10, pp 397-426.

30 Ludwig Chr.; Casey, W.H. *J. Coll Interf Sci* **1995,** *178*, 176.

31 Casey, W.H.; Ludwig, Chr. in: *Chemical Weathering Rates of Silicate Minerals*. Reviews in Mineralogy (White, A. F. and Brantley, S. L. eds.), Washington, D. C., 1995, Vol. 31,; Chapter 3, p. 87-114.

32 Rotzinger, F. P. *J. Am. Chem. Soc.* **1996,** *118*, 6760.

33 Crimp, S. J.; Spiccia, L.; Krouse, H. R.; Swaddle, T. W. *Inorg. Chem.* **1994,** *33*, 465.

34 Żutič, V.; Stumm, W. *Geochim. Cosmochim. Acta* **1984,** *48*, 1493.

35 Pulfer, K.; Schindler, P. W.; Westall, J. C.; Grauer, R. *J. Coll. Interface Sci.* **1984,** *101*, 554.

36 Katakis, D.; Gordon, G. *Mechanisms of Inorganic Reactions*. Wiley-Interscience. New York, 1987, 384 p.

37 Fox, M.; Whitesell, J. K. *Organic Chemistry*, 2nd edition. Jones and Bartlett. Sudbury, MA, 1997.

Chapter 13

Reactivity of Dissolved Mn(III) Complexes and Mn(IV) Species with Reductants: Mn Redox Chemistry Without a Dissolution Step?

George W. Luther, III, David T. Ruppel, and Caroline Burkhard

College of Marine Studies, University of Delaware, Lewes, DE 19958

Mn(III)-carboxylic acid complexes are stable to Mn(III) disproportionation to Mn(IV) and Mn(II) but decompose in a pH dependent intramolecular redox reaction that produces Mn^{2+} and CO_2. Mn(III) complexes derived from organic ligands without carboxyl groups and which are unsaturated are more stable to intramolecular redox reactions. However, all Mn(III) complexes tested react quickly with sulfide. Soluble MnO_2 is a polymeric form of 50 nm size or less with a characteristic UV-VIS spectrum that can be used to follow reaction progress. This form of MnO_2 is used to assess the empirical rate law for the reaction of MnO_2 with oxalate. The rate law is complex but is consistent with previous studies of this reaction using solid phase MnO_2. The ability to follow kinetics of the reaction with a soluble form of MnO_2 confirms the formation of a MnO_2-oxalate complex which then undergoes inner sphere redox reactions similar to Mn(III)-carboxylic acid complexes with subsequent release of Mn^{2+} and CO_2 to solution.

Solid manganese(III,IV) oxides and (oxy)hydroxides are significant oxidants of organic matter in the environment (1-2). Although high oxidation state *soluble* manganese compounds are well known in the chemical (3-7) and biochemical literature (8-10), they have not been explored and reported in the environmental literature until recently (11). Among the high oxidation state Mn compounds are Mn(III) complexes with organic ligands (3-7) and a polymeric MnO_2 phase (12) of about 50 nm size which can pass through the filters (0.2 and 0.4 μm) commonly used in environmental research. These phases have not been studied for their electron acceptor behavior as the solid phases have been. Only Kostka et al (11) have reported on the chemical and microbial reactivity of a Mn(III)-pyrophosphate complex under environmental conditions because the pyrophosphate complex has been detected in soils. This is in contrast to a host of studies which have shown that Fe(III) is almost 100% complexed to organic ligands in seawater (e.g.; 13-14).

Soluble Mn(III,IV) compounds should be of extreme importance at interfaces of oxic and anoxic zones and in suboxic (anoxic and non-sulfidic) zones where organic matter decomposition is intense and results in dissolved organic matter that can chelate metals (15-18). These zones are found in a variety of marine and freshwater sedimentary environments and in the stagnant water columns of the Black Sea, Chesapeake Bay and lakes. Because phytoplankton contain in their photosynthetic apparatus a Mn(III)/Mn(IV) cluster molecule that splits H_2O to O_2 (10), it is possible that these enzymes can be released during particulate organic matter decomposition and oxidize reduced species [e.g., H_2S, organic acids] in the presence or absence of O_2. Mn(III) complexes are also important in lignin degradation (8-9), and white rot fungi contain a peroxidase that uses Mn(III) as an oxidant that functions in the absence of light.

The objective of this paper is to begin to document the reactivity of Mn(III) organic complexes and polymeric/soluble MnO_2 with common natural reductants such as sulfide and oxalate. The Mn(III) complexes can be readily prepared (3-7). Some are quite stable; however, complexes with carboxylate binding groups decompose with CO_2 generation at low pH (3-7). Mn(III) complexes can be monitored by UV-VIS spectroscopy since many Mn(III) complexes exhibit color and by electrochemistry since many Mn(III) complexes reduce to Mn(II) in a voltammetric experiment (6). A soluble form of MnO_2 can be prepared by the reaction of permanganate with a reductant such as thiosulfate (12); an homogeneous solution results which coagulates and precipitates slowly on increasing the ionic strength of the solution. Perez-Benito et al (12) termed this form of MnO_2 a soluble colloid of 50 nm (500Å) size or less but we will term it soluble and polymeric for simplicity because it is not affected by gravitational settling. This soluble form of MnO_2 is not electroactive (this work) but has color, absorbs light in the UV-VIS region and follows Beer's law (19, 20). These properties can allow for the determination of the order of reaction in terms of the concentration of MnO_2 even though the MnO_2 is polymeric. Thus, knowledge of the surface area of MnO_2 is not necessary to specify kinetics.

EXPERIMENTAL

Mn(III) complexes were prepared by dissolving manganese(III) triacetate ($Mn(ac)_3$ from Alfa Products, Inc.) in an excess of the desired ligand at the ligand's natural pH in water or by adding base to the ligand to obtain the desired pH. Stock solutions were 1 mM in Mn(III) and diluted for experiments. Reactions were carried out in 0.1 M NaCl or in seawater at 20°C. The ligands initially tested to produce Mn(III) complexes included malonate, tartrate, the buffer bis-tris, oxalate, catechol, porphyrin, EDTA, oxalate, pyrophosphate, citrate, triethanolamine, 8-hydroxyquinoline, and were prepared from ACS reagent grade material. The complexes show stability to disproportionation of Mn(III) to Mn(II) and Mn(IV) in air because of the high ligand content. They are less stable at low pH and result in oxidation of the organic ligand by decarboxylation (3-5) or semiquinone / quinone (6) formation and Mn(II) formation. The compounds which have poly-hydroxy groups (with and without carboxylic acid groups) such as bis-tris, citrate and tartrate are more stable to oxidation because they are unsaturated with -OH groups involved with binding Mn(6). Detailed experiments centered on bis-tris and malonate complexes with Mn(III) because they showed a broad range of stability. The Mn(III) bis-tris, pyrophosphate and citrate complexes were more stable to disproportionation and redox processes at neutral pH than all others tested and could be stored for days to months. Bis-tris [2-[Bis(2-hydroxyethyl)imino]-2-(hydroxymethyl)-1,3-propanediol] has 5 hydroxyl groups and does not have carboxylic acid, aromatic or other unsaturated groups. Its stability as well as pyrophosphate's is related to the chelate effect and its saturated nature.

Mn(III)-malonate has an absorption peak centered at 266 nm (9; ϵ = 11,500). Mn(III)bis-tris as well as many other Mn(III) complexes (6,7) has a voltammetric peak for the reduction of Mn(III) to Mn(II) at a Hg electrode near Ep = -1.10. These spectroscopic and voltammmetric parameters were used to follow Mn(III) reactivity in a reaction vessel or directly in a voltammetric cell after the addition of a reductant.

Soluble MnO_2 was first detected in the oxidation of organic compounds by MnO_4^- using UV-VIS spectrophotometry (19). Laboratory synthesis is easily achieved using the method of Perez-Benito et al (12) by the stoichiometric reaction of permanganate with thiosulfate (eq. 1). A solution of 1 mM is stable to coagulation/precipitation and remained

$$8\ MnO_4^- + 3\ S_2O_3^{2-} + 2\ H^+ \dashrightarrow 8\ MnO_2 + 6\ SO_4^{2-} + H_2O \qquad (1)$$

so for several months. The polymeric MnO_2 consists of roughly spherical particles with a radius of 500 Å (50,000 pm or 50 nm) and we determined that the particles are NOT electroactive at the mercury electrode. The material has a negative electrostatic charge and is precipitated as more cations are added, and divalent cations increase the rate of precipitation more than monovalent cations (12). The UV-VIS characteristics are that Beer's law is followed over the pH range used in this study (4-8); at lower pH, the increase in ionic strength causes coagulation and precipitation. Polymeric and soluble MnO_2 is a possible product of chemical and bacterial Mn(II) oxidation in fresh waters such as lakes.

An Analytical Instrument Systems, Inc. (AIS) model DLK-1000 diode array VIS spectrophotometer was used to measure the spectra of Mn(III) complexes and soluble MnO_2; five scans were recorded and overlaid to give the spectrum. MnO_2 reactions were followed in a vessel in a manner similar to Xyla et al (21) at 20 °C. In some experiments, a 1 cm cuvette was used while monitoring the 400-450 nm region. In others, a 5 cm fiber optic dip probe with a 450 nm filter attached to a Brinkmann model PC800 colorimeter was used to measure MnO_2 absorbance. All UV-VIS experiments over the 400-450 nm region showed the same kinetics for MnO_2. A Metrohm model 716 DMS titrino in pH stat mode was used to maintain reactions at constant pH of 4, 5 or 6 (within ±0.02). For MnO_2 reactions with oxalate, oxalate was added to the reaction vessel after the oxalate was buffered to the desired pH by adding the appropriate amount of sodium hydroxide. MnO_2 was then added to the oxalate solution and the pH adjusted with the aid of the pH stat system; initial oxalate concentrations were always five-fold or greater in excess to initial MnO_2 concentrations. Thus, the ionic strength of a given solution was primarily due to the oxalate with a minor contribution from the cations and anions present from the MnO_2 synthesis of eq. 1. The volume change over the entire reaction was less than 0.25% based on the acid added to maintain pH constant. The MnO_2 concentrations used in this study ranged from 0.01 mM to 0.100 mM which are representative of dissolved Mn concentrations in sedimentary porewaters (22) and in the water column of the Black Sea (18) and Chesapeake Bay (15,17). Mn(II) production was confirmed with voltammetric measurements. No Mn(III)-oxalate complex was detected in solution by UV-VIS measurements.

Voltammetric measurements were performed with an AIS model DLK-100 voltammetric analyzer and a standard three electrode system consisting of a SCE reference electrode with salt bridge, a Pt counter electrode and a working electrode. In some experiments, a solid state Au/Hg microelectrode was used as the working electrode and in others an EG&G model 303A static dropping mercury electrode was interfaced to the analyzer. Both the Mn(III) to Mn(II) reduction and the Mn(II) to Mn(0) reduction onto the Hg electrode could be monitored simultaneously.

RESULTS AND DISCUSSION

Stability of Mn(III) complexes (internal metal -ligand redox reactions). The stability of a variety of Mn(III) complexes was tested by preparing the Mn(III) complex while varying pH and ionic strength. Most complexes exhibited stability to Mn(III) disproportionation in solutions that were made in air as well as purged with Ar or N_2. However, in some cases, the ligand decomposed with liberation of Mn(II) and CO_2 because of intramolecular electron transfer from the ligand to Mn(III). This was expected for ligands which bind metals with only carboxyl groups (1-5). Decomposition of the Mn(III)-malonate complex by intramolecular electron transfer, as observed by the loss of absorbance of the 266 nm UV-VIS peak (9), occurred for ligand to metal concentration ratios from 10 to 1000, and the variation in the rate constant for decomposition was only a factor of two over this range of concentration ratios at a constant pH. Plots of ln[Mn(malonate)]$_2^-$ vs time were linear ($r^2 \geq 0.99$) and showed first order decomposition of the $[Mn(malonate)_2]^-$ at a ligand to metal concentration of 10; similar results were obtained in air and N_2 purged solutions. **Figure 1** shows the complex increases in stability with pH since the first order rate constant for decomposition of the complex decreases with increasing pH. The pK_{a1} and pK_{a2} are 2.83 and 5.69 for malonate, respectively and the stability of the Mn(III) complex appears related to the protonation of the malonate because the kinetic stability becomes stronger and constant at pH ≥ 6 (Fig. 1). This behavior is expected based on the known chemistry of Mn(III) complexes (3-5). At higher pH, MnL_3 complexes are expected in solution, but on lowering pH complexes of form $MnL_2(H_2O)_2$ are expected (3, 6). Duke (3) showed that the decomposition of the Mn(III)-oxalate complex, $[Mn(ox)_2(H_2O)_2]^-$ to Mn^{2+} and CO_2 at low pH followed pseudo first order kinetics with a rate law given in eq. 2. The malonic acid complex with Mn(III) follows a similar rate law. These $MnL_2(H_2O)_2$

$$-d[Mn(III)] / dt = k_{obs} \{[Mn(ox)_2(H_2O)_2]^-\} \quad (2)$$

complexes are very labile relative to MnL_3 complexes because water is a monodentate ligand and Mn(III) is a high spin, labile metal cation [**I**, d^4, $t_{2g}{}^3e_g{}^1$] in octahedral

(I)

$$\underline{\uparrow}\ \underline{\ \ } \quad e_g^* \quad (x^2\text{-}y^2, z^2)$$

$$\underline{\uparrow}\ \underline{\uparrow}\ \underline{\uparrow} \quad t_{2g} \quad (xz, yz, xy)$$

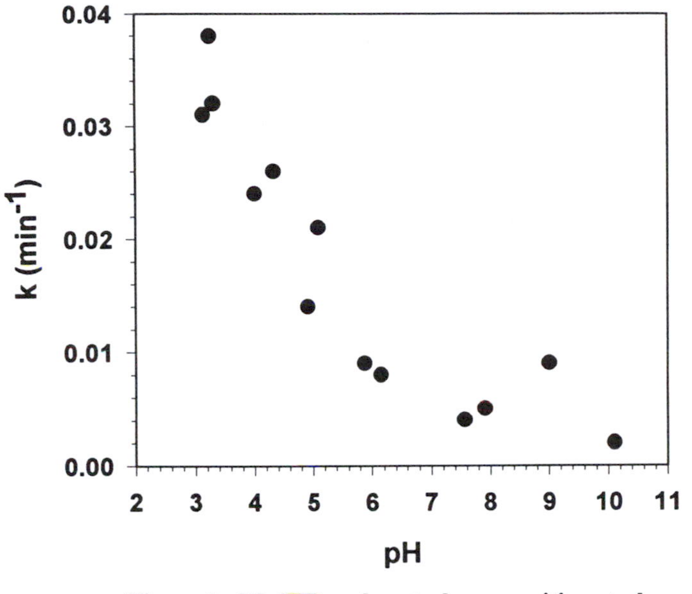

Figure 1. Mn(III) malonate decomposition study.
Mn(III) = 1mM; malonate = 10 mM.

geometry (23) which undergoes water exchange rapidly. Lability facilitates reactivity for inner sphere electron transfer processes and lowering pH enhances metal-ligand bond dissociation (23) as well as the redox reactions shown in Figure 1. Mn(III) is labile because one of the LUMO orbitals (e_g*) is not occupied and a geometry change towards square planar results: this is an example of the Jahn-Teller effect with long metal-ligand bonds that can dissociate on the z axis. As a consequence of its electron configuration, an one electron transfer with free radical generation for the organic reductant is expected in Mn(III) redox reactions (21, 24). Photochemical enhancement is slight for Mn(III) and Mn(IV) compounds because the LUMO orbital(s) is of σ symmetry (21).

These complexes as well as many of the others containing carboxylate groups (e.g., oxalate) showed similar stability in ionic strengths of 0.1 M NaCl and in seawater (I=0.7). However, Mn(III)-porphyrin complexes in particular decompose as ionic strength increases.

The Mn(III)bis-tris complex was also prepared and studied over the pH range 5-9. There was no change in the concentration of the complex over days as determined by monitoring the voltammetric peak at -1.1 V vs. SCE. We attribute the stability of this complex to the fact that bis-tris has no unsaturation or carboxylate groups that bind to Mn(III).

Mn(III) complexes with sulfide. Addition of ten-fold excess of sulfide to Mn(III) complexes resulted in the pseudo-first order reduction of Mn(III) to Mn(II) which could be easily followed by the decrease in the UV-VIS signal for the Mn(III) complex or by the decrease in the current of the Mn(III) to Mn(II) peak (**Figure 2**). The pseudo-first order rate constants, k_{OBS}, for sulfide reduction of Mn(III) complexes were greater than the k_{OBS} for internal redox with ligands described above by 6 (e.g., malonate) to > 500 (e.g., bis-tris). For the Mn(III)bis-tris complex, the rate of Mn(III) reduction did not vary over the pH range 5-9 and the ligand to metal ratio range of 10:1 to 1000:1, and the k_{OBS} for the reduction of Mn(III)bis-tris to Mn(II) by sulfide was determined to be 8.5 min^{-1}. These results indicate that the sulfide directly attacks the labile Mn(III)bis-tris complex in an inner sphere process (23) by a direct σ bonding interaction between the filled p orbital of the sulfide and the LUMO of Mn(III) which is of σ symmetry (e_g*). Electron transfer then follows with free radical generation.

Polymeric and soluble MnO_2 stability. Before testing the reactivity of polymeric and soluble MnO_2 with oxalate, we investigated its ability to coagulate and precipitate in NaCl and seawater solutions. Perez-Benito et al (12) indicated that increasing the ionic strength of solutions and increasing the amount of divalent cations relative to monovalent cations in solution induced precipitation. **Figure 3** shows the loss of MnO_2 measured by loss of absorbance when 0.100 mM of MnO_2 is added to varying NaCl concentrations with a final NaCl concentration of 0.01 to 0.04 M. The MnO_2 is almost totally precipitated at 30 hrs. For solutions of 0.01 mM MnO_2 and 0.05 M NaCl, 20 % of the MnO_2 is still in solution after 65 hrs. (data not shown). **Figure 3** shows that seawater of 0.5 (I=0.01 M) and 1.0 salinity units (I=0.02 M) has the same effect as 0.100 mM MnO_2 added to 0.02 M NaCl; i.e, almost total coagulation and precipitation occurs by 30 hrs. Seawater contains significant quantities of Ca^{2+} and Mg^{2+} which accelerate MnO_2 precipitation in accordance with the results of Perez-Benito et al (12).

Polymeric MnO_2 and oxalate reactivity. Based on these results, we investigated the well known reaction of oxalate, a microbial metabolite (24), and polymeric/soluble MnO_2 to compare with previous solid phase MnO_2 studies over the pH range 4-6 (21, 24) and to attempt to determine if Mn(III)-oxalate is an intermediate. Perez-Benito and Arias (25) have also performed this reaction but in 0.5 M acetic acid/acetate buffer solutions to maintain pH; their work showed that the acetate buffer could also reduce the MnO_2 over the pH region 4.3-5.1. To our knowledge, the reaction order for all reactants has not been reported; however, the reaction order for proton and other reductants with solid phase MnO_2 has been reported in some instances (24,26-28). To date the MnO_2 concentration dependence is generally unknown because most previous studies used solid MnO_2 phases; thus, soluble MnO_2 provides a way to estimate the reaction order.

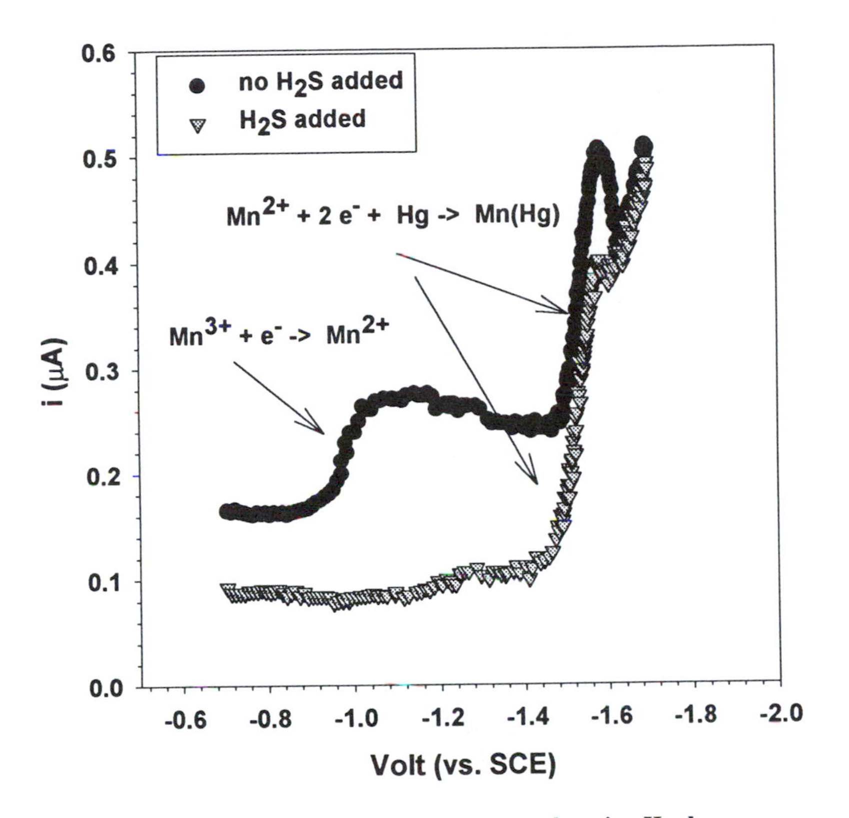

Figure 2. Linear sweep voltammograms at a hanging Hg drop electrode for Mn(III)bis-tris in seawater before and after addition of 1/2 an equivalent of sulfide at pH =7. The Mn(III) to Mn(II) reduction peak at -1.05 V vs. SCE decreases after sulfide addition. The total current for the Mn(III) reduction plus the Mn(II) to Mn(0) reduction before sulfide addition (peak at -1.57 V vs SCE) equals the total current for Mn(II) to Mn(0) after sulfide addition indicating that Mn(II) is the final product.

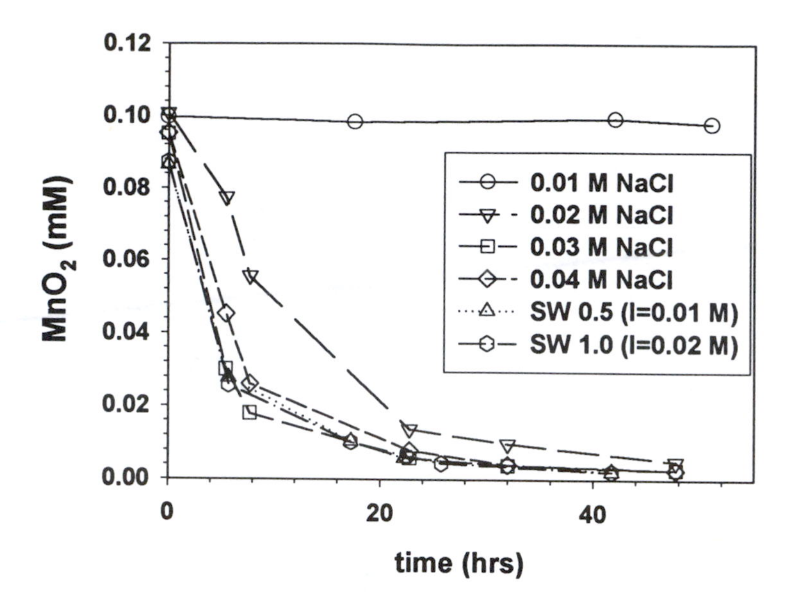

Figure 3. **MnO_2 (0.100 mM) precipitation induced by NaCl and seawater (SW).**

The overall reaction of oxalate with solid phase pyrolusite, β-MnO_2 has the following stoichiometry as observed by Xyla et al (21; eq. 3). In our reactions with soluble

$$4\,H^+ + C_2O_4^{2-} + MnO_2 \dashrightarrow 2CO_2 + Mn^{2+} + 2\,H_2O \qquad (3)$$

MnO_2 we found the same stoichiometry. **Figures 4-6** show representative kinetic studies at pH 4,5 and 6 for the reaction of 0.0200 and 0.0400 mM MnO_2 with 0.200 and 0.400 mM oxalate. These reactions are complete within 40 min. and are 60 times or more faster than the aggregation and precipitation reactions with salinity shown in **Figure 3**. Pseudo-first order kinetics are clearly observed and plots of ln[MnO_2] vs time were linear for the duration of the reaction with r^2 values of 0.99 or better. In the reaction of hydroquinone with solid phase MnO_2, Stone (26) showed linear behavior for ln{$[MnO_x]_0$ -[Mn(II)] / $[MnO_x]_0$} vs time plots only over the first 50 % of the reaction where ln{$[MnO_x]_0$ - [Mn(II)]/$[MnO_x]_0$} is a proxy for ln[MnO_2]. He attributed the non-linear behavior at longer

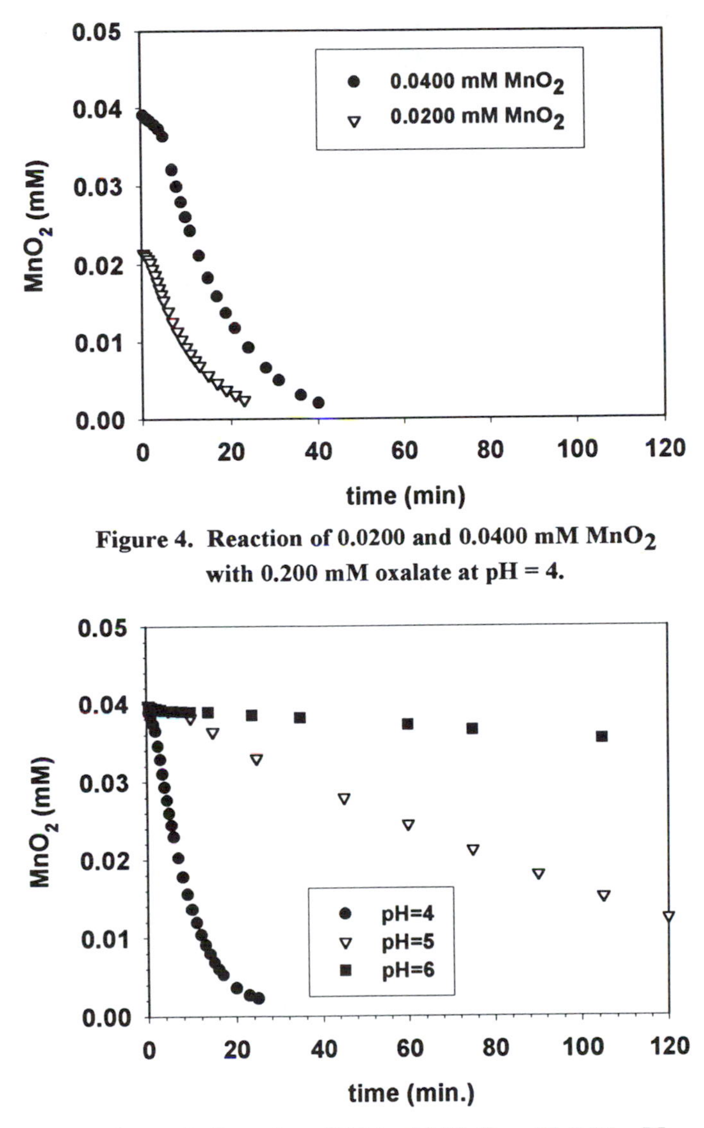

Figure 4. Reaction of 0.0200 and 0.0400 mM MnO_2 with 0.200 mM oxalate at pH = 4.

Figure 5. Reaction of 0.04 mM MnO_2 with 0.04 mM oxalate at pH 4, 5 and 6.

reaction times to alteration of the surface area and chemical characteristics of the solid oxide by extensive dissolution. This does not appear to be a problem with the polymeric MnO_2 which indicates it is reacting as a soluble entity.

Figures 5 and 6 show data for pH = 4 to 6 and demonstrate that the reaction is very slow at high pH in agreement with solid phase MnO_2 studies with a variety of organic reductants (21,26-28). **Figure 6b** shows the first 10 min. of the reactions in **Figure 6a**. At pH = 4, it is clear that a complex is forming prior to reaction since there is no decrease in the MnO_2 in the first 1.5 min for the reaction of 0.400 mM oxalate with 0.0400 mM MnO_2 and the first 3.0 min for the reaction of 0.0200 mM oxalate with 0.0400 mM MnO_2. These data show that a precursor complex of MnO_2 with oxalate is forming in accordance with previous solid phase studies (21, 24).

To determine the order of reaction for all the reactants in eq. 3, we used the initial rates of reaction calculated for the first 10 % loss of $[MnO_2]$ but after the precursor complex had formed. The rates were calculated as previously described (24,26-28) and representative data are in **Table 1**.

Table 1. Representative initial rate data used to determine the order of reaction for the reactants in eq. 3.

pH=4	**Initial rate (μM min^{-1})**
0.0400 mM MnO_2 with 0.400 mM oxalate	3.60
0.0200 mM MnO_2 with 0.200 mM oxalate	1.60
0.0400 mM MnO_2 with 0.200 mM oxalate	2.15
pH = 5	
0.0400 mM MnO_2 with 0.400 mM oxalate	0.355
0.0400 mM MnO_2 with 0.200 mM oxalate	0.254
pH = 6	
0.0400 mM MnO_2 with 0.400 mM oxalate	0.038

The empirical rate law (eq. 4) determined for eq. 3 is

$$\text{rate} = -\, d[MnO_2] / dt = k\, [MnO_2]^{0.37}\, [\text{oxalate}]^{0.75}\, [H^+]^{1.00} \qquad (4)$$

The rate law shows that the reaction is complex with an overall reaction order of approximately 2 and a second order rate constant, k, of 5.4×10^2 M^{-1} min^{-1}. Interestingly, the rate law shows that the order for oxalate is twice that of MnO_2 and suggests that a Mn-oxalate complex of unknown identity is forming in the reaction.

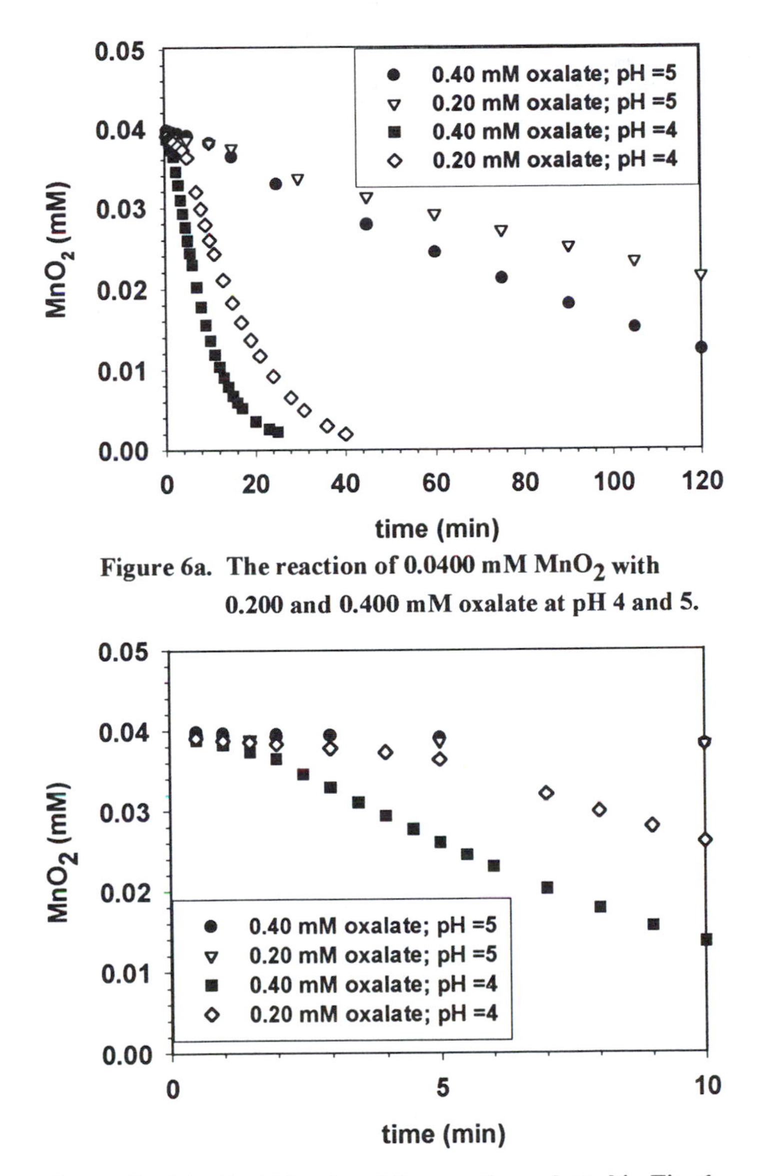

Figure 6a. The reaction of 0.0400 mM MnO_2 with 0.200 and 0.400 mM oxalate at pH 4 and 5.

Figure 6b. The first 10 min. of the reactions plotted in Fig. 6a.

The results in Figs. 5-6 indicate that acidity enhances the reaction in a manner similar to the intramolecular redox reaction of Mn(III) organic complexes and reductive dissolution of solid phase MnO_2 (21). However, the MnO_2 reactions are significantly slower than the intramolecular redox reactions because a precursor complex phase forms before noticeable redox (Fig. 6a). An inner sphere redox process is likely for oxalate reduction of soluble MnO_2 in accordance with the solid phase MnO_2 reaction in which surface complexation occurs (20, 23). Mn(IV) is an inert metal cation [**II**, d^3, t_{2g}^3] in

$$\text{(II)} \qquad \begin{array}{lll} \_\_ \;\; \_\_ & e_g^* & (x^2\text{-}y^2,\ z^2) \\ \underline{\uparrow}\;\; \underline{\uparrow}\;\; \underline{\uparrow} & t_{2g} & (xz,\ yz,\ xy) \end{array}$$

octahedral symmetry that does not undergo ligand exchange easily because inertness retards reactivity (23). Thus acidity is required to form Mn-OH and Mn-(H_2O) species which can dissociate from the Mn(IV) during attack by oxalate. Because the LUMO orbitals (e_g^*) are totally unoccupied and unpairing of electrons from the oxalate is not required during Mn(IV) reduction, two electrons could be transferred directly to Mn(IV) (21, 29) if the oxalate could bind directly to the Mn(IV). However, the complex nature of the rate law indicates that more than one mechanism may occur simultaneously; thus we cannot indicate whether two one-electron processes or one two-electron transfer may be operational.

An important question in determining the nature of the electron transfer step(s) is whether Mn(III) forms as an intermediate as Mn(IV) reduces to Mn(II). UV-VIS data in **Figure 7** demonstrate that it is not possible to distinguish or measure a Mn(III)-oxalate complex in solution during the reaction. The spectrum of soluble MnO_2 in oxalate solution does not show the characteristic spectrum of the Mn(III)-oxalate complex with a peak centered at 450 nm; it shows the spectrum of soluble MnO_2. Mn(III)-oxalate can form and decompose quickly to Mn^{2+} and CO_2 (3). Because the internal redox reaction of Mn(III)-oxalate is fast (3), not enough of the complex may build up as an intermediate to detect under our reaction conditions. Thus, we cannot distinguish between a single two-electron step or two one-electron steps.

However, our data do provide further support for the mechanism of MnO_2 reduction initially proposed by Stone for hydroquinone (26). The reaction scheme involves four steps but we have combined the release and detachment of Mn^{2+} and CO_2 into one step for simplicity (eqs. 5-7).

Precursor complex formation without reduction:

$$MnO_2 + C_2O_4^{2-} \underset{k_{-1}}{\overset{k_1}{\rightleftharpoons}} (MnO_2,\ C_2O_4^{2-}) \tag{5}$$

Electron transfer within the precursor complex

$$(MnO_2,\ C_2O_4^{2-}) \underset{k_{-2}}{\overset{k_2}{\rightleftharpoons}} (Mn^{3+},\ C_2O_4^{-}) + (Mn^{2+},\ C_2O_4) \tag{6}$$

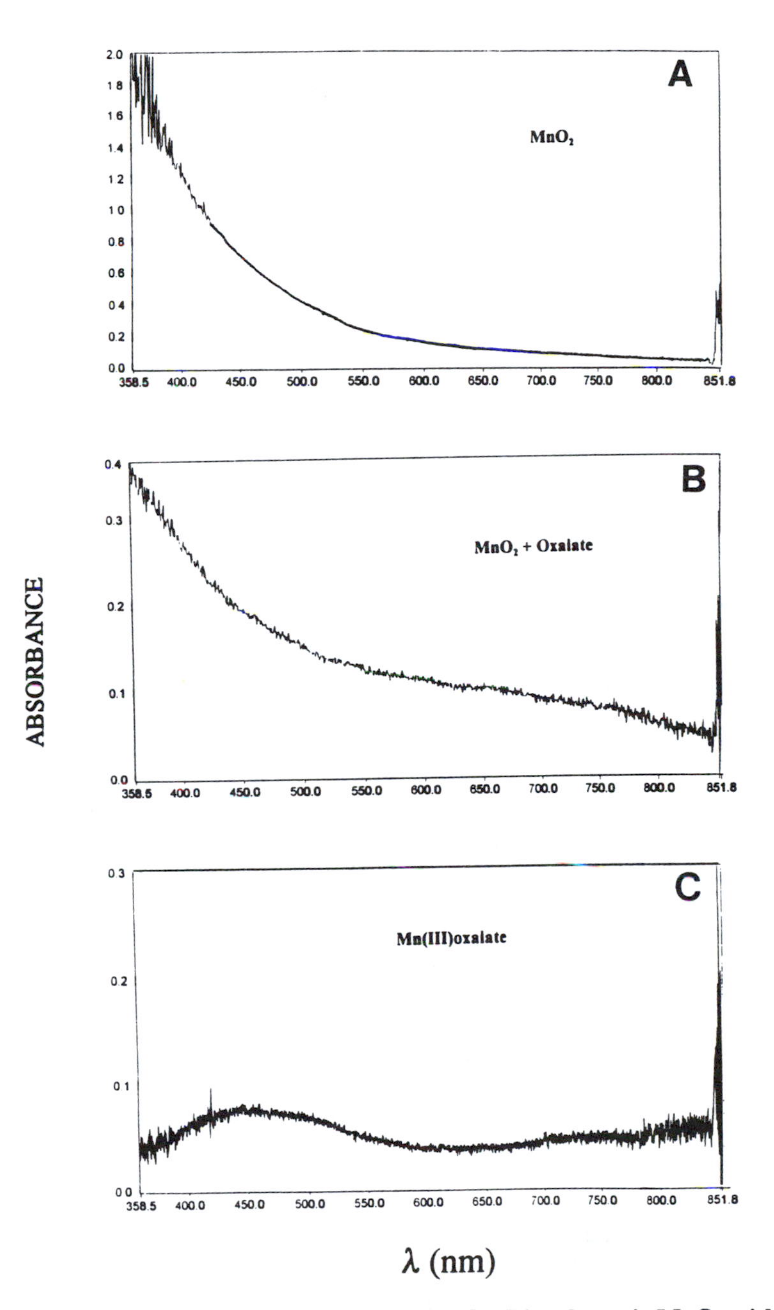

Figure 7. UV-VIS spectra for (A) polymeric MnO_2, (B) polymeric MnO_2 with oxalate and (C) $Mn(oxalate)_2^-$. Both (B) and (C) decompose as the spectrum is obtained.

Release and detachment of Mn^{2+} and CO_2

$$(Mn^{3+}, C_2O_4^{-}) + (Mn^{2+}, C_2O_4) \rightarrow Mn^{2+} + 2\ CO_2 \quad (7)$$

Figure 6b clearly shows that a precursor complex forms without significant reduction of polymeric MnO_2 and release of Mn^{2+} in the first couple of minutes of the reaction which is then followed by fast reduction and release of Mn^{2+}. The release of Mn^{2+} to solution is clearly dependent on the formation of a precursor complex and agrees with the kinetic expression given by Stone (26) for the dissolution rate of solid phase MnO_2 (eq. 8; modified

$$d[Mn^{2+}] / dt = - d[MnO_2] / dt = k_2 [(MnO_2, C_2O_4^{2-})] \quad (8)$$

for oxalate). Interestingly, eq. 8 is similar to the pseudo first order rate equation for the decomposition of the Mn(III)-oxalate complex, $[Mn(ox)_2(H_2O)_2]^-$, to Mn^{2+} and CO_2 first proposed by Duke (3; see eq. 2) in 1947. These data also indicate that electron transfer occurs in the complex which agrees with the data on Mn(III)-carboxylic acid complexes described above (an intramolecular inner sphere redox process) and which we observed for Mn(III)-oxalate. Unfortunately it is not possible to distinguish Mn^{3+} from Mn^{2+} in the electron transfer step (eq. 5) because Mn(III)-oxalate was not detectable in solution; thus, both Mn^{3+} and Mn^{2+} may form in the electron transfer step (eq. 6) with subsequent detachment of Mn(III)-oxalate radical and/or Mn^{2+} and CO_2. In addition, it is possible that Mn^{2+} and Mn^{3+} on the surface of MnO_2 can be oxidized by the MnO_2 in an autocatalytic inner sphere process before detachment; the original Mn^{2+} and Mn^{3+} then become Mn^{3+} and MnO_2 which are reduced by the oxalate. The complexity of the rate law (eq. 4) is consistent with multiple electron transfer reaction pathways for eq. 6. Our data do not support a buildup of the Mn(III)-oxalate complex, $[Mn(ox)_2(H_2O)_2]^-$, in solution which has been observed in sulfuric acid solutions of permanganate, Mn^{2+} and oxalate at $pH < 2$ (30).

Conclusions

Mn(III) complexes with ligands containing only carboxyl groups are not stable to intramolecular redox processes which result in Mn(II) and CO_2 formation; these reactions are pH dependent. Mn(III) organic complexes that do not have carboxyl groups and other unsaturated groups appear stable to intramolecular redox reactions and to Mn(III) disproportionation. All Mn(III) complexes tested react quickly with sulfide and are hypothesized to be important oxidants in organic matter decomposition zones where O_2 and H_2S are not present in significant quantities such as the suboxic zone of the Black Sea and sediments.

The empirical rate law for the reaction of oxalate with soluble and polymeric MnO_2 over the pH range of 4-6 is non-integer for both oxalate and MnO_2 which indicates that there are several electron transfer reaction pathways possible. Oxalate forms a precursor complex with MnO_2 prior to reaction as observed by almost constant MnO_2 absorbance after the initial addition of MnO_2 to the oxalate. After precursor complex formation, MnO_2 reacts with formation of Mn(II) and CO_2. MnO_2 acts as a "dissolved" species and it is not clear that there is a discrete dissolution step although the precursor complex may result in break-up of the polymeric MnO_2 with possible formation of Mn(III)-oxalate. However, Mn(III)-oxalate is not measureable as an intermediate by UV-VIS. These results do not allow us to distinguish whether the electron transfer step(s) may be a single two electron process or two discrete one-electron steps which may involve other autocatalytic processes. Polymeric MnO_2 is a potential strong oxidant in the environment which may be transported by physical processes because of its soluble character. Microbial catalysis of organic matter decomposition in the environment should also be very important with a polymeric species since less energy would be required to break-up the polymer for reaction compared to solid phase MnO_2.

The use of soluble and polymeric MnO_2 allows for direct analysis of MnO_2 as a reactant. In previous studies (21, 26-28), the solution had to be filtered to remove solid MnO_2 in order to determine soluble Mn^{2+}. Any solid MnO_2 remaining in the reaction vessel was calculated by the difference between the initial solid phase concentration and the Mn^{2+}.

Acknowledgments

This work was supported by a grant from the National Science Foundation (DEB-9612293).

Literature Cited

1. Froelich P. N., Klinkhammer G. P., Bender M. L., Luedtke N. A., Heath G. R., Cullen D., Dauphin P. Hammond D., Hartman B., and Maynard V., *Geochim. Cosmochim. Acta* **1979**, *43*, 1075-1090.
2. Burdige D. J.; Nealson K. H. *Geomicrobiol. J.* **1986**, *4*, 361-387.
3. Duke, F. *J. Amer. Chem. Soc.* **1947**, *69*, 2885-2888.
4. Hamm, R. E.; Suwyn, M. A. *Inorg. Chem.* **1967**, *6*, 139-142.
5. Li, X.; Pecoraro, V. L. *Inorg. Chem.* **1989**, *28*, 3403-3410.
6. Magers, K. D.; Smith, C. G.; Sawyer, D. T. *Inorg, Chem.* **1978**, *17*, 515-523.
7. Richens, D. T.; Smith, C. G.; Sawyer, D. T. *Inorg. Chem.* **1979**, *18*, 706-712.
8. Gold, M. H.; Wariishi, H.; Valli, K. In *Biocatalysis in agricultural biotechnology*; J. R. Whitaker and P. E. Sonnet; Eds.. ACS Symposium Series 1989; *Vol. 389*, 127-136.
9. Popp, J. L.; Kalyanaraman, B.; Kirk, T. K. *Biochemistry* **1990**, *29*, 10475-10480.
10. Larson, E. J.; Pecoraro, V. L. In *Manganese Redox Enzymes*; Ed. Pecoraro, V. L. VCH: New York, 1992; Ch. 1, 1-28.
11. Kostka J. E.; Luther,III, G. W.; Nealson, K. H. *Geochim. Cosmochim. Acta* **1995**, *59*, 885-894.
12. Perez-Benito, J. F.; Brillas, E.; Pouplana, R. *Inorg. Chem.* **1989**, *28*, 390-392.
13. Rue, E. L.; Bruland, K. W. *Mar. Chem.* **1995**, *50*, 117-138.

14. Wu, J.; Luther, III, G. W. *Mar. Chem.* **1995**, *50*, 159-177.
15. Luther, III, G. W.; Ferdelman, T.; Tsamakis, E. *Estuaries* **1988**, *11*, 281-285.
16. Luther, III, G. W.; Church, T. M.; Powell, D. *Deep Sea Research* **1991**, *38* (Suppl. 2A), S1121-S1137.
17. Millero, F. J. *Est. Coast. Shelf Sci.* **1991**, *33*, 521-527.
18. Murray J.W.; Codispoti, L. A.; Friederich, G. E. In *Aquatic Chemistry: Interfacial and Interspecies Processes*; C.P. Huang, C.R. O'Melia and J.J. Morgan; Eds. American Chemical Society: 1995; pp. 157-176.
19. Wiberg, K.; Deutsch; C. J.; Roček, J. *J. Amer. Chem. Soc.* **1973**, *95*, 3033-3035.
20. Perez-Benito, J. F.; Arias, C. *J. Colloid Interface Sci.* **1992**, *152*, 70-84.
21. Xyla, A. G.; Sulzberger, B.; Luther, III, G. W.; Hering, J. G.; Van Cappellen, P.; Stumm, W. *Langmuir* **1992**, *8*, 95-103.
22. Luther, III, G. W.; Sundby, B.; Lewis, B. L.; Brendel, P. J.; Silverberg, N. *Geochim. Cosmochim. Acta* **1997**, *61*, 4043-4052.
23. Shriver, D. F.; Atkins, P.; Langford, C. H. *Inorganic Chemistry* (2nd. Ed.), W. H. Freeman and Co., New York, 1994.
24. Stone, A. T.; Morgan, J. J. *Environ. Sci. Tech.* **1984**, *18*, 450-456.
25. Perez-Benito, J. F.; Arias, C.; Amat, E. *J. Colloid Interface Sci.* **1996**, *177*, 288-297.
26. Stone, A. T. *Environ. Sci. Tech.* **1987**, *21*, 979-988.
27. Stone, A. T. *Geochim Cosmochim. Acta* **1987**, *51*, 919-925.
28. Laha, S.; Luthy, R. G. *Environ. Sci. Tech.* **1990**, *24*, 363-373.
29. Luther, III, G. W. In *Aquatic Chemical Kinetics*; Ed. W. Stumm; John Wiley and Sons: New York, 1990, pp. 173-198.
30. Pimienta, V.; Lavabre, D.; Levy, G.; Micheau, J. C. *J. Phys. Chem.* **1994**, *98*, 13294-13299.

Photochemical and Microbially-Mediated Processes

Chapter 14

Degradation of Tetraphenylboron at Hydrated Smectite Surfaces Studied by Time Resolved IR and X-ray Absorption Spectroscopies

D. B. Hunter[1], W. P. Gates[1,3], P. M. Bertsch[1], and K. M. Kemner[2]

[1]Advanced Analytical Canter for Environmental Sciences, University of Georgia, Savannah River Ecology Laboratory, Drawer E, Aiken SC 29802
[2]Environmental Research Division, Argonne National Laboratory, Argonne, IL 60439

The surface catalyzed redox reactivity of an organoboron probe molecule (tetraphenylboron, or TPB) with fully hydrated clay mineral surfaces is described wherein one or both of two degradation pathways can occur. Attenuated total reflectance Fourier transform infrared (ATR-IR) spectroscopy can quantitatively measure, *in situ*, both acid hydrolytic and oxidative degradation of TPB as controlled by the exchange cation identity and the structural octahedral Fe content of the clay mineral. Oxidation of TPB at smectite surfaces is directly attributable to octahedral Fe(III) in the clay structure. The concomitant reduction of Fe(III) during TPB oxidation can be analytically measured by X-ray absorption near edge structure (XANES) spectroscopy. The resultant distortion of the clay mineral structure during the reduction of structural Fe could be probed directly by extended X-ray absorption fine structure (EXAFS) spectroscopy. The lack of evidence for clay mineral dissolution during the reaction with TPB permits defining the clay as a catalyst. The combination of these three *in situ* spectroscopic techniques to measure the reaction in real time, provides powerful insight into the reactivity of clay mineral surfaces.

The fate of organic contaminants in soils and sediments is of primary concern in environmental science. The capacity to which soil constituents can potentially react with organic contaminants may profoundly impact assessments of risks associated with specific contaminants and their degradation products. In particular, clay mineral surfaces are known to facilitate oxidation/reduction, acid/base, polymerization, and hydrolysis reactions at the mineral-aqueous interface (1, 2). Since these reactions are occurring on or at a hydrated mineral surface, non-invasive spectroscopic analytical methods are the preferred choice to accurately ascertain the reactant products and to monitor reactions in real time, in order to determine the role of the mineral surface in the reaction. Additionally, the *in situ* methods employed allow us to monitor the ultimate changes in the physico-chemical properties of the minerals.

An example where understanding the fate of potential organic contaminants in terrestrial and aquatic environments is important is the current use of Na-

[3]Current address: CSIRO Land and Water, Glen Osmond, SA 5064, Australia.

tetraphenylboron anion (TPB) at the Savannah River Site (SRS), a U.S. Department of Energy nuclear materials processing facility, located near Aiken, SC. Na-TPB is used in the precipitation of high-level cesium-containing liquid radioactive wastes at the SRS Defense Waste Processing Facility. An estimated 200,000 kg per year over 30 years will be needed in the vitrification of thirty-five years of accumulated high level waste at the SRS, thus requiring the first large scale industrial production of this chemical (3). Although TPB is relatively non-toxic to soil microorganisms and plants, the same is not true for its degradation products. For instance, one degradation product, diphenylboric acid (DPBA), has been shown to inhibit nitrification processes controlled by soil microorganisms (4, Bertsch, Univ. of Georgia, unpublished results). Such large-scale industrial usage increases the potential for accidental releases of TPB to the environment and necessitates understanding the behavior and ultimate fate of this chemical in soil and aquatic systems. This chapter will illustrate how *in situ* surface-probe spectroscopies can be used to provide fundamental understanding of TPB degradation and the role that mineral surfaces play in the process. While TPB is a specific organic molecule, a complete understanding of its reactivity at smectite surfaces may provide insight regarding other, more toxic organic contaminants.

Two forms of acidity predominate at mineral surfaces : Lewis acidity and Brønsted acidity. Lewis acidity arises from an electron pair transfer facilitated by secondary and tertiary structures of minerals, e.g., cations located in the structure of the mineral or at crystal edges (1, 5). The associated Lewis base is operationally defined as an adsorbate at the mineral surface. A mineral surface can behave as a Lewis acid by accepting electrons from adsorbed organic molecules (1, 6) where biotic degradation can be discounted (4, 6, 7). One example of the Lewis acidity is the reduction of structural Mn(III, IV) during the oxidation of phenolic compounds at manganese oxide surfaces (8, 9). Another example is the ring cleavage of pyrragol, a precursor in the formation of humic polymers (2, 8), facilitated by the presence of Fe(III) in the structure of nontronite, the Fe-rich end member phyllosilicate (10).

Brønsted acidity of mineral surfaces arises from proton transfer when exchangeable cations induce dissociation of interlayer water. In general, Brønsted acidity increases with increasing cation charge:radius ratio. That clay minerals also can behave as strong Brønsted acids is primarily attributable to the presence of acidic exchangeable cations (1, 6), such as Al^{3+}. Examples of the Brønsted acidity of mineral surfaces include protonation of ammonia and the formation of pyridinium ions upon sorption at acidic Brønsted acid sites (12).

Both Lewis and Brønsted acid degradation pathways have previously been described for TPB (4, 13). The first pathway is a two-electron transfer oxidation reaction producing diphenylboric acid (DPBA) and biphenyl (BP). We will define the degradation of TPB to DPBA as occurring at Lewis acid sites (14) at hydrated clay surfaces as shown in Figure 1A. The second pathway is an acidic (H^+) degradation to triphenylboron (TriPB), which subsequently decomposes in the presence of O_2 to phenyl diphenylborinate. Diphenylborinate then undergoes hydrolysis to yield phenol and DPBA. We will define the acidic degradation of TPB to TriPB as occurring at Brønsted acid sites on the mineral surface as shown in Figure 1B.

The Brønsted and Lewis degradation reactions depicted in Figure 1 are diagrammed schematically in Figure 2 with respect to an Fe-containing 2:1 phyllosilicate mineral surface. Smectites are hydrated 2:1 layered silicates (phyllosilicates) composed of an (Fe, Al, Mg) octahedral sheet linked between two (Si, Al) tetrahedral sheets by oxygen ligands (11). Most smectites contain appreciable amounts of octahedral Fe(III) (15) and are effected by redox reactions with the aqueous phase (15, 16, 17). Typically, little dissolution of the phyllosilicate occurs during redox reactions (18) as opposed to Fe(III) hydrous oxides, where dissolution and transformation are a predominant result of redox reactions (15, 19, 20). This chapter will highlight the capabilities of three distinct time-resolved spectroscopies to probe the reactions of TPB at phyllosilicate

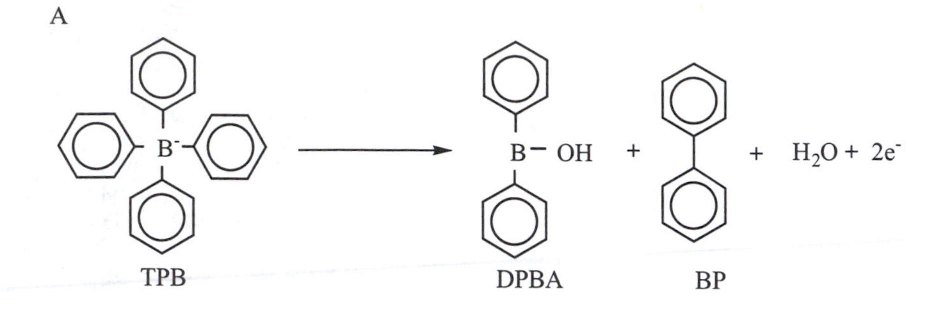

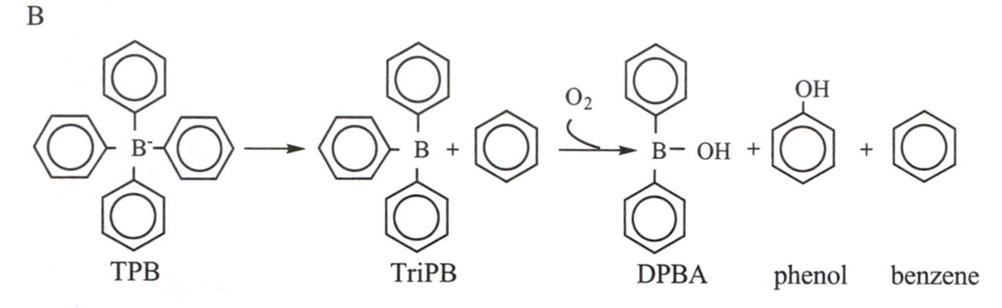

Figure 1. Primary reaction pathways for Lewis (A) and Brønsted (B) degradation of tetraphenylboron (TPB). Short-lived intermediates, e.g., radicals, have been excluded as they are not observed by the spectroscopic methods employed.

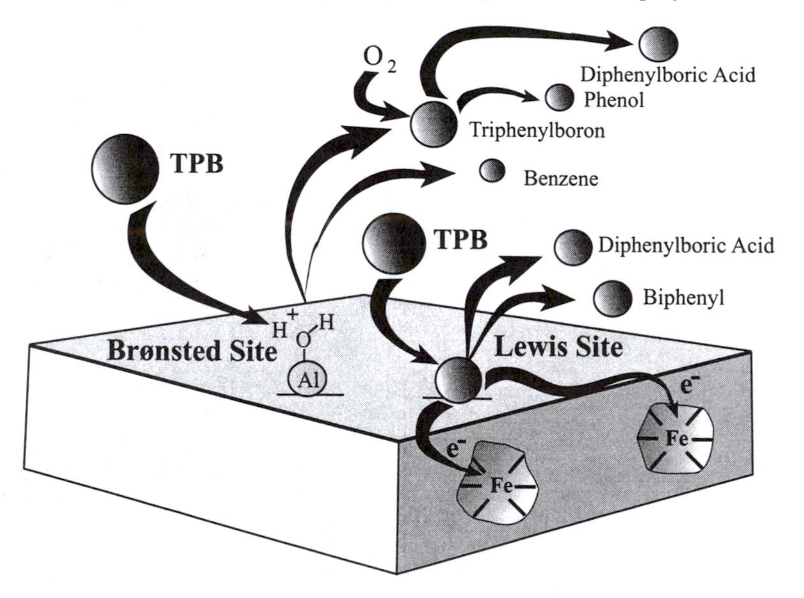

Figure 2. Representation of multiple reaction pathways of TPB catalyzed degradation at smectite surfaces that were monitored directly by non-invasive *in situ* spectroscopic techniques.

surfaces: 1) infrared (IR) spectroscopy to monitor the multiple degradation pathways of the organo-boron; 2) X-ray absorption near edge structure (XANES) to track the associated reduction of structural Fe in the clay mineral during oxidation of TPB at the Lewis site; and 3) extended X-ray absorption fine structure (EXAFS) to probe mineral structural changes resulting from the reduction of octahedral Fe(III).

IR Analysis: Fate of TPB

Infrared spectroscopy has previously been shown to be a powerful tool for studying reactions at surface acid sites using a variety of surface-sorbed probe molecules (12, 13, 21, 22). For example, the protonation of NH_3 to NH_4^+ when adsorbed at Brønsted sites can be clearly distinguished from NH_3 retained at Lewis sites using IR (12). Similarly, the formation of pyridinium ions upon the adsorption of pyridine at Brønsted sites is spectrally distinct from pyridine sorbed at Lewis sites (12). We have taken advantage of IR spectroscopy to identify and quantify the degradation products, TriPB and DPBA, as pathway-specific probes of TPB reactivity at respective Brønsted and Lewis acid sites of phyllosilicates (13).

From these studies, we can describe the reactive site(s) on hydrated reference clay minerals where either the electron acceptor (Lewis acid) or the proton donor (Brønsted acid) pathway can be detailed. Three reference clay minerals were chosen from the smectite family with known structural iron contents varying from ~nil to high Fe substitution (hectorite, SHCa-1; montmorillonite, SWy-1 and SWy-2; and nontronite, NG-1, respectively) (15). Their respective abilities to facilitate degradation of TPB to DPBA via a Lewis acid pathway were then observed. The degradation of TPB to TriPB could be examined by varying the Brønsted acidity of the predominant exchangeable cation (Na, Ca or Al) as well as the relative water content of the organo-clay system.

The application of IR spectroscopy to organic sorbates has been primarily limited to gas-phase adsorption kinetics and gas-phase catalysis (21). The usefulness of IR for investigating sorption/desorption processes *in situ*, has been demonstrated for the selectivity of conformer adsorption at mineral surfaces (22). Strong absorption of H_2O vibrational modes by IR radiation has been a major hindrance in the application of IR spectroscopy to study organic sorption at aqueous-mineral interfaces. Attenuated total reflectance-IR (ATR-IR) spectroscopy and the use of D_2O in the aqueous phase minimizes the water absorption problem. Figure 3 details the ATR setup used in this study. Clay pastes were loaded into Teflon plaques and clamped to both sides of a vertical ATR prism. Silicon sealant around the edges of the plaques prevented water evaporation during extended data collection times (up to 2 days). The area of the D_2O bending mode at 1250 cm^{-1} was measured to ensure that the pastes did not dehydrate in the purged environment of the optical bench. The ZnSe prism was placed in the vertical ATR cell adjusted to 45° angle of incidence. This geometry resulted in a calculated depth of penetration of 0.73 μm into the sample by the evanescent wave at 1580 cm^{-1} with an overall effective pathlength of 12.2 μm (23). ATR-IR is ideally suited for such a system with both strongly absorbing (both water and clay) and weakly absorbing (relatively dilute adsorbed organic compounds) matrices. The very short effective pathlengths achieved minimize the signal of the clay and water, but a strong signal to noise ratio can be achieved for the sorbed organic compounds from the multiple reflections off both sides of the prism as the beam passes down the crystal. In this study, identification of TPB and its reaction products could be performed in a spectral region made free from interferring bands associated with H_2O (the H_2O bending mode at 1640 cm^{-1}) by using D_2O as a solvent (bending mode is shifted to 1250 cm^{-1} for D_2O).

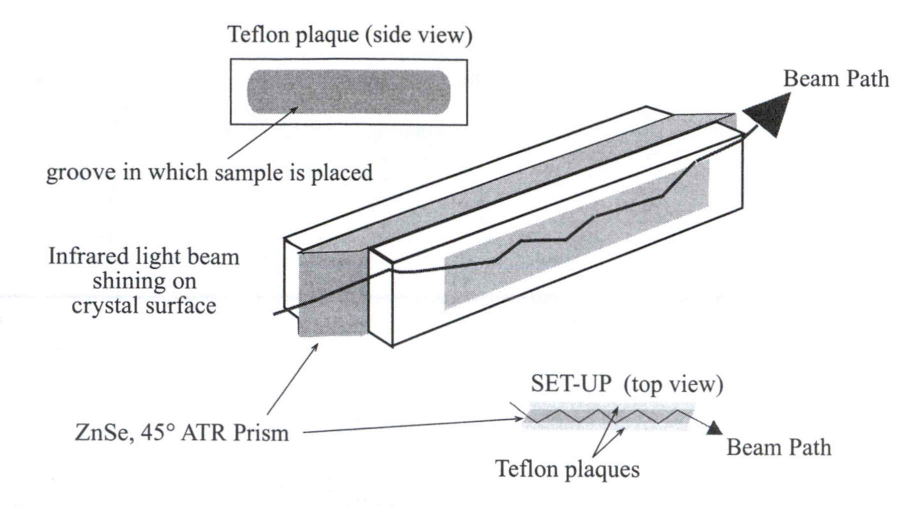

Figure 3. Schematic diagram of the vertical attenuated total reflectance setup. Many more reflections occur as the IR beam passes down the prism than are illustrated (see text).

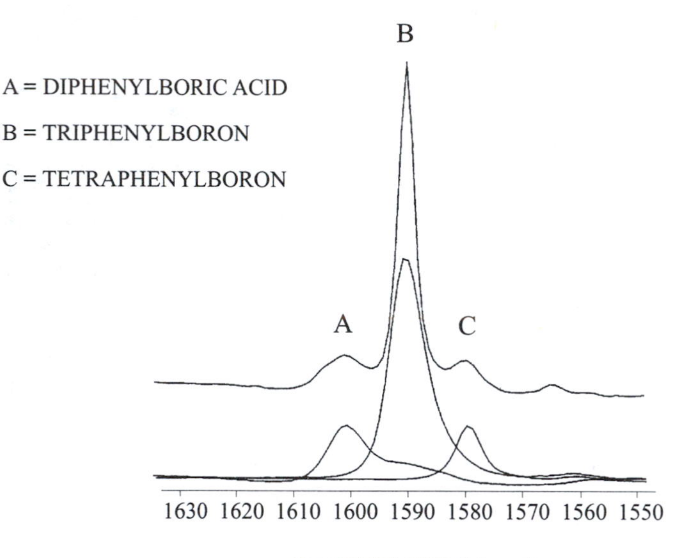

Figure 4. Bands of interest for quantitative determination of major mother and daughter products in the reaction of TPB at clay surfaces. Top trace is taken during the reaction of TPB on Al-exchanged SWy-1 and lower trace is an overlay of TPB (C), DiPB (A) and TriPB (B) aqueous references.

In the fully hydrated clay paste, the v_4 modes (24) of the surface-adsorbed organo-borate species exhibited a small 5 cm^{-1} shift compared to the aqueous references (Figure 4). Spectra of adsorbed molecules tend to resemble those seen in the liquid state, but it has been observed that surface interactions can shift band frequencies to either higher or lower values (12). Surface interactions can induce dipoles where none would otherwise exist and give rise to dipole oscillations in vibrations that would not normally exhibit them, thus, resulting in the appearance of bands not normally seen in the IR spectrum of aqueous references (12). The contrast of very intense and medium intense vibrational modes for TriPB, TPB and DPBA, as well as the excellent agreement with the HPLC results (13), strongly supports the peak assignments of chemical species despite their proximity and wavelength shift.

Lewis degradation pathway of TPB.

An example of time-resolved IR spectra of tetraphenylboron degradation on Na-exchanged SWy-1 demonstrates the appearance of the band associated only with DPBA over a 30-hour period (Figure 5). The time-resolved spectra of Na-NG-1, Ca-SWy-1 were qualitatively similar to Na-SWy-1 in that only DPBA was observed to form as a daughter product. Replicates of the clay pastes were also sampled at regular time intervals by extraction of the organo-boron compounds with acetonitrile, and these extractions were analyzed by HPLC (4, 13). HPLC (data not shown) provided a separate corroborating method to confirm the IR spectral interpretations as well as quantitative determinations (Figure 6).

IR was found to be a more reproducible method for studying the surface facilitated degradation rates of TPB than extraction/HPLC techniques (13). Only end products in the degradative reactions were considered in this study. Short-lived intermediates, such as phenyl diphenylborinate, which forms during the hydrolysis of TriPB to DPBA, were neglected. Potentially, IR is better able to observe the presence of these intermediates *in situ* , if they are present in sufficient concentration, as compared to a method requiring time consuming replicated extractions involving significant sample manipulation. In contrast, HPLC has better detection limits (approximately 10^3 times better) and is better able to speciate other reaction products such as biphenyl and phenol that may be obscured by the many other absorption bands present in the IR spectra of aqueous clay-TPB pastes. Both approaches gave the same fundamental information which was sufficient to isolate and model the degradation pathways (13). Although it was possible to successfully model the reactions, it should be noted that the model is insensitive to unique solutions because of multiple exponential terms. The goodness-of-fit of the described model does not indicate that the reactions are truly first-order or that other reaction schemes would not be as appropriate.

The IR spectra of TPB added to Na-hectorite (SHCa-1) remained unchanged over 40 h. Hectorite represents a low Fe end member in the smectite family with less than 0.01 mole% Fe(III) based on the number of Fe(III) cations per $O_{20}(OH)_4$ / total number of cations per $O_{20}(OH)_4$ (25). The specific role of structural Fe(III) in smectites in the Lewis acid degradation pathway of TPB is emphasized by the fact that when the Fe oxides associated with SHCa-1 (2% by total mass) were present, the hectorite suspension yeilded no reaction (13). TPB was reactive on Na-exchanged montmorillonite (SWy-1) and nontronite (NG-1) which had 10.1 and 98 mole% of Fe(III). The stability of TPB in the SHCa-1 clay suspension also clearly demonstrates that microbial degradation was not a factor in these experiments, which agrees with earlier studies (4).

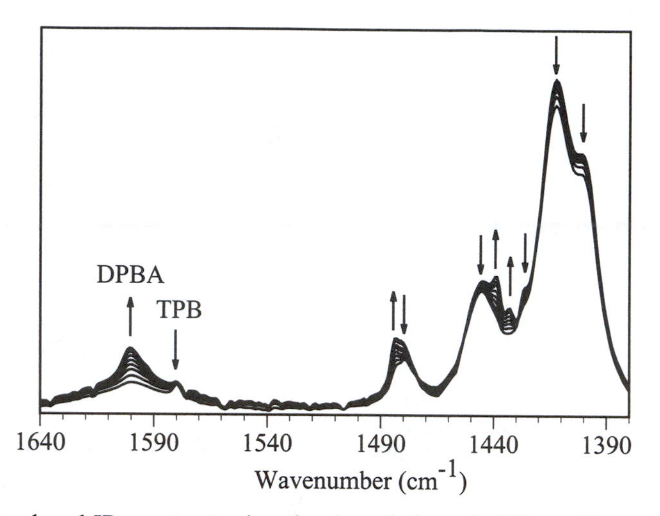

Figure 5. Overlayed IR spectra tracing the degradation of TPB on Na-exchanged montmorillonite (SWy-1) collected in 5-hour intervals over 40 hours. Arrows indicate appearance or loss of spectral bands with time of reaction.

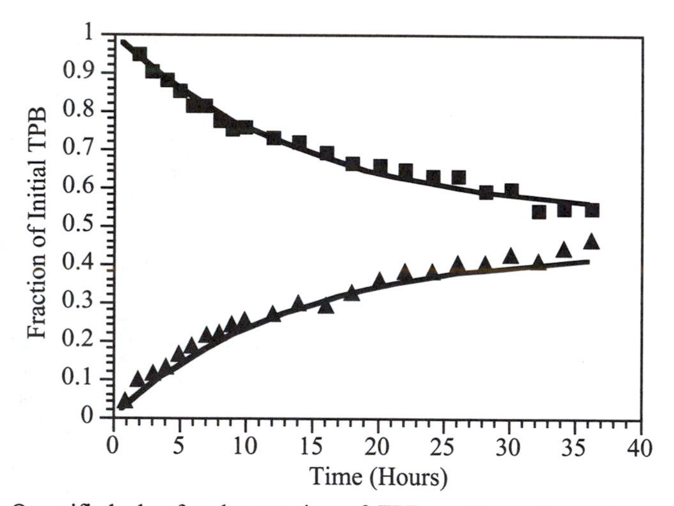

Figure 6. Quantified plot for the reaction of TPB on Na-exchanged montmorillonite (SWy-1) measured from IR spectra. Squares indicate loss of TPB determined by the band at 1580 cm^{-1}, and diamonds indicate production of DPBA determined by the band at 1600 cm^{-1}. The solid line is the fit from equations (1)-(3) with k_1=0.1 hr^{-1}.

Brønsted degradation pathway of TPB.

The Al-SHCa-1 spectra shown in Figure 7 demonstrate the initial production and subsequent loss of TriPB at 1590 cm^{-1} are observed, along with the eventual production of DPBA at 1600 cm^{-1}. Since TPB on a Na-exchanged SHCa-1 is not expected to degrade via the Lewis pathway, we infer that TPB degradation on the Al-SHCa-1 surface is via the Brønsted degradation pathway only. These results are quantified in Figure 8 where a rapid degradation of TPB to TriPB is observed. At longer reaction times TriPB degraded to DPBA.

Simultaneous Lewis and Brønsted degradation pathway of TPB.

When Al-exchanged SWy-1 clay is reacted with TPB, two bands associated with DPBA and TriPB arise over a similar time period (Figure 9). The spectra reveal the slow loss of TriPB after a very rapid initial appearance. Quantification of the reaction of TPB with Al-SWy-1 revealed that both TriPB and DPBA were formed as products, indicating that both Brønsted and Lewis reaction pathways are initially operating simultaneously (Figure 10). Note, however, that as TriPB degrades, additional formation of DPBA was observed, indicating that the second step in the Brønsted reaction continued for extended time periods. TriPB degrades to DPBA in the presence of oxygen (Figure 1B) and the increased production of DPBA in the Al-SWy-1 system after 5 hours results from this reaction, since the Lewis pathway would be insignificant following depletion of TPB.

Reaction Kinetics

We have demonstrated that either or both the Lewis and Brønsted pathways can occur under controlled conditions, specifically, when the mineral contains structural Fe(III) and/or the exchange cation is Al^{3+}. It is also known, that TriPB degrades ultimately to DPBA in the presence of oxygen (14). If O_2 is not limiting in the experimental setup, a simple, overall reaction scheme for the oxidative degradation of TPB at clay mineral surfaces can be described (13):

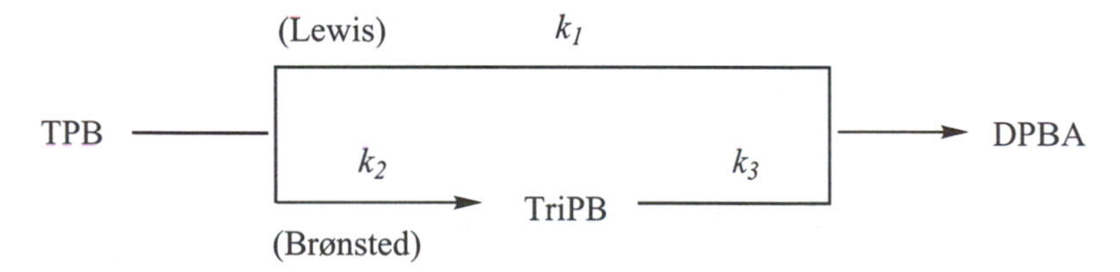

where k_i=1,2,3 are proposed first-order rate coefficients. Only the Lewis pathway operates in the Na/Ca-SWy-1 and Na-NG-1 systems (Figure 1A); only the Brønsted pathway operates in the Al-SHCa-1 system (Figure 1B); both pathways operate in the Al-SWy-1 system. First-order rate equations based on the above model can be solved (26) to predict the concentrations of, respectively, TPB, DPBA and TriPB as a function of time. If first-order kinetics are applied, then:

$$C_{TPB}(t) = C_{TPB}(0)e^{-k_T t} \quad (1)$$

$$C_{TriPB}(t) = C_{TriPB}(0)e^{-k_3 t} + \frac{k_2 k_3 C_{TPB}(0)(e^{-k_T t} - e^{-k_3 t})}{k_3 - k_T} \quad (2)$$

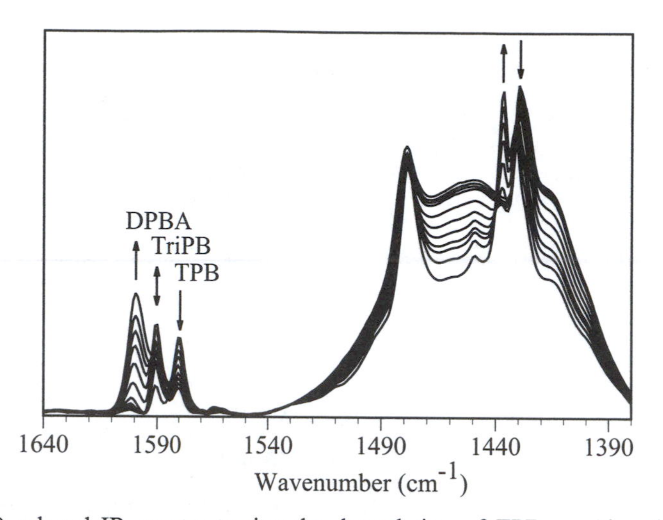

Figure 7. Overlayed IR spectra tracing the degradation of TPB on Al-exchanged hectorite (SHCa-1) over 30 hours. Arrows indicate appearance or loss of spectral bands with time; double ended arrows indicate inital increase followed by decrease of peak intentsity.

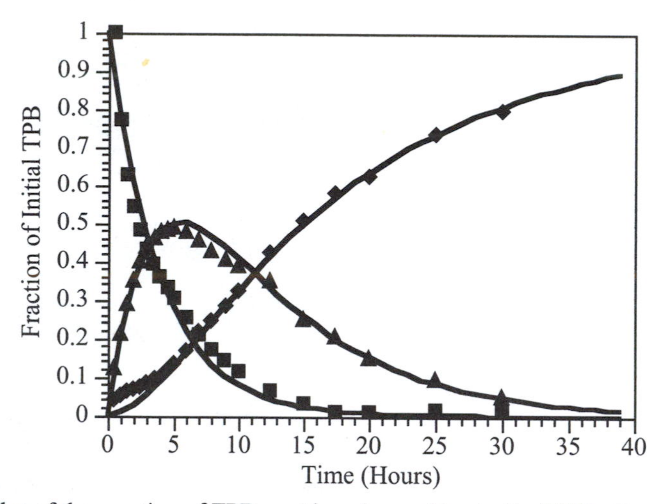

Figure 8. Plot of the reaction of TPB on Al-exchanged hectorite (SHCa-1) measured from IR spectra. Squares indicate loss of TPB determined by the band at 1580 cm^{-1}, triangles indicate the inital production and subsequent loss of TriTB determined by the band at 1590 cm^{-1}, and diamonds indicate the production of DPBA determined by the band at 1600 cm^{-1}. The solid lines represent the fit to equations (1)-(3) with k_2=0.253 hr^{-1}, k_3=0.013 hr^{-1}.

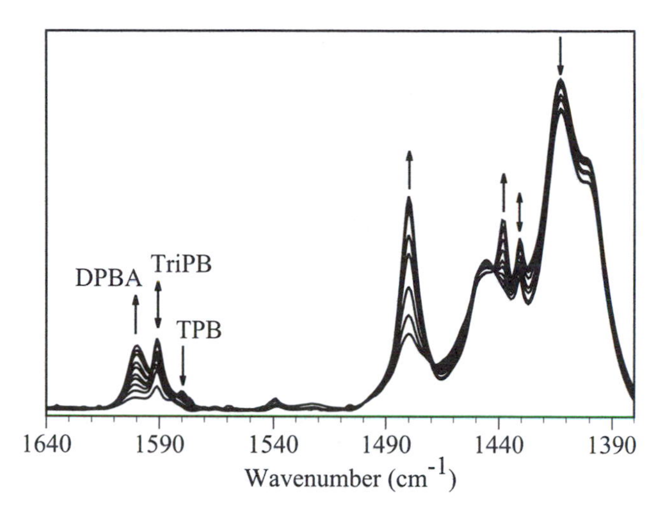

Figure 9. IR spectra following the degradation of TPB on Al-exchanged montmorillonite (SWy-1) in 5-hour intervals for 24 hours. Arrows indicate appearance or loss of spectral bands with time; double ended arrows indicate initial appearance followed by loss of band intensity.

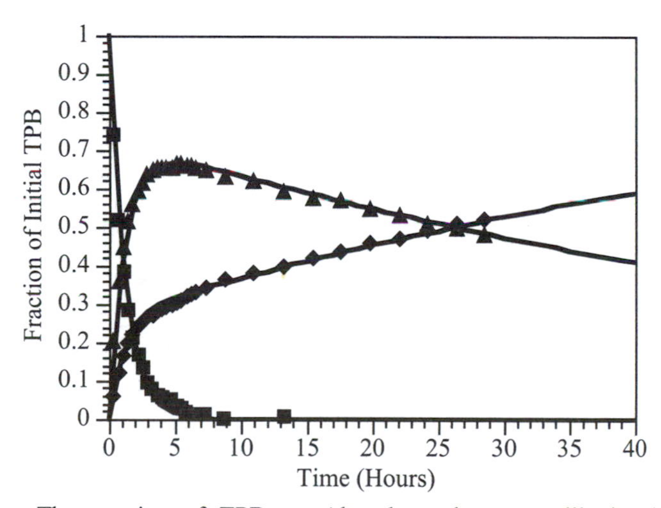

Figure 10. The reaction of TPB on Al-exchanged montmorillonite (SWy-1) quantitatively measured from IR spectra. Squares indicate loss of TPB determined by the band at 1580 cm^{-1}, triangles indicate appearance and loss of TriPB by the band at 1590 cm^{-1}, and diamonds indicate production of DPBA determined by the band at 1600 cm^{-1}. The solid lines represent the fit to equations (1)-(3) with k_1=0.225 hr^{-1}, k_2=0.575 hr^{-1}, k_3=0.015 hr^{-1}.

$$C_{DPBA}(t) = C_{DPBA}(0) + C_{TriPB}(0)(1 - e^{-k_3 t}) + \frac{C_{TPB}(0)k_1(1 - e^{-k_T t})}{k_T} + \frac{k_2 k_3 C_{TPB}(0)(1 - e^{-k_T t})}{k_T(k_3 - k_T)} - \frac{k_2 C_{TPB}(0)(1 - e^{-k_3 t})}{k_3 - k_T} \quad (3)$$

where $k_T = k_1 + k_2$, $C_i(0)$ is the initial concentration of i=TPB, DPBA or TriPB and $C_i(t)$ are their respective concentrations at time, t. The solid lines in Figures 6, 8, and 10 show the non-linear regression best fits based on the above first-order reaction model and assumptions. In Figure 8, k_1 was assumed to be zero for TPB degradation on Al-SHCa-1 since the Lewis pathway was not operative on Na-SHCa-1. In general, the model can effectively fit the data.

Effect of Water Content on Reaction Kinetics

Mills et al. (4), reported that the moisture content of field soils affected the degradation reaction of TPB. The reaction was observed to occur only at relatively low water contents, well below the field capacity of the soils. Therefore, a systematic study of the TPB-clay system was conducted to further delineate the role of water content in governing the reactivity of the clay surface for TPB. Treatments ranged from 500 to 2000 mg g^{-1} water added to Na-exchanged montmorillonite as depicted in Figure 12. The overall reaction efficiency dropped with increasing water content consistent with the observations of Mills et al . (4). The lowest moisture content (500 mg g^{-1} water to clay) reported here is still sufficiently high for the occurrence of multilayer water, both in the interlayer and external surfaces of the clay (27). This suggests that at the lower water contents used, TPB can outcompete water and react preferentially at the clay surface. These results are in contrast to other reactive organics at clay surfaces, such as benzene and nitrobenzene, where greater than monolayer water contents impede the reactivity of the clay surface by preventing direct sorption of the reactant to the surface (28). In addition, these results are in general agreement with sorption from aqueous solution of other large polar (or weakly polar) aromatics, such as benzidene and its derivatives (29). However, it appears that there is a specific range of water content where maximum surface sorption of TPB and, thus, reactivity occurs.

Evidence that TPB still reacts at the clay surface at high relative water contents is indicated by the striking loss in intensity of the phenyl bands of TPB observed in the initial stages of the reaction. The stacked plots in Figure 11 provide direct evidence for a surface-sorbed species of TPB as well as an interaction of TPB with the surface: note the changes in solution TPB phenyl bands (top trace) compared to 500 mg g^{-1} water content (second from bottom trace). The bulk TPB has C-C stretching bands associated with the phenyl rings at 1480 and 1430 cm^{-1}, while surface sorption of TPB resulted in a shift of the 1480 cm^{-1} band to 1450 cm^{-1}, and both a shift and splitting of the 1430 cm^{-1} band to 1410 and 1400 cm^{-1}. An increase in water content in the TPB-clay paste resulted in an increase in bulk phase TPB, and increased the intensity of TPB phenyl and surface adsorbed TPB bands. However, the relative amount of surface-sorbed TPB did not change substantially, albeit the band near 1450 cm^{-1} appears to broaden at the high frequency side. While this may be interpreted as evidence of TPB sorption at the edges of the clay crystals (5), the Brønsted acid pathway coupled to exchangeable Al^{3+} in the interlayer of hectorite (previous sections), and the Lewis acid degradation pathway coupled to reduction of Fe(III) in the structure of the montmorillonite (next sections), indicate that the sorption sites are most probably the basal planes.

The IR data presented in Figure 11 may provide an explanation of the apparent anomaly that an anion (TPB) would sorb to the negatively charged surface of smectites.

These data suggest a van der Waals interaction between polar phenyl rings upon sorption to the clay surface. Such interactions have been documented in other phenyl containing compounds (5, 29). McBride and Johnson (29) provided evidence of changes in the conformation of benzidene ions upon sorption from aqueous solution to smectite surfaces. The conformational changes of sorbed benzidene are believed to be responsible for the electron transfer and blue color of benzidene-smectite complexes (29). The splitting of the 1430 cm^{-1} band for solution TPB into two distinct bands at 1410 and 1400 cm^{-1} (Figure 11) indicate the possibility of a similar change in conformation for sorbed TPB. Although TPB appears to be capable of directly adsorbing at the basal planes, apparently by van der Waals interactions between the phenyl groups and the clay surface, high water contents (e.g., 2000 mg g^{-1}) can still impact sorption and the resulting reactivity (Figures 11 and 12).

The solid lines in Figure 12 represent the nonlinear best fits of the IR absorption data (Figure 11) to the proposed pseudo-first-order reaction kinetics. It is interesting to note that in each instance, the rate constant was essentially unchanged ($k= 0.1\ hr^{-1}$). Hence, the amount of water did not affect the rate of TPB reaction, but rather the quantity of TPB that was able to react. In other words, increasing water contents decreased the efficiency of the Lewis pathway of TPB degradation at the clay surface, probably by inhibiting the adsorption of TPB and the formation of the redox couple. This demonstrates that further insights can be elucidated by *in situ* spectroscopic techniques through manipulation of physical factors affecting the reaction.

Coupling of TPB Oxidation to Structural Fe(III) Reduction in Smectites

A color change observed for nontronite from red to blue-green during TPB degradation (13) can be attributed to the reduction of structural Fe(III) in nontronite. The blue-green color has been identified as an intervalence charge transfer (IT) transition and is attributed to the presence of mixed valent Fe [Fe(III)-O-Fe(II)] couples in the octahedral sheet (30). Since the Lewis degradation of TPB yields two electrons (14), we assume that two Fe(III) are reduced during the Lewis acid reaction on montmorillonite and nontronite. Since the Lewis acid pathway did not proceed on Fe-free hectorite, it is reasonable to assume that octahedral Fe in the structure of nontronite and montmorillonite is the electron acceptor as has been argued previously (31, 32). It is already well established that electrons readily pass from edge and basal surfaces to octahedral Fe (33). However, the behavior of the IT transition in nontronites (30, 34) and the relatively low structural Fe(III) content in montmorillonites indicate that the reduction of structural Fe in smectites proceeds from the basal surface rather than along edges.

Reduction of Fe(III) to Fe(II) in nontronite results in a decrease in specific surface area which has been attributed to the collapse of partially or fully expanded layers to unexpanded layers with increased layer charge (17, 18). Such textural changes might explain why the reaction could not be modeled with simple first-order equations as was done with montmorillonite. Additionally, octahedral Fe(III) reduction results in changes in the structure of the tetrahedral sheet in montmorillonites (35, 36) and the octahedral sheet of nontronites (37, 38). It is possible that changes in surface structure and layer charge during the course of reaction affected the reaction rate more at nontronite surfaces than at montmorillonite surfaces because of the greater Fe content in nontronites. For montmorillonite, the electron transfer is assumed to be almost entirely to octahedral $Fe\text{-}O_2\text{-}Al$ sites, because of the lower quantity of Fe in SWy-1 than NG-1.

The IR spectral data of the organic component of TPB-clay reactions on different Fe-containing clays indirectly infers the role of structural Fe(III) in the Lewis reaction pathway. Although the octahedral Fe-OH deformation bands in the IR spectra of Fe-containing smectites (data not shown) support the hypothesis of structural Fe reduction, the relevant bands are ill-resolved and difficult to interpret analytically. The application

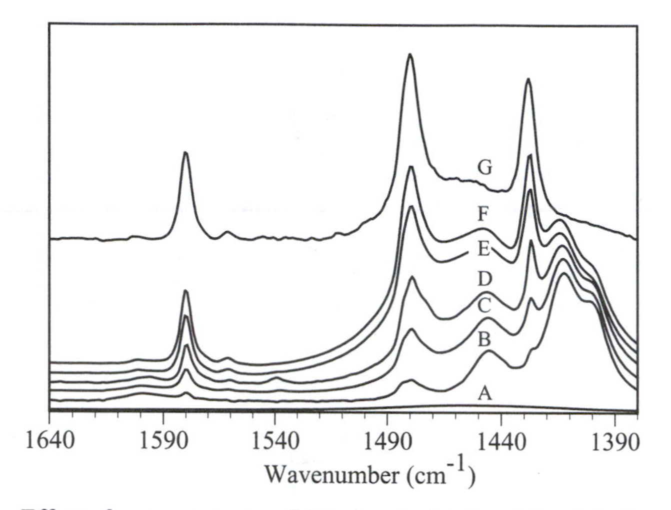

Figure 11. Effect of water content on TPB phenyl vibrational bands in the presence of SWy-2 preceeding the reaction. The spectra from bottom to top are: Na-SWy-2+D_2O (A); 50% (B); 75% (C), 100% (D); 150% (E); 200% (F) loading of TPB/D_2O solution. Top trace is TPB/D_2O solution without SWy-2 present (G).

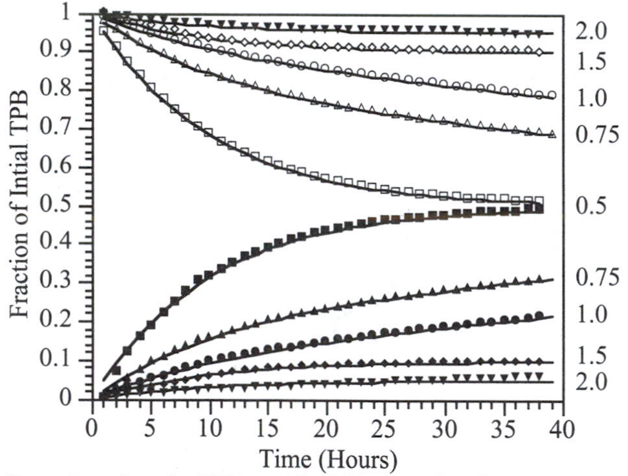

Figure 12. Reaction plots for TPB on Na-SWy-2 under the same water loadings of Figure 11. Open symbols show loss of TPB; closed symbols show production of DPBA. Right hand axis shows fractional water loading ($g\ g^{-1}$) for each trace.

of X-ray spectroscopy can provide another spectroscopic probe of the *in situ* reduction of Fe in real time.

XANES as an *in situ* Probe for Fe(II)/Fe(III)

X-ray absorption spectroscopy (XAS) is an attractive technique for *in situ* oxidation state determination and speciation analyses. Recent studies of X-ray absorption near edge structure (XANES) demonstrate that oxidation states can be determined from the precise energy of the X-ray absorption edge and associated absorption resonances (39, 40, 41). Waychunas et al. (39), reported that the Fe K absorption edge energy was systematically higher for ferric bearing minerals compared to ferrous bearing minerals. These reports indicate that the observed energy shift is approximately 2 eV per unit increase in oxidation state. However, the precise energies of these features also depend on matrix effects so that matrix-matched standards must be used for quantitative applications (42).

In the TPB-smectite system, temporal studies alleviate matrix issues because the same sample is being probed over time. Reference samples were prepared by reducing clay samples with dithionite and washing excess salts under inert atmospheric conditions (43). Wet chemical determinations of Fe(II) and total Fe were determined by digesting the reduced smectites in the presence of excess tris 1,10-phenanthroline and recording the absorbance of the Fe^{2+}-phenanthroline complex at 510 nm (44). XANES was conducted on dry powdered subsamples stored under inert conditions within days of the chemical analyses. The wet chemical results were used to standardize the edge shift of the XANES spectra for Fe(II)/Fe(III) content (45).

The TPB degradation kinetic studies were reproduced at beamline X-26A at the National Synchrotron Light Source, Brookhaven National Laboratory, Upton NY, under its normal configuration using a channel-cut 111 Si-monchromator and an 8:1 elliptical focusing mirror. High quality spectra could be acquired within 15 minutes by measuring the Fe Kα fluorescence intensity with an energy dispersive Si(Li) detector (45).

Results indicated that up to 50% of the structural Fe in montmorillonite could be reduced during the course of the reaction with TPB at a 500 mg g^{-1} water content. After 12-15 hours, the samples began to reoxidize, thus reaction kinetics were only fit to the first 10 hours of the reaction. The time resolved series of XANES spectra are presented in Figure 13A and the fraction of Fe (III) remaining following reaction of TPB with SWy-2 is plotted in Figure 13B. The solid line in Figure 13B is the best fit to the nonlinear first order reaction kinetics (equation 1). The data were fit with first order rate constant: k_e=0.207 h^{-1}. This is approximately twice the rate measured for the degradation of TPB by IR analysis. Since the oxidation of TPB yields two electrons, the rate constant, k_e, should be related to k_T by a factor of 2, $k_e=2k_T$. The XANES data for the reduction of Fe are thus consistent with the oxidation of TPB measured by IR.

Reacted samples remeasured 3 months later showed that all of the reduced Fe had subsequently reoxidized to Fe(III). Previous studies have demonstrated that little clay dissolution occurs during redox cycles of phyllosilicate clays (18, 36, 44). This permits the clay to be defined as a catalyst in the truest sense. The clay facilitates the oxidation of TPB but the clay itself is not destroyed. The redox of Fe in smectites is essentially reversible (15, 38), hence the reactivity of this system is in contrast to the widely studied Fe-oxide and Mn-oxide facilitated reactions in which the oxide typically reductively dissolves and transforms during the reaction (19, 20).

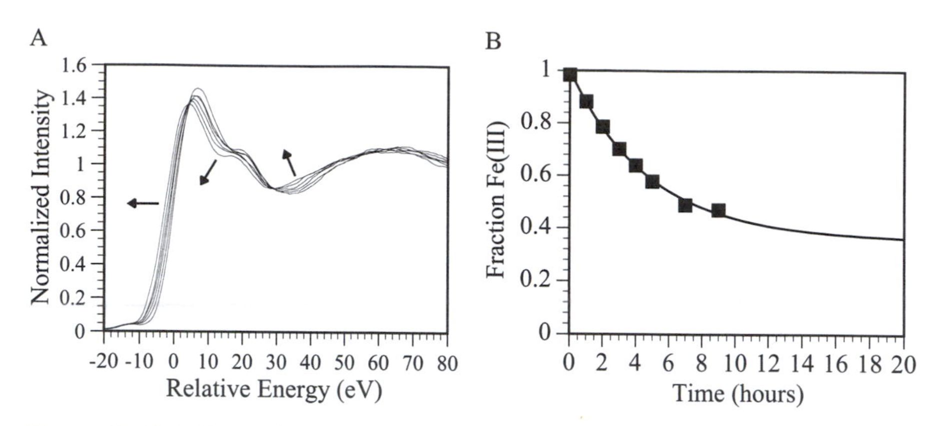

Figure 13. XANES analysis of the Lewis acid oxidative degradation of TPB on Na-exchanged montmorillonite (SWy-2). (A) The change in the Fe near edge structure over 10 hr. The arrows indicate the direction of change in spectral features. (B) The reduction of Fe (III) measured from the edge positions in (A).

Effect of Fe Reduction on Clay Structure: EXAFS

Because dissolution is negligible (18, 36, 44), most physical, chemical and structural properties of smectites are essentially reversible following reduction and oxidation reactions of octahedral Fe (38). The relative reversibility of these processes suggests modifications in the structure during redox reactions rather than reductive dissolution (35, 36, 37, 38). Indeed, the layered structure of dioctahedral smectites (Figure 14A) can allow for significant distortions in bond lengths and bond angles as a function of charge and composition (46, 47, 48), as well as reduction of structural Fe (36). Extended X-ray absorption fine structure (EXAFS) is an excellent spectroscopic tool to probe the local structure of specific elements within minerals. (For a detailed discussion of EXAFS see 49) In this study we have employed EXAFS to investigate the change in the average local environment of Fe in these clays on reaction with TPB to discern structural modifications when Fe is reduced from III to II valence states.

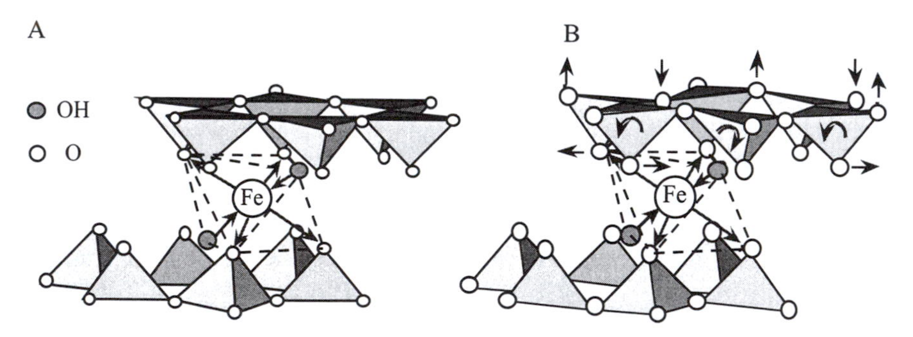

Figure 14. (A) Polyhedral representation of montmorillonite showing the linkage of the tetrahedral sheet with the octahedral sheet. (B) Effective change in the average Fe-O bondlength upon reduction of Fe(III) to Fe(II) results in the distortion of the local clay crystal structure.

EXAFS spectroscopy was conducted on beamline X23-A2 at the National Synchrotron Light Source, Brookhaven National Laboratory, Upton NY with an electron beam energy of 2.54 GeV and stored currents between 180 and 250 mA. The X23A2 monochromator uses 311 Si crystals in a fixed exit geometry. Because the energy of the transmitted harmonic from the 311 reflection (beyond the 7 KeV primary) equals 21 KeV, and no significant elemental concentrations with Z>43 occurs in these clays, no harmonic rejection was used during these experiments. Measurements on the SWy-1 clays were made at room temperature in the fluorescence mode using a Mn filter-Soller slit combination. (50). High quality EXAFS scans were measured every 20 minutes (51).

Data were analyzed using the MACXAFS package (52) following established EXAFS analysis procedures (53). The pre-edge background absorption was determined from a linear fit to the data roughly 100-50 eV below the absorption edge energy and then extrapolated over the entire energy range of the spectrum. The K-edge absorption was then isolated by subtracting the extrapolated pre-edge background from the post-edge absorption signal. The smoothly varying atomic absorption was determined by fitting the post-edge data with a cubic-spline function. The absorption spectra were then step-normalized by dividing by the amplitude of the absorption edge. The location of the absorption edge (E_0) for the clays was determined as the maximum of the first derivative. Initial energy calibration was performed by assigning the maximum of the derivative of the data from an Fe foil to be 7112.0 eV. All time-dependent scans for each clay were analyzed in the same manner to remove any possible distortions of the EXAFS results due to analysis procedures.

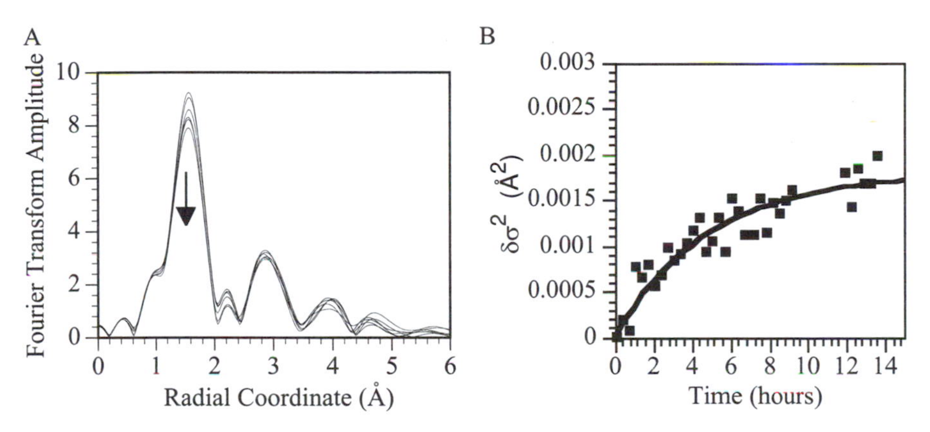

Figure 15. (A) Fourier transform of single scan EXAFS data collected every 20 minutes over 12 hours during reaction with TPB with Na-SWy-2. The arrow shows(B) The change in Debye Waller factor ($\delta\sigma^2$) during the reaction period.

The Fourier transform of the EXAFS spectra displayed in Figure 15A reveal significant changes in the average local structure of Fe in montmorillonite upon its reduction by reaction with TPB. In an initial attempt at fitting the data, the Fe coordination number was fixed at six, and the radial distance and EXAFS Debye-Waller factor were allowed to vary. Results from this fitting scenario yielded a slight increase in the average Fe-O bondlength ($\Delta r = 0.01 \pm 0.0075$ Å) and a significant increase in the EXAFS Debye-Waller factor (0.002 ± 0.00075 Å^2) relative to their values at time = 0. A plot of the change in the EXAFS Debye-Waller factor versus time is shown in Fig. 15B. The solid line in this figure is the best exponential fit to the data and indicates a reaction rate constant of k=0.202 hr^{-1}. This value is in agreement with the rate constant determined from the XANES data described earlier.

One possible explanation for this increase in the EXAFS Debye-Waller factor could be that there is an increase in the spread of Fe-O bondlengths about their initial average value, i.e. the shorter Fe-O bonds become shorter and the longer Fe-O bonds become longer. Note, however, that this fitting scenario does not take into account the possibility of two different average Fe-O bondlengths for Fe, depending upon whether Fe is in the II or III valence state.

Different Fe-O bondlengths in layered silicates, (depending on the oxidation state of the Fe) have been demonstrated previously (54). To investigate this possibility, the filtered data were fit with 2 Fe-O subshells, one with the Fe-O bondlength constrained to equal 2.00 Å (to represent the average local environment of Fe(III)) and the other with the Fe-O bondlength constrained to equal 2.10 Å (to represent the average local environment of Fe(II)). Results for this subshell fitting scenario of the coordination number values indicated an increase in the percent of Fe(II) with time. The best exponential fit to the time-dependent change in percent of Fe(II) in the clay resulted in a time constant equal to approximately 0.2 hr^{-1}. This value is in complete agreement with the previously-discussed XANES results and the EXAFS Debye-Waller factor results. A more detailed discussion of these EXAFS results is presented elsewhere (51).

The effect of an increase of the average Fe(II)-O bondlength of ~0.1 Å is illustrated in Figure 14B. Because of the linkage through the apical oxygen of the tetrahedral and octahedral sheets, any change in the octahedral cation-oxygen ligand bondlength would result in tilting of silica tetrahedra to maintain structural integrity (9). This hypothesis is consistent with the results obtained by other indirect spectroscopic techniques such as FTIR (35, 37) and the interpretation of ^{29}Si MAS NMR (36).

Summary

We have described the utilitiy of three *in situ* spectroscopies for examining heterogeneous surface reactions of sorbed organics in real time (Figure 2). Using tetraphenylboron as our probe molecule, we have shown that IR is capable of monitoring multiple reaction pathways of the reactant and daughter products involving both direct adsorption and hydrolysis mechanisms, respectively. XANES spectroscopy, in combination with wet chemical analyses, can quantify the reduction of octahedral Fe in the catalyst (clay mineral) structure. EXAFS, constrained by relative Fe(II)/Fe(III) populations derived from XANES, can yield detailed information on local structural changes within the reactive mineral surface. This information is critical to understanding the dynamic role that mineral surfaces play in facilitating heterogeneous reactions of organic species.

Acknowledgments

This research was supported by Financial Assistance Award Number DE-FC09-96SR18546 from the U.S. Department of Energy to the University of Georgia Research Foundation and by ERDA/WSRC subcontract AA46420T. The work of KMK was supported by the U. S. Department of Energy, Office of Energy Research, Office of Health and Environmental Research, under contract W-31-109-Eng-38.

Literature Cited

1. Laszlo, P. *Science* . **1987**, *235*, 1473.
2. Wang, M. C.; Huang, P. M. *Clays Clay Miner*. **1989**, *37*, 525.
3. Bertsch, P. M.; *Borax Rev*. **1991**, *9*, 16.
4. Mills, G. L.; Kaplan, D.; Schwind, D.; Adriano, D. *J. Environ. Qual*. **1990**, *19*, 135.

5. Norris, J.; Giese, R.F.; Van Oss, C.J.; Costanzo, P.M. *Clays Clay Miner.* **1992**, *40*, 327.
6. Vouldrias, E. A.; Reinhard, M. In *Geochemical Processes at Mineral Surfaces.*; Davis, J. A.; Hayes, K. F., Ed.; ACS Symposium Series 323; American Chemical Society: Washington, DC 1986, pp 462.
7. Mills, G. L.; Schwind, D. *Environ. Toxicol. Chem.* **1990**. 9, 569.
8. Ukrainczyk, L.; McBride, M. B. *Clays Clay Miner.* **1992**. *40*, 1992.
9. Stone, A. T.; Morgan, J. J. *Environ. Sci. Technol.* **1984**, *18*, 617.
10. Wang, M. C. *Clays Clay Miner.* **1991**, *39*, 202.
11. Brindley, G. W. In *Crystal Structures of Clay Minerals and Their X-ray Identification.*; Brindley, G. W.; Brown, G., Ed.; Mineralogical Society: London; **1984**, Chapter 2, pp 170.
12. Peri, J. B. In *Catalysis Science and Technology* ; Anderson, J. R.; Boudart, M., Ed.; Springer Verlag, New York **1984**, Vol 5, Chapter 3.
13. Hunter, D.B.; Bertsch, P.M. 1994. *Environ. Sci. Technol.* **1994**, *28*, 686-691.
14. Geske, D. H. *J. Phys. Chem.* **1959**, *63*, 1062.
15. Stucki, J.W. In *Iron in Soils and Clay Minerals.* J.W. Stucki, Goodman, B.A.; Schwertmann, U. Ed. D. Reidel, Drodretecht, The Netherlands; **1988**, Chapter 17, pp 625-675.
16. Stucki, J.W.; Low, P.F.; Roth, C.B.; Golden, D.C. *Clays Clay Miner.* **1984**, *32*, 357.
17. Lear, P. R.; Stucki, J. W *Clays Clay Miner.* **1989**, *37*, 547.
18. Stucki, J.W. ; Golden, D.C.; Roth, C.B. *Clays Clay Miner.* **1984**, *32*, 350.
19. Herring, J.G.; Stumm,W. P. In *Reviews in Mineralogy*; Hochella, (Jr), M.F.; White, A.F., Ed.; Mineral-Water Interface Chemistry. Mineralogical Society of America. Washington DC. 1990, Vol. 23; pp 427.
20. Ryan J.N.; Gschwend, P.M. *Environ. Sci. Technol.* **1994**, *28*, 1717.
21. Johnston, C. T.; Tipton, T.; Trabue, S. L.; Erickson, C.; Stone, D. A. *Environ. Sci. Technol.* **1992**, *26*, 382.
22. Aochi, Y. O.; Farmer, W. J.; Shawney, B. L. *Environ. Sci. Technol.* **1992**, *26*, 329.
23. Griffiths, P. R.; deHaseth, J. A.; *Fourier Transform Infrared Spectroscopy*; John Wiley and Son: NY, 1986, Chapter 5.
24. Costa, G., Camus, A., Marsich, N., Gatti, L. J. *Organometallic Chem.* **1967**, *8*, 339.
25. Newman, A. C. D.; Brown, G. In *Chemistry of Clays and Clay Minerals*; Newman, A. C. D.,Ed.; John Wiley & Son, NY, 1987, Chapter 1.
26. Lasaga, A. C. In *Kinetics of Geochemical Processes* ; Lasaga, A. C.; Kirkpatrick, R. J., Ed.; Mineralogical Society of America: Washington, DC 1981, Chapter 1.
27. Johnston, C.T.; Sposito, G ; Erickson, C. *Clays Clay Miner.* **1992**, *40*, 722.
28. Yarif, S.; Russell, J.D.; Farmer, V.C. *Israel. J. Chemistry.* **1966**, *4*, 201.
29. McBride, M.B.; Johnson, M.G. *Clays Clay Miner.* **1986**, *34*, 686.
30. Lear, P. R.; Stucki, J. W. *Clays Clay Miner.* **1987**, *35*, 373.
31. Ainsworth, C. C.; McVeety, B. D.; Smith, S. C.; Zachara, J. M *Clays Clay Miner.* **1991**, *39*, 416.
32. McBride, M.B. *Clays Clay Miner* **. 1979**, *27*, 224.
33. Tennakoon, D. T. B., Thomas, J. M., Tricker, M. J. *J. Chemica l Soc., Dalton Trans.*, **1974**, 2211.
34. Komadel, P.; Lear, P.L.; Stucki, J.W. *Clays Clay Miner* . **1990**, *38*, 203.
35. Rozenson, I.; Heller-Kallai, L. *Clays Clay Miner.* **1976**, *24*, 271.
36. Gates, W.P.; Stucki, J.W.; Kirkpatrick, R.J. *Phys. Chem. Miner.* **1996**, *23*, 535.
37. Stucki, J.W.; Roth, C.B. *Clays Clay Miner.* **1976**, *24*, 293.
38. Komadel P.; Madejová, J.; Stucki, J.W. *Clays Clay Miner.* **1995**, *43*, 105.

39. Waychunas, G.A.; Brown, G.E., Jr.; Apted, M.J. *Phys. Chem. Miner.* **1983**, *10*, 1.
40. Sutton, S.R.; Jones, K.W.; Gordon, B.; Rivers, M.L.; Bajt, S.; Smith, J.V. *Geochim. Cosmchim. Acta* **1993**, *57*, 461.
41. Bajt, S.; Sutton, S.R.; Delaney, J.S. *Geochim. Cosmchim. Acta.* **1994**, *58*, 5209.
42. Manceau, A.; Gates, W.P. *Clays Clay Miner* **. 1997**, *45*, 448.
43. Stucki, J.W. ; Golden, D.C.; Roth, C.B. *Clays Clay Miner.* **1984**, *32*, 191.
44. Komadel, P.; Stucki, J.W. *Clays Clay Miner* **. 1988**, 36, 379-381.
45. Gates, W.P.; Hunter, D.B.; Nuessle, P.R.; Bertsch, P.M. *Journal des Physiques IV - France.* **1997**, *7*, C2785.
46. Newman, A. C. D. In *Chemistry of Clays and Clay Minerals*; Newman, A.C.D. Ed.; John Wiley & Son, NY, 1987, Chapter 5.
47. Weiss, C.A.; Altaner, S.P.; Kirkpatrick, R.J. *Am. Mineral.* **1987**, *72*, 935.
48. Sanz, J.; Robert, J.L. *Phys. Chem. Miner.* **1992**, *19*, 39.
49. Koningsberger D. C.; Prins R. *X-Ray Absorption: Principles, Applications, Techniques of EXAFS, SEXAFS and XANES*, Wiley, New York, **1988**.
50. Stern, E.A.; Heald S.M. *Rev. Sci Instrum.* **1979**, *50*, 1579.
51. Kemner, K.M.; Hunter, D.B.; Gates, W.P.; Pratt, S.; Bertsch, P.M. *J. Phys. Chem.* in review
52. Bouldin, C.; Furenlid, L.; Elam, T. *Physica B* **1995**, *208 & 209*, 190,
53. Sayers, D.E.; Bunker, B.A. In *X-Ray Absorption: Principles, Applications, Techniques of EXAFS, SEXAFS, and XANES*; Koningsberger, D.C.; Prins, R., Ed.; Wiley, NY, 1988**,** chapter 6.
54. Manceau, A.; Bonnin, D.; Kaiser, P.; Frétigney. *Phys. Chem. Miner.* **1988**, *16*, 180.

Chapter 15

The Role of Oxides in Reduction Reactions at the Metal–Water Interface

Michelle M. Scherer[1], Barbara A. Balko[2] and Paul G. Tratnyek[1]

[1]Department of Environmental Science and Engineering, Oregon Graduate Institute of Science and Technology, Portland, OR 97291
[2]Department of Chemistry, Lewis and Clark College, Portland, OR 97219

The oxide layer that lies at the iron-water interface under environmental conditions can influence the reduction of solutes by acting as a passive film, semiconductor, or coordinating surface. As a passive film, oxides may inhibit reaction by providing a physical barrier between the underlying metal and dissolved oxidants. Sustained reduction of solutes requires localized defects in the passive film (e.g., pits), or some mechanism for transferring electrons through the oxide. In the semiconductor model, conduction band electrons from the oxide may contribute to solute reduction, but electron hopping (resonance tunneling) appears to be more important due to the high population of localized states in oxides formed under environmental conditions. Ultimately, electron transfer to the solute must occur via a precursor complex at the oxide-water interface. For dehalogenation of chlorinated aliphatic compounds on an iron oxide surface, a surface complexation model suggests that the outer-sphere precursor complex is weak and partially displaced by common environmental ligands.

Zero-valent metals such as iron, tin, and zinc, are moderately strong reducing agents that react with a variety of oxidants, including halogenated alkanes and alkenes, nitro aromatic compounds, nitrate, and chromate. Although these reactions have been known for a long time, the recent application of granular iron metal (Fe^0) to remediation of contaminated groundwater has sparked new interest in their chemistry *(1, 2)*. The majority of new work in this area concerns reduction of chlorinated aliphatic compounds, so we haved focused our analysis on this class of reactions. The analysis, however, should be broadly applicable to contaminant reactions in heterogeneous environmental systems.

Under environmental conditions, the Fe^0-H_2O interface has a surface layer of corrosion products that develops due to the thermodynamic instability of Fe^0 in the presence of water. Long-term batch and column studies have shown that this layer evolves with time into a complex mixture of amorphous iron oxides, iron oxide salts, and other mineral precipitates *(2-5)*. Because this material lies at the metal-water interface, it must, in some manner, mediate the reduction of contaminants by the underlying metal. Understanding the mechanism by which metals reduce contaminants in the presence of a substantial layer of oxides is one of the critical, remaining challenges for researchers in this field. The goal of the following analysis is to develop a conceptual framework for this problem by reevaluating recent data in terms of established models for the structure and reactivity of metal oxides.

Chemical Background

Reduction reactions in Fe^0-H_2O systems are heterogeneous reactions driven by the oxidation of Fe^0 (equation 1) or the oxidation of surface-bound Fe^{2+} ($Fe^{2+}_{(sur)}$) (equation 2). In the absence of oxygen, these oxidations are coupled with the reduction of interfacial H_2O *(6-8)*:

$$Fe^0 \rightleftharpoons Fe^{2+} + 2\,e^- \tag{1}$$

$$Fe^{2+}_{(sur)} \rightleftharpoons Fe^{3+}_{(sur)} + e^- \tag{2}$$

$$2\,H_2O + 2\,e^- \rightleftharpoons H_2(g) + 2\,OH^- \tag{3}$$

In principle, Fe^0, $Fe^{2+}_{(sur)}$, and H_2 can contribute to the reduction of contaminants *(6)*. However, for chlorinated aliphatic compounds (RX), several lines of evidence suggest that most of the observed reduction is due to reaction with Fe^0 or $Fe^{2+}_{(sur)}$, and little, if any, involves H_2 (or H• or H^-) at near neutral pH values *(9, 10)*. The predominant degradation pathway for RX appears to be reductive dechlorination,

$$RX + H^+ + 2\,e^- \rightarrow RH + X^- \tag{4}$$

although variable amounts of reductive elimination have been reported *(11, 12)*. Both pathways presumably involve heterogeneous electron transfer (ET) from the iron metal or iron oxide to RX. The ET mechanism requires formation of a precursor complex, which appears to be weak and outer sphere for typical organic contaminants *(13)*.

In the absence of an oxide layer, it appears that reduction of RX involves primarily ET from Fe^0 (i.e., equations 1 and 4). An oxide free iron metal surface can be achieved in electrochemically-controlled laboratory systems *(9)*, and may apply where localized corrosion occurs at defects in the oxide surface layer (see section on The Oxide as a Physical Barrier). Under environmental conditions, however, the layer of iron oxides that covers the Fe^0 surface will contain $Fe^{2+}_{(sur)}$. Therefore, the observed reduction of RX, may reflect ET from $Fe^{2+}_{(sur)}$ as well as Fe^0 (equations 1, 2, and 4). For the purposes of the following discussion, $Fe^{2+}_{(sur)}$ may be adsorbed or structural iron *(14)*, as long as it is available for reaction at the oxide-water interface.

Mass Transport to the Interface

Heterogeneous reactions can be broken down into mass transport and surface reaction steps. In the case portrayed in Figure 1a, mass transport involves diffusion of dissolved RX through the stagnant water layer to the surface, and surface reaction involves precursor complex formation and electron transfer. Controlled experiments conducted with a rotating disk electrode clearly show that mass transport has a negligible effect on the overall reduction rate of carbon tetrachloride (CCl_4) on oxide-free Fe^0 *(9)*. Over the range of conditions of interest in environmental applications, however, both steps could influence the reaction rate, so it is not surprising that some evidence for mass transport effects has been reported *(summarized in 9)*. The surface reaction step appears to be the dominant factor for chlorinated aliphatics, however, because reported reduction rates vary widely (over four orders of magnitude) with chemical structure *(15)*, while there is only a small (less than one order of magnitude) variation in their diffusion coefficients *(9)*.

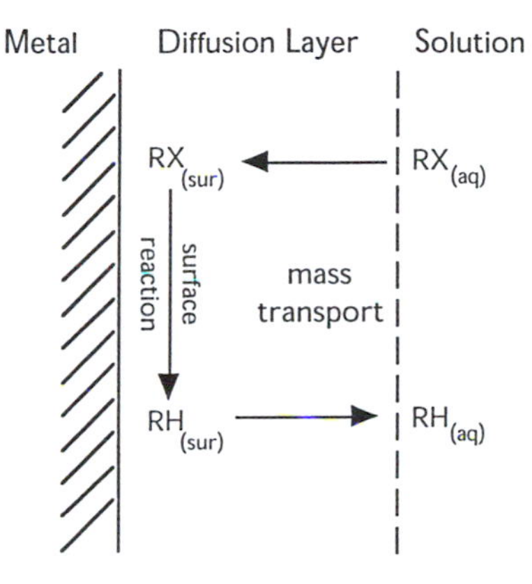

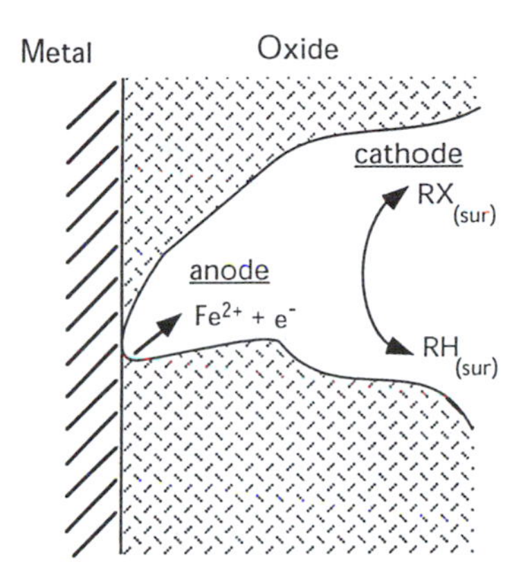

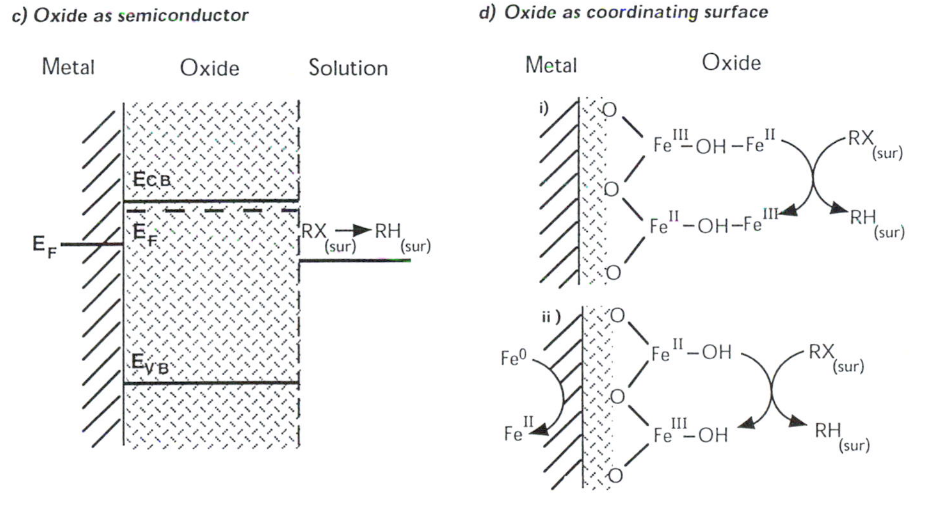

Figure 1. Conceptual models of processes that may be involved in reduction of chlorinated aliphatics (RX → RH) at the iron-oxide-water interface: (a) bare electrode (b) passivated electrode, (c) semiconductor surface, and (d) coordinating surface. E_F, E_{CB}, and E_{VB} refer to the Fermi, conduction band, and valence band energies, respectively.

The Oxide-Water Interface

In the presence of an oxide layer, there are several mechanisms by which electrons may be transferred from the underlying Fe^0 or $Fe^{2+}_{(sur)}$ to the adsorbed RX. The major mechanisms of ET are represented by three conceptual models in which the oxide layer serves as (i) an inert physical barrier, where electrons must come directly from the metal through defects, such as pits or grain boundaries in the oxide layer (Figure 1b); (ii) a semiconductor, where electrons are transported through the oxide layer via the oxide conduction band, impurity bands, or localized states (Figure 1c); and (iii) a surface consisting of a finite number of coordination sites, where reduction may occur by metal-to-ligand charge transfer (Figure 1d). These models are not necessarily exclusive of one another; in fact, it is likely that the observed dehalogenation of RX by Fe^0 is affected by the oxide layer acting in all three ways. However, it is necessary to have a conceptual framework that clearly distinguishes between each mode of ET before their relative importance can be assessed from experimental data.

Composition of the Oxide. All three modes of reactivity for oxides on Fe^0 (Figure 1) are influenced by the composition of the oxide layer. The composition of this layer can be complex and variable, but spectroscopic analysis of electrochemically generated oxide films (passive films) have generally confirmed the two-layer model originally proposed by Nagayama and Cohen *(16)*, in which the film consists of an inner layer of magnetite (Fe_3O_4) and an outer layer of maghemite (γ-Fe_2O_3) *(17, 18)*. Other models that have been postulated include: (i) less discrete layers of amorphous oxides (similar in structure to $Fe(OH)_2$, Fe_3O_4, and γ-Fe_2O_3) *(19, 20)*; (ii) one layer of a mixed-valent iron oxide in which the proportion of Fe^{2+} decreases exponentially from the metal-oxide contact to the oxide-water interface *(21)*; or (iii) a layer consisting of a spinel structure similiar to Fe_3O_4 and γ-Fe_2O_3 with random cation vacancies and interstitials *(22)*. The amorphous, non-crystalline character of these materials has been attributed to the presence of coordinated water molecules *(23, 24)*.

The oxide layer that forms on the iron used in remediation applications, however, may differ from models developed to describe passive films because the iron used is an impure, recycled material that is manufactured primarily for use as a conditioner in building materials. Spectroscopic analyses of these materials show that they consist of a complex mixture of crystalline phases *(25-27)*. Recently, Raman spectra obtained on iron particles from Master Builder Inc. (Cleveland, OH) and Peerless Metal Powder & Abrasives (Detroit, MI) revealed maghemite, magnetite, and hematite (α-Fe_2O_3) on samples analyzed as received *(27)*.

With exposure to aqueous solutions, the original air-dried oxides on iron metal rehydrate and new phases may be formed by diagenesis or precipitation. For example, scanning electron microscopy (SEM) with energy dispersive X-ray spectroscopy (EDS) on iron particles (Aldrich) exposed to carbonate buffer (15 mM) and nitrobenzene for several days revealed the formation of siderite ($FeCO_3$) *(28)*. Siderite was also identified in a column of iron filings (VWR Scientific) exposed to carbonate buffered water and trichloroethene *(5)*. Peerless iron that had been exposed to groundwater from the Borden aquifer for several weeks showed disappearance of maghemite, an abundance of magnetite, and the formation of a new phase identified by Raman spectroscopy as green rust *(27)*. Green rusts are mixed-valent iron oxyhydroxide salts whose interlayers contain anions such as chloride, sulfate, and carbonate *(29-31)*. Green rusts have been identified in natural soil like environments *(32)*, and are commonly found as an aqueous corrosion products of iron *(33)*. More relevant is the greenish-blue precipitate observed on iron particle surfaces in a column exposed to high concentrations of CCl_4 in our laboratory *(3)*. SEM on grains removed from this column revealed hexagonal crystals (Figure 2) that are consistent with those of green rust *(33)*.

Thus, it appears likely that the oxide layer on Fe^0 evolves from a predominantly Fe(III) phase under air-formed conditions to a mixed-valent or pure Fe(II) phase under the highly-reducing conditions found in groundwater remediation applications. Therefore, the oxide formed on Fe^0 during long-term dechlorination experiments is probably best modeled as a mixed-valent or pure Fe(II) phase such as magnetite, siderite, or green rust, rather than a pure Fe(III) phase such as maghemite, hematite, or goethite. Additional solid phases, due to precipitation of sulfides and carbonates, may be important under some environmental conditions, but their role in the reduction of contaminants by Fe^0 is not addressed in the following discussion.

The Oxide as a Physical Barrier. The oxide surface layer can be interpreted as a physical barrier to mass transport (similar to the role of an oxide film on a passivated electrode) that inhibits ET from the iron metal to RX *(4)*. If the oxide layer is a non-conductive physical barrier, then ET may occur from the metal to the dissolved substrate through defects such as pits or grain boundaries. Pitting and crevice (grain) corrosion are localized forms of corrosion, where rupture of the oxide layer introduces new paths for diffusion (i.e., pore diffusion) and new "catalytic" dissolution pathways *(34, 35)*.

Pitting is typically initiated where an aggressive anion, such as chloride, stimulates dissolution at a defect or inclusion in the surface of the oxide *(36, 37)*. Once pit growth is established, it usually continues until the oxide film is perforated, exposing bare metal (Figure 1b). The new, longer diffusion path to the bottom of the pit restricts transport of aqueous oxidants (e.g., O_2 or RX) from the bulk solution, which creates a concentration gradient in the pit that is known as a differential aeration cell (in reference to pitting corrosion in the presence oxygen) *(38, 39)*. Alone, the oxidant concentration gradient that develops across the pit is not enough to sustain a significantly increased corrosion rate, but two additional effects are involved. First, a pH gradient develops as hydrolysis of Fe^{2+} at the bottom of the pit releases hydrogen ions which make the bottom of the pit more acidic *(37)*. Acidic conditions at the bottom of the pit prevent an oxide layer from reforming. At the mouth of the pit, relatively alkaline conditions and Fe^{2+} diffusing from below favor repassivation along the edges *(38)*.

Although the microscopic mechanism of pit initiation and oxide breakdown is still not fully understood *(40, 41)*, the macroscopic behavior of enhanced local dissolution and diffusion of dissolved metal ions can be described using current-potential (i-E) curves (Figure 3). The solution conditions in a pit create two distinct electrochemical cells. At the bottom of the pit, the oxidation half-reaction is acidic dissolution of Fe^0 (equation 1), which is balanced primarily by reduction of water to hydrogen gas (equation 3). The second cell is at the mouth of the pit, where the half-reactions are dissolution at a passivated iron metal surface (alkaline conditions) and reduction of water or stronger oxidants such as O_2 or RX.

If the pitting mechanism is involved in the reduction of RX by Fe^0, then (in the absence of O_2 and presence of RX) the major reduction half-reaction at the mouth of the pit will be dehalogenation of RX (equation 4) *(9)*. At equilibrium,

Figure 2. Scanning electron micrograph showing precipitates on an Fe^0 surface after long-term exposure (>1 year) to CCl_4 dissolved in deionized water *(3)*. (magnification: x3000)

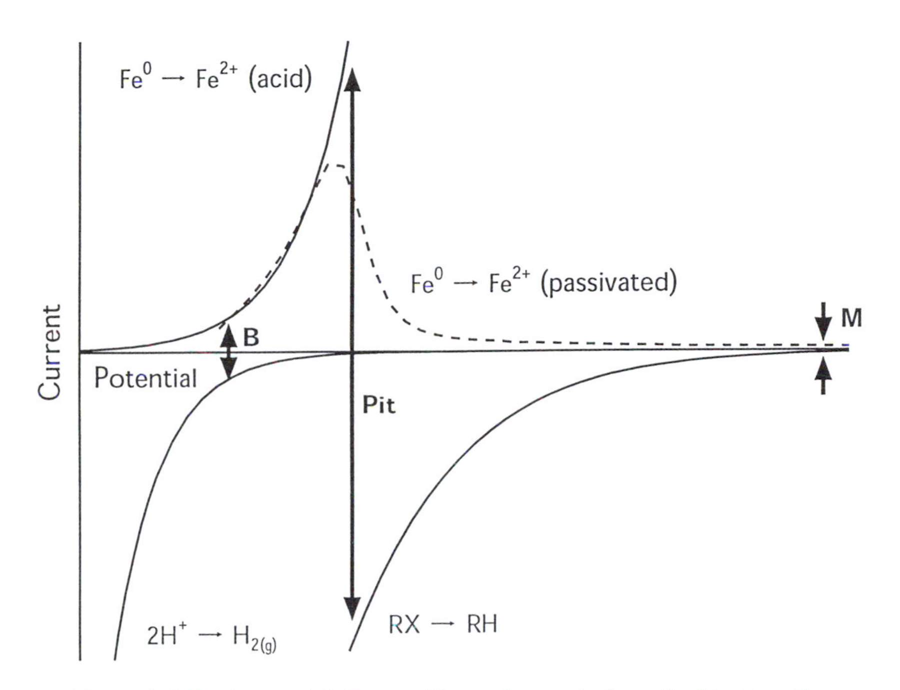

Figure 3. Mixed potential diagram illustrating controls on the kinetics of corrosion at a pitted, oxide-covered metal. The potential range is from -700 to +300 mV/NHE. Arrows: (B) corrosion current at the bottom of the pit, controlled by $Fe^0 \rightarrow Fe^{2+}$ (acid) and $2H^+ \rightarrow H_2$; (M) corrosion current at the mouth of the pit, controlled by the partial currents for $Fe^0 \rightarrow Fe^{2+}$ (passivated) and $RX \rightarrow RH$; (Pit) corrosion current for the short-circuited pit, controlled by $Fe^0 \rightarrow Fe^{2+}$ (acid) and $RX \rightarrow RH$. The three solid curves are generated using the Tafel equation and exchange current densities and Tafel slopes from reference *(9)*. The dashed curve was measured at 5 mV s^{-1} in pH 8.4 borate buffer, using methods described in reference *(9)*.

anodic and cathodic reactions must be balanced in each cell, and these conditions determine the net corrosion rate in each case. Figure 3 illustrates this principle for conditions at the bottom of a pit (labeled B) and for the mouth of a pit (labeled M) using i-E curves. The acceleration of corrosion associated with pitting results because the proximity of the two cells creates a coupled cell where acidic iron dissolution at the bottom of the pit and reduction of a strong oxidant at the mouth of the pit become the controlling anodic and cathodic processes. If pitting contributes to reduction of RX by Fe^0, the dehalogenation rate will equal the net corrosion rate of the short-circuited cell (labeled Pit in Figure 3).

Since pitting corrosion is greatly accelerated by aggresive anions such as chloride *(42, 43)*, and chloride is a product of dechlorination (equation 4), it is possible that dechlorination might favor further dechlorination if enough chloride accumulates to provide auto-catalysis. At least two studies have found effects consistent with the possibility of an auto-catalytic link between dechlorination and pitting corrosion: Heland *et al.* reported a 60% increase in CCl_4 reduction rates by Fe^0 with increased contact time in batch systems *(44)*, and Johnson *et al.* showed that large concentrations of added chloride (up to 60 mM) increased the rate of CCl_4 degradation as much as four-fold *(13)*. However, the gradual increase in rate with time that would be expected from a strongly auto-catalytic process has rarely been observed, and most groups have reported consistently first-order degradation kinetics *(5, 45)*, or gradually decreasing rates with time (i.e., tailing) *(2, 46-48)*.

Although there is little evidence for auto-catalysis in dechlorination by Fe^0, it is still possible that localized corrosion contributes to the remediation of contaminants in environmental applications. Various investigators have postulated that localized corrosion contributes through increased surface area *(44)* and creation of corrosion cell domains *(49-51)*. The corrosion cell model works on the same principle as the electrochemical model described above (Figure 3), but invokes additional effects such as the reduction of protons as the major cathodic reaction, and the creation of an electrical double layer between the anode and cathode that permits transport due to electrical migration as well as diffusion. Although many aspects of these models are plausible, there are not yet any data that specifically support them, and a study that systematically addresses the role of localized corrosion in remediation applications of Fe^0 remains to be done.

The Oxide as a Semiconductor. In addition to direct reduction of RX via pits or other defects in the oxide layer, it is possible that reduction occurs indirectly, by transport of charge through the oxide from Fe^0 to RX. Possible charge carriers within the oxide include: (i) anion and cation vacancies (lattice sites where ions are absent), (ii) anion and cation interstitials (sites where ions are imbedded between lattice sites), and (iii) electrons and holes *(52)*. In this section, we will focus on electrons as the charge carriers and examine the transport of electrons through the oxide layer, treating the oxide as a semiconductor.

Most iron oxides are semiconductors (Table 1). Specifically, this means that they exhibit a bandgap (E_{BG}) where no energy levels are found, between the conduction band (energy levels unoccupied by electrons) and the valence band (energy levels occupied by electrons) *(53-55)*. The oxide layer that forms on Fe^0 can also be considered a semiconductor. For example, passive films composed of magnetite, maghemite, and other mixed-valent (hydr)oxides (see section on Composition of the Oxide) have been modeled as semiconducting Fe_2O_3 that is doped with Fe(II) *(56)*.

Long term exposure of granular Fe^0 to the highly-reducing conditions found in groundwater remediation applications, however, is likely to produce oxides that are mixed-valent. These oxides will exhibit metallic conductivity, because mixed oxidation states impart mobility of electrons between atoms *(53)*. For example, magnetite is mixed-valent, has a very small band gap (Table 1), and exhibits conductivity that is typical of a metal. The same should be true for the green rusts. (Green rusts, however, do not have well-defined crystal structures *(57)*, so their electronic structure does not appear to have been investigated.) Green rusts are of particular interest because of recent reports that they form on Fe^0 under environmental conditions *(3, 27)*.

Table I. Selected electronic properties of iron oxides

Oxide	Formula	E_{CB} (eV)[1]	E_F (eV)[1]	E_{BG} (eV)
Fe(II) only				
Siderite	$FeCO_3$			
Pyrite	FeS_2	0.17 *(58)*		1.0 *(58)*
Mixed Fe(II), Fe(III)				
Magnetite	Fe_3O_4			0.11 *(59)*
Green Rust I	[2]			
Green Rust II	[3]			
Fe(III) only				
Ferrihydrite	$Fe_5HO_8 \cdot 4H_2O$			
Hematite	α-Fe_2O_3	-0.245 *(60)*, 0.242 *(61)*	-0.225 *(57)*[5], -0.28 *(62)*, -0.17 *(59)*, -0.285 *(57)*[4]	2.02 *(57)*, 2.2 *(59)*
Maghemite	γ-Fe_2O_3		0.315 *(57)*[5]	2.03 *(57)*
Goethite	α-FeOOH		0.135 *(57)*[5]	2.10 *(57)*
Akaganeite	β-FeOOH		0.835 *(57)*[5]	2.12 *(57)*
Lepidocrocite	γ-FeOOH		0.705 *(57)*[5]	2.06 *(57)*
Feroxyhyte	δ'-FeOOH		0.625 *(57)*[5]	1.94 *(57)*

[1]At pH 7 versus NHE. [2]$3Fe(OH)_2 \cdot Fe(OH)_2Cl \cdot n\, H_2O$ where $3 \geq n \geq 2$ *(63)*. [3]$[Fe(II)_4\, Fe(III)_2\, (OH)_{12}]^{2+} \cdot [SO_4 \cdot 2H_2O]^{2-}$ *(64)*. [4]Single crystal. [5]Quasi-Fermi energy level for electrons. This is the Fermi energy level under illumination (nonequilibrium conditions). For n-type semiconductors where electrons are the majority carriers, the quasi-Fermi energy level is approximately equal to the Fermi energy level *(65)*.

The details of ET involving semiconductors are determined by their electron energy levels. Important energy levels include: (i) the upper edge of the valence band (E_{VB}); (ii) the lower edge of the conduction band (E_{CB}); and (iii) the Fermi energy (E_F), which is the electrochemical potential of electrons in the bulk semiconductor. Table 1 is a compilation of these energies for the iron oxides. Passive films on Fe^0 have a bandgap of ~1.9 eV and a flatband potential of -0.3 to 0.0 V (vs. NHE at pH 7) *(66-69)*. (The flatband potential is equivalent to the Fermi energy before ET in or out of the semiconductor that produces a space charge region *(55)*.) Figure 1c summarizes these energy levels in a band diagram for a single crystal semiconductor that has been adapted for the Fe^0-oxide-water system *(70)*.

A fully-accurate band diagram for oxides present on grains of Fe^0 used in remediation applications, however, would be more complicated than the version portrayed in Figure 1c, because the oxide layer that forms on granular Fe^0 lacks long-range order. Passive films on Fe^0 are typically only a few monolayers thick, so long-range order normal to the surface cannot develop. Thicker oxide layers tend to be amorphous because a more coherent film would be strained and break up *(55, 71-74)*. While thick oxide layers will have bands equivalent to the conduction and valence bands found in single crystal semiconductors (shown in Figure 1c), additional bands, which result from a large density of impurities or ions in a different oxidation state, may also be present *(55, 75)*. In addition, there will be a much greater density of localized states (energy levels that exist in the bandgap) than is found in single crystal semiconductors. These localized states result from the lack of identical lattice sites throughout the oxide *(55, 71, 73-75)*. Impurity bands and localized states are important because they can serve as electron donors in ET *(68, 76)*, and because they are likely to be abundant in the oxide layers that develop during remediation applications of Fe^0.

The band diagram in Figure 1c is also simplified with respect to electric field effects. The differences in energy levels of E_F(metal), E_F(oxide), and E^0(RX/R•) will produce electric fields in the oxide layer, and these fields can affect ET. For example, at equilibrium, there will be a potential barrier for injecting electrons from Fe^0 to the conduction band of the iron oxide, because E_F(oxide) is more negative than E_F(metal) *(77)*. Similarly, because E_F(oxide) can be more negative than E^0(RX/R•), the oxide layer may be in depletion: i.e., there may be a space charge region in the oxide layer that acts as a barrier to electron transport from the conduction band to RX. It should be noted that this barrier develops as equilibrium is established between the semiconductor and RX because ET occurs from the semiconductor to RX until E_F(oxide) = E^0(RX) *(54)*. This barrier may not be fully established, however, because the rate at which equilibration occurs depends on the rate of ET between the oxide and redox couples in solution *(54)*, and this appears to be fairly slow for iron-oxide-RX systems *(9, 15)*.

Although the semiconductor band model shown in Figure 1c is a simplification of the iron-oxide-solution system, it does help reveal some of the factors controlling reduction of chlorinated aliphatic compounds by passivated iron. For ET from Fe^0 through the oxide layer to RX to be thermodynamically favorable, the energy of electrons in the metal and/or the oxide must be more negative than E^0(redox). In addition, the likelihood of ET from the semiconductor to the electrolyte is greater if there is good overlap between electron energy levels in the metal or semiconductor and the unoccupied electron energy levels of the redox couple in solution *(55, 75)*.

If ET through the oxide originates from the metal, then E_F(metal) will determine the driving force for reaction. For the reduction of CCl_4 by passivated iron, E_F(metal) is more negative than $E^0(CCl_4/{\bullet}CCl_3)$ *(70)* so electrons could originate from Fe^0. ET through the oxide may occur by tunneling (direct or resonance) *(55, 73, 78, 79)*. Direct tunneling of electrons from the underlying Fe^0 through the oxide layer is only possible for thin films of oxide (~10 Å). Resonance tunneling, in which the electrons tunnel or "hop" between localized states in the oxide, is more probable for the oxide layer that forms on grains of Fe^0 in remediation applications *(55, 73, 74, 78, 79)*.

If, however, ET from the oxide layer is not influence by the underlying Fe^0, the driving force will be determined by E_{CB} of the oxide, E_F(oxide), or the energies of localized states or impurity bands. For the reduction of CCl_4 by passivated iron, E_{CB} and E_F(oxide) are both more negative than $E^0(CCl_4/{\bullet}CCl_3)$ *(70)*. The energy levels of localized states and impurity bands are unknown, but presumably there will be a significant number of localized states with energies sufficiently negative to drive reduction of most chlorinated aliphatic compounds. Thus, the electrons involved in ET to RX could originate from the conduction band or localized states in the oxide. If a space charge barrier exists (which we believe to be the case for passivated iron in the presence of aqueous CCl_4), conduction band electrons will have to tunnel through the barrier either directly or via localized states *(55, 71, 74, 76, 80)*. There is evidence that ET to dissolved ferrous cyanide can occur via tunneling from the conduction band through the space charge region of a passivated iron electrode *(81, 82)*. However, our studies have shown that ET from granular Fe^0 to aqueous CCl_4 does not occur via the oxide conduction band *(70)*. Thus, in groundwater remediation applications, where there are likely to be high concentrations of Fe(II) ions in the oxide layer, ET is more likely to occur via resonance tunneling from the metal.

The Oxide as a Coordinating Surface. In addition to direct ET from zero-valent metal to the adsorbed contaminant (Figures 1a-b) or indirect ET through the oxide film (Figure 1c), there are two possibilities for reduction of RX at the surface of the oxide film (Figure 1d): (i) electrons may be donated from surface-bound Fe(II) sites created by adsorption of Fe^{2+} from solution *(83)*, or (ii) electrons may originate from the underlying metal, creating Fe(II) sites within the oxide lattice ("structural" Fe(II) *(84)*), which eventually result in surface sites that reduce RX. Reactive surface Fe(II) sites have been postulated in a recent study of the reduction of nitro aromatic compounds in suspensions of Fe(III) minerals in the presence of dissolved Fe(II) *(83)*. Reduction via ET from surface reactive sites is further supported by additional studies showing the reduction of chlorinated aliphatics *(7, 85-89)* and inorganic groundwater contaminants in the presence of iron oxides (alone, no Fe^0) *(90, 91)*.

In the presence of a finite number of sites, the rate of dehalogenation should increase with increasing concentration of RX until there are no more vacant surface sites, producing the hyperbolic profile often attributed to site saturation effects in heterogeneous systems *(92)*. Site saturation behavior has been observed for the reduction of CCl_4 by Fe^0 *(13, 15, 93)*, and the reduction of tetrachloroethene by zinc metal in laboratory batch experiments *(12)*. It has also been suggested that the effects of initial contaminant concentration and competing adsorbates (such as chromate or EDTA) on the kinetics of reduction of chlorinated aliphatic compounds might reflect competition for surface coordination sites *(12, 13)*.

Adsorbate competition for a limited number of sites may be modeled using surface complexation models (SCMs). The SCMs extend equilibrium solution chemistry to include surface chemical species. Fundamentally, these models are based on two main assumptions: (i) adsorption occurs at coordinating sites with specific functional groups, and (ii) equilibrium can be described using mass law equations with an electrostatic correction derived from electric double layer theory *(94-96)*. Combining the mass law equations with a mole balance on the total number of sites results in a set of simultaneous equations that can be solved numerically. In the following analysis, we derive a simplified SCM (with no electrostatics) that describes the site saturation behavior observed in Fe^0 systems. Additional applications of the SCM to remediation by Fe^0 include: (i) evaluating the role of adsorption of Fe^{2+} to the iron oxide surface, and (ii) understanding the effect of site competition from co-contaminants and naturally occurring ligands.

SCMs have been used to model metal oxide dissolution by treating the dissolution rate as proportional to the number of sites occupied by ligands that promote dissolution *(97-100)*. The oxide surface is modeled as a finite number of surface functional groups that act as possible dissolution sites. The following derivation is based on a similiar hypothesis: that the dehalogenation rate is proportional to the number of sites occupied by the chlorinated aliphatic compound.

As discussed earlier, it is reasonable to assume the surface is a mixed Fe(II)\Fe(III) oxide, sometimes approaching a pure Fe(II) oxide under highly reducing conditions (Figure 1d). The basic kinetic formulation involves adsorption (or association) of the dissolved species with the surface and subsequent ET to form products:

$$\equiv FeOH^0 + RX \underset{k_2}{\overset{k_1}{\rightleftharpoons}} \equiv FeOH\text{-}RX \overset{k_{RX}}{\rightarrow} \text{Products} \quad (5)$$

The rate equation, assuming a steady-state population of the precusor complex ≡FeOH-RX, is:

$$-\frac{d[RX]}{dt} = -k_{RX}\rho_a[\equiv FeOH\text{-}RX] \quad (6)$$

where ρ_a is the surface area concentration of the solid ($m^2 L^{-1}$), which serves to convert surface concentrations ($mol\ m^{-2}$) to aqueous concentrations ($mol\ L^{-1}$). The rate equation can be modified to include additional terms that represent adsorption to non-reactive sites *(101, 102)*, or solubility enhancements in the presence of surfactants *(103)* and dissolved organic matter, as the significance of these processes becomes better understood. The concentration of the precusor complex can be found by substituting the mass law equations (from Table 2) into a mole balance equation for the total number of surface sites, $[\equiv Fe_{TOT}]$:

$$[\equiv Fe_{TOT}] = [\equiv FeOH^0] + [\equiv FeL] + [\equiv FeOH\text{-}RX] \quad (7)$$

As written, equation 7 does not include adsorption of aqueous metals (such as Fe^{2+}), formation of ternary complexes, or formation of binuclear complexes. These terms have been neglected only because there is not yet a three-dimensional data set collected as a function of pH, ionic strength, and ligand concentration with which to evaluate the importance of such a large number of complexes and adsorptive sinks. When the data become available, the model can be revised to include the sum of all plausible metal-centered complexes.

Substitution of the mass law equations from Table II into equation 7 gives a general competition model describing the rate of dehalogenation by iron in the presence of a competing ligand:

$$-\frac{d[\mathrm{RX}]}{dt} = \frac{k_{RX}K_{RX}[\equiv \mathrm{Fe_{TOT}}][\mathrm{RX}]}{1+K_{RX}[\mathrm{RX}]+K_L[\mathrm{HL}]} \qquad (8)$$

In the absence of any competition (e.g., a single RX with no ligands present), equation 8 simplifies to:

$$-\frac{d[\mathrm{RX}]}{dt} = \frac{k_{RX}K_{RX}[\equiv \mathrm{Fe_{TOT}}][\mathrm{RX}]}{1+K_{RX}[\mathrm{RX}]} \qquad (9)$$

Equation 9 is a hyperbolic relationship, similar to the Michaelis-Menton equation derived for enzyme kinetics *(104)*, the Langmuir equation as applied to adsorption on soils *(105)*, and an adaptation of these models for dechlorination by Fe^0 that we published previously *(13)*. As such, all four models are capable of describing site saturation phenomena commonly found in heterogenous systems; however, only the new model (equations 8 and 9) explicitly distinguishes thermodynamically-related parameters from the kinetic constants.

The non-competitive SCM (equation 9) can be used to estimate the equilibrium association of RX with the surface (K_{RX}), and the surface rate constant (k_{RX}), by modeling previously published site saturation data for systems containing Fe^0 (Figure 4). To do this, we estimated the total number of sites using

$$[\equiv \mathrm{Fe_{TOT}}] = \rho_a \mathrm{N_s} / \mathrm{A} \qquad (10)$$

where N_s is the number of sites per unit area and A is Avogadro's number. Experimental values of N_s for common iron oxides range from 2 -16 sites nm^{-2} *(106)*, with an average calculated surface density of 5 sites nm^{-2} *(95)*. Using reported Fe^0 surface area concentrations (ρ_a, typically measured using BET adsorption techniques) and an estimated N_s of 5 sites nm^{-2}, the total number of sites was estimated for two different Fe^0 systems (Table 3). K_{RX} and k_{RX} were then determined as fitting coefficients by nonlinear least-squares parameter adjustment with equation 9. The arrows in Figure 4 provide graphical interpretation of K_{RX} and k_{RX} by indicating the maximum rate of dechlorination (equivalent to V_{max} in the Michaelis-Menton formulation *(13)*) and the conditional equilibrium constant for the association of RX with the surface (equivalent to the reciprocal of $K_{1/2}$, which is the concentration of RX at $V_{max}/2$ in the Michaelis-Menton formulation *(13)*).

Even though the two systems contain different types of Fe^0, the estimated values of K_{RX} are the same within five percent (Table 3). Since CCl_4 is most likely driven to the surface by a weak hydrophobic interaction *(13)*, partitioning of CCl_4 to

Table II. Mass action equations for the surface complexation model

Reaction	Equilibrium constant
Protonation Reactions	
$\equiv FeOH_2 \rightleftharpoons \equiv FeOH^0 + H^+$	$K_{a1} = [\equiv FeOH^0][H^+] / [\equiv FeOH_2]$
$\equiv FeOH^0 \rightleftharpoons \equiv FeO^- + H^+$	$K_{a2} = [\equiv FeO^-][H^+] / [\equiv FeOH^0]$
Ligand Exchange Reactions	
$\equiv FeOH^0 + HL \rightleftharpoons \equiv FeL + H_2O$	$K_{L1} = [\equiv FeL] / [\equiv FeOH^0][HL]$
$2 \equiv FeOH^0 + H_2L \rightleftharpoons \equiv Fe_2L + H_2O$	$K_{L2} = [\equiv Fe_2L] / [\equiv FeOH^0][H_2L]$
Hydrophobic Partitioning	
$\equiv FeOH^0 + RX \rightleftharpoons \equiv FeOH^0\text{-}RX$	$K_{RX} = [\equiv FeOH^0\text{-}RX] / [\equiv FeOH^0][RX]$
Metal Partitioning	
$\equiv FeOH^0 + Fe^{2+} \rightleftharpoons \equiv FeOFe^+ + H^+$	$K_{M1} = [\equiv FeOFe^+][H^+] / [\equiv FeOH^0][Fe^{2+}]$
$2 \equiv FeOH^0 + Fe^{2+} \rightleftharpoons \equiv (FeO)_2Fe + 2H^+$	$K_{M2} = [\equiv (FeO_2)Fe][H^+]^2 / [\equiv FeOH^0][Fe^{2+}]$
Ternary Complex Formation	
$\equiv FeOH^0 + Fe^{2+} + L^- \rightleftharpoons \equiv FeOFeL + H^+$	$K_{T1} = [\equiv FeOFeL][H^+] / [\equiv FeOH^0][Fe^{2+}][L^-]$
$\equiv FeOH^0 + L^- + Fe^{2+} \rightleftharpoons \equiv FeLFe^+ + OH^-$	$K_{T2} = [\equiv FeLFe^+][OH^-] / [\equiv FeOH^0][Fe^{2+}][L^-]$

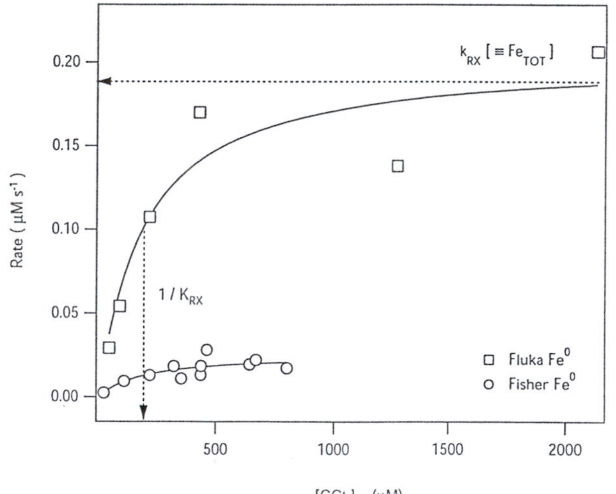

Figure 4. Effect of initial CCl_4 concentration, $[CCl_4]_0$, on the rate of CCl_4 reduction by Fe^0, in two well-mixed anaerobic batch systems. Squares: Fluka Fe^0 turnings ($\rho_a = 0.79\ m^2\ L^{-1}$) in deionized water buffered with 10 mM HEPES at pH 6.2 *(13)*. Circles: Fisher Fe^0 filings ($\rho_a = 1.02\ m^2\ L^{-1}$) in unbuffered deionized water *(15, 93)*. Solid curves are drawn from the non-competitive SCM (equation 8) using fitted values of K_{RX} and k_{RX} (given in Table 3).

the surface should be independent of iron type. On the other hand, values for the surface reaction rate constant (k_{RX}) differ by a factor of ten. This can be attributed to differences in intrinsic reactivity of the sites on each metal. Similar results and conclusions were obtained previously from the same data using a less explicit model *(13)*.

Table III. SCM Model parameters for iron metal systems.

Iron-RX System	$[\equiv Fe_{TOT}]$ (μM)	K_{RX} (μM^{-1})	k_{RX} (s^{-1})
[a]Fluka Fe^0 and CCl_4	6.6	$(5.3 \pm 2.6) \times 10^{-3}$	$(3.1 \pm 0.41) \times 10^{-2}$
[b]Fisher Fe^0 and CCl_4	8.5	$(5.4 \pm 4.4) \times 10^{-3}$	$(3.0 \pm 0.76) \times 10^{-3}$

[a] Ref. *(13)*. [b] Ref. *(15, 93)*.

The competitive SCM (equation 8) can be used to model the rate of dehalogenation in the presence of competing ligands. Given the weak interaction of RX with the surface, complex formation by strong ligands will block sites, thereby decreasing the rate of dehalogenation. This effect has been observed for the reduction of CCl_4 in the presence of acetate, ascorbate, and catechol *(13)*. However, as shown in Figure 5, the observed inhibition is less than that which is predicted from the competitive SCM (equation 8) using the values in Table III. The implication of this result is that dehalogenation rates approach a plateau at 50-70% of non-competitive rates, rather than declining rapidly toward zero as predicted by the model. For this to be the case, there must be two (or more) mechanisms of reaction, at least one of which is unaffected by ligand competition. There is not yet enough data on the effect of ligands to be certain of this interpretation, but several mechanisms that might not be sensitive to ligand competition can be suggested based on what has already been discussed. One possibility is formation of ternary complexes (FeOFeL) or metal bound complexes ($FeOFe^+$) that permit or even facilitate ET through bridging ligands. An example of the latter was observed in the catalytic dissolution of hematite in the presence of Fe(II) and oxalate *(107)*. Other possibilities include pitting of the surface oxide layer or tunneling across the oxide-water interface (see section on The Oxide as a Semiconductor).

Implications for Environmental and Engineering Applications

The treatment of environmental contaminants with zero-valent metals is still a new field, but one that has fundamental features in common with a variety of other chemical technologies. These include environmental applications such as (i) "redox manipulation" to form in-situ treatment walls of structural ferrous iron *(108)*, (ii) high pressure grinding (ball milling) to degrade chlorinated benzenes *(109)*, (iii) high temperature oxidation on metal oxides *(110)*, and (iv) oxidation by photocatalysis using semiconducting oxides *(111)*, as well as non-environmental applications like (v) chemical synthesis with dissolving metals *(112)* and Grignard reagents *(113)*, (vi) controlling material damage by corrosion of metals and metal oxides *(38)*, and (vii) high pressure lubrication with chlorinated hydrocarbons *(114)*. Although our

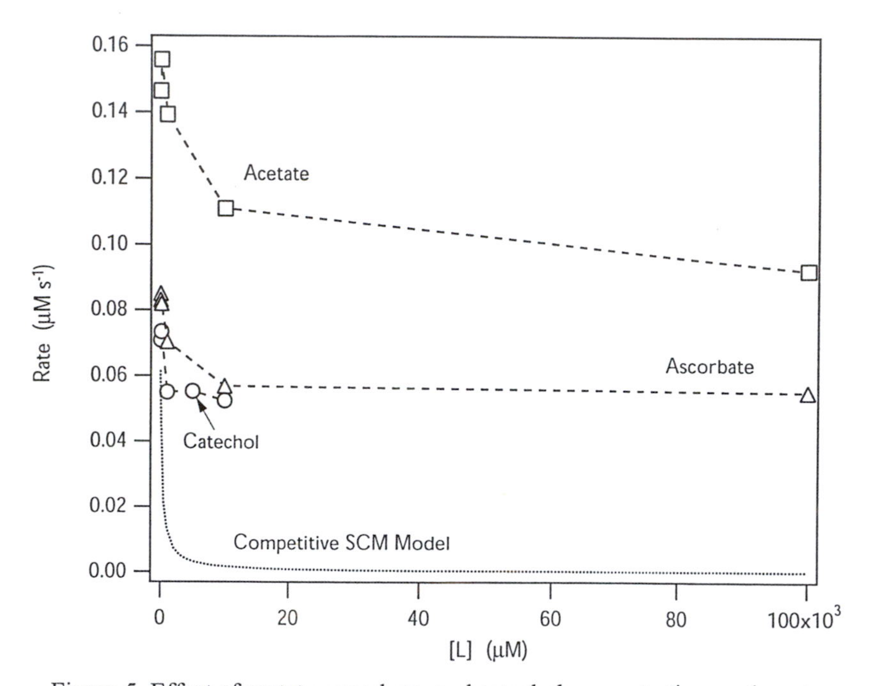

Figure 5. Effect of acetate, ascorbate, and catechol concentration on the rate of CCl_4 dechlorination by Fluka Fe^0 *(13)*. Same conditions as in Figure 4 and Table III. The dotted curve is calculated from the competitive SCM (equation 9) using values in Table III for Fluka Fe^0 and assuming $K_L = K_{RX}$.

primary concern in this paper has been with contaminant degradation by Fe^0, we have tried to perserve generality in our perspective throughout.

In an effort to reach a comprehensive general picture of the mechanism of contaminant degradation in metal-oxide-water systems, we have tried to balance our treatment of the oxide in its various roles as a physical barrier, a semiconductor, and a coordinating surface. Each model provides its own useful insights into the treatability of contaminants with zero-valent metals. For example: (i) the physical barrier posed by the oxide layer may be overcome by reversing the strategies used by corrosion engineers to protect metals from corrosion, (ii) the semiconducting properties of the oxide layer may respond to manipulations that are equivalent to doping, and (iii) the coordination properties of the surface may be altered in predictable ways by various types of ligands and surfactants. To develop these ideas further, it will help to clearly distinguish the processes involved in each model (as we have tried to do in Figs. 1b-d).

Distinguishing the passive film, semiconductor, and coordinating surface models does not make them exclusive, however, and additional insights are to be gained by examining their common elements. For example, (de)passivation involves changes in the oxide layer that can be caused by surface complexation *(100)*, photoinduced electrochemical reactions on passive films can be attributed to semiconducting properties of the film *(72)*, and doping of a semiconductor may be initiated by ligand attachment (i.e., surface complexation) *(115)*. One consequence of the former, is that surface complexation by corrosion inhibitors can inhibit the degradation of chlorinated solvents on aluminum *(116)*. A consequence of the latter, is that defects in coordination structure of semiconducting oxides can enhance photoreactivity *(117)*. With respect to the remediation of RX by Fe^0, there are many elements common to all three models discussed in this review.

The benefits of unifying specialized models of the oxide-water interface into a more comprehensive general picture have been argued previously for geochemistry and corrosion science *(34, 100, 118)*. Organic substances play a small role in these fields, mainly as ligands whose fate is not of primary interest. Perhaps the impact of future research in remediation with zero-valent metals will be to develop the largely unexplored disciplinary-interface between organic chemistry (contaminant degradation) and corrosion science (passive films as physical barriers), solid state physics (semiconducting oxides), and geochemistry (surface complexation at the mineral-water interface).

Legend of Symbols

ρ_a Surface area concentration ($m^2\ L^{-1}$)
k_1 Forward association rate constant ($L\ mol^{-1}\ s^{-1}$)
k_2 Backward association rate constant (s^{-1})
k_{RX} Rate constant for reduction of RX at the surface (s^{-1})
K_{RX} Conditional equilibrium constant for association of RX with surface ($L\ mol^{-1}$)
N_s Number of sites per unit area ($mol\ m^{-2}$)
A Avogadro's number (6.02×10^{23} sites mol^{-1})
E_F Fermi energy (eV)
E_{CB} Conduction band energy (eV)
E_{VB} Valence band energy (eV)
E_{BG} Band gap energy (eV)

Acknowledgments

Acknowledgement is made to the donors of The Petroleum Research Fund, administered by the ACS, for the primary support of this research through Type AC award number 29995-AC5 and two supplemental awards. Additional support was provided by the University Consortium Solvents-In-Groundwater Research Programme, and the Murdock Trust. In addition, the authors gratefully acknowledge J. Westall (Oregon State University) and B. Fish (Oregon Graduate Institute) for their insightful comments on this manuscript, and T. Johnson (Oregon Graduate Institute) for use of Figure 2.

Literature Cited

1. Tratnyek, P. G. *Chem. Ind. (London)* **1996**, 499-503.
2. Gillham, R. W.; O'Hannesin, S. F. *Ground Water* **1994**, *32*, 958-967.
3. Johnson, T. L.; Tratnyek, P. G. *Proceedings of the 33rd Hanford Symposium on Health & the Environment. In-Situ Remediation: Scientific Basis for Current and Future Technologies,* Pasco, WA, Battelle Pacific Northwest Laboratories, 1994; Vol. 2, pp. 931-947.
4. Agrawal, A.; Tratnyek, P. G. *Environ. Sci. Technol.* **1996**, *30*, 153-160.
5. Mackenzie, P. D.; Baghel, S. S.; Eykholt, G. R.; Horney, D. P.; Salvo, J. J.; Sivavec, T. M. *209th National Meeting,* Anaheim, CA, American Chemical Society, Preprint Extended Abstracts, Division of Environmental Chemistry, 1995; Vol. 35, No. 1, pp. 796-799.
6. Matheson, L. J.; Tratnyek, P. G. *Environ. Sci. Technol.* **1994**, *28*, 2045-2053.
7. Sivavec, T. M.; Horney, D. P. *213th National Meeting,* San Francisco, CA, American Chemical Society, Preprint Extended Abstracts, Division of Environmental Chemistry, 1997; Vol. 37, No. 1, pp. 115-117.
8. Sivavec, T. M.; Horney, D. P. *209th National Meeting,* Anaheim, CA, American Chemical Society, Preprint Extended Abstracts, Division of Environmental Chemistry, 1995; Vol. 35, No. 1, pp. 695-698.
9. Scherer, M. M.; Westall, J. C.; Ziomek-Moroz, M.; Tratnyek, P. G. *Environ. Sci. Technol.* **1997**, *31*, 2385-2391.
10. Johnson, T. L.; Tratnyek, P. G. *209th National Meeting,* Anaheim, CA, American Chemical Society, Preprint Extended Abstracts, Division of Environmental Chemistry, 1995; Vol. 35, No. 1, pp. 699-701.
11. Roberts, A. L.; Totten, L. A.; Arnold, W. A.; Burris, D. R.; Campbell, T. J. *Environ. Sci. Technol.* **1996**, *30*, 2654-2659.
12. Arnold, W. A.; Roberts, A. L. *213th National Meeting,* San Francisco, CA, American Chemical Society, Preprint Extended Abstracts, Division of Environmental Chemistry, 1997; Vol. 37, No. 1, pp. 76-77.
13. Johnson, T. L.; Fish, W.; Gorby, Y. A.; Tratnyek, P. G. *J. Contam. Hydrol.* **1998**, *29(4)*, 377-396.
14. Haderlein, S. B.; Pecher, K. In *Kinetics and Mechanism of Reactions at the Mineral/Water Interface*; Grundl, T.; Sparks, D., Ed.; American Chemical Society: this volume.
15. Johnson, T. L.; Scherer, M. M.; Tratnyek, P. G. *Environ. Sci. Technol.* **1996**, *30*, 2634-2640.
16. Nagayama, M.; Cohen, M. *J. Electrochem. Soc.* **1962**, *109*, 781-790.

17. Davenport, A. J.; Bardwell, J. A.; Vitus, C. M. *J. Electrochem. Soc.* **1995**, *142*, 721-724.
18. Davenport, A. J.; Sansone, M. *J. Electrochem. Soc.* **1995**, *142*, 725-730.
19. Oblonsky, L. J.; Devine, T. M. *Corr. Sci.* **1995**, *37*, 17-41.
20. Gui, J.; Devine, T. M. *Corr. Sci.* **1991**, *32*, 1105-1124.
21. Cahan, B. D.; Chen, C.-T. *J. Electrochem. Soc.* **1982**, *129*, 921-925.
22. Toney, M. F.; Davenport, A. J.; Oblonsky, L. J.; Ryan, M. P.; Vitus, C. M. *Phys. Rev. Let.* **1997**, *79*, 4282-4285.
23. O'Grady, W. E. *J. Electrochem. Soc.* **1980**, *127*, 555-563.
24. Bockris, J. O. M. *Corr. Sci.* **1989**, *29*, 291-312.
25. Sivavec, T. M.; Horney, D. P.; Baghel, S. S. *Emerging Technologies in Hazardous Waste Management VII, Extended Abstracts for the Special Symposium,* Atlanta, GA, Industrial & Engineering Chemistry Division, American Chemical Society, 1995; Vol. pp. 42-45.
26. Pratt, A. R.; Blowes, D. W.; Ptacek, C. J. *Environ. Sci. Technol.* **1997**, *31*, 2492.
27. Odziemkowski, M. S.; Gillham, R. W. *213th National Meeting,* San Francisco, CA, American Chemical Society, Preprint Extended Abstracts, Division of Environmental Chemistry, 1997; Vol. 37, No. 1, pp. 177-180.
28. Agrawal, A.; Tratnyek, P. G.; Stoffyn-Egli, P.; Liang, L. *209th National Meeting,* Anaheim, CA, American Chemical Society, Preprint Extended Abstracts, Division of Environmental Chemistry, 1995; Vol. 35, No. 1, pp. 720-723.
29. Taylor, R. M. *Clay Miner.* **1980**, *15*, 369-382.
30. Schwertmann, U.; Cornell, R. M. *Iron Oxides in the Laboratory*; VCH: Weinheim, 1991.
31. Hansen, H. C. B.; Borggaard, O. K.; Sørensen, J. *Geochim. Cosmochim. Acta* **1994**, *58*, 2599-2608.
32. Koch, C. B.; Moerup, S. *Clay Miner.* **1991**, *26*, 577-82.
33. Kassim, J.; Baird, T.; Fryer, J. R. *Corr. Sci.* **1982**, *22*, 147-158.
34. Sato, N. *Corrosion* **1989**, *45*, 354-368.
35. Sukamto, J. P. H.; Smyrl, W. H.; Casillas, N.; Al-Odan, M.; James, P.; Jin, W.; Douglas, L. *Mat. Sci. Eng.* **1995**, *A198*, 196.
36. Sato, N. *J. Electrochem. Soc.* **1982**, *129*, 255-260.
37. Sato, N. *J. Electrochem. Soc.* **1982**, *129*, 260-264.
38. Jones, D. A. *Principles and Prevention of Corrosion*; Macmillan: New York, 1992.
39. Roe, F. L.; A., L.; Funk, T. *Corr. Sci.* **1996**, *52*, 744-752.
40. Pou, T. E.; Murphy, O. J.; Young, V.; Bockris, J. O. M. *J. Electrochem. Soc.* **1984**, *131*, 1243-1251.
41. Asanuma, M.; Aogaki, R. *J. Chem. Phys.* **1997**, *106*, 9938-9943.
42. Abdel, M. S.; Wahdan, M. H. *Br. Corrosion J.* **1981**, *16*, 205-211.
43. Bardwell, J. A.; Fraser, J. W.; MacDougall, B.; Graham, M. M. *J. Electrochem. Soc.* **1992**, *139*, 366-370.
44. Helland, B. R.; Alvarez, P. J. J.; Schnoor, J. L. *J. Haz. Mat.* **1995**, *41*, 205-216.
45. Orth, S., W.; Gillham, R. W. *Environ. Sci. Technol.* **1996**, *30*, 66-71.
46. Lipczynska-Kochany, E.; Harms, S.; Milburn, R.; Sprah, G.; Nadarajah, N. *Chemosphere* **1994**, *29*, 1477-1489.
47. Schreier, C. G.; Reinhard, M. *Chemosphere* **1994**, *29*, 1743-1753.

48. Campbell, T. J.; Burris, D. R.; Roberts, A. L.; Wells, J. R. *Environ. Toxicol. Chem.* **1997**, *16*, 625-630.
49. Powell, R. M.; Puls, R. W.; Hightower, S. K.; Sabatini, D. A. *Environ. Sci. Technol.* **1995**, *29*, 1913-1922.
50. Powell, R. M.; Puls, R. W. *Environ. Sci. Technol.* **1997**, *31*, 2244-2251.
51. Khudenko, B. M. *Wat. Sci. Technol.* **1991**, *23*, 1873-1881.
52. Battaglia, V.; Newman, J. *J. Electrochem. Soc.* **1995**, *142*, 1423-1430.
53. Cox, P. A. *Transition Metal Oxides: An Introduction to their Electronic Structure and Properties*; Oxford: New York, 1997; Vol. 27.
54. Finklea, H. O. In *Semiconductor Electrodes*; Finklea, H. O., Ed.; Elsevier: Amsterdam, 1988; pp 1-42.
55. Morrison, S. R. *Electrochemistry at Semiconductor and Oxidized Metal Electrodes*; Plenum: New York, 1980.
56. Schmuki, P.; Büchler, M.; Virtanen, S.; Böhni, H.; Muller, R.; Gauckler, L. J. *J. Electrochem. Soc.* **1995**, *142*, 3336-3342.
57. Leland, J. K.; Bard, A. J. *J. Phys. Chem.* **1987**, *91*, 5076-5083.
58. Wei, D.; Osses-Asare, K. *J. Electrochem. Soc.* **1996**, *143*, 3192-3198.
59. Anderman, M.; Kennedy, J. H. In *Semiconductor Electrodes*; Finklea, H. O., Ed.; Elsevier: Amsterdam, 1988; pp 147-202.
60. Kormann, C.; Bahnemann, D. W.; Hoffmann, M. R. *J. Photochem. Photobiol., A. Chem.* **1989**, *48*, 161-169.
61. Hoffmann, M. R. In *Aquatic Chemical Kinetics*; Stumm, W., Ed.; Wiley: New York, 1990; pp 71-111.
62. Dimitrijevíc, N. M.; Savíc, D.; Mícíc, O. I.; Nozik, A. J. *J. Phys. Chem.* **1984**, *88*, 4278-4283.
63. Refait, P.; Génin, J.-M. R. *Corr. Sci.* **1994**, *36*, 55-65.
64. Génin, J.-M. R.; Olowe, A. A.; Refait, P.; Simon, L. *Corr. Sci.* **1996**, *38*, 1751-1762.
65. Pierret, R. F. *Semiconductor Fundamentals*; Addison-Wesley: New York, 1988; Vol. 1.
66. Schmuki, P.; Böhni, H. *Electrochim. Acta* **1995**, *40*, 775-783.
67. Abrantes, L. M.; Peter, L. M. *J. Electroanal. Chem.* **1983**, *150*, 593-601.
68. Stimming, U.; Schultze, J. S. *Ber. Bunsenges. Phys. Chem.* **1976**, *80*, 1297-1302.
69. Wilhelm, S. M.; Yun, K. S.; Ballenger, L. W.; Hackerman, N. *J. Electrochem. Soc.* **1979**, *126*, 419-424.
70. Balko, B. A.; Tratnyek, P. G. *J. Phys. Chem. B,* **1998**, *102, 1459-1465.*
71. Searson, P. C.; Latanision, R. M.; Stimming, U. *J. Electrochem. Soc.* **1988**, *135*, 1358-1363.
72. Stimming, U. *Electrochim. Acta* **1986**, *31*, 415-429.
73. Stimming, U. *Langmuir* **1987**, *3*, 423-428.
74. Newmark, A. R.; Stimming, U. *Electrochim. Acta* **1987**, *32*, 1217-1221.
75. Gerischer, H. *Corr. Sci.* **1989**, *29*, 191-195.
76. Schmickler, W. *Ber. Bunsenges. Phys. Chem.* **1978**, *82*, 477-487.
77. Hummel, R. E. *Electronic Properties of Materials;* 2nd ed.; Springer: New York, 1993.
78. Leiva, E.; Meyer, P.; Schmickler, W. *Corr. Sci.* **1989**, *29*, 225-236.
79. Schmickler, W. *J. Electroanal. Chem.* **1977**, *82*, 65-80.
80. Stimming, U.; Schultze, J. W. *Electrochem. Acta* **1979**, *24*, 858-869.

81. Meisterjahn, P.; Schultze, J. W.; Siemensmeyer, B.; Stimming, U.; Dean, M. H. *Chem. Phys.* **1990**, *141*, 131-141.
82. Schultze, J. W.; Stimming, U. *Z. Phys. Chem. N. F.* **1975**, *98*, 285-302.
83. Klausen, J.; Trüber, S. P.; Haderlein, S. B.; Schwarzenbach, R. P. *Environ. Sci. Technol.* **1995**, *29*, 2396-2404.
84. Stucki, J. W. *Nato Adv. Sciences Inst. Ser., Ser. C* **1988**, *217*, 625-75.
85. Kriegman-King, M. R.; Reinhard, M. In *Organic Substances and Sediments in Water*; Baker, R., Ed.; Lewis: MI, 1991; Vol. 2; pp 349-364.
86. Kriegman-King, M. R.; Reinhard, M. *Environ. Sci. Technol.* **1992**, *26*, 2198-2206.
87. Kriegman-King, M. R.; Reinhard, M. *Environ. Sci. Technol.* **1994**, *28*, 692-700.
88. Pecher, K.; Haderlein, S. B.; Schwarzenbach, R. P. *213th National Meeting,* San Francisco, CA, American Chemical Society, Preprint Extended Abstracts, Division of Environmental Chemistry, 1997; Vol. 37, No. 1, pp. 185-187.
89. Butler, E. C.; Hayes, K. F. *213th National Meeting,* San Francisco, CA, American Chemical Society, Preprint Extended Abstracts, Division of Environmental Chemistry, 1997; Vol. 37, pp. 113-115.
90. Peterson, M. L.; White, A. F.; Brown, G. E.; Parks, G. A. *Environ. Sci. Technol.* **1997**, *31*, 1573-1576.
91. Hansen, H. C. B.; Koch, C. B.; Nancke-Kroge, H.; Borggaard, O. K.; Sørensen, J. *Environ. Sci. Technol.* **1996**, *30*, 2053-2056.
92. Zepp, R. G.; Wolfe, N. L. In *Aquatic Surface Chemistry: Chemical Processes at the Particle-Water Interface*; Stumm, W., Ed.; Wiley: New York, 1987; pp 423-455.
93. Scherer, M. M.; Tratnyek, P. G. *209th National Meeting,* Anaheim, CA, American Chemical Society, Preprint Extended Abstracts, Division of Environmental Chemistry, 1995; Vol. 35, No. 1, pp. 805-806.
94. Morel, F. M. M.; Hering, J. G. *Principles and Applications of Aquatic Chemistry*; Wiley: New York, 1993.
95. Stumm, W. *Chemistry of the Solid-Water Interface: Processes at the Mineral-Water and Particle-Water Interface of Natural Systems*; Wiley: New York, 1992.
96. Dzombak, D. A.; Morel, F. M. M. *Surface Complexation Modeling: Hydrous Ferric Oxide*; John Wiley & Sons: New York, 1990.
97. Furrer, G.; Stumm, W. *Geochim. Cosmochim. Acta* **1986**, *50*, 1847-1860.
98. Wieland, E.; Wehrli, B.; Stumm, W. *Geochim. Cosmochim. Acta* **1988**, *52*, 1969-1981.
99. Zinder, B.; Furrer, G.; Stumm, W. *Geochim. Cosmochim. Acta* **1986**, *50*, 1861-1869.
100. Stumm, W. *Colloids and Surfaces. A: Physicochemical and Engineering Aspects* **1997**, *120*, 143-166.
101. Burris, D. R.; Campbell, T. J.; Manoranjan, V. S. *Environ. Sci. Technol.* **1995**, *29*, 2850-2855.
102. Allen-King, R. M.; Halket, R. M.; Burris, D. R. *Environ. Toxicol. Chem.* **1997**, *16*, 424-429.
103. Bizzigotti, G. O.; Reynolds, D. A.; Kueper, B. H. *Environ. Sci. Technol.* **1997**, *31*, 472-478.

104. Huennekens, F. M.; Chance, B. In *Investigation of Rates and Mechanisms of Reactions*; Friess, S. L.; Lewis, E. S.; Weissberger, A., Ed.; Interscience: New York, 1963; Vol. 8, Part 2; pp 1231-1314.
105. Hingston, F. J.; Posner, A. M.; Quirk, J. P. *Discuss. Faraday Soc.* **1971**, *52*, 334-342.
106. Davis, J. A.; Kent, D. B. In *Mineral-Water Interface Geochemistry*; Hochella, J., M.F.; White, A. F., Ed.; Mineralogical Society of America: 1990; Vol. 23; pp 177-260.
107. Wehrli, B.; Sulzberger, B.; Stumm, W. *Chem. Geol.* **1989**, *78*, 167-179.
108. Fruchter, J. S.; Amonette, J. E.; Cole, C. R.; Gorby, Y. A.; Humphrey, M. D.; Istok, J. D.; Spane, F. A.; Szecsody, J. E.; Teel, S. S.; Vermeul, V. R.; Williams, M. D.; Yabusaki, S. B. In Situ Redox Manipulation Field Injection Test Report - Hanford 100-H Area, Pacific Northwest National Laboratory, 1996.
109. Loiselle, S.; Branca, M.; Mulas, G.; Cocco, G. *Environ. Sci. Technol.* **1997**, *31*, 261-265.
110. Morlando, R.; Manahan, S. E.; Larsen, D. W. *Environ. Sci. Technol.* **1997**, *31*, 409-415.
111. Choi, W. Y.; Hoffmann, M. R. *Environ. Sci. Technol.* **1997**, *31*, 89-95.
112. Hudlicky, M. *Reductions in Organic Chemistry*; Ellis Horwood: Chichester, 1984.
113. Lai, Y.-H. *Synthesis* **1981**, 585-604.
114. Smentkowski, V. S.; Cheng, C. C.; Yates, J. T. J. *Langmuir* **1990**, *6*, 147-158.
115. Bockris, J. O. M.; Khan, S. U. M. *Surface Electrochemistry. A Molecular Level Approach*; Plenum: New York, 1993.
116. Archer, W. L. *Ind. Eng. Chem. Prod. Res. Dev.* **1982**, *21*, 670-672.
117. Choi, W.; Termin, A.; Hoffman, M. R. *J. Phys. Chem.* **1994**, *98*, 13669-13679.
118. Blesa, M. A.; Morando, P. J.; Regazzoni, A. E. *Chemical Dissolution of Metal Oxides*; CRC Press: Ann Arbor, 1994.

Chapter 16

The Reduction of Aqueous Metal Species on the Surfaces of Fe(II)-Containing Oxides: The Role of Surface Passivation

Art F. White[1] and Maria L. Peterson[2,3]

[1]U.S. Geological Survey, Menlo Park, CA 94025
[2]Department of Geological and Environmental Sciences, Stanford University, Stanford, CA 94305
[3]ECK Technologies, Hayward, CA 94545

The reduction of aqueous transition metal species at the surfaces of Fe(II)- containing oxides has important ramifications in predicting the transport behavior in ground water aquifers. Experimental studies using mineral suspensions and electrodes demonstrate that structural Fe(II) heterogeneously reduces aqueous ferric, cupric, vanadate and chromate ions on magnetite and ilmenite surfaces. The rates of metal reduction on natural oxides is strongly dependent on the extent of surface passivation and redox conditions in the weathering environment. Synchrotron studies show that surface oxidation of Fe(II)-containing oxide minerals decreases their capacity for Cr(VI) reduction at hazardous waste disposal sites.

The solubility, sorption and transport of aqueous metal species in the environment are strongly dependent on their valence state. Changes in valance states can occur via electron transfer mechanisms between aqueous metal species and electrochemically reactive mineral substrates. One class of such heterogeneous redox reactions involves the reduction of metal species during the weathering and oxidation of Fe(II)-containing oxides and silicates (*1*). These minerals are common in many aquifers and, therefore, may be important in controlling the chemical behavior of natural or contaminant metal species dissolved in groundwater.

Reduction of Fe(III) occurs on the surfaces of basalt and Fe(II) containing silicates (*2, 3*). Reduction of Cr(VI) by biotite has been experimentally investigated (*4, 5*) and the oxidation of structural Fe(II) by interlayer Cu(II) has been observed in both natural (*6*) and experimentally reacted biotites (*7, 8*). The adsorption of Cr(VI) and partial reduction to Cr(III) also occurs on α-FeOOH that contains small amounts of Fe(II) (*9*). Recently the reduction of aqueous metal species including Fe(III), Cr

(VI), Cu(II) and V(IV) on the surfaces of magnetite and ilmenite have been extensively characterized (*10-15*).

Redox Reactions Involving Magnetite and Ilmenite

Magnetite and ilmenite are the most common Fe(II)-containing oxide minerals in the earth's crust. Magnetite and ilmenite asemiconductors and are capable of transferring electrons both within the solid state and also across the solid-liquid interface to adsorbed metal ions. These minerals are therefore potentially important in controlling heterogeneous redox reactions involving aqueous transition metals in groundwater systems. Oxidation of magnetite in the presence of water can be written as (*16*)

$$[Fe^{2+}\,Fe^{3+}{}_2]O_{4(magnetite)} + 2H^+ \rightarrow [Fe^{3+}{}_2]O_{3(maghemite)} + Fe^{2+} + H_2O. \quad (1)$$

Oxidation of magnetite in the presence of dissolved O_2 can be written as

$$3[Fe^{2+}Fe^{3+}{}_2]O_{4(magnetite)} + 1/2O_2 + 2H^+ \rightarrow 4[Fe^{3+}{}_2]O_{3(maghemite)} + Fe^{2+} + H_2O. \quad (2)$$

In both cases, maghemite, a structural polymorph of hematite, is the most common weathering product of magnetite (*17, 18*). A parallel reaction describing the commonly observed weathering of ilmenite to pseudorutile (*19*) is

$$3[Fe^{2+}Ti]O_{3(ilmenite)} + 2H^+ + 1/2O_2 \rightarrow Fe_2{}^{3+}Ti_3O_{9(pseudorutile)} + Fe^{2+} + H_2O. \quad (3)$$

Results of experimental studies of dissolution of magnetite and ilmenite are consistent with reactions predicted by equations 1-3. Due to reductive dissolution, Fe(II) and not Fe(III) is released to solution (*20, 21*). The rate of release of Fe(II) is linear with time and increases as a function of decreasing pH (Figure 1). Hydrogen ion is a reactant during dissolution as predicted by equations 1 - 3. Reaction of magnetite with acidic solutions ($pH < 3$) under oxic and anoxic conditions produce similar release rates for Fe(II), suggesting that the H^+ -driven reactions (equation 1) dominates over the O_2-driven reactions (equations 2 - 3). Comparisons of rates at pH >3 are not possible due to the oxidization and precipitation of Fe(II) in the presence of dissolved oxygen. Rates of ferrous ion release to anoxic solution can be described by the expression

$$N_{Fe} = k_o s_o t \quad (4)$$

where N_{Fe} (moles) is the mass of Fe released, k_o ($moles \cdot cm^{-2} \cdot s^{-1}$) is the rate constant at a specific pH and temperature, s_o (cm^2) is the surface area, and t (s) is time (*16*).

Coupled Half Cell Reactions. The oxidation of magnetite under anoxic conditions without the presence of reducible aqueous metal species consists of coupled anodic and cathodic half cell reactions (*16*)

$$3[Fe^{2+}Fe^{3+}{}_2]O_{4\,(magnetite)} \rightarrow 4[Fe^{3+}{}_2]O_{3\,(maghemite)} + Fe^{2+} + 2e^- \quad (5)$$

and

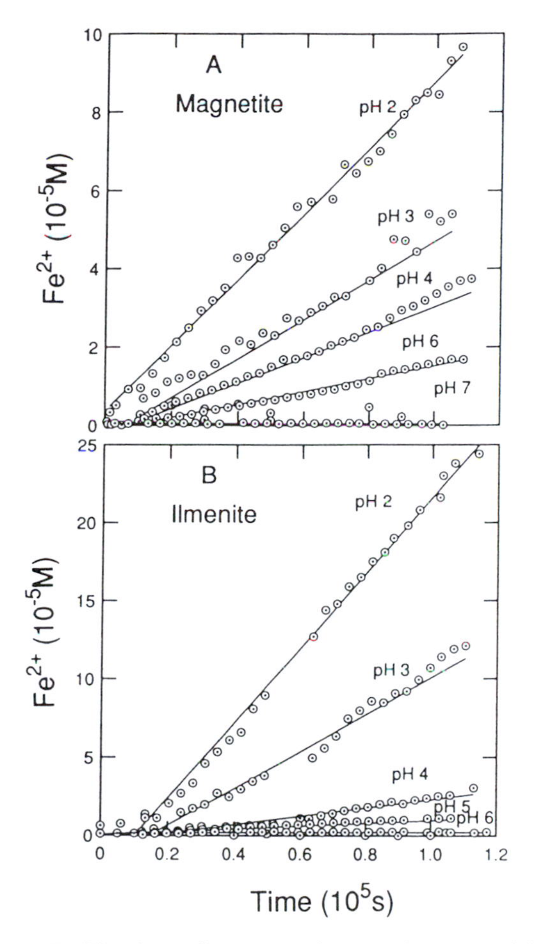

Figure 1. Aqueous Fe(II) release from natural magnetite (A) and ilmenite (B) as functions of time and pH at 25°C (adapted from ref. *16*).

$$[Fe^{2+}Fe^{3+}{}_2]O_{4\,(magnetite)} + 8\,H^+ + 2e^- \rightarrow 3Fe^{2+} + 4H_2O. \quad (6)$$

Equation 5 describes the oxidation of magnetite to form maghemite and equation 6 describes the reductive dissolution of magnetite (*22*). The decoupled half cell reactions describing ilmenite oxidation (equation 3) are

$$3[Fe^{2+}Ti]O_{3(ilmenite)} \rightarrow Fe_2{}^{3+}Ti_3O_{9(pseudorutile)} + 2e^- + Fe^{2+} \quad (7)$$

$$1/2O_2 + 2H^+ + 2e^- \rightarrow H_2O \quad (8)$$

The reaction of pure ilmenite does not involve reductive dissolution because the ideal structure contains only Fe(II). However most natural ilmenites represent partial solid solutions with hematite and contain significant amounts of Fe(III).

Electrochemical Studies. The mechanisms and kinetics associated with the decoupled reactions (equations 1 and 2) have been investigated by dynamic polarization of electrodes constructed from natural magnetite and ilmenite samples (*16*). Examples of resulting potentiodynamic scans are shown in Figure 2. The log of the absolute current density generated is plotted as a function of potential applied to the mineral electrodes relative to a calomel electrode. Increases in electrode current densities above approximately 1000 mV and below -200 mV corresponded to the respective oxidation and reduction of H_2O to form O_2 and H_2 .
Sharp current minimums in Figure 2 for magnetite (400 mV) and ilmenite (-40 mV) correspond to conditions in which no current is applied and the rates for the anodic and cathodic half cell reactions are equivalent and equal to equation 1. Polarization of the magnetite electrode to potentials less than 400 mV results in the dominance of the reductive dissolution of magnetite as described by equation 6. This reaction consumes electrons by reducing ferric atoms in the magnetite structure and releasing Fe(II) to solution.

Comparable current peaks are plotted in the inset (Figure 2) for polarization experiments at pH 1-4. The current axis is a linear scale rather than a log scale in this plot. The positive current density therefore corresponds to cathodic electrons. The intensity of this cathodic peak decreases from pH 1 to 3 and was not detected above pH 4. This pH effect is in concordance with the strong dependence of reductive dissolution on hydrogen ion activity (equation 6). Reductive dissolution has been intensively investigated under applied reducing potentials for synthetic magnetites that control corrosion inhibition in stainless steel (*21, 23*). Due to a lack of significant Fe(III), application of negative potentials, relative to E_o, does not result in a cathodic current peak at low pHs for ilmenite (Figure 2).

Above 400 mV, reactions at the magnetite electrode are dominated by the oxidation to maghemite with the concurrent release of electrons (anodic current) and Fe(II) to solution (equation 5). Increasing positive potentials accelerate this reaction. Above 400 mV, the current density remains relatively constant up to the potential at which H_2O begins to dissociate (Figure 2). Constant currents as functions of increasing positive potentials are commonly attributed in metal and metal oxide electrodes to passivation caused by the formation of unreactive oxidized surfaces. In magnetite, such passivation can be attributed to the formation of maghemite. The anodic reaction also passivates the ilmenite surface (equation 7), resulting in a relative constant current density from 150 mV up to a potential at which H_2O begins to dissociate (Figure 2). This passivation at positive potentials is similar to that observed for magnetite.

The application of positive potentials accelerates oxidation processes that occur at slower rates during natural weathering. Surface passivation can also be documented under open circuit conditions in which no external potential was applied to the oxide electrode (Figure 3). Experiments show that the potentials for freshly polished magnetite surfaces increase with time and decreased with solution pH (*16*). Steady state potentials were commonly approached in 1 to 2 hours after initial exposure to solution. Experiments in oxic solutions indicated that E_0 also increased with dissolved oxygen content. These increased potentials are caused by increasing surface oxidation.

Heterogeneous Reduction of Aqueous Transition Metals

In the presence of reducible aqueous metal species, additional reactions resulting in the oxidation of Fe(II) oxide surfaces can occur. Anodic half cell reactions describing the oxidation of Fe(II) on the surfaces of magnetite and ilmenite can be decoupled from the corresponding reductive half cell reactions (equations 6 and 8) and linked to the reduction of aqueous metal species at the mineral surfaces (*10*).

For an aqueous transition metal m, such reactions would be written as

$$3[Fe^{2+}Fe^{3+}{}_2]O_{4\,(magnetite)} + 2/n\ m^z \rightarrow 4([Fe^{3+}{}_2]O_{3\,(maghemite)} + Fe^{2+} + 2/n\ m^{z-n} \qquad (9)$$

and

$$3[Fe^{2+}Ti]O_{3(ilmenite)} + 2/n\ m^z \rightarrow Fe_2{}^{3+}Ti_3O_{9(pseudorutile)} + Fe^{2+} + 2/n\ m^{z-n} \qquad (10)$$

were z is the valance state and n is the charge transfer number. The above reactions involve the transport of both cations (Fe^{2+}) and electrons from reactive sites at or beneath the mineral surface to sites where reduction of aqueous metal species occurs. Free electrons cannot be transported effectively in solution, requiring that the metal be in close proximity to the mineral interface, most probably in an adsorbed state. In the presence of aqueous metals, the above reactions (equations 9 and 10) occur concurrently with and compete directly with the normal weathering reactions controlling the oxidation of magnetite and ilmenite (equations 1 - 3).

Heterogeneous reduction of several metal species including Fe(III), Cr(VI), Cu(II) and V(V) have been investigated (*10*). These reactions can be described for a generic Fe(II) containing oxide phase as

$$3[Fe^{2+}]_{oxide} + 2Fe^{3+} \rightarrow 2[Fe^{3+}]_{oxide} + 3\ Fe^{2+} \qquad (11)$$

$$9[Fe^{2+}]_{oxide} + Cr_2O_7{}^{2-} + 14\ H^+ \rightarrow 6[Fe^{3+}]_{oxide} + 3Fe^{2+} + 2Cr^{3+} + 7H_2O \qquad (12)$$

$$3[Fe^{2+}]_{oxide} + 2VO_2{}^+ + 4H^+ \rightarrow 2[Fe^{3+}]_{oxide} + Fe^{2+} + 2VO^{2+} + 2H_2O \qquad (13)$$

Examples of changes in aqueous metal concentrations in the presence of ilmenite at pH 3 are shown in Figure 4. The dotted horizontal lines correspond to the

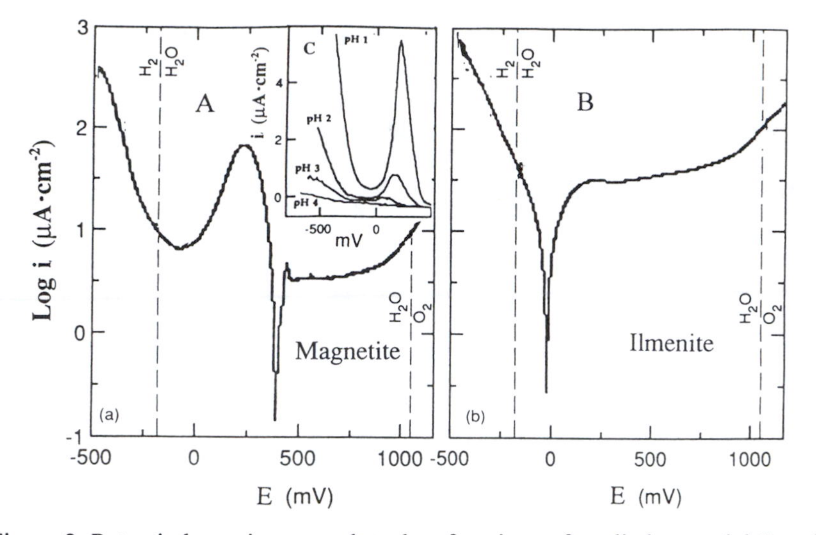

Figure 2. Potentiodynamic scans plotted as functions of applied potential E and measured current i for natural magnetite (A) and ilmenite (B) electrodes in anoxic solutions at pH 3. Inset is a detail of the current peak defining reductive dissolution of magnetite as a function of pH (adapted from ref. *16*)

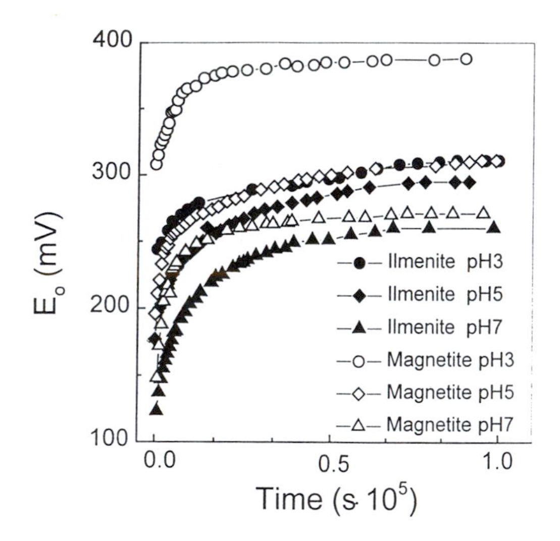

Figure 3. Self induced potentials (no applied current) as functions of time and pH for natural magnetite and ilmenite electrodes (adapted from ref. *16*)

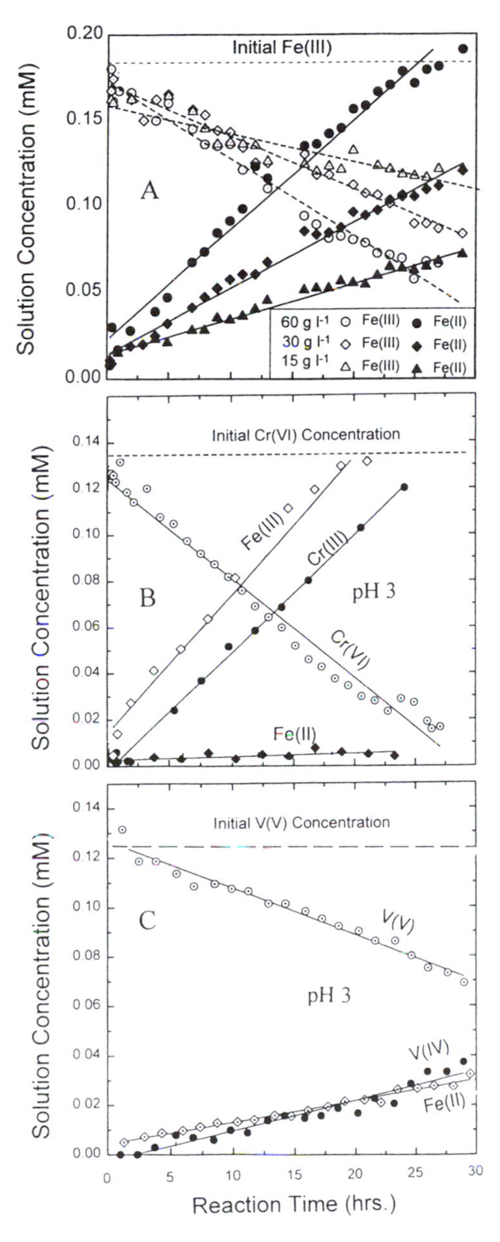

Figure 4. Time trends showing the rates of ferric (A), chromate (B) and vanadate (C) reduction in ilmenite suspensions at pH 3 at 25°C. Rates of ferric reduction increase with increasing suspended ilmenite (g l^{-1}). Solid lines are linear regression fits to data. Dashed lines are initial solution concentrations (adapted from ref. *10*).

initial solution concentration of the oxidized form of the transition metals i.e., Fe(III), Cr(VI) and V(V). With increasing time, these oxidized species are transformed to their reduced counterparts, Fe(II), Cr(III) and V(IV) at approximately linear rates. Additional experiments conducted at higher pH resulted in slower reduction rates and the precipitation of reduced forms of the metal hydroxides.

Role of Homogeneous Reactions. The half cell potential involving structural Fe(II) i.e. $[Fe^{2+}]_{oxide} \rightarrow [Fe^{3+}]_{oxide} + e^-$ is between -0.34 and -0.65 V making it a stronger reducing agent than aqueous Fe, i.e. Fe(II) → Fe(III) (-0.77 V) (*10*). Depending on the specific potential of the metal half cell reaction, homogeneous redox reactions also can occur which involve reduction of transition metals by aqueous Fe(II) produced by concurrent mineral dissolution (Eqns. 1 and 2). Such homogeneous reactions can be written in the examples of Cr(VI) and V(V) reduction as

$$3Fe^{2+} + 1/2Cr_2O_7^{2-} + 7H^+ \rightarrow 3Fe^{3+} + Cr^{3+} + 7/2H_2O \quad (E = 0.589\ V) \tag{14}$$

and

$$Fe^{2+} + VO_2^+ + 2H^+ \rightarrow Fe^{3+} + VO^{2+} + H_2O(E = -0.229\ V) \tag{15}$$

where the potentials are calculated for 1 mM metal solutions at pH 3. Ferrous ion is expected to homogeneously reduce aqueous chromate as indicated by the positive potentials in equation 14. This prediction is consistent with the increase in aqueous Fe(III) rather than Fe(II) during the oxidation of magnetite and ilmenite in acidic anoxic solutions (Figure 4B). Cr(VI) reduction in the presence of Fe(II) oxides occurs via three reactions: (1) Heterogeneous Cr(VI) reduction by direct electron transfer with structural Fe(II); (2) Homogeneous Cr(VI) reduction by Fe(II) released during the heterogeneous reaction (equation 9); and (3) Homogeneous Cr(VI) reduction by Fe(II) produced during dissolution of the ferrous oxide by H^+ or O_2 (equations 1-3) The stoichiometry and rates of Cr(VI) reduction in the first two reactions are directly linked. The last reaction is not linked to the first two reactions but is dependent on the weathering rates.

Cr(VI) reduction in the presence of hematite and biotite has been proposed to occur exclusively by homogeneous Fe(II) oxidation in solution (*4*). However for Fe(II) oxides, reaction stoichiometry requires that 3 Fe(III) ions be produced for every Cr(VI) ion reduced (equation 14) (*10*). The slopes for aqueous increases in Fe(III) and Cr(III) are nearly identical (Figure 4B) indicating that only one aqueous Fe(III) ion is being produced for every Cr(III) ion produced.

Equations 12 and 14 can be combined to describe Cr(VI) behavior by concurrent heterogeneous reaction and homogeneous oxidation of the released Fe(II) such that

$$3[Fe^{2+}]_{oxide} + 1/2Cr_2O_7^{2-} + 7H^+ \rightarrow 2[Fe^{3+}]_{oxide} + Fe^{3+} + Cr^{3+} + 7/2H_2O. \tag{16}$$

Solid state charge balance requires that one Fe(II) atom be expelled from the oxide structure to the aqueous solution for every 2 Fe(II) atoms oxidized in situ in the oxide structure. The one-to one correlation between Fe(III) and Cr(III) production in the experiments (Figure 4B) supports this reaction stoichiometry. This implies that approximately 2/3 of Cr(VI) is being reduced by heterogeneous electron transfer to the magnetite and ilmenite surfaces and 1/3 is being subsequently reduced in solution by release of Fe(II) and oxidation to Fe(III).

As shown in Figure 4C, Fe(II) and not Fe(III) coexists in the presence of V(V). The concurrent loss of V(V) and the production of V(IV) can not be attributed to homogeneous reduction in solution (equation 15) but must occur by heterogeneous reduction associated with direct electron transfer between aqueous V(V) and structural Fe(II) in the oxide (equation 13). Clearly the 1:2 stoichiometric relationship between V(IV) and Fe(II) production required by equation 13 is not satisfied by the experimental data in which an approximate 1:1 correlation is produced. As indicated, the total initial V concentration is not conservative and net V(IV) loss must be attributed to adsorption or precipitation.

Suppression of Reductive Dissolution Reactions. Dynamic polarization of mineral electrodes can be used to investigate the effects of aqueous metal species on solid state electrochemical reactions (*10*). Experiments on magnctite electrodes in the presence of varying concentrations of aqueous Cr(VI) (0 to 5.0 mM) at pH 3 are shown in Figure 5A. The sharp current density peak in the electrode response in a blank solution corresponds to the reductive dissolution of magnetite (equation 5) which is independent of the presence of aqueous metals. However, as the aqueous Cr(VI) concentrations are increased, the intensive of the peak decreases and is nearly absent in the presence of 5.0 mM Cr(VI). This is direct proof that the oxidation of aqueous Cr(VI) competes with and suppresses the reductive dissolution reaction. In contrast, increasing V(IV) concentrations have a much smaller impact on the reductive dissolution peak (Figure 5B), indicating that reduction of V(III) is less successful in competing with the reductive dissolution of magnetite. This effect is consistent with the fact that vanadate is a much weaker oxidizer in acid solutions than chromate.

Effects of aqueous metal concentrations on electrode potentials. Magnetite and ilmenite electrodes exhibit essentially no potential variations in the presence of increasing concentrations of metal cations with a single aqueous valence state under self induced (open circuit) conditions (*10*). This condition is apparent for plots of magnetite potential versus aqueous concentrations ($10^{-8.0}$ to 10^{-1} M) of Zn (II), Co (II), Mn (II) and Ni (II) at pH 3 (Figure 6A). These metals cannot be reduced to lower aqueous oxidation states and cannot serve as electron acceptors during magnetite oxidation. Likewise, increasing Fe(II) concentrations from 10^{-8} to 10^{-2} M at pH 3 and 4 produced only a 50 mV decrease in the self potential (Figure 6b). Like the other bivalent metal species discussed above, Fe(II) cannot act as an electron acceptor and therefore has a minimal impact on the solid state electrochemistry.

In contrast, increasing aqueous Fe(III) and Cu(II) concentrations from 10^{-8} to 10^{-1} M increased magnetite electrode potentials by several hundred mV (Figure 6B and 6C). At very low aqueous concentrations ($<10^{-6}$ M), the potentials are influenced only by differences in pH. At moderate concentrations, the electrode potentials increased with both decreasing pH and increasing aqueous concentrations. At concentrations $>10^{-3}$ M, the potentials are independent of pH and follow a log-linear relationship predicted by the Nernst equation. However, the slopes generated by the

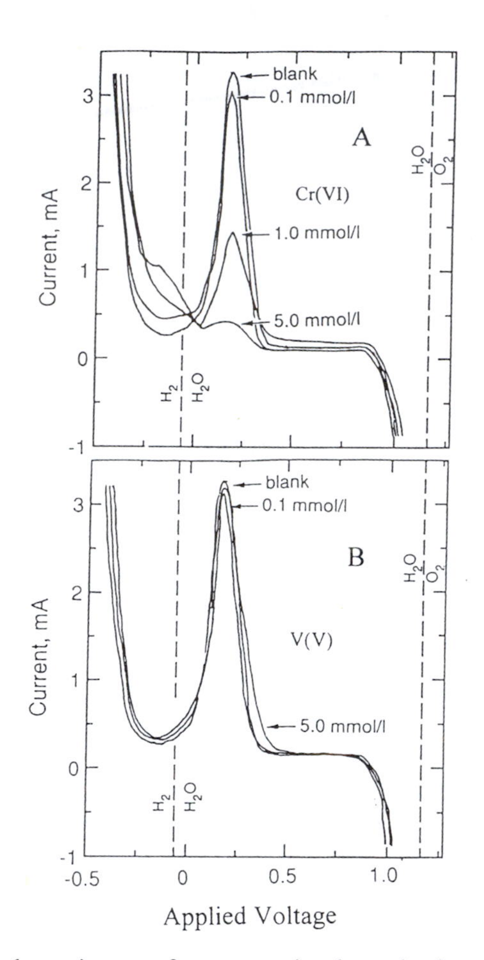

Figure 5. Potentiodynamic scans for a magnetite electrode plotted as functions of applied potential and measured current for anoxic pH 3 solutions at 25°C containing variable concentrations of (A) Cr(VI) and (B) V(V) (adapted from ref.*10*).

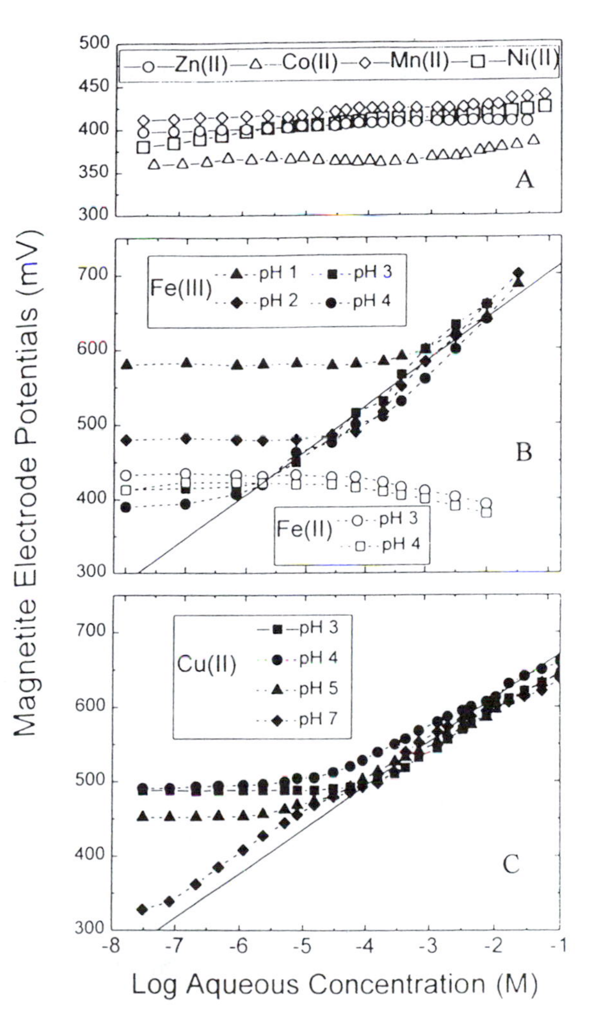

Figure 6. Self-induced potentials for magnetite electrodes in solutions at 25°C in the presence of (A) bivalent Zn, Co, Mn, and Ni at pH 3, (B) ferrous and ferric ions at differing pHs, and (C) cupric ion at differing pHs. Solid symbols represent reducible aqueous species and open symbols represent non-reducible species. Straight lines are linear regression fits to pH-independent regions of the data for Fe(III) and Cu(II) (adapted from ref. *10*).

electrode responses (Fe(III) = 0.073 and Cu(II) = 0.047) deviate from ideality (Fe = 0.059, Cu = 0.030). Magnetite and ilmenite electrodes produce non-linear positive potential responses to log concentration increases in aqueous Cr(VI) and V(V) species. The potentials are also sensitive to solution pH even at high aqueous metal concentrations. The significant increases in self potentials are indicative of surface oxidation.

Role of Surface Passivation

The preceding results were conducted under laboratory conditions using freshly prepared natural mineral surfaces that were reacted for relatively short time periods. In order to assess the effectiveness of natural Fe(II) oxides in reducing and immobilizing transition metal under natural aquifer conditions, several important parameters need to be assessed. These include the reductive capacity of the oxide minerals, the impact of surface passivation and the effects of competition and poisoning by of other aqueous species.

Reductive Capacities. The limitations of the reactive capacity of synthetic magnetites have been recently assessed (*13*). The results of two Cr(VI) redox experiments using magnetite are shown in Figure 7. Both experiments used the same solid/solution ratio of 20 g magnetite l^{-1} at pH 7. Short term experiments were performed using 2mM Cr(VI), a solution concentration the complete reduction of which is calculated to sorb at monolayer density on the magnetite surface. A much higher solution concentration of 50 mM Cr(VI) was used in a second experiment to determine the amount of Cr(VI) which could be reduced to Cr(III) before the surface would become passivated by depletion of Fe(II).

Cr(VI) uptake and reduction to Cr(III) was complete within 6 hours for the dilute solution. In this reaction at neutral pH, complete removal of Cr from solution was observed. In contrast, the long-term reaction showed Cr(VI) reduction in solution occurring over the first 24 hours, followed by steady state Cr (VI) that persisted upwards of 500 hours at about 80% of the initial concentration. In the short-term reaction, the amount of Cr(VI) reduced and sorbed or precipitated to the magnetite surface as Cr(III) corresponded to a surface coverage of 4.4 µmol m^{-2}. In the long term reaction, the initial Cr(VI) concentration was 25-fold higher and yet the steady state-state surface coverage increased by less than 5-fold to 21 µmol m^{-2}. The total possible amount of Cr(VI) based on complete reaction of all the Fe(II) in the stoichiometric formula is about 31 µmol m^{-2}. This discrepancy suggests that there is a limit to the capacity of magnetite to reduce Cr(VI) at neutral pH due to surface passivation.

Evidence for the Formation of a Passivation Layer. High resolution transmission electron microscopy (TEM) images of unreacted synthetic magnetite used in the preceding experiments (*13*) shows that the magnetite structure extends to the edges of the grains (Figure 8A). In contrast, magnetite grains which had reacted with 50 µM Cr(VI) for 3 hours to 5 weeks showed a surface layer which does not possess the bulk magnetite structure (Figure 8B). This surface reaction product is responsible for

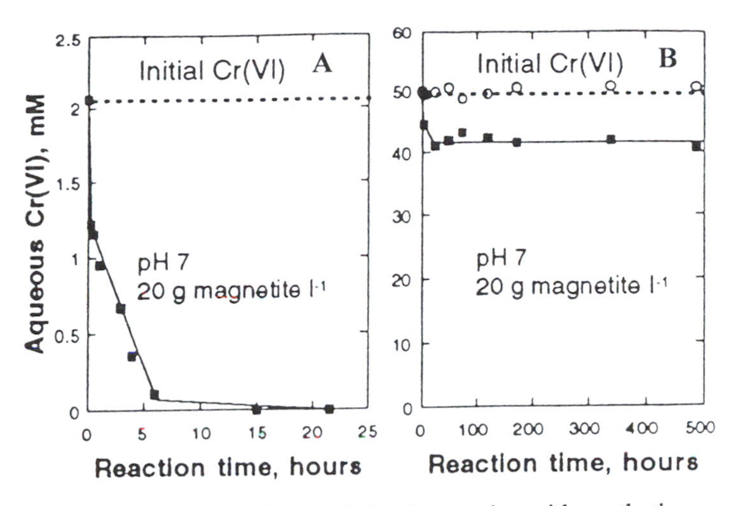

Figure 7. Removal of Cr(VI) from solution by reaction with synthetic magnetite for (A) short term and (B) long term reaction. Also shown are the initial Cr(VI) concentrations (dashed lines) and Cr(VI) concentrations in a blank solution with no magnetite (open circles)) (adapted from ref. *13*).

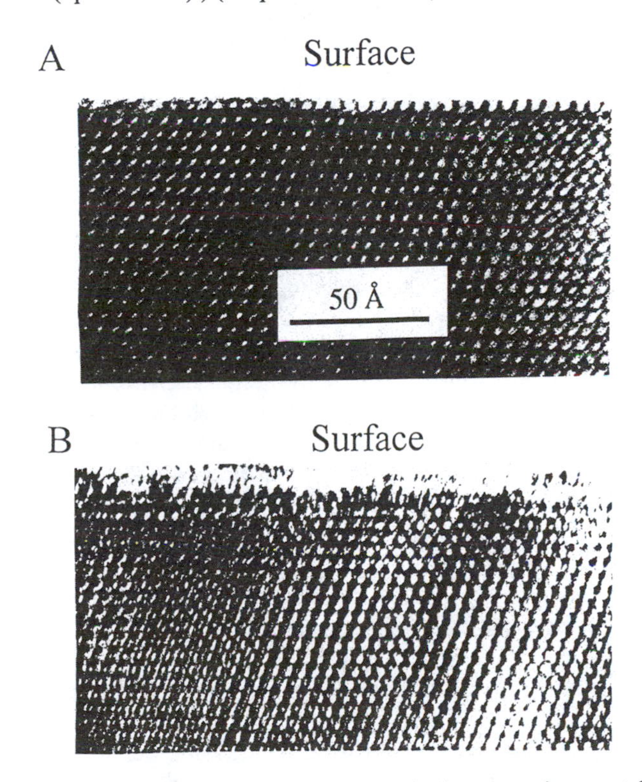

Figure 8. High resolution TEM images of synthetic magnetite crystals oriented along the <110> zone axis: (A) unreacted magnetite; (B) magnetite reacted for 5 weeks in 50 μM Cr(VI) at pH 7 (adapted from ref. (*13*).

passivating the magnetite surface and explains why magnetite does not fully oxidize to maghemite at pH 7 by reaction with Cr(VI).

The measured TEM thickness (10-20 Å) is equivalent to the range in maghemite thicknesses calculated from long term experiments at neutral pH without the presence of reducible metal species (*16*). These experimental data were fitted to a shrinking core model describing the formation of an oxidized maghemite rind on a magnetite grain. This model calculated a diffusion rate of Fe through the altered layer of approximately 10^{-12} to 10^{-13} m s^{-1}. This rate is consistent with inter-pore dissolution associated with the formation of 25% porosity created by the topotactic transformation of magnetite to maghemite. Based on the progressive thickening of the oxidized layer with time, this diffusion rate predicts residence times of 10^6 to 10^7 years which is consistent with the persistence of magnetite and ilmenite in the weathering record.

Experimental Metal Reduction Using Naturally Weathered Magnetite. Experiments have investigated the effects of natural weathering environments on the ability of natural magnetite to experimentally reduce aqueous Cr(VI) (*10*). Magnetite sand fractions were obtained from an oxidizing vadose zone alluvial deposit along the Los Angeles River (LA) and from an anoxic alluvial aquifer beneath the Kesterson Reservoir in the Central Valley of California. The Kesterson sample was obtained from 10 meters below the water table during coring beneath the reservoir. Detailed characterization of the groundwater chemistry indicated a reducing environment (dissolved $O_2 < 0.001$ mM and $Fe^{2+} > 0.1$ mM) (*24*).

The extent of Cr(VI) reduction by the LA River magnetite is minor (Figure 9A), decreasing from 0.060 to 0.045 mM over 24 hours. No measurable Fe(III) is produced in solution suggesting that even this Cr(VI) loss is not attributable to heterogeneous reduction reactions (equation 14). As summarized by (*25*), chromate is adsorbed at pHs below 4 on ferric oxides such as the secondary goethite and lepidocrocite which were detected by X-ray analysis in the LA sample.

In contrast, Cr(VI) loss from solution in the presence of the Kesterson magnetite is complete in less than 3 hours (Figure 9B). The rapid uptake is accompanied by the release of significant Fe to solution. During the initial stage of reaction, Fe(III) is the dominant Fe species in solution as expected in the presence of aqueous chromate. However as the concentration of chromate is depleted, aqueous Fe(III) decreases and is replaced by Fe(II) as the dominant aqueous Fe species. This trend in Fe speciation is attributed to the onset of heterogeneous Fe(III) reduction at the magnetite surface in the absence of Cr(VI) (equation 12) and the release of Fe(II) to solution by continued reaction of the magnetite surface with aqueous H^+ (equation 1). These results suggest that Fe(II) oxide surfaces remain electrochemically active in natural reducing conditions but that such reactivity may be significantly impeded by the formation of surface oxidation products in oxic conditions.

Field Evidence of Heterogeneous Metal Reduction by Fe(II) Oxides

Field studies have ascribed the loss of Cr(VI) in alluvial groundwater systems to the reduction to Cr(III) on magnetite surfaces (*26, 27*). The latter study, which involved the injection of a Cr(VI) tracer into a well field at Cape Cod, Massachusetts demonstrated the strong attenuation of Cr(VI) in the suboxic zone. Breakthrough

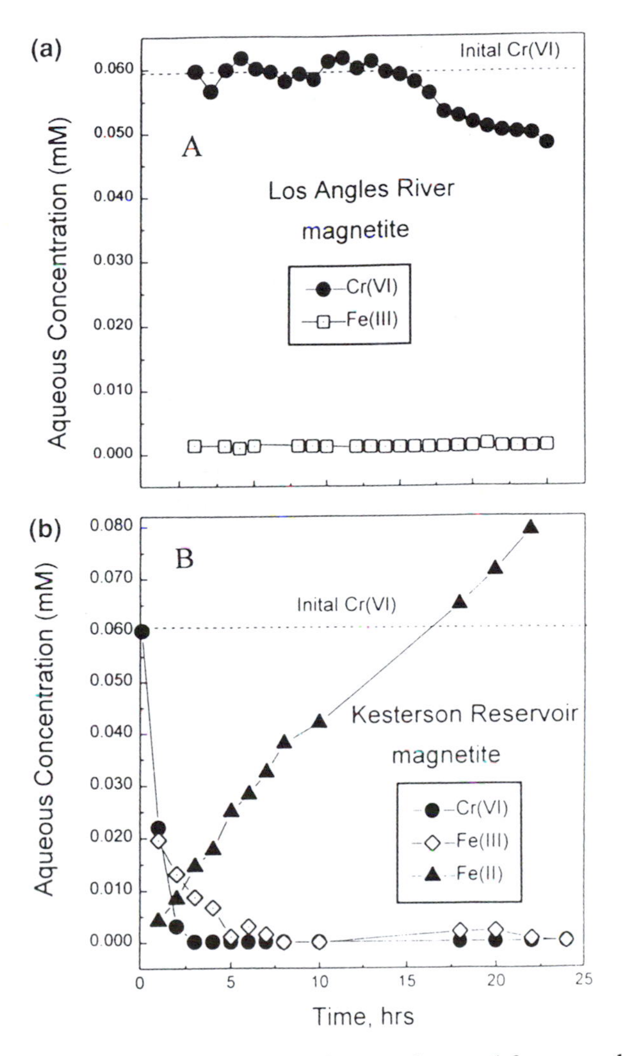

Figure 9. Time trends showing rates of chromate loss and ferrous and ferric ion production at 25°C in pH 3 suspensions of magnetites naturally weathered under (A) oxic vadose conditions (LA River) and (B) anoxic groundwater conditions (Kesterson Reservoir). Horizontal dashed lines are initial Na_2CrO_4 concentrations (0.06 mM) (adapted from ref. *10*).

curves for Cr(total), Cr(VI) and the conservation Br tracer indicated only the Cr(VI) species was mobile but that aqueous concentrations were strongly attenuated with increasing distance down gradient and with depth in the tracer cloud. Concurrent laboratory characterization demonstrated that Cr(VI) reduction was much faster in the fine size fraction of the aquifer material relative to the coarse sand (*27*). Citing low organic contents of the aquifer, these workers invoked Fe(II) containing minerals such as magnetite as the principal pathway for Cr(VI) reduction.

The lack of detailed characterization of the surface chemistry associated with heterogeneous redox invoked in these previous studies was addressed in studies of two waste water contaminant sites, a Cr plating facility near Keyport WA where a corroded disposal pipe allowed seepage of Cr(VI) effluent into the surrounding soil and a chemical waste landfill containing several surface disposal pits for chromium and other chemical waste solutions, in use from the 1960 to 1980 at Sandia National Laboratory, Albuquerque, NM (*14*). The soil mineralogy of both sites is dominated by silicate minerals with accessory magnetite. Sequential extraction and bomb digestion of the soils indicated that the Keyport soils contained 400 to 1200 $\mu g\ g^{-1}$ total Cr and the soils in the Sandia pits contained up to 20,000 $\mu g\ g^{-1}$.

The valence state and chemical nature of solid state chromium associated with separated silicate and magnetite fractions of the soils was investigated by comparing soil samples with Cr model compounds using X-ray absorption fine structure (XAFS) spectroscopy (*14*). At synchrotron energies in the chromium K absorption edge region, the dominant scattering process is multiple scattering involving outgoing and backscattered photoelectrons between absorber Cr and surrounding atoms. This accounts for most of the fine structure in the near-edge energy region (Figure 10). In addition, a prominent pre-edge peak at 5993.5 eV occurs when Cr(VI) is present, due to a *1s* to *3d* bound-state transition. The intensity of this pre-edge peak is proportional to the Cr(VI):total Cr ratio. Figure 10 shows the increase in pre-edge peak intensity with increasing Cr(VI):total Cr for physical mixtures of Cr_2O_3 and $Na_2CrO_4.4H_2O$. The pre-edge feature was calibrated based on background-subtracted peak height and fit with a second-order polynomial (Fig 1; $R = 0.998$) (*14*).

When the pre-edge height calibration is applied to the normalized pre-edge peaks of Cr-containing soils from Keyport, the results showed that Cr(VI) in the non-magnetic silicate fraction comprised up to 6.4% of the total Cr. Keyport magnetites contain no measurable Cr(VI) and their pre-edge features are similar to pre-edge features of the Cr(III) model. These results are consistent with the explanation that mixed oxidation state Cr waste reacted with the magnetite grains was reduced to Cr(III) by electron transfer from Fe(II) in the magnetite whereas the Cr waste reacted with the silicates remained dominantly unchanged in oxidation state.

Based on pre-edge peak height, the Sandia soils were highly variable in Cr(VI)/Cr(III) ratios but contained significantly more Cr(VI) (2 - 67%) than did the Keyport samples (0 - 6.4%). A comparison of the nonmagnetic and magnetite fractions from the Sandia 1980s pit showed little difference in pre-edge heights. This suggests that either the amount of Cr(VI) introduced into the pit exceeded the reductive capacity of the minor amounts of magnetite present and/or the surfaces of the magnetite in this pit have become oxidized and passivated and no longer capable of Cr(VI) reduction. Also an abundance of Pb was present on the mineral surfaces,

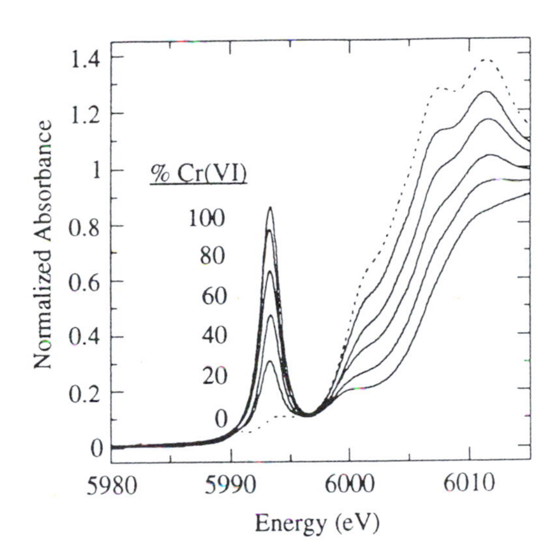

Figure 10. Normalized XAFS pre-edge region of mixtures of Cr(III) + Cr(VI) in mol % ratios of Cr_2O_3 and $Na_2CrO_4.4H_2O$. The 100% Cr(III) end member (Cr_2O_3) spectrum is dotted to illustrate the relationship between edge position, edge shape, pre-edge height and Cr(VI):Cr(III) ratio. Increasing pre-edge height reflects increasing Cr(VI) content (adapted from ref. *14*).

suggesting that precipitation of Cr(VI) as $CrPbO_4$ predominated over reduction of Cr(VI). In contrast, the magnetite fraction from the 1970s pit exhibited significantly lower portion of Cr(VI) than the feldspar+quartz function based on pre-edge peak height. As in the case of the Keyport soils, this deficiency suggests that a significant amount of the Cr(VI) has been reduced to Cr(III). These data are the first direct evidence of the extent of metal reduction on a Fe(II) containing oxide under field conditions.

Summary

The reduction on aqueous transition metal species at the surfaces of Fe(II)-containing oxides is defined in terms of heterogeneous redox reactions that occur concurrently with oxidation and weathering of the mineral surfaces. Electrochemical measurements made on mineral electrodes document that such reactions are coupled half cells in which reductive dissolution can be decoupled and substituted for reduction of aqueous species. This is confirmed experimentally in aqueous/mineral suspensions in which Cr(VI), V(V) Fe(III) and Cu(II) are reduced to lower valance states.

The effectiveness of natural Fe(II) oxides in reducing and immobilizing transition metals under natural aquifer conditions is dependent on the reductive capacity of the Fe(II) oxide minerals, the impact of surface passivation, and the effects of competition of other aqueous species. Studies of Cr(VI) reduction on natural magnetite samples indicate that the weathering environment is critical in determining the extent of passivation, with anoxic ground water conditions maintaining reactivity and oxic vadose environments promoting passivation. Application of synchrotron radiation techniques permit quantitative characterization of the portion of residual Cr(VI) on Fe(II) oxide surfaces and thus the effects of natural geochemical conditions on surface reduction.

Literature Cited

1. A. F. White, *in* "Mineral-Water Interface Geochemistry" (M. F. Hochella and A. F. White, eds.), Reviews in Mineralogy Vol. 23, p. 467, 1990.
2. A. F. White, A. Yee, and S. Flexser, *Chem. Geol.* **49**, 73 (1985).
3. A. F. White and A. Yee, *Geochim. Cosmochim. Acta* **49**, 1263 (1985).
4. L. E. Eary and D. Rai, *Am. J. Sci.* **289**, 180 (1989).
5. E. S. Ilton and D. R. Veblen, *Geochim. Cosmochim. Acta* **58**, 2777 (1994).
6. E. S. Ilton and D. R. Veblen, *Econ. Geol.* **88**, 885 (1993).
7. M. Sayin, *Clay Clay Miner.* **30**, 287 (1982).
8. D. Earley III, M. D. Dyar, I. E.S., and A. A. Granthem, *Geochim. Cosmochim. Acta* **59**, 2423 (1995).
9. G. Bidogilio, P. N. Gibson, M. O'Gorman, and K. J. Roberts, *Geochim. Cosmochim. Acta* **57**, 2389 (1993).
10. A. F. White and M. L. Peterson, *Geochim. Cosmochim. Acta* **60**, 3799 (1996).
11. M. L. Peterson, G. E. Brown, and G. A. Parks, *Coll. Surf. Anal.* **107**, 77 (1996).

12. M. L. Peterson, G. E. Brown, and G. A. Parks, *Mat. Res. Soc. Proc.* **432**, 75 (1997).
13. M. L. Peterson, A. F. White, G. E. Brown, and G. A. Parks, *Enviro. Sci. and Tech.* **31**, 1573 (1997).
14. M. L. Peterson, G. E. Brown, G. A. Parks, and C. L. Stein, *Geochim. Cosmochim. Acta* **61**, in press (1997).
15. M. L. Peterson, G. E. Brown, and G. A. Parks, *J. de Physique* **7**, C2-781 (1997).
16. A. F. White, *Geochim. Cosmochim. Acta* **58**, 1859 (1994).
17. J.-P. Jolivet and E. Tronc, *J. Coll. Inter. Sci.* **125**, 688 (1987).
18. J. F. Banfield, P. J. Wasilewski, and D. R. Veblen, *Am. Mineral.* **79**, 654 (1994).
19. J. N. Ryan and P. M. Gschwend, *Geochim. Cosmochim. Acta* **56**, 1507 (1992).
20. P. D. Allen and N. A. Hampson, *J. Electroanalytical Chem.y* **99**, 299 (1979).
21. P. D. Allen and N. A. Hampson, *J. Electroanalytical Chem.* **111**, 223 (1980).
22. M. A. Blesa, H. A. Mainovich, E. C. Baumgartner, and A. J. G. Maroto, *Inorg. Chem.* **26**, 3713 (1987).
23. N. Valverde, *Beriche der Busen-Gesellschaft* **80**, 333 (1976).
24. A. F. White, S. M. Benson, A. W. Yee, H. A. Wollenberg, and S. Flexser, *Water Resources Res.* **27**, 1085 (1991).
25. D. A. Dzombak and F. M. M. Morel, "Surface Complexation Modeling-Hydrous Ferric Oxide," p. 392. John Wiley and Sons, New York, 1990.
26. K. G. Stollenwek and D. B. Grove, *J. Enviro. Qual.* **14**, 396 (1985).
27. N. J. Anderson, B. A. Bolto, and L. Pawlowski, *Nuclear and Chem. Waste Management* **5**, 125 (1984).

Chapter 17

Pollutant Reduction in Heterogeneous Fe(II)–Fe(III) Systems

Stefan B. Haderlein[1,3] and Klaus Pecher[2]

[1]Ralph M. Parsons Laboratory, Department of Civil and Environmental Engineering, Massachusetts Institute of Technology, Cambridge, MA 02139
[2]Department of Hydrology, University of Bayreuth, D-95440 Bayreuth, Germany

Several classes of priority pollutants can undergo abiotic reductive transformation reactions in anoxic aqueous environments. Such reactions may be environmentally benign since they often lead to products which are more bioavailable and more easily to degrade. In certain cases, however, the products of such reactions are more toxic than the parent compounds. Thus, knowledge of the reductants and processes involved in the reductive transformation of pollutants is essential to evaluate the risk of pollution and the need for remediation of contaminated sites.

In recent years, increasing evidence for the participation of ferrous iron in reductive transformation of pollutants became available through laboratory and field studies. The pollutants studied include halogenated solvents (e.g., (*1-3*)), nitroaromatic compounds (*4-6*), nitrite (*7, 8*), nitrate (*9*), chromate (*10, 11*), selenate (*12*), and pertechnetate (*13*). A common feature of most of these laboratory studies is that ferrous iron species present at mineral surfaces were much more reactive than dissolved Fe(II) present in the background electrolyte. These findings can be rationalized by inspection of Figure 1. Shown are the reduction potentials in aqueous solution, $E_h^0(w)$, of some environmentally relevant iron redox couples as well as of selected organic pollutants (for details see (*14, 15*)). As can be seen, iron is an extremely versatile redox sensitive element. The redox potentials of geochemically important forms of iron vary by more than 1000 mV, depending on which species are involved. Species which stabilize Fe(III) formed in the reaction tend to exhibit the lowest redox potentials. Aqueous Fe(II) is a poor reductant compared to minerals containing structural iron such as magnetite or iron silicates which can reduce polyhalogenated alkanes, nitroaromatic compounds , or aromatic azo compounds under standard environmental conditions. Since oxidation of iron(II) to iron(III) involves the transfer of only one electron, the reduction potential of a given iron(III)/iron(II) couple can be directly applied to evaluate the relative reactivities of iron(II) species in various chemical environments. However, this consideration is only applicable if the actual electron transfer from Fe(II) to a pollutant is at least in part rate limiting. This can not always be assumed, particularly for heterogeneous reactions, where in the course of the reaction the properties of the reductant (i.e., the surface) is changed (see below). Although minerals containing structural Fe(II), e.g., phyllosilicates, were found to

[3]Current address: EAWAG/ETH Zurich, CH-8600 Dübendorf, Switzerland (Fax: +41–1–823 5471; e-mail: haderlein@eawag.ch).

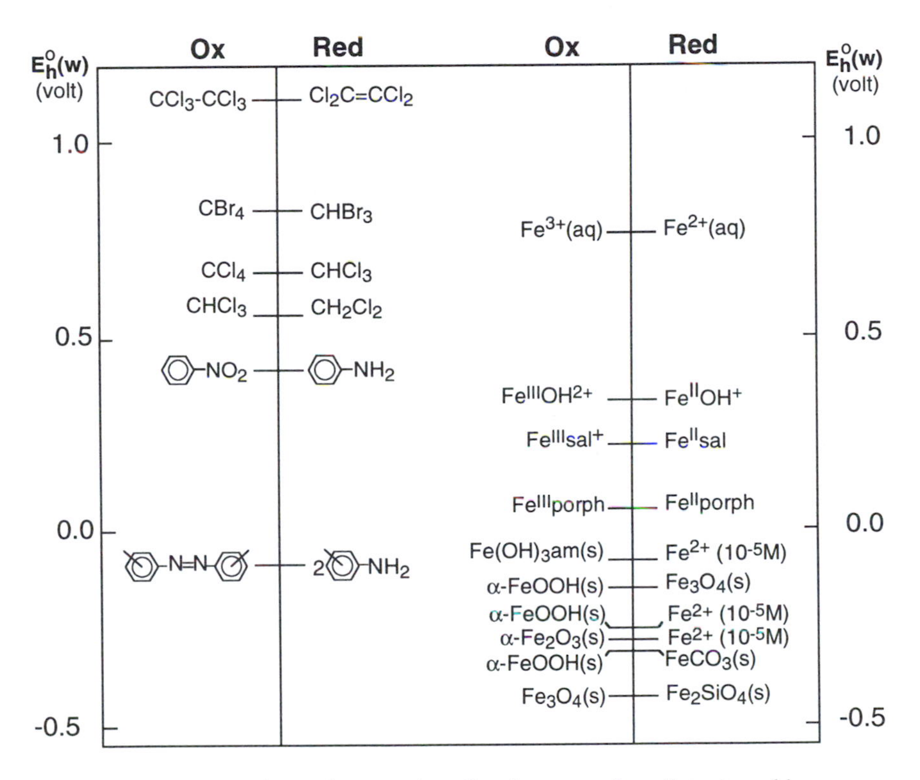

Figure 1. Representative redox couples of various organic pollutants and iron species. Given are standard potentials at pH = 7 and molar concentrations of the reactants but at standard environmental concentrations of the major anions involved: $[HCO_3^-] = [Cl^-] = 10^{-3}$ M; $[Br^-] = 10^{-5}$ M; (s) = solid. Adapted from (*44*).

reduce organic substances such as CCl_4 (*16*), the rate constants for such reactions typically are very low. Obviously, the reactivity of the various Fe(II) species cannot be predicted from their E_h^0(w) values alone. Other processes than the electron-transfer, such as sorption, precipitation or dissolution reactions, may be rate-limiting and must be considered as well to evaluate the reactivity of Fe(II) species.

The purpose of this chapter is to give an overview of the chemical and biological processes that control the reactivity of Fe(II) in heterogeneous aqueous systems with respect to pollutant transformation. To this end, we will evaluate data collected in various laboratory systems as well as field studies. Two classes of model compounds with complementary properties will be used to monitor the reactivity of Fe(II) species in the various systems. Nitroaromatic compounds (NACs) primarily served to characterize the systems in terms of mass and electron balances. Reduction of NACs by Fe(II) species results in only a few major products (aromatic amines and hydroxylamines) which can be easily quantified by standard HPLC-UV methods in the low μM range. Polyhalogenated aliphatic compounds (PHAs) were used if little perturbation of the systems in terms of electron transfer to the organic substrates was crucial. Reduction of PHAs requires fewer electrons than nitro reduction and PHAs can be quantified by standard GC-ECD methods in the low ppb range.

Factors Controlling the Reactivity of Ferrous Iron in Heterogeneous Aqueous Systems

Figure 2a shows a typical example of the reactivity of ferrous iron at ambient pH in (anoxic) aqueous solutions or suspensions containing Fe(III) bearing minerals, respectively. Significant reactivity of aqueous Fe(II) was observed neither for NACs (*17*) nor for PHAs (*18*). Addition of Fe(III)(hydr)oxides or magnetite to anoxic solutions of Fe(II) resulted in rapid reduction of the organic substrates. Apparently, uptake of Fe(II) by such surfaces created very reactive Fe(II) species.

Systems containing dissolved Fe(II) and iron oxy-hydroxide minerals showed a high reactivity as long as sufficient aqueous Fe(II) was available to replace Fe(II) surface species that had reacted with NACs (Figure 2b). During the entire course of the reaction, the number of electrons transferred to NACs matched the consumption of aqueous Fe(II). Thus, reactive Fe(II) species formed at these surfaces can be regenerated rapidly by uptake of Fe(II) from solution, once they have been consumed by oxidation of pollutants.

Although the reduction of NACs with most Fe(II)-containing minerals thermodynamically is highly favorable at neutral pH (Figure 1), virtually no transformation was observed in suspensions of magnetite without added aqueous Fe(II). Long-term reaction rates of iron sulfides or phyllosilicates such as biotite and vermiculite for CCl_4 or NACs in aqueous suspension were also found to be very slow at near-neutral pH values (*16, 19-22*). Because the oxidative dissolution of such minerals by organic pollutants under environmental conditions is a relatively slow process, it seems rather unlikely that such reactions can compete with the significant rates of formation and regeneration of reactive, surface-bound Fe(II). In the following, the various factors that control the reactivity of Fe(II) at iron oxy-hydroxide surfaces will be reviewed.

pH Value. The pH of the system is a master variable that is related to several factors affecting the reactivity of heterogeneous Fe(II)/Fe(III) systems. With increasing pH values, observed pseudo-first-order reaction rates with respect to pollutant transformation increased for both NACs (Figure 3a) and PHAs (*18*). Apparently, uptake of Fe(II) from aqueous solution by iron oxy-hydroxide surfaces, which is a prerequisite for formation of reactive Fe(II) surface species, is strongly pH dependent (Figure 3b and (*23*)). However, other important system variables also change with pH, namely the redox potential and the speciation of both aqueous Fe(II) and surface sites of the

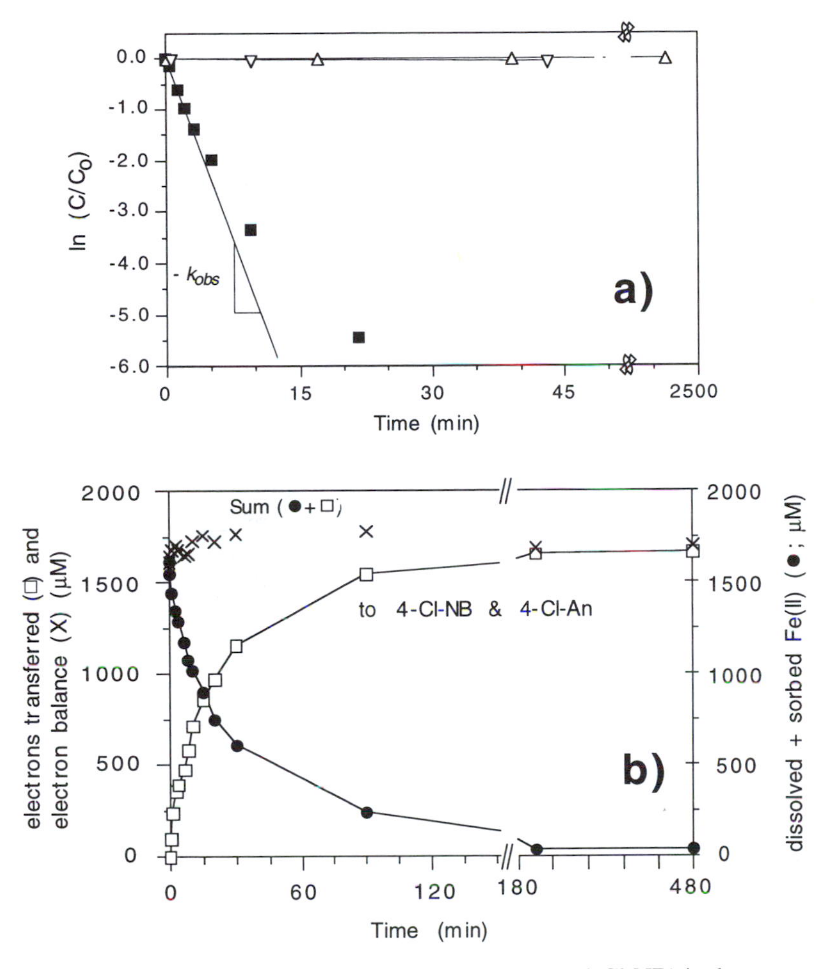

Figure 2. a) Reduction of 50 μM 4-chloro nitrobenzene (4-Cl-NB) in the presence of 17 m^2L^{-1} magnetite and an initial concentration of 2.3 mM Fe(II) at pH 7.0 and 25 °C (■). The rate law deviates from pseudo-first-order behavior for longer observation times. 4-Cl-NB was not reduced significantly in suspensions of magnetite without dissolved Fe(II) (∇) or in solutions of Fe(II) without magnetite (Δ); (adapted from (*17*)) b) Electron balance (×) for the reduction of 4-Cl-NB repeatedly added at times t = 0, 6, 17 min to a suspension containing 11.2 m^2L^{-1} magnetite and an initial concentration of dissolved Fe(II) of 1.6 mM at pH = 7.75 and T = 25 °C; (adapted from (*6*)).

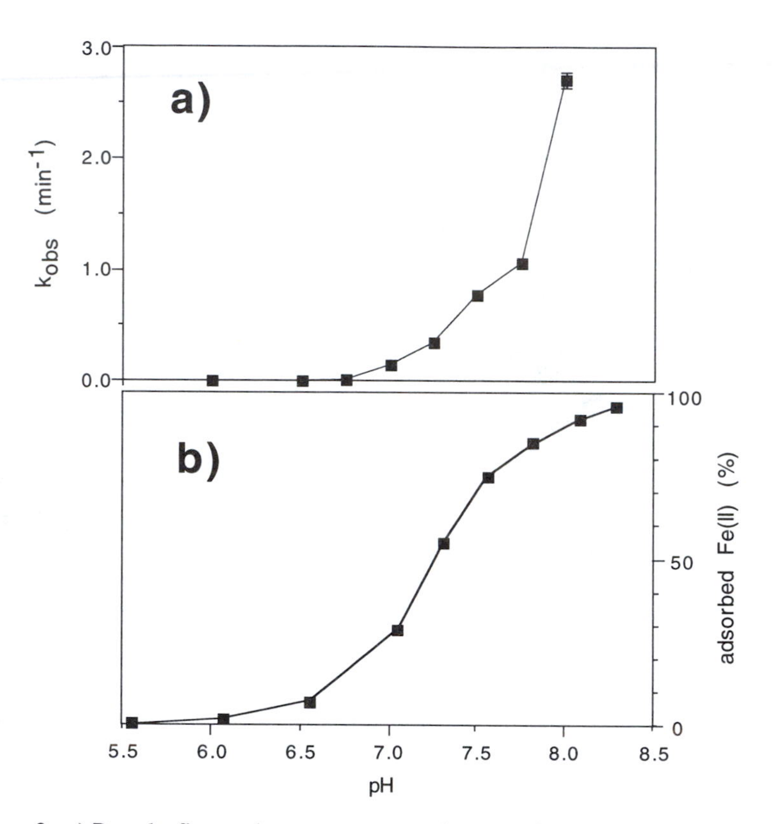

Figure 3. a) Pseudo-first-order rate constants, k_{obs},as a function of pH for the reduction of 50 μM nitrobenzene in suspensions of 11.2 m^2L^{-1} magnetite and an initial concentration of 1.5 mM Fe(II). b) Adsorption of 0.2 mM Fe(II) onto 0.55 gL^{-1} magnetite as a function of pH (adapted from (*17*)).

minerals. Moreover, different types of reactive Fe(II) species may be formed at surfaces, depending on the amount of Fe(II) uptake. Thus, we will try to discuss the reactivity of the system by considering variations in only one of these variables at a time.

During reduction of nitrobenzenes in heterogeneous Fe(II)/Fe(III) systems, a pseudo-first-order rate law was obtained only for the initial phase of the experiments, followed by a decrease in reaction rates (Figure 2a). Such behavior may indicate that the most reactive Fe(II) species were depleted and less reactive species and/or the changing properties of the system controlled the reaction rates at later stages of the experiments. The further evaluation of the reactivity of heterogeneous Fe(II)/Fe(III) systems will be based on the reduction of polyhalogenated alkanes. In contrast to NACs, which take up six electrons per nitro group during the reduction to the respective amines, reduction of hexachloroethane and dibromodichloromethane to tetrachloroethene and monobromdichloromethane, respectively, consumes only two electrons per molecule. Furthermore, reduction of these PHAs can be studied at much lower concentrations. The reduction of PHAs by surface-associated Fe(II) caused sufficiently low perturbation of the system properties to result in pseudo-first-order reduction kinetics during the entire course of the reaction (*18*).

Remodeling of Fe(II) at Iron(hydr)oxides with Time. The reactivity of Fe(II) in suspensions of goethite strongly depended on how long the suspended minerals had been exposed to dissolved Fe(II) before the organic substrate was spiked to the systems. Figure 4 shows normalized reaction rates, k'_{norm}, for two PHAs as a function of this equilibration time, t_{eq}. Reaction rates increased with increasing t_{eq} and somehow levelled off after about 20 h. This behavior was independent of the pollutant studied since it was observed for both PHAs and NACs. It demonstrates that Fe(II) bound to the oxide surface alters its reactivity with time and that during the course of interaction of ferrous iron with the goethite surface various reactive Fe(II) species are formed. Such remodeling of surface-bound Fe(II) species is corroborated by results of experiments where Fe(II) was desorbed from goethite as a function of equilibration time of Fe(II) and goethite. Figure 5 shows results of such experiments where phenanthroline was used to desorb Fe(II) (*24*)). The non desorbable fraction of surface-bound ferrous iron increased with equilibration time, t_{eq}. until a plateau was reached after about 20 h. Thus, under appropriate reaction conditions (pH, surface density of Fe(II)), more strongly bound Fe(II) species are being formed with increasing equilibration time, consistent with a remodeling of the surface in terms of formation of surface precipitates or surface clusters. Coughlin & Stone (*23*) also reported considerable uptake of Fe(II) by goethite after 18 h equilibration time even under acidic conditions (pH 4 - 5.5). Similar to our study they found that a significant fraction of surface-bound ferrous iron was not desorbable, even after treatment with HNO_3 for 28 h.

Surface Density of Fe(II)-Species. Figure 6 shows the rate constants for the reduction of dibromodichloromethane in suspensions containing goethite and Fe(II) as a function of total ferrous iron present and pre-equilibration time of Fe(II) with the surface. A strong dependence of pseudo-first-order reaction rates on total ferrous iron concentrations was observed for long pre-equilibration times (t_{eq} > 30 h) which provides further evidence that surface species of Fe(II) formed after prolonged contact of ferrous iron with iron(hydr)oxide surfaces are most reactive. Experiments such as shown in Figure 6 do not allow one to calculate second-order rate constants as it is remains unclear which species or fraction(s) of surface-bound Fe(II) is involved in the reaction.

The increase of reduction rates with higher concentrations of $Fe(II)_{tot}$ may also be due to a decreasing reduction potential of the system. Assuming that the redox potential of the Fe(II)/Fe(III) couple in homogeneous solution represents the redox

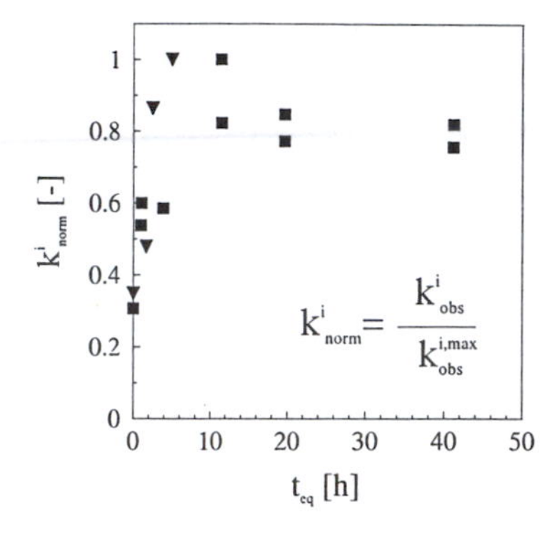

Figure 4. Dependence of normalized pseudo-first-order rate constants (k^i_{norm}) on pre-equilibration time of Fe^{2+} with goethite (t_{eq}(pH 7.2, $Fe(II)_{tot}$=1 mM, I=20 mM, T=25°C, 25 $m^2\ L^{-1}$ goethite, (▼): $C_0(C_2Cl_6)$=0.4 µM, (■): $C_0(CCl_2Br_2)$=5 µM); adapted from (*1*).

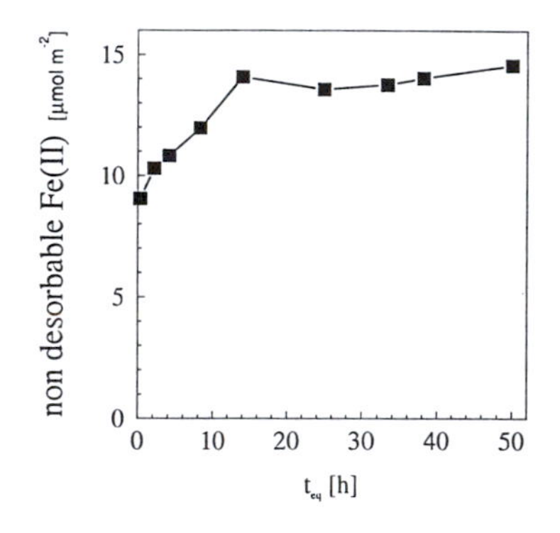

Figure 5. Non-desorbable surface bound Fe(II) as a function of contact time of Fe(II) with suspended goethite (pH 7.2, $Fe(II)_{tot}$=1 mM, I=20 mM, T=25°C, 25 $m^2\ L^{-1}$ goethite); adapted from(*24*).

potential of the system reasonably well and that the ratios of aqueous Fe(II)/Fe(III) for the two sets of experiments shown in Figure 6 are similar at the respective $Fe(II)_{tot}$ concentrations, the results suggest that changes in the redox potential were not a major factor for the increase in reaction rates with $Fe(II)_{tot}$.

Interaction of Fe(II) with Iron(hydr)oxides - Formation and Characterization of Reactive Sites. Adsorption kinetics of Fe(II) to iron oxide minerals was studied under strictly anoxic conditions (glove box). Kinetics of adsorption at Fe(II) surface saturation > 60% exhibited two distinct features (shown in Figure 7 for several iron oxides at pH = 7.2). In an almost immediate first step, a certain fraction of the ferrous iron added was bound to the oxides. The different amounts of initially adsorbed Fe(II) for the various minerals may be due to differences in electrostatic forces (different pH_{pzc} of the minerals) as well as morphologic features (differences in crystallographic planes, kinks, steps, edges, etc.). In a second step, the fraction of bound ferrous iron slowly increased and levelled off after about 30 h. At given $Fe(II)_{tot}$, this behavior was significant for goethite at pH values > 6.0 (Figure 8). The suspensions showed a distinct change in color from bright yellow to blue-green at pH > 7.6, which may indicate the formation of green rust phases during the experiments (*8, 25*).

Two-step adsorption kinetics of metal cations at iron oxides has been observed for several metals (*26-28*) and mainly attributed to interaction of the metal cations with different sites of the metal oxide surface (*29*). Diffusion of cations into micro- or macropores can be ruled out for synthetic goethites (*30*) due to an essential lack of porosity as determined by t-plot analysis (*31*). However, diffusion processes may be important for other oxides, especially amorphous iron oxides or oxides with certain coatings and oxide surfaces which have been subject to substantial dissolution processes. Thus, other processes than diffusion must have caused the two different kinetic domains of Fe(II) sorption to iron oxide surfaces.

Several authors proposed slow electron transfer reactions, induced by adsorption of Fe(II) (*32-36*), at the interface of spinel-like iron oxides containing Fe(II) and Fe(III) and also provided spectroscopic evidence for product formation (*37*). However, these results must be carefully evaluated since contamination of the systems with molecular oxygen cannot be excluded, especially in the early studies. Thus, the experimental conditions and systems reported in this literature are not directly comparable to our study where predominantly non mixed-valence iron oxides were investigated under strict exclusion of molecular oxygen. Direct electron transfer between adsorbed Fe(II) and structural Fe(III) at the goethite surface seems unlikely as Fe(III) within the crystal structure is stabilized by neighbouring oxygen ligands.

An alternative explanation for the observed slow kinetics of Fe(II) uptake is the formation of surface precipitates. The greenish color developed in goethite suspensions at pH > 7.6 may indicate Fe(II)/Fe(III)-precipitates, possibly green rust phases. Traces of Fe(III), a prerequisite for the formation of such mixed valence iron phases, may either have been present as impurity in the Fe(II) spike solutions or due to dissolution of the iron(hydr)oxides. Incorporation of divalent metal cations into goethite or hematite has recently been shown during aging of hydrous iron oxides (*38*). The hypothesized Fe(II)/Fe(III) surface precipitates, which seem to be responsible for the high reactivity of systems studied at high pH values, are currently analyzed by surface spectroscopic techniques.

Despite the missing spectroscopic characterization of Fe(II) surface precipitates in such systems, analysis of adsorption isotherms of ferrous iron on several iron oxides (*24*) supports this hypothesis. Modeling of Fe(II) adsorption isotherms considering surface precipitation reactions (*39*) gave excellent fits to adsorption data (Figure 9). The model used allows for a transition from surface complexation to surface precipitation, which becomes relevant at surface saturations exceeding 20% (*26*). Fitted surface site density data of the studied iron oxides (Table I) agree with experimental data. Furthermore, the solubility product fitted for a hypothetical $Fe(OH)_2(s)$

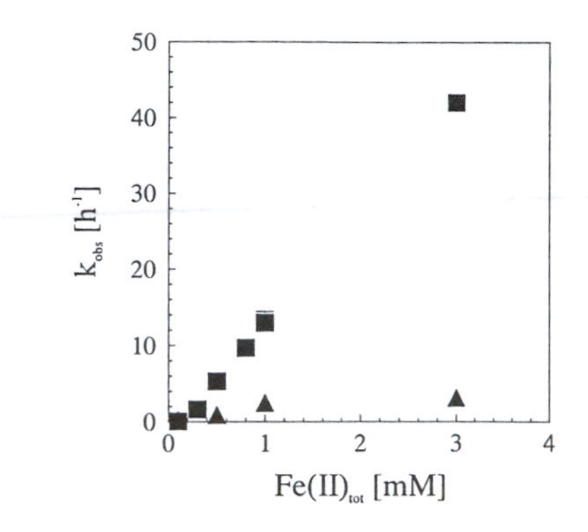

Figure 6. Pseudo-first-order rate constant of transformation of CCl_2Br_2 in suspensions of goethite (25 m^2 L^{-1}) as a function of concentrations of $Fe(II)_{tot}$ (pH=7.2, I=20 mM, T=25°C, $C_0(CCl_2Br_2)$=0.5 μM, (■): t_{eq}>30 h, (▲):t_{eq}=0 h); adapted from (*1*).

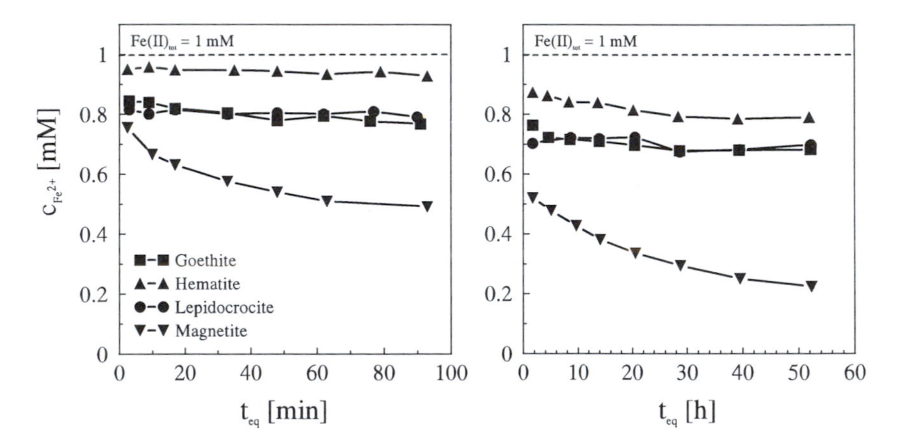

Figure 7. Uptake kinetics of dissolved Fe(II) by different iron oxides in anoxic aqueous supensions (pH 7.2, 25 m^2 L^{-1}, I=20 mM, T=25°C, $Fe(II)_{sol}$= dissolved concentration of ferrous iron). The difference between $Fe(II)_{tot}$ and $Fe(II)_{sol}$ gives surface-bound ferrous iron. Adapted from (*1*).

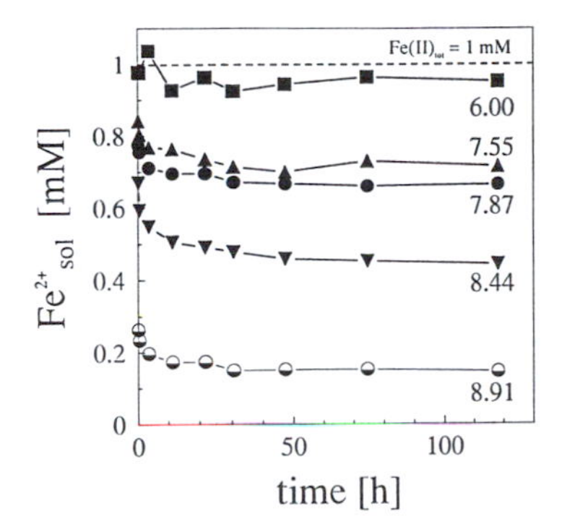

Figure 8. Uptake kinetics of dissolved Fe(II) by goethite in anoxic aqueous supensions at different pH values (25 m^2 L^{-1}, I=20 mM, T=25°C, $Fe(II)_{sol}$= dissolved concentration of ferrous iron). The difference between $Fe(II)_{tot}$ and $Fe(II)_{sol}$ gives surface-bound ferrous iron. Adapted from (*24*).

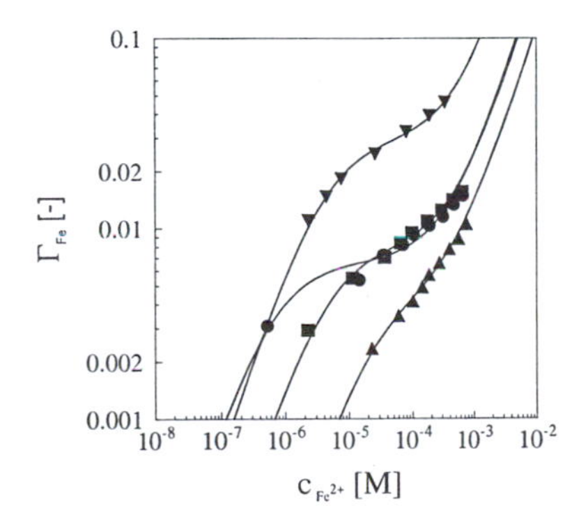

Figure 9. Isotherm fits of the surface precipitation model for the sorption of ferrous iron on iron oxides: (▼) magnetite, (■) goethite, (●) lepidocrocite, (▲) hematite (for parameter values see Table I, $Fe(II)_{sol}$ = concentration of dissolved ferrous iron, Γ_{Fe}= concentration of surface-bound ferrous iron per total concentration of iron oxide, pH 7.2, I=20 mM, T=25°C, 25 m^2 L^{-1}, t_{eq}=15 min); adapted from (*1*).

precipitate closely matched the experimental value reported by Baes & Mesmer (*40*), independent of the iron oxide used. However, we are aware that this model is a first approximation and needs to be refined in terms of pH-dependent isotherms and inclusion of measured stoichiometries of the surface precipitates. Surface precipitation may be neglected at lower total ferrous iron concentrations due to low surface coverage of sorbed Fe(II) (*41*).

Table I. Calculated Parameter Values of Isotherm Fits [a] of the Surface Precipitation Model [b] for Uptake of Fe(II) by Various Iron(hydr)oxides.

Mineral	K_{ads} [c]	K_{spM} [d]	S_T [e]
magnetite	0.014	$4.7\ 10^{-12}$	$3.9\ 10^{-4}$
goethite	0.013	$1.5\ 10^{-13}$	$1.3\ 10^{-4}$
hematite	0.0026	$2.5\ 10^{-13}$	$4.7\ 10^{-5}$
lepidocrocite	0.096	$1.4\ 10^{-13}$	$1.06\ 10^{-4}$

[a] shown in Figure 9; [b] for details see Ref. (*39*); [c] K_{ads}=equilibrium coefficient for adsorption; [d] K_{spM}=solubility product of the metal hydroxide formed; [e] S_T=surface site density [M].

Significance of Surface-Bound Fe(II) for Pollutant Reduction in Natural Systems

In natural anoxic matrices, several biogeochemical redox processes may take place in parallel, giving rise to several potential reductants for subsurface contaminants. Such species may be present in aqueous solution or at solid surfaces, which complicates chemical analyses. Under such complex redox conditions, the traditional approach to predicting the fate of reactive pollutants often fails due to inadequate system characterization. The use of reactive tracers as *in situ* probes may help to identify the predominant reductants in anoxic natural systems (*42*).

Rügge et al. (*43*) demonstrated how *in situ* information on reductive transformation processes in anaerobic aquifers can be obtained using NACs as reactive tracers. NACs can be used to identify the predominant reductants since the effects of NAC substituents on reduction rates are characteristic for various important subsurface reductants (*44*). Furthermore, depending on the pattern of substituents, reduction rates of NACs by a given species may vary by orders of magnitude. This allows one to study reactions at very different time scales by choosing an appropriate set of substituted NACs as probe compounds.

The reactivity pattern of five nitroaromatic probe compounds was studied in an anaerobic leachate plume of a landfill at Grindsted, Denmark, by means of an injection experiment in the plume area, by *in situ* microcosm experiments installed along a flow line in the plume, and by laboratory batch experiments containing groundwater and aquifer sediment. The *in situ* reactivity of the compounds in the aquifer was compared to their reactivity in well defined model systems to delineate processes and reductants that were active in the leachate plume. The redox conditions in the plume were typical for complex redox environments with several processes occurring simultaneously (Figure 10a). Closest to the landfill, methane production and sulfate reduction prevailed, followed by sulfate- and iron reduction. Generally, the groundwater was rich in dissolved or colloidal organic matter (DOM = 50-120 mg C_{org} L^{-1}) and in dissolved inorganic Fe(II) (>1.4 mM). Since reduction of NACs by dissolved H_2S, Fe(II),(*45*) and Fe(II) bound to DOM relatively slow, the major reductants for NACs expected under these conditions include microorganisms, reduced DOM, and Fe(II) associated with iron(hydr)oxide surfaces.

In the aquifer, dinitrobenzenes generally were reduced at very high rates and without any noticeable lag phase. These compounds were always reduced first within the mixture of the NACs studied. These results and the very similar pattern of NAC reduction under unchanged and microbially deactivated conditions suggest that NAC reduction was dominated by abiotic reactions rather than biotransformation. The presence of aquifer sediment strongly enhanced the reduction of all NACs compared to filtrated groundwater. Throughout the plume, a very similar pattern of NAC reduction was found, indicating that the same type of reactions occurred in the entire anaerobic part of the aquifer.

The reactivity pattern of the NACs within the aquifer resembled very much the pattern observed in model systems where surface-bound Fe(II) species were the reductants (Figure 10b). In both systems, the range of relative reactivities and the sequence of NAC reduction were similar. Also, similar competition effects among NACs present in mixtures occurred in the two systems which indicates that the reactive sites present in the aquifer had a similar affinity spectrum for NACs than reactive, surface-bound Fe(II) species. If DOM was the predominant reductant in the leachate plume, the NAC mixture was expected to be reduced without competition effects. Also, dinitrobenzenes were expected to react 2-3 orders of magnitude faster than mono-NACs, and 4-CH_3-NB should have been reduced at the lowest rate (Figure 10b). The results suggest that ferrous iron adsorbed to Fe(III)(hydr)oxides was the dominating reductant of the NACs throughout the plume.

Formation and Regeneration of Reactive Fe(II) Surface Sites in Natural Systems. As has been addressed in the discussion of Figure 2b, reactive Fe(II) surface sites at Fe(III)-containing minerals are formed and can be regenerated by adsorption of dissolved Fe(II) from solution. However, such sites can also be formed by microbial activity. Heijman et al. (*4*) and Heijman (*46*) investigated the formation and reactivity of Fe(II) surface sites in microbially active laboratory columns containing sandy aquifer material and a model sediment consisting of pure quartz sand coated with amorphous FeOOH, respectively. The columns were operated under ferrogenic conditions and reduced NACs quantitatively to the corresponding anilines. It was shown that reduction of NACs occurred exclusively by abiotic reaction with surface-bound iron(II) species as discussed above. The formation of reactive iron(II) species was coupled to oxidation of organic material by dissimilatory iron reducing bacteria and mediated the transfer electrons from the organic substrate of the bacteria to the pollutants (Figure 11). In contrast to sterile Fe(II)/Fe(III) batch systems, under the prevailing conditions, the NACs studied were *all reduced at the same rates.* Within the columns virtually no ferrous iron appeared in aqueous solution. Under such conditions, the regeneration of the reactive sites (i.e., the surface-bound iron(II) species) was not possible by re-adsorption of Fe(II) from a reservoir in solution and thus was entirely controlled by the activity of the iron reducing bacteria. Factors that stimulated the activity of iron-reducing microorganisms (e.g., addition of electron donors such as acetate) increased the rates of pollutant reduction.

The conditions prevailing in the these column experiments may be regarded as an extreme case since reducing conditions in natural porous media often involve the presence of significant amounts of dissolved ferrous iron due to chemical and microbial reduction of iron(hydr)oxide minerals or dissolution of Fe(II)-containing minerals. Thus, in natural anoxic media, re-adsorption of Fe(II) from solution will compete with direct microbial regeneration of reactive Fe(II) surface sites.

Throughout the studied part of the Grindsted aquifer, the concentrations of aqueous Fe(II) were very high (about 1.5 mM), providing a large reservoir of Fe(II) bulk reductants. Under these conditions, the (re)generation of reactive Fe(II) species at the minerals proceeded predominately by adsorption of aqueous Fe(II) to mineral

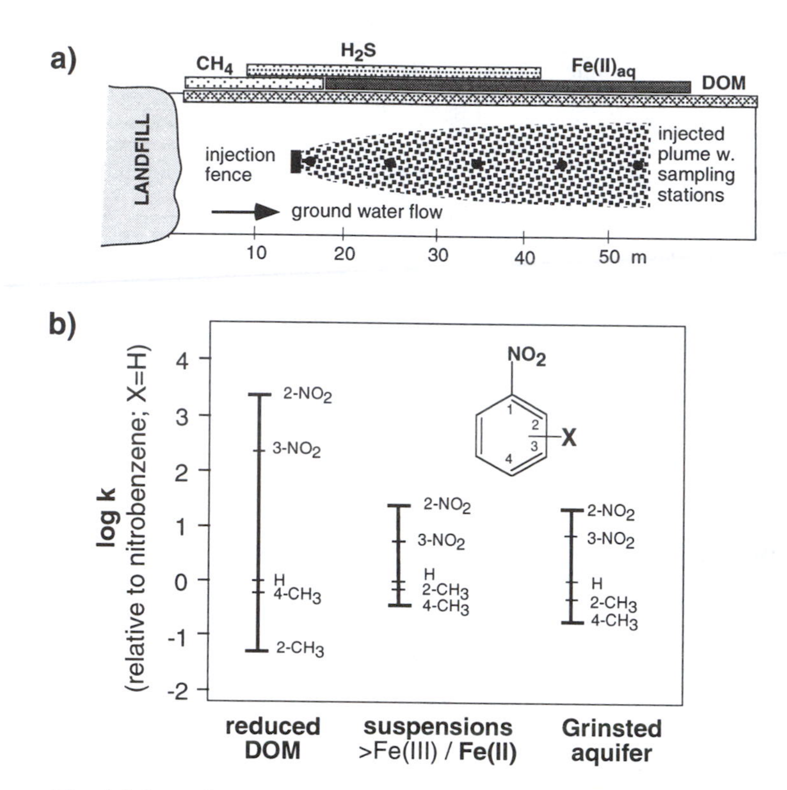

Figure 10. a) Schematic representation of the field injection experiment of Rügge et al. (*43*) conducted in the anaerobic part of a landfill leachate plume. The reactivity patterns of nitroaromatic tracers injected in the aquifer were used to characterize the active reductants present. b) Comparison of the reactivity of the tracers within the aquifer and in model systems exhibiting only NOM or surface-bound Fe(II) as reductants.

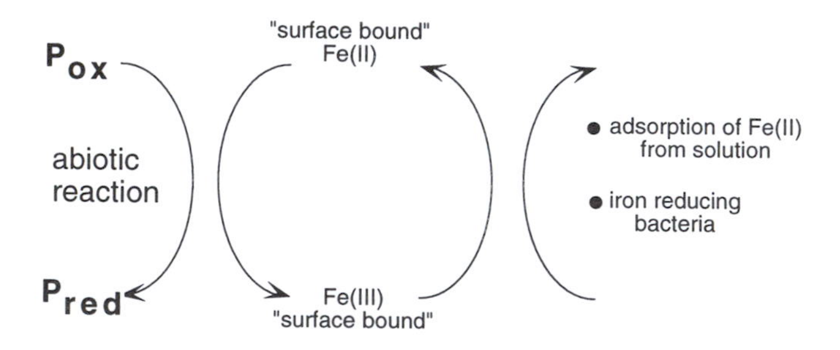

Figure 11. General scheme for reduction of pollutants by reactive Fe(II) surface sites and processes that regenerate such reductants.

surfaces as reflected by a similar reactivity of untreated and microbially inhibited samples.

Nielsen et al. (*47*) studied the transformation of various xenobiotic organic compounds including NACs and CCl_4 in the anaerobic plume the Vejen landfill, Denmark. In contrast to the Grindsted aquifer, the Vejen aquifer exhibited considerable variability with respect to the concentrations of aqueous ferrous iron throughout. In the presence of high concentrations of dissolved Fe(II) their results were similar to those obtained in the Grindsted aquifer. However, further out in the plume, i.e., at lower concentrations of dissolved Fe(II), inhibition of microbial activity by adding biocides to *in situ* microcosms significantly slowed down the reduction rate constants. Their conclusion was, that biological processes or chemical processes coupled to biological activity were important in the reduction of nitrobenzene under these conditions, while chemical processes were responsible for the reduction in the more reduced part of the plume where Fe(II) was more abundant. This is in good accordance with our discussion on the interplay of microbial iron reduction and adsorption of aqueous Fe(II) in regenerating reactive Fe(II) surface sites at Fe(III)minerals.

Conclusions

Ferrous iron associated with iron(III)(hydr)oxide surfaces is a versatile and powerful reductant for many reducible organic pollutants of environmental concern. Evidence is presented for the importance of such Fe(II)/Fe(III) species in the overall transformation rates of reducible pollutants in anoxic aquifers. In contrast to the reactivity of minerals containing structural Fe(II), which is often very low once the reactive surface sites are exhausted, the high reactivity of surface-bound Fe(II) can be maintained over long periods of time since such species may continuously be regenerated by uptake of Fe(II) from solution or by the activity of iron reducing microorganisms. As dissolved Fe(II) and iron(III)(hydr)oxides are almost ubiquitously present in anoxic environments, abiotic reduction of pollutants by such species generally may be significant in reducing subsurface systems.

The type and the reactivity of Fe(II) species formed at Fe(III)(hydr)oxide surfaces are subject to various environmental factors. The most important variables include solution pH, surface charge and morphologic properties of the supporting Fe(III) mineral, and degree of surface saturation with respect to Fe(II). Generally, reaction conditions favouring surface precipitation, i.e., high surface saturation of oxides with Fe(II) and high pH values, speed up kinetics of reduction reactions and extent the range of reducible substrates. Various macroscopic evidence suggests that surface clusters of Fe(II) or mixed valence iron precipitates such as green rusts may be responsible for the high reactivity of these systems. This hypothesis currently is further investigated by surface sensitive spectroscopic techniques.

Groundwater remediation using metal iron as a bulk reductant for pollutants has become an emerging technique in recent years. However, the processes and the nature of the actual reductants that lead to the observed high reduction rates and low substrate specificity of metal iron systems are not yet fully resolved. In such systems, reaction conditions prevail that favor the surface precipitation of Fe(II) species (i.e., high pH values, high surface saturation of oxides with Fe(II)). In fact, spectroscopic studies on corrosion the of steel recently showed that metastabile mixed valence iron coatings are formed as precursors to stabile Fe(III)oxide coatings of the metal (*48-50*). Thus, Fe(II) present at Fe(III)(hydr)oxides may play a pivotal role for the abiotic reduction of pollutants in natural reducing matrices as well as in technical groundwater remediation systems.

Acknowledgments

Financial support was provided by the Swiss National Science Foundation (SNF grant No. 8220-46517) and by the German Research Council (DFG grant No. Pe 581/1-1).

Literature Cited

(1) Pecher, K.; Haderlein, S.B.; Schwarzenbach, R.P. In: *213th ACS National Meeting, Symposium on Redox Reactions in Natural and Engineered Aqueous Systems*; ACS, Division of Environmental Chemistry, Ed.; Preprints of Papers 37; 1997, 185-187.
(2) Sivavec, T.M.; Horney, D.P. In: *213th ACS National Meeting, Symposium on Redox Reactions in Natural and Engineered Aqueous Systems*; ACS, Division of Environmental Chemistry, Ed.; Preprints of Papers 37; 1997, 115-117.
(3) Rügge, K.; Bjerg, P.L.; Christensen, T.H. *Water Resources Research* **1997**, submited.
(4) Heijman, C.G.; Grieder, E.; Holliger, C.; Schwarzenbach, R.P. *Environ. Sci. Technol.* **1995**, *29*, 775-783.
(5) Heijman, C.G.; Holliger, C.; Glaus, M.A.; Schwarzenbach, R.P.; Zeyer, J. *Appl. Environ. Microbiol.* **1993**, *59*, 4350-4353.
(6) Klausen, J. Ph.D. Thesis, ETH Zurich, 1995.
(7) Sørensen, J.; Thorling, L. *Geochchim. Cosmochim. Acta* **1991**, *55*, 1289-1294.
(8) Hansen, H.C.B.; Borgaard, O.K.; Sørensen, J. *Geochim. Cosmochim. Acta* **1994**, *58*, 2599-2608.
(9) Ottley, C.J.; Davison, W.; Edmunds, W.M. *Geochim. Cosmochim. Acta* **1997**, *61*, 1819-1828.
(10) Buerge, I.J.; Hug, S.J. *Environ. Sci. Technol.* **1997**, *31*, 1426-1432.
(11) Sedlak, D.L.; Chan, P.G. *Geochim. Cosmochim. Acta* **1997**, *61*, 2185-2192.
(12) Myreni, S.C.B. et al. *Science* **1997**, *278*, 1106ff.
(13) Cui, D.; Eriksen, T.E. *Environ. Sci. Technol.* **1996**, *30*, 2259-2262.
(14) Schwarzenbach, R.P.; Gschwend, P.M.; Imboden, D.M. *Environmental Organic Chemistry*; John Wiley & Sons: New York, 1993.
(15) Stumm, W. *Chemistry of the Solid-Water Interface*; Wiley: New York, 1992.
(16) Kriegmann-King, M.R.; Reinhard, M. *Environ. Sci. Technol.* **1992**, 2198-2206.
(17) Klausen, J.; Tröber, S.P.; Haderlein, S.B.; Schwarzenbach, R.P. *Environ. Sci. Technol.* **1995**, *29*, 2396-2404.
(18) Pecher, K.; Haderlein, S.B.; Schwarzenbach, R.P. *Environ. Sci. Technol.* **1998**, in preparation.
(19) Lanz, K.; Schwarzenbach, R.P.1988. unpublished results. EAWAG (CH).,
(20) Yu, S.Y.; Bailey, G.W. *J. Environ. Qual.* **1992**, *21*, 86 - 94.
(21) Kriegmann-King, M.R.; Reinhard, M. *Environ. Sci. Technol.* **1994**, *28*, 692-700.
(22) Assaf-Anid, N.; Lin, K.-Y.; Mahony, J. In: *213th ACS National Meeting, Symposium on Redox Reactions in Natural and Engineered Aqueous Systems*; ACS, Division of Environmental Chemistry, Ed.; Preprints of Papers 37; 1997, 194-195.
(23) Coughlin, B.R.; Stone, A.T. *Environ. Sci. Technol.* **1995**, *29*, 2445-2455.
(24) Pecher, K.; Schwarzenbach, R.P. *Geochim. Cosmochim. Acta* **1998**, in preparation.
(25) Hansen, H.C.B.; Koch, C.B.; Nancke-Kroh, H.; Borggaard, O.K.; Sørensen, J. *Environ. Sci. Technol.* **1995**, *30*, 2053-2056.
(26) Dzombak, D.A.; Morel, F.M.M. *Surface complexation modeling. Hydrous ferric oxide.*; John Wiley & Sons: New York, 1990.

(27) Kinniburgh, D.G.; Jackson, M.L.; Anderson, A.A.; Rubin, A.J. *Cation adsorption by hydrous metal oxides and clays/ Adsorption of inorganics at solid-liquid interfaces.*; Ann Arbor Science: Ann Arbor (MI), 1981.
(28) Sposito, G. In *Geochemical Processes at Mineral Surfaces*; ACS Symposium Series, 323, Davis, J. A.; K. F. Hayes, Ed.; American Chemical Society: Washington, 1986. 217-228.
(29) Hachiya, K.; Sasaki, M.; Ikeda, T.; Mikami, N.; Yasunaga, T. *J. Phys. Chem.* **1984**, *88*, 27-31.
(30) Cornell, R.M.; Schwertmann, U. *The Iron Oxides*; VCH: Weinheim, 1996.
(31) De Boer, J.; Lippens, B.; Linsen, B.; Broekhiff, J.; van den Heuvel, A.; Osinga, T. *J. Colloid Interf. Sci.* **1966**, *21*, 405-414.
(32) Jolivet, J.-P.; Tronc, E. *J. Coll. Interf. Sci.* **1988**, *125*, 688-701.
(33) Tronc, E.; Jolivet, J.-P. *Ads. Sci. Technol.* **1984**, *1*,
(34) Tronc, E.; Jolivet, J.-P.; Lefebvre, J.; Massart, R. *J. Chem. Soc. Faraday Trans.* **1984**, *80*, 2619-2629.
(35) Tronc, E.; Belleville, P.; Jolivet, J.-P.; Livage, J. *Langmuir* **1992**, *8*, 313-319.
(36) Tamaura, Y.; Ito, K.; Katsura, T. *J. Chem. Soc. Dalton Trans.* **1983**, 189-194.
(37) Tronc, E.; Jolivet, J.P.; Belleville, P.; Livage, J. *Hyperfine Interact.* **1989**, *46*, 637-643.
(38) Ford, R.G.; Bertsch, P.M.; Farley, K.J. *Environ. Sci. Technol.* **1997**, *31*, 2028-2033.
(39) Farley, K.J.; Dzombak, D.A.; Morel, F.M.M. *J. Coll. Interf. Sci.* **1985**, *106*, 226-242.
(40) Baes Jr., C.F.; Mesmer, R.E. *The Hydrolysis of Cations.*; John Wiley & Sons: New York, 1976.
(41) Zhang, Y.; Charlet, L.; Schindler, P.W. *Colloids Surf.* **1992**, *63*, 259-268.
(42) Tratnyek, P.G.; Smolen, J.M.; Weber, E.J. In: *213th ACS National Meeting, Symposium on Redox Reactions in Natural and Engineered Aqueous Systems*; ACS, Division of Environmental Chemistry, Ed.; Preprints of Papers 37; 1997, 119-121.
(43) Rügge, K.; Hofstetter, T.; Haderlein, S.B., et al. *Environ. Sci. Technol.* **1998**, *32(1)*, in press.
(44) Haderlein, S.B.; Schwarzenbach, R.P. In *Biodegradation of Nitroaromatic Compounds*; Spain, J., Ed.; Plenum: New York, 1995. Chapter 12, 199-225.
(45) Colon, D.; Weber, E.J.; Anderson, J.L. In: *213th ACS National Meeting, Symposium on Redox Reactions in Natural and Engineered Aqueous Systems*; ACS, Division of Environmental Chemistry, Ed.; Preprints of Papers 37; 1997, 193-194.
(46) Heijman, C.G. Ph.D. Thesis, ETH Zurich, 1995.
(47) Nielsen, P.H.; Bjarnadottir, H.; Winter, P.L. *J. Contam. Hydrol.* **1995**, *20*, 51-66.
(48) Abdelmoula, M.; Refait, P.; Drissi, S.H.; Mihe, J.P.; Génin, J.-M.R. *Corrosion Science* **1996**, *38*, 626-633.
(49) Novakova, A.A.; Gendler, T.S.; Manyurova, N.D.; Turishcheva, A. *Corrosion Science* **1997**, *39*, 1585-1594.
(50) Réfait, P.H.; Drissi, S.H.; Pytkiewicz, J.; Génin, J.M.R. *Corrosion Science* **1997**, *39*, 1699-1709.

Chapter 18

Auto-Inhibition of Oxide Mineral Reductive Capacity Toward Co(II)EDTA

Scott Fendorf[1], Phillip M. Jardine[2], David L. Taylor[2], Scott C. Brooks[2], and Elizabeth A. Rochette[1]

[1]Soil Science Division, University of Idaho, Moscow, ID 83844
[2]Environmental Science Division, Oak Ridge National Laboratory, Oak Ridge, TN

Subsurface migration of ^{60}Co has been attributed to organic chelating agents and oxidation of $Co(II)EDTA^{2-}$ to $Co(III)EDTA^{-}$ by mineral surfaces. Although the oxidized product ($Co(III)EDTA^{-}$) has been detected in solution, a reduced species has not been measured. As a result, fate and transport mechanisms involving ^{60}Co remain ill-defined. Accordingly, the objective of this research was to determine redox changes in the solid-phase oxidants, β-MnO_2 and $Fe(OH)_3 \cdot nH_2O$, during reaction with $Co(II)EDTA^{2-}$. Time-resolved changes in the surface composition of β-MnO_2 was accomplished in hydrodynamic systems using XANES spectroscopy. The transport of $Co(II)EDTA^{2-}$ through packed beds of β -MnO_2 resulted in a decrease in structural Mn(IV) and an increase in Mn(III) as a Mn_2O_3-like phase. As the quantity of Mn_2O_3 increased, the production of $Co(III)EDTA^{-}$ decreased. Thus, it appears that the surface association of Mn_2O_3 produced from the oxidation of $Co(II)EDTA^{2-}$ impedes the redox reaction. We were unable to detect surface structural alterations on $Fe(OH)_3 \cdot nH_2O$ upon reacting with $Co(II)EDTA^{2-}$.

The migration of radionuclides away from low-level waste disposal facilities has been documented over a broad geographical range from Oak Ridge, TN (1) to West Valley, NY (2), to the Canadian burial grounds at the Chalk River Nuclear Labs (3). The enhanced migration in groundwater of ^{60}Co at ORNL (1,4) and Chalk River (3) has been attributed to the presence of synthetic organic chelating agents, such as ethylenediaminetetraacetic acid (EDTA), that increase aqueous concentrations of cobalt. The stability of Co-EDTA complexes, and thus the fate and transport of ^{60}Co in the subsurface, is strongly dependent on the oxidation state of cobalt (log $K_{Co(II)EDTA}$ = 18.2; log $K_{Co(III)EDTA}$ = 39.8). For example, Szescody *et al.* (5) have demonstrated that in batch experiments ferric iron (Fe(III)) in ferrihydrite ($Fe(OH)_3 \cdot nH_2O$) can displace Co(II) from $Co(II)EDTA^{2-}$, resulting in the formation of aqueous $Fe(III)EDTA^-$ (log $K_{Fe(III)EDTA}$ = 27.7); cobalt is then released into solution followed by re-adsorption on the ferrihydrite surface. However, in flow-through columns, which result in finite residence times, the oxidation of $Co(II)EDTA^{2-}$ to $Co(III)EDTA^-$ dominate the reactive transport of cobalt (6); the formation of $Fe(III)EDTA^-$ was a reaction of minor importance. The results of the column experiments strongly suggested that Fe(III) served as the oxidant in the ferrihydrite-$Co(II)EDTA^{2-}$ system, representing the first report that Fe(III) can oxidize $Co(II)EDTA^{2-}$ to $Co(III)EDTA^-$. In addition, the experimental results indicated that the reduced iron remained in the solid phase and was not released to solution.

Other mineral phases that exhibit strong control on the environmental distribution of cobalt are the hydrous oxides of manganese; the strong geochemical association between manganese and cobalt has long been recognized (for example see reference 7). The specific affinity of Co for Mn oxides has been attributed to the oxidation of Co(II) to Co(III) at the mineral-solution interface and the subsequent retention of Co(III) on the Mn-mineral (8-11). The oxidized species Co(III) has been detected at the surface of Mn-oxides using X-ray photoelectron spectroscopy (11). Thus, we know that Mn-oxides can oxidize Co(II) to Co(III); when complexed by EDTA, however, Co(III) desorbs from the surface rather than forming a sorbate layer.

$Co(III)EDTA^-$ was formed during transport of $Co(II)EDTA^{2-}$ through columns packed with pyrolusite (ß-MnO_2)-coated quartz sand (12). Although the oxidized species, $Co(III)EDTA^-$, was detected the identity of the reduced product remained unknown. While it has been suggested that Mn^{2+} is the reduced product (8,11), it has also been speculated that a Mn(III) solid phase is formed. Jardine and Taylor (12) reported the reversible loss of the oxidative potential of MnO_2 with continual exposure to $Co(II)EDTA^{2-}$, and they demonstrated that the accumulation of Mn^{2+} on the oxide surface did not explain the loss in oxidative potential. Rather, they hypothesized that the formation of a thin layer of Mn_2O_3 on the MnO_2 surface resulted in the loss of oxidative capacity. This is consistent with x-ray photoelectron spectroscopy which indicated the production of Mn(III) after birnessite was reacted with Co(II) (13). The production of Mn(III) is also noted in the oxidation of Mn(II) as either a transient intermediate phase, dominantly Mn_3O_4 or β-MnOOH, or as a stable end product, γ-MnOOH (14-17).

Although the oxidized species $Co(III)EDTA^-$ has been detected after the association of $Co(II)EDTA^{2-}$ with Fe- and Mn-oxides, the identity of the reduced product remains the object of speculation. The positive identification of the reduced product is required to define the Eh and pH conditions that favor the formation of the less stable $Co(II)EDTA^{2-}$ versus $Co(III)EDTA^-$. Furthermore, without knowledge of the reduced products, there is an incomplete mechanistic understanding of the governing reactions, thus limiting the generic applicability of the observations. The governing reactions are implicitly heterogeneous--involving the interaction of the solution and solid phases; an understanding of the aqueous phase alone will not suffice. Because the reduced products appear to remain tightly associated with the solid phase, their detection will rely on the use of surface spectroscopic techniques. In response to this need, we used X-ray absorption fine structure spectroscopy to measure changes in Mn and Fe surface speciation upon exposure to $Co(II)EDTA^{2-}$ under hydrodynamic conditions. This approach allowed for *in situ* measurements of Mn and Fe chemical states.

Our objectives for this research were to: (1) use XANES spectroscopy to quantify subtle changes in solid-state geochemistry after hydrodynamic flow; (2) determine if geochemical reduction of structural Mn(IV) or Fe(III) in oxide minerals occurred by surface-associated $Co(II)EDTA^{2-}$; (3) identify the source of Co(II) oxidation inhibition in pyrolusite and ferrihydrite media.

Materials and Methods

Reduction of Fe- and Mn-oxide minerals by $Co(II)EDTA^{2-}$ is believed to be a surface-mediated phenomena that affects only the outer molecular layer(s) of the mineral. XANES spectroscopy is inherently a bulk technique that is not surface sensitive without special experimental configurations such as glancing incidence angles; alternatively, surface sensitivity can be gained by measuring electron yields. Neither of these approaches is suitable for performing surface sensitive XANES measurements in hydrodynamic experiments using packed mineral beds. We circumvented this limitation be synthesizing our reactive solids as thin coatings (only several monolayers thick) on an inert, high surface area substrate. If the oxide coating is too thick, distinguishing the reduced elements from the majority of oxidized elements is futile using this approach. Therefore, we adopted the following technique that provides surface sensitivity to prepare our Mn- and Fe-oxide solid phase materials for XANES analysis.

Solid-Phase Preparation. To obtain surface sensitivity, both manganese- and iron-(hydr)oxide minerals were synthesized as thin coatings on an inert, high-surface area silica substrate.

Pyrolusite-coated Silica. High surface area silica (SiO_2) with a grain size <44 μm (silt sized and smaller) was obtained from TIMCO Corp. Using measured packing bulk densities of the silica, a quantity of Mn that allowed for multiple valence states to be identified without self-absorption effects was calculated; this value was set at 0.25 mg Mn/cm^2 cell area. Pyrolusite (ß-MnO_2) was prepared on the silica surface at 1.5x and 15x the optimal amount calculated for XANES analysis, which resulted in a surface coverage of 0.10 and 1.0 wt% MnO_2 (0.06 and 0.6 wt% Mn), respectively. For the lower surface coverage, a 50% w/w reagent grade $Mn(NO_3)_2$ solution was diluted with deionized water to make a solution weighing about 7% of the silica used. After mixing well with the silica, this amount of solution just barely dampened the silica, thus dispersing the $Mn(NO_3)_2$. To ensure that the pyrolusite formed a uniform coverage on the silica surface, the mixture was slowly rotated in Teflon vessels at 100°C until the liquid had evaporated. The mixture was baked further at 160°C for 3 days to ensure that pyrolusite had formed. The formation of pyrolusite using this procedure was confirmed using x-ray diffraction and XANES spectroscopy. When the coating process was complete, residual nitrate was washed from the pyrolusite-coated silica using reagent water.

Ferrihydrite-coated Silica. High surface area silica was also used as a template for precipitating ferrihydrite ($Fe(OH)_3{\bullet}nH_2O$). We used a quantity of Fe equal to 0.29 mg Fe/cm^2 cell area to minimize self-absorption but still provide adequate signal to noise. Ferrihydrite was prepared on the silica surface at 1.5x and 15x the optimal rate calculated for XANES analysis. This resulted in a surface coverage of 0.128% and 1.28% ferrihydrite, respectively. Ferric chloride, hexahydrate ($FeCl_3{\bullet}6H_2O$) was dissolved in carbonate-free reagent water, and low carbonate NaOH was slowly added to the vigorously stirred solution until pH 3.5 was achieved. Using an HPLC advanced gradient pump, helium- purged 0.5 N NaOH was added to the rapidly stirring $FeCl_3$ solution at a rate of 0.5 mL/min through small diameter PEEK tubing to avoid localized areas of high pH. The percent NaOH added and the flow rate were decreased as the $FeCl_3$ solution pH slowly increased to 7.5. The solution was aged overnight. The next day the $FeCl_3$ solution was readjusted to pH 7.5 and an appropriate mass of silica was added to the solution. The mixture was slowly agitated several times during a period of 8 hours. Gentle agitation is essential since ferrihydrite will easily abrade off the silica. The mixture was allowed to settle overnight into three distinct layers. The total solution level in the container was noted, and the top clear layer was decanted. The silica and dark Fe layer were mixed together, and 0.1 mM NaCl (pH ~7) was added to replace the solution that was removed. The suspension was mixed and aged overnight. The last three steps were repeated twice more, and the pH of the dark Fe layer was adjusted to 6.5-7.0 using dilute NaOH prior to mixing. Finally, the well-mixed suspension was poured into a Pyrex baking dish and dried using convective air displacement at 25° C. The dry ferrihydrite-coated silica was washed with reagent water prior to use.

Dynamic Flow Experiments. Time-dependent reduction of the Mn- and Fe-oxide coated silica by $Co(II)EDTA^{2-}$ was investigated using a miscible displacement technique. Columns were wet packed to a uniform bulk density using the approach detailed in Jardine and Taylor (12) and Brooks et al. (6). To obtain time-resolved changes in Mn states, a series of identical columns were packed with pyrolusite coated silica. The reductant, 0.20 mM $Co(II)EDTA^{2-}$, a nonreactive Br^- tracer, and a background electrolyte of 5.0 mM $CaCl_2$ adjusted to pH 6.5, were passed through the columns using a continuous steady-state pulse. Columns were sacrificed at known times during the course of the redox reaction for analysis by XANES spectroscopy. At the designated times, the solid phase material was quickly (≈ 30 s) extracted from the column, quickly packed into a sample holder, and sealed with Mylar tape; XANES spectroscopy was then performed on the solid-phase as described below. Solutions were collected using an automated fraction collector and CoEDTA was speciated using ion chromatographic methods (18)

X-ray Absorption Fine Structure Spectroscopy. X-ray absorption near-edge structure (XANES) spectroscopy was performed on beamlines 4-1 and 4-2 (beamline 4 is an 8-pole wiggler) at the Stanford Synchrotron Radiation Laboratory (SSRL), running under dedicated conditions. The ring operated at 3 GeV with a current ranging from ~ 100 mA to ~ 50 mA. Energy selection was accomplished using a Si(111) monochromator, with a focused beam on 4-2 and an unfocused beam on 4-1. XANES spectra were recorded by fluorescent x-ray production using a wide-angle ionization chamber; incident and transmitted intensities were also measured with in-line ionization chambers. Scattered primary radiation was restricted from entering the fluorescent detector by placing a Cr filter for Mn, or a Mn filter for Fe, in front of the detector. XANES spectra were recorded over the energy range of -100 to +200 eV about the K-edge of the respective absorber (Mn or Fe).

No changes in Fe states were observed during the course of reaction, as will be discussed in the next section, and consequently methods for Fe XANES analysis are not discussed further. In contrast, distinct changes in Mn states were noted during the reaction of pyrolusite with $Co(II)EDTA^{2-}$.

To quantify the Mn species in the solid-phase, XANES spectra were acquired over the energy range 6300 eV to 6950 eV; step sizes of 0.25 eV were used over the range 6530 to 6620 eV followed by 1 eV steps from 6620 to 6950 eV. Between 3 and 5 individual spectra were averaged for each time period analyzed. Spectra acquired by these means allowed for Mn quantification using a linear combination of known compounds. A series of manganese minerals, primarily those of Mn(III), were used to reconstruct the unknown spectra: ß-MnO_2, α-Mn_2O_3, γ-Mn_2O_3, γ-MnOOH, β-MnOOH, Mn_3O_4, and $MnCO_3$. Pyrolusite (ß-MnO_2) was obtained from Fisher Scientific; bixbyite (α-Mn_2O_3) and rhodochrosite ($MnCO_3$) were obtained from STREM chemicals. Hausmannite (Mn_3O_4), feitkechtite (β-MnOOH), γ-Mn_2O_3, and manganite (γ-MnOOH) were synthesized by modified procedures of Hem (14) and Bricker (19). X-ray diffraction and electron microscopy were used to confirm the mineral identity and purity.

Standard spectra were recorded from finely ground mineral powders that were well dispersed in a polystyrene resin (20); the total quantity of Mn was restricted to give a change in optical depth of less than 0.1 across the absorption edge. On the basis of previous studies indicating that Mn(II) did explain the poisoning of the reaction between Co and MnO_2 (12), we only used a single Mn(II) bearing phase to reconstruct the spectra. If Mn(II) was deemed a more important constituent, one should certainly use a more elaborate series of Mn(II) standards including $Mn(OH)_2$ and Mn(II)-MnO_2 (Mn(II) adsorbed on the Mn(IV) mineral).

Both unknown and standard spectra were acquired and treated in an identical manner. The background from averaged spectra was removed using a low-order polynomial function. The total absorption was then normalized to a uniform atomic-cross section and the spectra calibrated relative to metallic manganese (inflection point = 6539.0 eV). First derivatives of the spectra were used to enhance line-shapes and were obtained using a Savitzky-Golay algorithm with a smoothing function at 3%. A linear combination of the standard first-derivative spectra was then fit to the unknown using a least-squares routine to minimize the error between the unknown and 'reconstructed' spectrum. The reconstruction is summarized:

$$\mu(E) = \sum a_i \, \mu_i(E \pm \Delta E) \qquad (1)$$

where $\mu(E)$ is the reconstructed spectrum (absorption as a function of energy), a_i is the amplitude of the standard i^{th}-species, and $\mu_i(E)$ is the absorption spectrum of the ith-standard. Although spectra were calibrated to elemental manganese, a small energy offset (ΔE, constrained within 0.5 eV) is used to account for limitations in calibration precision (21). Our approach is similar to that used to quantify sulfur and selenium species in complex media using XANES spectroscopy (21-23). We did not explicitly correct for self-absorption dampening because both the unknown and standard specimens contained Mn quantities that were well dispersed within a non-Mn media and contained sufficiently small quantities of the absorber so that self-absorption is negligible.

Results and Discussion

The oxidation of $Co(II)EDTA^{2-}$ to $Co(III)EDTA^{-}$ by pyrolusite-coated silica under dynamic flow conditions exhibited breakthrough characteristics (Figure 1) that were consistent with earlier research (Jardine and Taylor, 1995). The oxidation was initially rapid and slowed with continual exposure of $Co(II)EDTA^{2-}$ to the Mn-oxide. Jardine and Taylor (12) showed that the loss of oxidative potential was not caused by the accumulation of Mn^{2+} on the surface and postulated that an intermediate Mn(III)-oxide solid phase, Mn_2O_3, was formed that impeded the redox reaction. XANES spectroscopy was used to determine the speciation of the solid-phase as a function of reaction time in order to identify the inhibition mechanism.

Time-resolved studies of manganese solid-phase transformations were performed by sacrificing column experiments at known times for analysis by XANES spectroscopy. Possible contributions from solution species can be removed by rinsing the extruded material or by spectral subtraction. However, previous studies have noted that no soluble Mn products are formed from the reaction of β-MnO_2 with $Co(II)EDTA^{2-}$ (12); therefore, we neither rinsed the entrained solutions from the reacted material nor did we subtract solution contributions spectrally.

Speciation of Mn solids in extruded columns was accomplished using XANES spectroscopy by fitting a series of reference spectra (Figures 2 and 3) to the solid-phase products of β-MnO_2 reacted with $Co(II)EDTA^{2-}$ (Figure 4). An extensive series of Mn(III) containing reference materials and one Mn(II) reference, along with the parent Mn(IV) bearing oxide, were used to fit the unknown spectra. The Mn(III) standards reflect Mn(III) minerals that have been observed in oxidative and reductive transformations of Mn; only α-MnOOH (groutite) was omitted from this comprehensive list of mineral standards because it has less frequently been observed. We focused on Mn(III) bearing solids because Mn(II) was previously demonstrate to not influence the reaction (12).

The main-edge positions of the Mn-minerals vary with changes in oxidation state as does the scattering region above the edge (the EXAFS) with

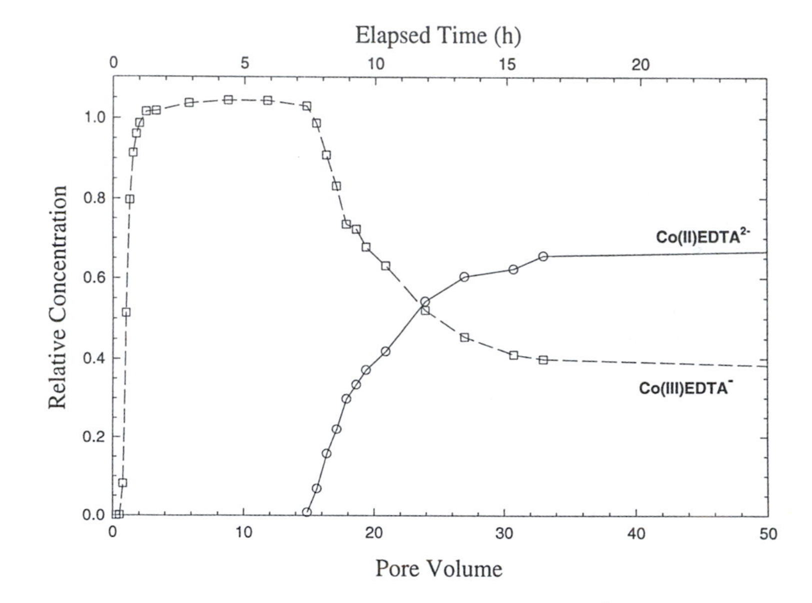

Figure 1. Reduced effluent concentrations of $Co(II)EDTA^{2-}$ and $Co(III)EDTA^{-}$ as a function of time and pore volume following the continuous injection of $Co(II)EDTA^{2-}$ through a packed bed of 0.1% MnO_2 coated silica (pH=6.5 and q=3.8 cm/h).

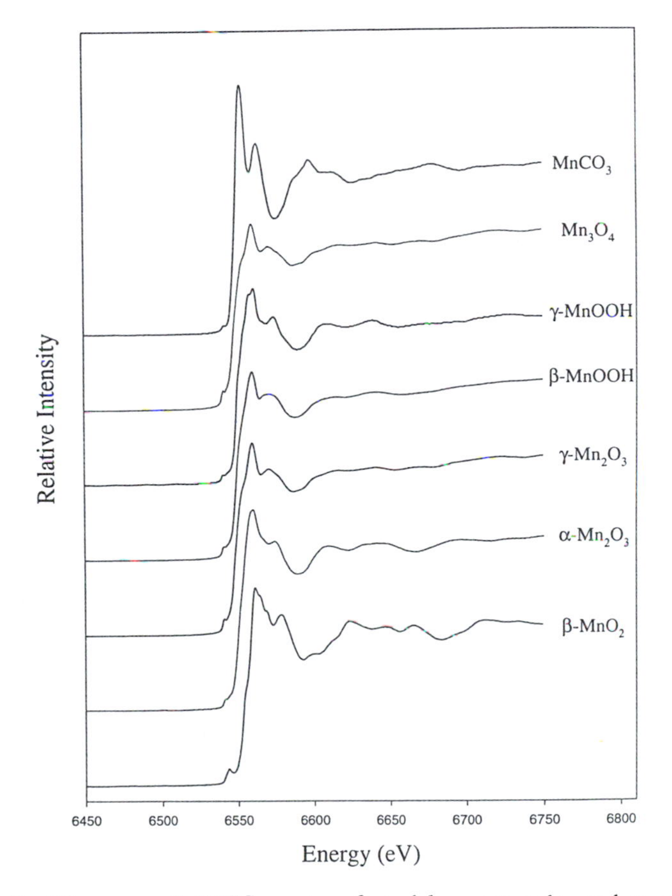

Figure 2. Manganese XANES spectra of model compounds used as a linear combination to fit unknown spectra of β-MnO_2 reacted with Co(II)$EDTA^{2-}$.

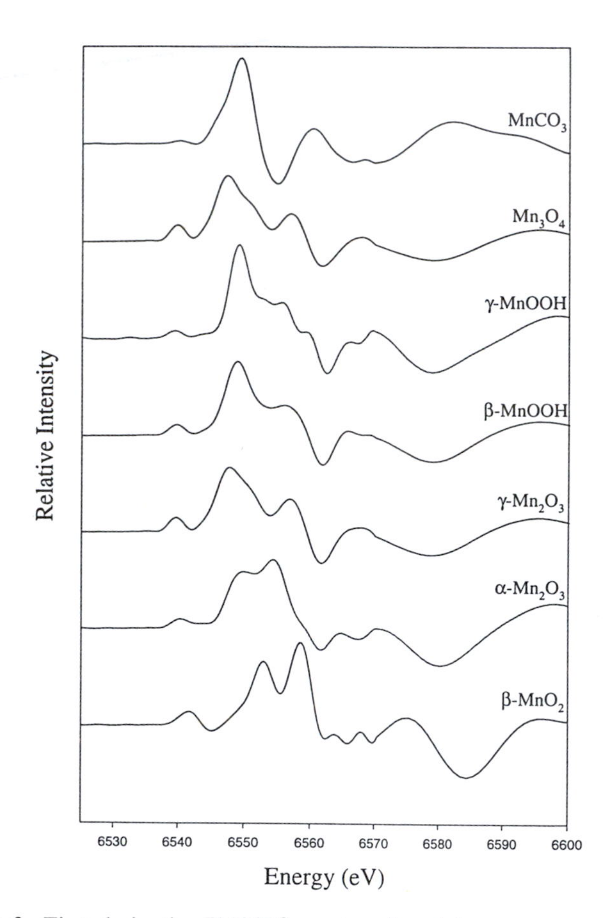

Figure 3. First-derivative XANES spectra of model manganese minerals.

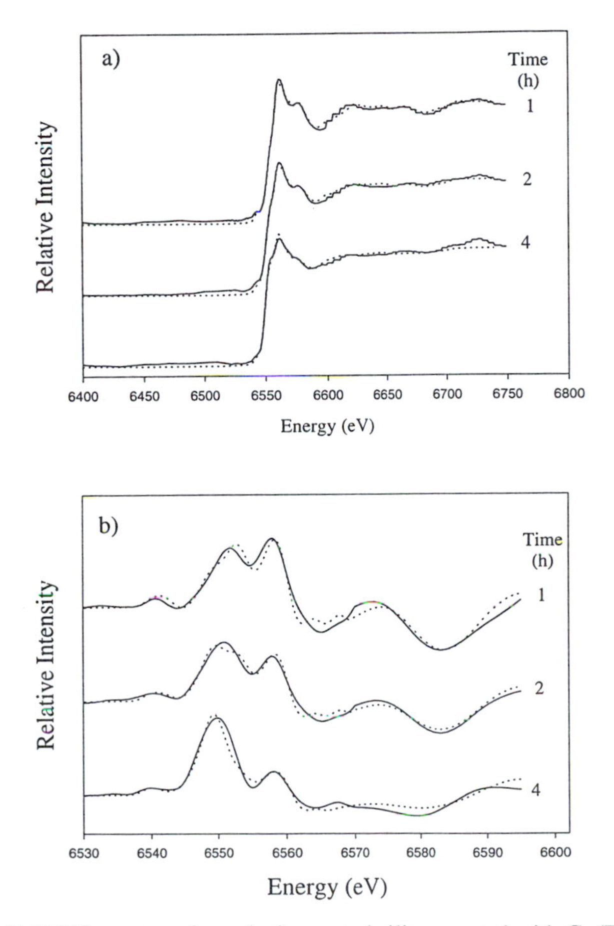

Figure 4. XANES spectra of pyrolusite coated silica reacted with Co(II)EDTA^{2-}: (a) experimental curves and (b) corresponding first-derivative experimental curves (solid lines) and associated fits (dashed-line). Fitted functions were constructed using a linear combination of the standard spectra described in Table 1.

changes in local structure (Figure 2). First-derivative curves of a limited spectral range (ca 6530 to 6600 eV) encompassing the edge region clearly indicate different chemical states for each of the standard compounds. Moreover, the distinct line-shapes of each model compound allows for speciation of the solid through robust fitting procedures using a linear combination of the standards. The single exception is the similarity between γ-Mn_2O_3 and Mn_3O_4.; however, as discussed below, neither of these phases contributed to the unknown XANES spectra. Therefore, our XANES analysis is not limited to discerning oxidation states but extends to speciation of the reacted β-MnO_2 material.

Figure 4 presents the experimental Mn XANES spectra for pyrolusite reacted with a continuous pulse of $Co(II)EDTA^{2-}$ for 1, 2, and 4 h. It is apparent that the main-edge position (Figure 4a) and inflection point (Figure 4b) of the Mn absorption edge shifts to lower energies with increased reaction time, indicative of an increase in more reduced states. Consistent with this qualitative view, spectral fitting revealed that α-Mn_2O_3 increased dramatically with increased reaction time (Table 1; Figure 4b). Spectral contributions of each standard used in the fitting are provided in Table 1. After 1 h of reaction with $Co(II)EDTA^{2-}$, greater than 80% of the Mn remains as pyrolusite. However, after exposure to a continuous pulse of the reductant for 2 h, half of the Mn(IV) has converted to α-Mn_2O_3; and after 4 h of reaction α–Mn_2O_3 is the dominant (67%) solid. Manganese(II) is a minor product, comprising less than 15% of the solid-phase Mn after 4 h of reaction (Table 1; Figure 4); additionally, no soluble Mn products are detected from the reaction of $Co(II)EDTA^{2-}$ with β-MnO_2. The small quantities of Mn(II) produced and the apparent lack of influence on the reductive capacity of pyrolusite are consistent with the findings of Jardine and Taylor (12). The peak production of $Co(III)EDTA^-$ occurs after about 1 h of flow and then decreases to a pseudo steady-state that is less than half the maximum value (see Figure 1). The decrease in Mn-oxide reductive capacity correlates with the increased quantity of Mn(III).

It therefore appears that the surface association of an α-Mn_2O_3-like phase, which is produced from the oxidation of $Co(II)EDTA^{2-}$, impedes the redox reaction and limits the quantity of $Co(III)EDTA^-$ produced. A secondary steady-state in $Co(III)EDTA^-$ level is then attained from the slower oxidation rate of $Co(II)EDTA^{2-}$ by α -Mn_2O_3--or from a limited exposure of β-MnO_2 (Figure 1). Although we cannot directly document the spatial distribution of the Mn_2O_3 on the β-MnO_2 with this procedure, it is probable that the Mn_2O_3 layer formed on the β-MnO_2 surface is well distributed over the substrate. This hypothesis is based on the initial thin films of pyrolusite coated on silica coupled with the fact that nearly 70% of the Mn is converted to a α-Mn_2O_3 phase.

The success of the experiments documented above was primarily the result of the method used to prepare the MnO_2 solid phase. By preparing a thin coating of the MnO_2 mineral on a high surface area substrate (i.e. silt sized and smaller silica), it is possible to detect Mn reduction products in the presence of the original oxidized mineral. Because the redox reaction is surface mediated, too much of the

Table I. Spectral contribution of manganese minerals (percent species) in β - MnO_2 reacted with Co(II)EDTA^{2-} for varying lengths of flow periods

2	50	ND	ND	41	ND	ND	9
4	19	ND	ND	67	ND	ND	14

†ND: Not detected.

Mn(IV) mineral will mask the detection of the Mn(III) or Mn(II) reduction products. This was the case for experiments performed with a 10x higher concentration of MnO_2 coated silica (1.0% vs. 0.1% coating) and Co(II)EDTA^{2-} (2.0 mM vs. 0.2 mM). Even though the ratio of the solid phase Mn and the reductant were constant, Mn reduction products were not detected due to the overwhelming signal of the Mn(IV) (data not shown). These sets of experiments add credibility to the notion that oxidation of Co(II)EDTA^{2-} by Mn-oxides is a surface-mediated process that is restricted to the outer molecular layers of the mineral structure.

Fe-oxide Studies. The oxidation of Co(II)EDTA^{2-} to Co(III)EDTA^{-} by the ferrihydrite-coated silica under dynamic flow conditions exhibited breakthrough characteristics (data not shown) that were consistent with earlier research (6). The reaction quickly reached a steady-state condition that was pH and flow rate dependent. Previous studies by Brooks et al. (6) showed that in the absence of oxygen, the amount of oxidation slowly decreased with time, suggesting that in the absence of oxygen, the amount of oxidation strongly depended on the amount of Fe(III) originally in the flow field. Brooks et al. (6) postulated that Fe(III) served as the oxidant in the ferrihydrite-Co(II)EDTA system with subsequent formation of solid phase Fe(II). XANES spectroscopy was used to determine the speciation of the solid phase as a function of reaction time in order to identify the reaction mechanism.

Relative to Mn-oxides, Fe(III) bearing minerals have a low propensity to oxidize Co(II)EDTA^{2-}. As a consequence, one would expect lower quantities of Fe(II) to result on the surface of the substrate oxidant. This was observed for experiments performed at pH=7, as no appreciable Fe(II) was noted on ferrihydrite-reacted Co(II)EDTA^{2-}. These results were consistent with Brooks et al. (6) who observed only a small portion of Co(II)EDTA^{2-} being oxidized by Fe-oxides at this pH. Therefore, we attempted to optimize the conditions favoring the oxidation of Co(II)EDTA^{2-} by ferrihydrite by performing experiments at a lower pH (pH=4.8 vs. 7.0) and a slower flow rate (1.0 vs. 3.0 mL h^{-}). Time-resolved XANES analysis was inconclusive and indicated that no more that 5% of the structural Fe existed as Fe(II), which is essentially the lower limit of our detection. Greater sensitivity of Fe states may reveal that indeed Fe(II) is retained on the mineral surface and correlates with the slight decrease in reactivity of ferrihydrite toward Co(II)EDTA^{2-}, but our analysis failed to provide such evidence.

Summary and Conclusions

X-ray Absorption Near Edge Spectroscopy (XANES) was performed to provide direct observations of contaminant-induced reduction of structural Mn(IV) in oxide minerals. Our approach used time-resolved spectroscopy to assess dynamic geochemical processes during saturated flow through porous media. The experiments were performed directly in the X-ray beam where the changes in solid-phase speciation were measured after designated reaction times. XANES analysis showed that the transport of $Co(II)EDTA^{2-}$ through Mn-oxide mineral beds was characterized by a decrease in structural Mn(IV) accompanied by an increase in structural Mn(III)--a α-Mn_2O_3-like phase developed on the β-MnO_2 surface and diminished the reaction rate. Surface Mn(II) was not observed in the reacted materials. As the quantity of Mn(III) increased, the production of $Co(III)EDTA^-$ decreased, which is important from an environmental perspective since $Co(III)EDTA^-$ is very stable and highly mobile.

Acknowledgments

This research was funded by the Laboratory Directed Research and Development Program of the Oak Ridge National Laboratory and the Subsurface Science Program of the Office of Biological and Environmental Research, U.S. Department of Energy, under contract DE-AC05-96OR22464 with Lockheed Martin Energy Corporation. The authors appreciate the efforts of Dr. Frank Wobber, the contract officer for the Department of Energy, who partially supported this work. We also gratefully acknowledge the staff and scientists at the Stanford Synchrotron Radiation Laboratory (SSRL) for their help in conducting the XANES analyses. SSRL is operated by the Department of Energy, Office of Basic Energy Sciences. The SSRL Biotechnology Program is supported by the National Institutes of Health, National Center for Research Resources, Biomedical Technology Program, and by the Department of Energy, Office of Biological and Environmental Research.

Literature Cited

1. Olsen, C. R.; Lowry, P. D.; Lee, S. Y.; Larsen, I. L.; Cutshall, N. H. *Geochim. Cosmochim. Acta* **1986**, *50*, 593-607.
2. Francis, A. J.; Iden, C. R.; Nine, B. J.; Chang, C. K. *Nucl. Technol.* **1980**, *50*, 158-163.
3. Killey, R. W. D.; McHugh, J. O.; Champ, D. R.; Cooper, E. L.; Young, J. L. *Environ. Sci. Technol.* **1984**, *18*, 148-157.
4. Means, J. L.; Crerar, D. A.; Borscik, M. P.; Duguid, J. O. *Geochim. Cosmochim. Acta* **1978**, *42*, 1763-1773.
5. Szecsody, J. E.; Zachara, J. M.; Bruckhart, P. L. *Environ. Sci. Technol.* **1994**, *28*, 1706-1716.
6. Brooks, S. C.; Taylor, D. L.; Jardine, P. M. *Geochim. Cosmochim. Acta* **1996**, *60*, 1899-1908.

7. Burns, R. G. *Geochim. Cosmochim. Acta* **1976**, *40*, 95-102.
8. Murray, D. J.; Healy, T. W.; Fuerstenau, D. W. In *Adsorption from Aqueous Solution;* Weber, W. J. and Matijevic, E., Eds.; Advances in Chemistry Series; 1968, 79, 74-81.
9. Hem, J. D. *Chem. Geol.* **1978**, *21*, 199-218.
10. Hem, J. D. In *Particulates in Water;* Kavanaugh, M. C. and Leckie, J. O., Eds.; Advances in Chemistry Series; **1980**, *189*, 45-72.
11. Murray, D. J.; Dillard, J. G. *Geochim. Cosmochim. Acta* **1979**, *43*, 781-787.
12. Jardine, P. M.; Taylor, D. L. *Geochim. Cosmochim. Acta* **1995**, *59*, 4193-4203.
13. Crowther, D. L.; Dillard, J. G.; Murray, J. W. *Geochim. Cosmochim. Acta* **1983**, *47*, 1399-1403.
14. Hem, J. D. *Geochim. Cosmochim. Acta* **1981**, *45*, 1369-1374.
15. Hem., J. D.; Roberson, C. E.; Fournier, R. B. *Water Resour. Res.* **1982**, *18*, 563-570.
16. Hem, J. D.; Lind, C. J. *Geochim. Cosmochim. Acta* **1983**, *47,* 2037-2046.
17. Murray, J. W.; Dillard, J. G.; Giovanoli, R.; Moers, H.; Stumm, W. *Geochim. Cosmochim. Acta* **1985**, *49*, 463-470.
18. Taylor, D. L.; Jardine, P. M. *J. Environ. Qual.* **1995**, *24*, 789-792.
19. Bricker, O. *Am. Mineralogist* **1965**, *50*, 1296-1355.
20. Marcus, M. A.; Flood, W. *Rev. Sci. Instrum.* **1991**, *62*, 839-840.
21. Waldo, G. S.; Carlson, R. M. K.; Moldowan, J. M.; Peters, D. E.; Penner-Hahn, J. E. *Geochim. Cosmochim. Acta* **1991**, *55*, 801-814.
22. Vairavamurthy, A.; Zhou, W.; Eglinton, T.; Manowitz, B. *Geochim. Cosmochim. Acta* **1994**, *58*, 4681-4687.
23. Pickering, I. J.; Brown, G. E., Jr.; Tokunaga, T. K.. *Environ. Sci. Technol.* **1995**, *29*, 2456-2459.

Photochemical and Microbially Mediated Processes

Chapter 19

Adsorption and Sensitization Effects in Photocatalytic Degradation of Trace Contaminants

T. David Waite[1], Stephan J. Hug[2], and Andrew J. Feitz[1]

[1]School of Civil and Environmental Engineering, The University of New South Wales, Sydney, New South Wales 2052, Australia
[2]Swiss Federal Institute for Environmental Science and Technology (EAWAG), CH-8600 Duebendorf, Switzerland

While many aqueous contaminants are recognized to be amenable to semiconductor-catalyzed photodegradation, studies have generally been performed under well-defined conditions free of the complexities of natural waters. In this chapter, results of recently reported studies into the semiconductor-catalyzed photodegradation of the blue-green algal toxin microcystin-LR in the presence of additional algal exudates are summarized and a proposed conceptual model for toxin degradation described. The veracity of the proposed model is tested by determining species concentrations for assumed initial conditions and kinetic constants. Very similar dependencies and trends to those observed in the laboratory studies were obtained suggesting that adsorption and sensitization effects are critical. In particular, under conditions where the contaminant adsorbs strongly to semiconductor surface sites, the primary degradation step appears to involve reaction between surface-located long-lived organic radicals and adsorbed trace contaminant. These long-lived radicals may also diffuse into solution where they may react with dissolved contaminant under conditions where contaminant adsorption is not significant.

Many organic contaminants present in natural waters have been shown to be susceptible to oxidative degradation by semiconductor-catalyzed photolysis and considerable progress has been made in elucidating the kinetics and mechanism of degradation of a wide range of individual compounds of environmental concern through detailed laboratory investigations (*1-3*). Trace contaminants, however, typically occur in natural waters in the presence of a wide range of other constituents, many of which are present at concentrations orders of magnitude higher than the contaminant. These other constituents, which may include humic and fulvic acids and bacterial and algal exudates, are likely to be surface active compounds which, like the specific trace contaminant of concern, may also be susceptible to semiconductor-mediated photo-transformation. While it is obviously critical to understand the transformation dynamics operating in such composite systems, very little progress has been made because of their inherent complexity.

We have recently reported the results of TiO_2-catalysed photodegradation studies in one such composite system involving the breakdown of a blue-green algal toxin in the presence of a large excess of ill-defined algal exudate (Feitz, A.J.; Waite, T.D.; Jones, G.J.; Boyden, B.H.; Orr, P.T. *Environ. Sci. Technol.*, submitted). For these studies, exudate containing approximately 3 wt% of the highly toxic hepatotoxin microcystin-LR (*4, 5*) (for which the structure is shown in Figure 1 and abbreviated below as MLR) was separated from a cyanobacterial bloom *of Microcystis aeruginosa* and the removal of MLR from solution was investigated under a wide range of system conditions. Results of dark adsorption studies of both the toxin and algal exudate (abbreviated below as AE) to the semiconductor particles and of light-enhanced removal of the toxin from solution as a function of system pH provided the basis for a qualitative model of the major processes operating in this system.

In this chapter, we examine the proposed model in more detail and develop quantitative solutions for the concentrations of constituents as a function of time in a hypothetical system similar to that examined in the laboratory. The insights developed from the results of such modeling should assist in developing a firmer understanding of the relative importance of processes such as adsorption, sensitization as well as competition effects in the semiconductor-catalyzed degradation of contaminants in complex aqueous environments.

Laboratory Investigations of Microcystin-LR TiO_2-Catalysed Photodegradation

Experimental Approach. As described in detail by Feitz et al. (*Environ. Sci. Technol.*; submitted) near-UV lamp experiments were performed in a water-jacketed borosilicate glass vessel incorporating a quartz window for illumination. Near-UV light was supplied by a 100 W high-pressure Hg arc lamp with a 365 nm band pass filter. The titanium dioxide solution consisted of Milli-Q water and Tioxide Australia TiO_2 (principally anatase) with an average crystal size of approximately 100 nm and a surface area of 10 m^2g^{-1}. The temperature of the reaction vessel was kept constant at 20.0°C using a recirculating refrigerated water bath and pH was adjusted with 5% HNO_3 or 0.1 M NaOH depending on the required pH before injection of the microcystin-LR/algal exudate mixture. One hour dark adsorption experiments were conducted in the same reaction vessel and under similar conditions to the photodegradation studies.

The algal extract used contained 73 μg/mL of microcystin-LR and a dissolved organic carbon content of 2200 μg/mL. The organic matter present in addition to microcystin-LR was presumably a complex mixture of algal exudates.

This crude mixture was used in all experiments rather than pure microcystin-LR both because of the very high cost of pure microcystin-LR and because this is the mixture that would be released from a naturally lysing population of cyanobacteria. Previous studies (*6*) have shown that other cyanobacterial exudates enhance the rate of microcystin photooxidation but these rates are small compared with the rate of TiO_2 catalyzed photooxidation.

High Performance Liquid Chromatography was used to determine microcystin-LR concentrations in both adsorption and photodegradation studies. The presence of possible breakdown products of photodegraded MLR toxin was investigated by GC-MS analysis. Details of these techniques are given elsewhere (Feitz, A.J.; Waite, T.D.; Jones, G.J.; Boyden, B.H.; Orr, P.T. *Environ. Sci. Technol.*, submitted).

Results. The concentrations of MLR remaining in pH 6.4 solutions over time in the absence and presence of TiO_2 and in the absence and presence of light are shown in Figure 2. It is clear that essentially no degradation occurs in the absence of either the catalyst or light while an initial rate of degradation of MLR of 6×10^{-11} M/s is observed in the presence of both catalyst and light and, under such conditions, the concentration of MLR is reduced from 85 nM to 1 nM in approximately 30 minutes. These results support the hypothesis that a TiO_2 catalyzed photodegradation process accounts for the reduction in concentration of microcystin-LR and suggests that homogeneous photodegradation of MLR (as previously noted by Tsuji et al. (*6*)) is not a significant process under these conditions.

Effect of pH. As can be seen from Figure 3, the rate of degradation increases with decreasing pH down to a pH of around 3 below which the toxin appears to be slightly more stable. An exponential decrease in MLR concentration with time is observed under acidic pH conditions. Under alkaline conditions, significantly slower initial rates of MLR decay are observed than under acidic conditions. In these high pH cases, an initial lag in degradation is observed with a subsequent increase in rate of MLR decay.

The dramatic effects of pH are clearly shown in Figure 4 where initial rates of photodegradation (assuming first-order in all cases) are plotted as a function of suspension pH. Significantly, the pH dependence of the initial MLR photodegradation rate is approximately mirrored by the extent of MLR adsorption to TiO_2 as a function of pH. Under alkaline conditions, where toxin adsorption to TiO_2 appears to be minimal, the initial degradation rates are essentially zero though, as noted above, significant degradation is observed at later times (Figure 3). Feitz et al. (*Environ. Sci. Technol.*; submitted) have shown that a simple surface complexation model adequately describes the effect of pH on MLR adsorption to TiO_2 (solid line in Figure 4).

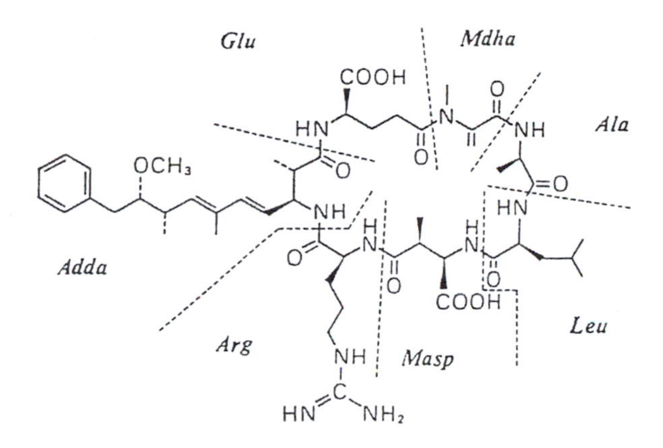

Figure 1. Structure of the blue-green algal toxin microcystin-LR. Besides the two variable L-amino acids, leucine and arginine, the microcystin contains three D-amino acids (glutamic acid, alanine and methylaspartic acid) and two unusual amino acids, N-methyl-dehydroalanine (Mdha) and 3-amino-9-methoxy-2,6,8-trimethyl-10-phenyl-deca-4,6-dienoic acid (Adda) (see Ref. 4).

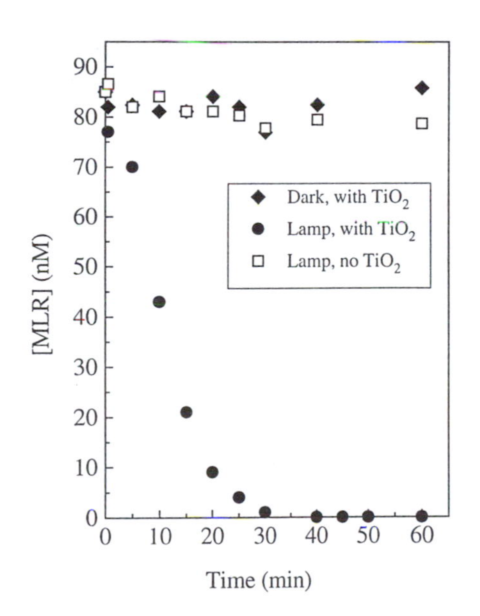

Figure 2. Loss of microcystin-LR from solution at pH 6.4 over time in the presence and absence of added TiO_2 and in the dark and 365 nm light. All runs at initial toxin concentration of 85 nM and TiO_2 loading of 1 g/L except for Lamp with TiO_2 run where $[TiO_2]$ = 0.5 g/L (optimised catalyst concentration). (Adapted from Feitz et al.; *Environ. Sci. Technol.*; submitted).

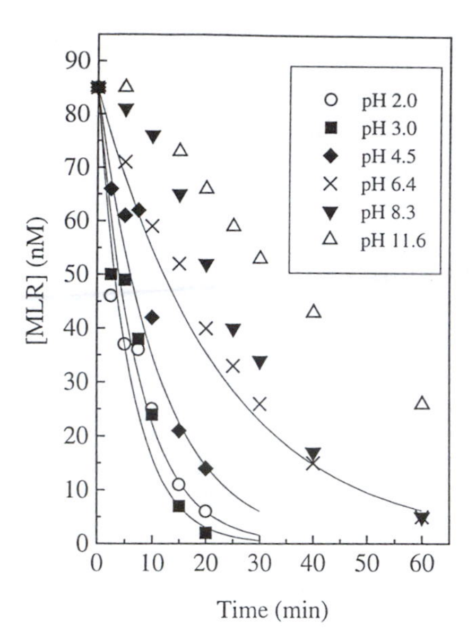

Figure 3. Loss of microcystin-LR from solutions of various pH. All runs at initial toxin concentration of 85 nM and TiO_2 loading of 1 g/L. Solid lines represent exponential functions of best fit to data for pH 2.0 to 6.4. (Reproduced from Feitz et al., *Environ. Sci. Technol.*; submitted).

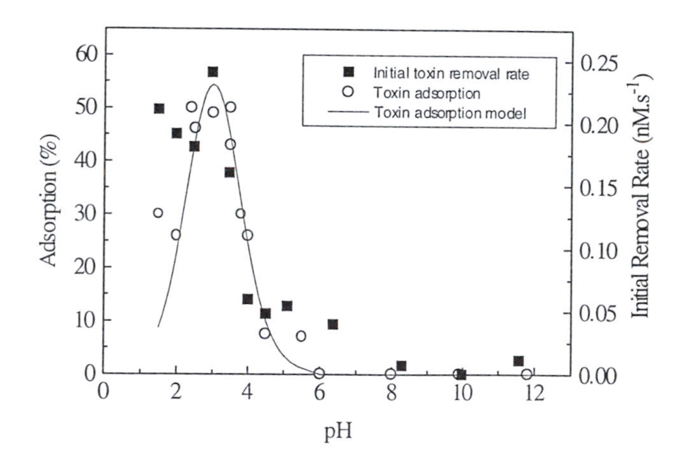

Figure 4. Percentage of 85 nM microcystin-LR removed at various pH after one hour contact time with 1 g/L TiO_2 in the dark. The solid line represents the percentage of 85 nM MLR adsorption to 1 g/L TiO_2 as a function of pH predicted using a surface complexation model. The initial removal rate of microcystin-LR from solution in the light is also shown as a function of pH. (Reproduced from Feitz et al., *Environ. Sci. Technol.*; submitted).

While not always justified (*7*), it has been common under conditions where adsorption is the (rapid) precursor to a rate controlling surface-mediated degradative step, to assume that the initial rate of contaminant removal is dependent upon both an equilibrium adsorbed contaminant concentration and the concentration of surface-located oxidizing species (*1, 3*). Under these conditions, an expression for the initial rate of removal of microcystin-LR may be written as:

$$\frac{-\,d[MLR]}{dt} = k\,.[>TiMLRH]\,.[oxidising\ species] \qquad (1)$$

where >TiMLRH represents a protonated toxin species at the titanium oxide surface. While little is known about the structure of this surface-located species, formation of an inner-sphere surface complex would appear possible.

If the concentration of the oxidizing species rapidly attains some steady state concentration, this expression may be simplified to the pseudo-first order form

$$\frac{-\,d[MLR]}{dt} = k'\,.[>TiMLRH] \qquad (2)$$

Thus, in accord with the results shown in Figure 4, a dependence is to be expected between the rate of toxin removal and the concentration of adsorbed toxin.

In reality of course, contaminant adsorption is unlikely to be instantaneous and assumption that sorptive equilibrium is reached is likely to be flawed. The extent of departure of the actual kinetic behavior from that predicted by equation 2 will be dependent, at least in part, upon the relative rates of adsorption and degradation.

Effect of concentration of toxin (and other algal exudate). While the pH dependencies of MLR adsorption and initial toxin photodegradation rates mirror each other closely, the dependencies of extent of toxin adsorption and photodegradation behavior on added toxin concentration appear to be rather dissimilar. As shown in Figure 5, increasing uptake on increasing MLR concentration is initially observed but the extent of adsorption plateaus or "saturates" at higher toxin concentrations. This adsorption data is adequately described by a Langmuir isotherm expression; i.e.

$$[MLR]_{ads} = \frac{K_{ads}.[MLR]_{ads}^{max}.[MLR]_{soln}}{(1 + K_{ads}.[MLR]_{soln})} \qquad (3)$$

where $[MLR]_{soln}$ and $[MLR]_{ads}$ are the concentrations of toxin in solution and on the solid respectively after one hour contact with the solid, K_{ads} is the conditional equilibrium constant (for formation of an adsorbed toxin species) and $[MLR]_{ads}^{max}$ represents the maximum concentration of toxin that may be adsorbed to the TiO_2 surface (within the one hour equilibration time allowed). Estimates for K_{ads} of $10^{6.8}$ M^{-1} and for $[MLR]_{ads}^{max}$ of 128 nM (at pH 3.5) are obtained by linearization and regression of the data (Feitz et al., *Environ. Sci. Technol.*; submitted). Given that 1 $g.L^{-1}$ of TiO_2 has been used in these adsorption studies, this represents a maximum adsorption density of 128 nmoles of toxin per gram of TiO_2 at pH 3.5. A site density of 2.4 $sites.nm^{-2}$ has been reported for TiO_2 (*8*) thus, under the conditions of this study, no more than 0.3% of total available surface hydroxy sites (approx. 4.0 x 10^{-5} $moles.L^{-1}$) will be occupied by toxin molecules.

In comparison with the Langmuirian behavior of MLR adsorption, the rate of MLR loss from solution is observed to decrease on increasing the concentration of added toxin (and, concomitantly, other algal exudate). Thus, as shown in Figure 6, the pseudo-first order rate constants (k') obtained from best fit exponential functions to the pH 3.5 results exhibit a decrease from over 0.2 min^{-1} at an initial MLR concentration of 50 nM to around 0.04 min^{-1} for a toxin concentration of 400 nM.

Feitz et al. (*Environ. Sci. Technol.*; submitted) have proposed that these effects are related to competitive phenomenon involving the large amount of non-specific algal exudate present and the trace concentrations of toxin. As shown in Figure 7, this organic matter adsorbs significantly to the colloidal TiO_2 both at low and high pH. Rectangular hyperbolae may be fitted to this sorption data yielding conditional adsorption constants of $10^{2.3}$ $L.g^{-1}$ and $10^{2.1}$ $L.g^{-1}$ and maximum adsorption capacities of 2.1 $mg.L^{-1}$ and 1.7 $mg.L^{-1}$ at pH 3.5 and 8.6 respectively. These results indicate that, even at pH 3.5, only a fraction of the available TiO_2 surface sites will be occupied by algal exudate. Thus, if the algal exudate possesses an average molecular weight of 300, less than 17.5% of available sites will be occupied by exudate. This proportion reduces to less than 5.3% if the mean molecular weight of the exudate is 1000.

Qualitative Model of Microcystin-LR Removal from Solution

A generic conceptual model has been proposed by Feitz et al. (*Environ. Sci. Technol.*; submitted) to account for the results described above.

Both microcystin-LR and other algal exudate are considered to sorb reversibly to TiO_2 surface hydroxy groups; i.e.

$$> TiOH + MLRH^{-} + H^{+} \leftrightarrow > TiMLRH + H_2O \quad (4)$$

$$> TiOH + AEH^{-} + H^{+} \leftrightarrow > TiAEH + H_2O \quad (5)$$

While debate still exists concerning the nature of the primary oxidizing species on TiO_2 (*2, 9*), oxidation is assumed to occur via hydroxyl radicals which

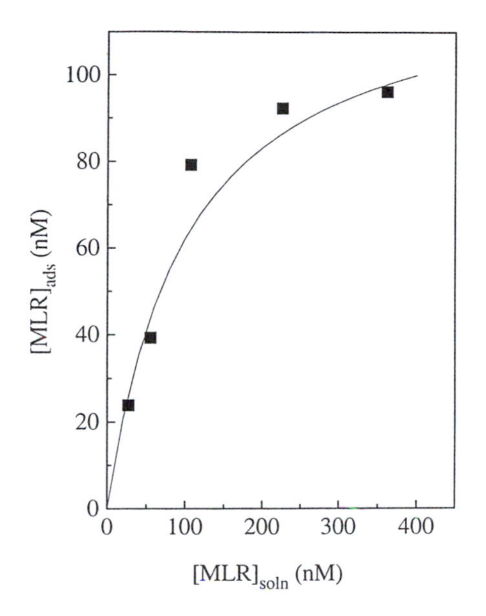

Figure 5. Concentration of MLR adsorbed to 1 g/L TiO_2 at pH 3.5 as a function of the solution concentration of the toxin. The solid line through the data represents the best fit as determined by linear transformation of equation 3 and regression of the data. (Reproduced from Feitz et al., *Environ. Sci. Technol.*; submitted).

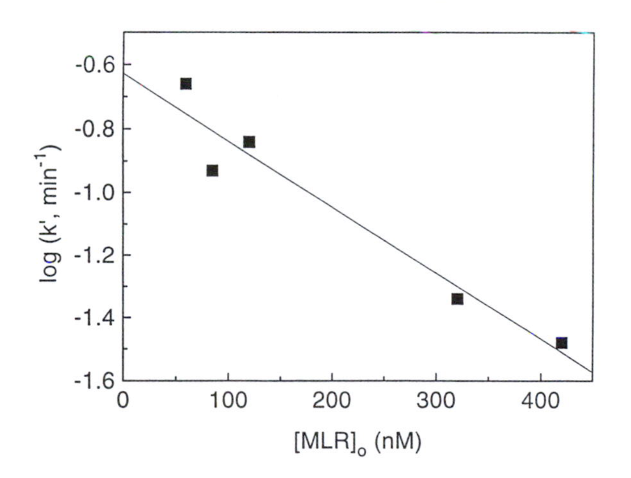

Figure 6. Change in pseudo-first order rate constant of microcystin-LR removal from solution at pH 3.5 as a function of initial MLR concentration. (Adapted from Feitz et al., *Environ. Sci. Technol.*; submitted).

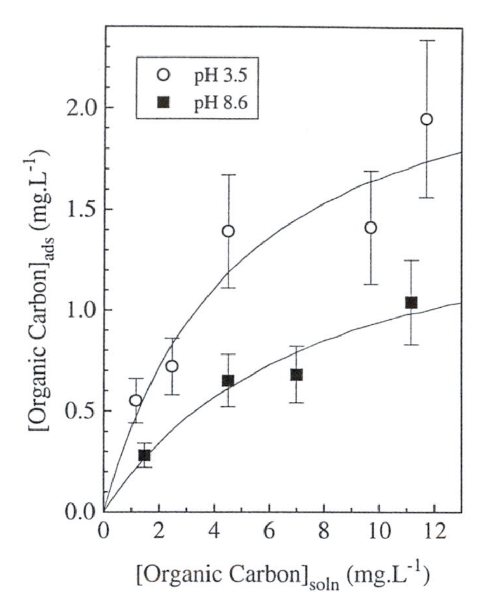

Figure 7. Adsorption isotherms for uptake of organic carbon (principally algal exudate) onto 1 g/L TiO_2 from solutions of pH 3.5 and 8.6. (Reproduced from Feitz et al., *Environ. Sci. Technol.*; submitted).

are trapped at the surface by titanol groups (*3, 9*) and represented here as >TiOH$\bullet$; i.e.

$$TiO_2 + h\nu \rightarrow \ > TiOH\bullet \tag{6}$$

The rate of recombination of these surface hydroxyl radicals with conduction band electrons is expected to be relatively slow (*3*) enabling interaction with other readily available (adsorbed) oxidizable species including toxin and algal exudate; i.e.

$$> TiOH\bullet \ + \ > TiMLRH \rightarrow \ > TiMLR\bullet^{-} \ + \ > TiOH_2^{+} \tag{7}$$

$$> TiOH\bullet \ + \ > TiAEH \rightarrow \ > TiAE\bullet^{-} \ + \ > TiOH_2^{+} \tag{8}$$

Note that while we have written equations 7 and 8 as H abstraction reactions, the processes could also occur by electron transfer however no information is available enabling differentiation between these two mechanisms. The algal exudate radicals so generated at the semiconductor surface might be expected to either react with other oxidizable species at the surface (such as adsorbed MLR) or desorb to solution:

$$> TiAE\bullet^{-} \ + \ > TiMLRH \rightarrow \ > TiMLR\bullet^{-} \ + \ > TiAEH \tag{9}$$

$$> TiAE\bullet^{-} \rightarrow \ > TiOH \ + \ AE\bullet^{-} \tag{10}$$

$$AE\bullet^{-} + O_2 \rightarrow AEO_2\bullet^{-} \tag{10a}$$

The observed decrease in toxin removal rate that is observed in the acidic pH range on increase in exudate concentrations (Figure 6) suggests that release of exudate radicals to solution (equation 10) is a more significant process than reaction with MLR at the surface (equation 9). While reaction of the released exudate radicals with oxygen is likely (equation 10a) the extent of algal exudate peroxyl radical formation is unknown.

MLR radicals generated at the surface (or in solution) are likely to be unstable and apparently readily decompose (on the basis of our analytical results). Feitz et al. (*Environ. Sci. Technol.*; submitted) have proposed that the relatively long-lived algal exudate radicals released to solution (which may include exudate peroxyl radicals) are deactivated by reaction with other readily oxidizable species such as dissolved toxin (equations 12 and 12a), by reaction with other components of the ill-defined algal exudate (equation 13) or simply by decomposition (equation 13a); i.e.

$$> TiMLR\bullet^{-} \rightarrow \ > TiOH \ + \ \text{MLR degradation products} \tag{11}$$

$$MLRH^{\cdot} + AE\bullet^{-} \rightarrow AE + MLR \text{ degradation products} \quad (12)$$

$$MLRH^{\cdot} + AEO_2\bullet^{-} \rightarrow AE + MLR \text{ degradation products} \quad (12a)$$

$$AE\bullet + AE\bullet \rightarrow AE \text{ degradation products} \quad (13)$$

$$AEO_2\bullet^{-} \rightarrow O_2\bullet^{-} + AE \text{ degradation products} \quad (13a)$$

While this conceptual model is highly simplified and contains many speculative components, it would appear to provide a useful basis for more detailed analysis of the photocatalysed degradation of trace contaminants in the presence of a large excess of other oxidizable material.

Approach to Quantitative Modeling of Semiconductor-Catalyzed Contaminant Photodegradation

In order to assess whether the proposed model described above is capable of generating results similar to those obtained by Feitz et al. (*Environ. Sci. Technol.*; submitted), we have defined a hypothetical system similar to that examined by Feitz et al. and varied critical parameters in order to simulate the conditions used. Concentrations of the various species present have then been determined (for the range of system conditions of interest) using the ACUCHEM computer program (*10*). The ACUCHEM input file used for these calculations is given in the Appendix and features of the hypothetical system used are described below.

Initial Concentrations of Toxin and Exudate. An initial contaminant (toxin) concentration of 10^{-7} M was used with the concentration of the ill-defined exudate set at 100 times this value. This approximately mirrors the initial conditions used by Feitz et al. (*Environ. Sci. Technol.*; submitted) if the average molecular weight of the exudate were 300 $gm.mole^{-1}$. These initial concentrations were used in all calculations except where the impact of changing exudate concentrations was examined. In this case, $[MLR]_o$ was varied between 5×10^{-8} M and 5×10^{-7} M and $[AE]_o$ varied between 5×10^{-6} M and 5×10^{-5} M.

Adsorption of Toxin and Exudate to TiO_2. The pH dependent tendency of the toxin and exudate to adsorb to the semiconductor surface was modeled by varying the conditional formation constants for toxin and exudate surface complexes. This was done by setting the forward reaction rates for formation of these surface complexes at fixed values and varying the back (desorption) reaction rates (see reactions 1 and 2, Appendix). The range of conditional formation constants used as well as the resulting (dark) equilibrium proportions of toxin and exudate adsorbed to the 3×10^{-5} M of available TiO_2 surface sites assumed in these calculations are given in Table I.

Table I. Proportion of toxin and exudate adsorbed[a] to TiO_2 for various assumed conditional formation constants for toxin and exudate surface complexes.

Case	$\log K^{cond}_{TiMLR}$	% MLR Adsorbed	$\log K^{cond}_{TiAE}$	% AE Adsorbed
1	6.70	58.1	4.00	21.7
2	6.22	32.1	3.81	15.9
3	5.70	12.6	3.70	12.6
4	4.70	1.4	3.60	10.4
5	3.70	0.1	3.52	8.8
6	2.70	0	3.45	7.7

[a] $[TiOH]_{total} = 3 \times 10^{-5}$ M, $[MLR]_o = 10^{-7}$ M and $[AE]_o = 10^{-5}$ M.

As is the case in the experimental investigation described above, extent of toxin adsorbed varies from over 50% to zero while the exudate exhibits less variation in extent of adsorption with a significant proportion adsorbed (7-21%) in all cases examined.

Production of Primary Oxidizing Species. As noted above, it is assumed here that the primary oxidizing species are surface-located hydroxyl radicals that are formed by the trapping of valence band holes by >TiOH surface groups. It is presumed that a steady-state concentration of oxidizing species will be established for a given TiO_2 loading and light intensity. In all calculations undertaken here, a steady-state concentration of surface-located hydroxyl radicals (represented as TiOHr in the ACUCHEM model) of 10^{-9} M is assumed.

Surface-located Radical Reactions. As proposed in the conceptual model above, a number of competing reactions involving either the primary oxidizing species (the surface-located hydroxyl radicals) or secondary, longer-lived products are envisaged to occur at the particle surface. Thus, fast and indiscriminate scavenging of surface hydroxyl radicals by the toxin and exudate is envisaged with proposed forward rate constants of 10^{15} $M^{-1}.s^{-1}$. These very high bimolecular rate constants reflect the high local concentrations of adsorbed species. With a surface area of 10 $m^2.L^{-1}$ and an estimated layer thickness of 10 nm in which adsorbed compounds are considered to be located, we obtain a volume of 10^{-4} L for adsorbed species. Thus, local surface concentrations must be on the order of 10^4 larger than if they are averaged over the suspension volume. Bimolecular reaction rate constants should reflect this by being, formally, a factor of 10^8 larger than for similar reactions in solution.

Given the large excess of non-specific exudate over toxin at the TiO_2 surface, it is expected that the bulk of the surface hydroxyl radicals will react with exudate rather than toxin. It is proposed that the exudate free radicals so produced, which are envisaged to be relatively long-lived, may either react with surface-bound toxin (with a bimolecular rate constant of 10^{13} $M^{-1}.s^{-1}$) or may be (relatively rapidly) released to solution (forward rate constant of 10^{6} s^{-1}. While not included, a third possible mode of deactivation of surface-located exudate free radicals involves interaction between the exudate free radicals and adsorbed exudate. Such a process may create a "second generation" of reducible exudate radicals at the surface but a lowered reactivity due to time delays and inefficiencies in energy transfer would be expected. Note that, given the likely insignificance of the direct oxidation of adsorbed MLR by hydroxyl radicals, reaction of toxin with exudate free radicals is expected to be the dominant MLR degradation pathway. It is proposed that toxin molecules rapidly leave the surface once oxidized (forward rate constant of 10^{9} s^{-1}).

Solution-phase Radical Reactions. A relatively slow reaction between soluble toxin molecules and released exudate radicals is envisaged (bimolecular rate constant of 10^{3} $M^{-1}.s^{-1}$). These radicals may also be scavenged by other exudate radicals (bimolecular rate constant of 10^{3} $M^{-1}.s^{-1}$ assumed) or the exudate itself (reaction not included). These solution phase processes will be particularly important under conditions (such as high pH) where the toxin molecules do not adsorb strongly to the semiconductor surface.

Results and Discussion of Quantitative Modeling

The results of application of the above model to our hypothetical system are shown in Figures 8 to 10.

Effect of Varying Extent of Toxin and Exudate Adsorption. Simulation of the effect of pH in modifying the extent of adsorption of both exudate and (particularly) toxin to the TiO_2 surface results in rapid (pseudo-first order) removal of toxin from solution under conditions where toxin adsorbs strongly to TiO_2 (Figure 8). In a manner identical to that observed in the "real" system (Figure 3), toxin is only degraded after a significant time lag under conditions where MLR adsorption to the semiconductor surface is weak or nonexistent.

A comparison of the extent of toxin removal under dark and light conditions for cases of high and low dark adsorption of toxin to the semiconductor surface is shown in Figure 9. For the "high MLR adsorption"` case (58.1% and 21.7% MLR and AE adsorbed to TiO_2 respectively), pseudo-first order removal of toxin from solution is observed with essentially all toxin degraded in the first 30 minutes. This result closely matches that observed by Feitz et al. (*Environ. Sci. Technol.*; submitted) at pH 3.5. In the "low adsorption" case (0% and 7.7% MLR

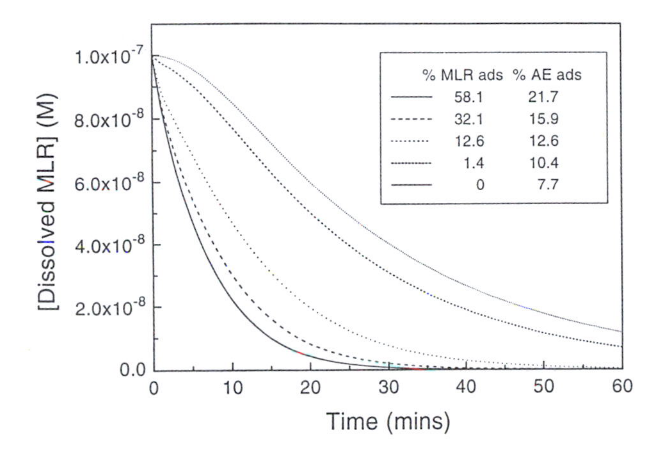

Figure 8. Effect of varying affinity of MLR and algal exudate for TiO_2 surface sites on rate of removal of toxin from solution. The cases modeled are identical to those defined in Table I and to the initial conditions and rate constants detailed in the Appendix.

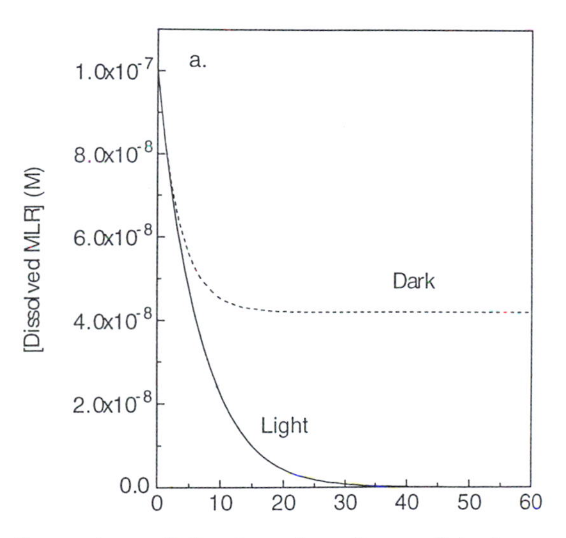

Figure 9. Comparison of the time dependence of toxin removal under conditions of a) strong and b) weak adsorption of toxin and algal exudate to the semiconductor as defined in Table I. a) 58.1% and 21.7% MLR and AE adsorbed to TiO_2 respectively (Case 1) and b) 0% and 7.7% MLR and AE adsorbed respectively (Case 6).

Continued on next page.

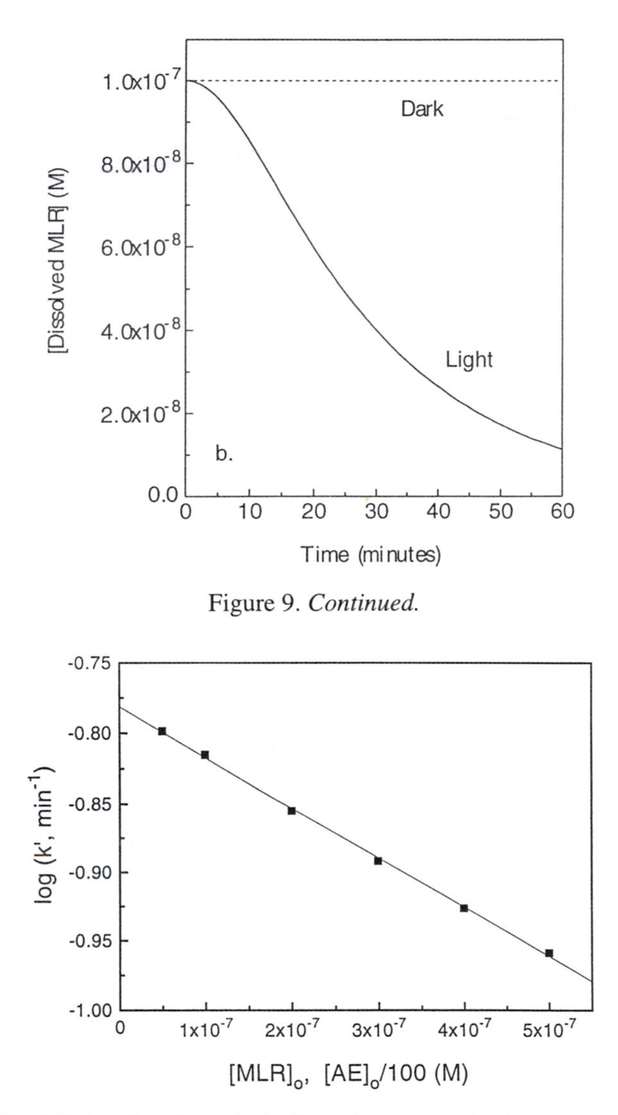

Figure 9. *Continued.*

Figure 10. Effect of toxin and algal exudate concentrations on pseudo-first order rate constant for removal of toxin from solution.

and AE adsorbed respectively), one hour of photolysis is required to degrade approximately 90% of the toxin.

Effect of Toxin and Exudate Concentrations. As found in studies on the crude toxin and exudate extract (Figure 6), increasing the concentration of these reactants in the hypothetical system results in a reduction in the pseudo-first order rate constant for toxin removal under conditions where adsorption to the TiO_2 surface is strong (Figure 10). This lower than proportional increase in toxin removal rate on increasing toxin and exudate concentrations presumably occurs because of the concentration dependence of surface-located exudate radical loss to solution (equation 10). That is, while increasing toxin and exudate concentrations results in an increase in rate of production of exudate radicals and a subsequent increase in rate of degradation of toxin (principally via reaction with exudate radicals), the increase is lower than might be expected since the rate of loss of exudate radicals to solution is also increased.

While the large number of parameters that might be varied in our hypothetical example renders too detailed an analysis superfluous, it is observed that the extent of reduction in the pseudo-first order rate constant for toxin removal on increasing MLR and AE concentrations is somewhat less than found in the laboratory investigations (Figure 10 cf. Figure 6). The greater extent of impact of increased toxin and exudate concentrations in the "real" compared to hypothetical cases may reflect the presence of additional concentration dependent pathways for deactivation of exudate free radicals that are unaccounted for in our relatively simple hypothetical system. As noted earlier, one such deactivation pathway may involve the scavenging of exudate free radicals by other exudate molecules adsorbed to the semiconductor surface. Despite this small difference, the results of modeling our hypothetical system with a relatively simple mechanistic model and a number of assumed (but not unreasonable) rate constants produces results that closely resemble those obtained in laboratory investigations.

Conclusions

Results of recently reported studies into the semiconductor-catalyzed photodegradation of the blue-green algal toxin microcystin-LR have been summarized in this chapter and a proposed conceptual model for removal of toxin from solutions containing high concentrations of additional ill-defined organic matter described. It is suggested that this model provides a useful generic framework for interpreting the semiconductor-catalyzed degradation of trace contaminants in complex natural waters.

In order to assess the veracity of the proposed conceptual model, a hypothetical example generally simulating the conditions used in the toxin study and involving assumed initial conditions and kinetic constants has been defined and quantitative solutions for all species present determined as a function of time.

Very similar dependencies and trends to those observed in the toxin/exudate laboratory studies were obtained suggesting that:

- adsorption of trace contaminant and "bulk" organic is an important precursor to rapid degradation of the adsorbed contaminant;
- the adsorbed ill-defined organic matter readily scavenges surface-generated hydroxyl radicals with the resultant production of relatively long-lived organic radicals;
- reaction between these long-lived radicals and adsorbed trace contaminant represents the major pathway for degradation of the adsorbed contaminant;
- increasing concentration of (ill-defined) organic matter at the semiconductor surface creates increased opportunities for deactivation of the oxidizing species (organic free radicals) present;
- these long-lived radicals may diffuse into solution where it is proposed they react (initially slowly due to the low reactant concentrations) with dissolved contaminant thus inducing contaminant degradation under conditions where adsorption of the contaminant to the semiconductor surface is not significant.

Acknowledgements

Financial support from DEET through the award of an APA-I scholarship to AJF is gratefully acknowledged as is the funding provided by the Hunter Water Corporation. Resources provided to TDW during a brief stay at EAWAG courtesy of Barbara Sulzberger were invaluable in producing this chapter, as were the facilities at The Australia Centre, University of Potsdam.

Literature Cited

1. Ollis, D.F.; Pelizzetti, E.; Serpone, N. In *Photocatalysis: Fundamentals and Applications*; Serpone, N.; Pelizzetti, E., Eds; John Wiley & Sons: New York 1989; pp 603-637.
2. Bahnemann, D.; Cunningham, J.; Fox, M.A.; Pelizzetti, E.; Pichat, P.; Serpone, N. In: *Aquatic and Surface Photochemistry* ; Helz, G.R., Zepp, R.G., Crosby, D.G., Eds.; Lewis Publishers: Boca Raton, 1994, p 216.
3. Hoffman, M.R.; Martin, S.T.; Choi, W.; Bahnemann, D.W. *Chem. Rev.* **1995**, 95, 69.
4. Carmichael, W.W.; Mahmood, N.A.; Hyde, E.G. In *Marine Toxins: Origin, Structure, and Molecular Pharmacology* . Hall, S., Strichartz, G, Eds; American Chemical Society: Washington, DC, 1990, p 87.
5. Nishiwaki-Matsushima, R.; Ohta, T.; Nishiwaki, S.; Suganama, M.; Kohyama, K.; Ishikawa, T.; Carmichael, W.W.; Fuuki, H. *J. Cancer Res. Clin. Oncol.* **1992**, 118, 420.
6. Tsuji, K.; Nalto, S.; Kondo, F.; Ishikawa, N.; Watanabe, M.F.; Suzuki, M.; Harada, K-H. *Environ. Sci. Technol.* **1994**, 28, 173.
7. Herrmann, V.M.; Boehm, H.P. *Z. Anorg. Allg. Chem.* **1969**, 368, 73.

8. Cunningham, J.; Al-Sayyed, G.; Srijaranai, S. In: *Aquatic and Surface Photochemistry*; Helz, G.R., Zepp, R.G., Crosby, D.G., Eds.; Lewis Publishers: Boca Raton, 1994, p 317.
9. Kesselman, J.M.; Weres, O.; Lewis, N.S.; Hoffmann, M.R. *J. Phys. Chem. B* **1997**, 101, 2637.
10. Braun, W.; Herron, J.T.; Kahaner, D. ACUCHEM Computer Program for Modelling Complex Reaction Systems; National Bureau of Standards: Gaithersburg, Maryland, 1986.

Appendix

ACUCHEM input file for hypothetical system involving TiO_2 catalysed photodegradation of trace contaminant (toxin, MLRH) in the presence of a higher concentration of organic material (algal exudate, AEH).

```
input
0010
;                                  MLRH Toxin Degradation
; Sept.13, 1997
; This model roughly reproduces dark adsorption of MLRH and photodegradation
; Surface toxin complex formation
4a, TiOH + MLRH = TiMLRH,1e2
4b, TiMLRH = TiOH + MLRH,  2.0e-3  6.0e-3  2.0e-2  2.0e-1  2.0e+0  2.0e+1
; Surface exudate complex formation
5a, TiOH + AEH = TiAEH,1e4
5b, TiAEH = TiOH + AEH, 1.0  1.5  2.0  2.5  3.0  3.5
;  Production of TiOHr from light absorption by bulk TiOH
06, =TiOHr,1e-9
; Surface toxin degradation
07, TiOHr+TiMLRH= TiMLRr,1e15            ; fast and indiscriminate reaction
08, TiOHr+TiAEH = TiAEr,1e15             ; of TiOHr with TiAEH and TiMLRH
09, TiAEr + TiMLRH = TiMLRr+TiAEH,1e13        ; necessary unless k6>100*k7
10, TiAEr = TiOH + AEr,1e6                    ; surface detachment of AEr
11, TiMLRr =TiOH, 1e9
; Solution toxin degradation
12, MLRH + AEr = AEH,1e3
13, AEr+AEr=,1e3                         ; necessary to prevent builup of AEr
end
; 2 sites/nm2  1.5g/l  10m2/g
TiOH, 30e-6                  ; with ca. 2 sites/nm2
AEH,10e-6                    ; always 100*c(MLRH) assuming MW=300g/mol
MLRH,100e-9                  ; 100ug/L with MW=1000g/mol
end
0.001
3600
end
&
6
MLRH
TiMLRH
AEH
AEr
TiOH
TiAEH
```

Chapter 20

Microbially Mediated Oxidative Precipitation Reactions

B. M. Tebo and L. M. He

Marine Biology Research Division and Center for Marine Biotechnology and Biomedicine, Scripps Institution of Oceanography, University of California at San Diego, La Jolla, CA 92093–0202

The oxidation of iron(II) and manganese(II) usually leads to the formation of insoluble metal oxides, hydroxides, and oxyhydroxides which are of wide-ranging environmental importance. Microorganisms catalyze the oxidation of Fe(II) and Mn(II), however, the mechanisms of metal oxidation, particularly in the case of Mn(II) oxidation, and the effect microorganisms have on the properties of the solid phases, are not well known. The state of knowledge concerning Fe and Mn biomineralization is briefly reviewed. Considerable progress has been made in recent years in understanding the mechanisms and products of Mn(II) oxidation by bacteria. Results obtained using classical and modern approaches, including biochemical, molecular biological, mineralogical and stable oxygen isotopes have provided new insights into bacterial Mn(II) oxidation.

Iron (Fe) and manganese (Mn) oxides (a collective term meant here to include oxides, hydroxides, and oxyhydroxides) are recognized as reactive mineral components in soils, sediments, and aquatic systems. They adsorb a variety of ions and participate in oxidation and reduction reactions with inorganic and organic species and compounds. Microorganisms, especially bacteria, are known to participate in Fe and Mn mineralization processes, however, the importance of these "iron bacteria" and "manganese bacteria" in environmental processes is often overlooked. In addition, because of the difficulty in working with many of these organisms (many of them have never been cultured or some grow only very slowly in culture) and the complex chemistry of Fe and Mn, there is a lack of knowledge concerning the physiology and biochemistry of these microbes. Recently, new insights into bacterial Mn(II) oxidation have been realized through modern approaches using molecular biological and stable isotopic methods. This paper presents general background information on Fe and Mn chemistry and microbiology and a more detailed focused description of recent results concerning a model Mn(II)-oxidizing bacterium, the marine *Bacillus* sp. strain SG-1.

Geochemistry of Fe and Mn and their minerals

Iron and Mn, the two most abundant transitional metals in the Earth's crust (Fe 5.1 wt%, Mn 0.12 wt%), bear many geochemical similarities. This is due in part to the different valence states in which both commonly can been found in the environment.

Iron is the most important metal in the universe. It is considered the most abundant element in the earth as a whole and it ranks fourth in abundance in the crust after oxygen (O), silicon (Si), and aluminum (Al). It occurs as a major or minor constituent of all mineral classes. Of the major chemical elements in the Earth's crust, Fe is unusual in that it occurs in two valence states, i.e., ferrous [Fe(II)] and ferric [Fe(III)], under environmental conditions. The fact that it may be oxidized or reduced in nature markedly influences its geochemical cycle. Fe(II) occurs as insoluble hydroxides, carbonates, phosphates or sulfides. Only under highly acidic or reducing conditions do significant amounts of soluble Fe(II) species occur. Fe(III), the oxidation state favored under oxidizing conditions, usually forms solid phase Fe oxides (oxides, hydroxides, and oxyhydroxides) in nature, although soluble Fe(III) can exist at acidic pHs or as organic complexes. Iron plays a critical role in the biosphere. In animals it acts in transporting oxygen from air or water to the body tissues. In green plants Fe is necessary for the formation of chlorophyll. Some chemolithoautotrophic bacteria possess an enzyme system that transfers electrons from ferrous iron to O_2, thus freeing them from dependence on organic matter as an energy source.

Manganese exists in a number of oxidation states, among which the II, III and IV oxidation states are of greatest environmental importance. Mn(II) can occur as insoluble carbonate or sulfide species or as soluble ions, the form available to organisms. Because the activation energy for Mn(II) oxidation is high, soluble Mn(II) can occur at fairly high concentrations in natural waters, even in the presence of O_2. Mn(III) is thermodynamically unstable and does not occur in soluble form except in the presence of strong complexing agents such as humic or other organic acids. Mn(III) and Mn(IV) primarily form insoluble oxides and oxyhydroxides. Mn is a major biogenic micronutrient required in the photosynthetic process and as an enzymatic cofactor.

Kinetics of Fe(II) and Mn(II) oxidation. The dissolved concentrations of Fe and Mn in soils and sediments is controlled by pH and Eh. These two master variables govern the dissolution of Fe and Mn oxides and the oxidation of dissolved Fe(II) and Mn(II) in natural environments. The oxidation of Fe(II) and Mn(II) by O_2 are thermodynamically favorable, but kinetically may be very slow. Thus, certain reduced species such as Fe^{2+} and Mn^{2+} are often present in surface or subsurface waters along with dissolved O_2, an indication that conditions of nonequilibrium are common. The oxidation of Fe(II) and subsequent precipitation of ferric hydroxide is rapid at neutral pH but the oxidation of Fe(II) is very slow below pH 6. The rate law of oxidation of dissolved Fe(II) is known to be (*1*)

$$\frac{-d[\mathrm{Fe(II)}]}{dt} = k[\mathrm{Fe(II)}][\mathrm{OH^-}]^2 p_{O_2}$$

where $k \approx 8.0 \times 10^{13}$ min^{-1} atm^{-1} $mole^{-2}$ $liter^{2}$ at 20°C. The rate of Fe(II) oxidation is first-order with respect to the concentrations of both Fe(II) and O_2 and second-order with respect to the OH^- ion. Thus, a 100-fold increase in the rate of reaction occurs for a unit increase in pH. Some cations (e.g., Cu^{2+} and Co^{2+}) in trace quantities as well as anions (e.g., HPO_4^{2-}) increase the reaction rate significantly. For a given pH, the rate increases about tenfold for a 15°C temperature increase. The oxidation of Fe(II) in seawater is much slower than in freshwater, indicating the effect of ionic strength on the rate.

Although Mn(II) in solution is thermodynamically unstable with respect to oxidation by O_2 at any pH higher than 4, the rate of uncatalyzed oxidation is not appreciable unless the pH is well above 8. The rate of Mn(II) oxidation may be expressed as

$$\frac{-d[\mathrm{Mn(II)}]}{dt} = k_0[\mathrm{Mn(II)}] + k[\mathrm{Mn(II)}][\mathrm{MnO_2}][\mathrm{OH^-}]^2 p\mathrm{O_2}.$$

That the rate depends on MnO_2, the product of Mn(II) oxidation, indicates that the reaction is autocatalytic (*1*). As Mn oxides precipitate, oxidation accelerates in response to the increased surface available for selective adsorption of Mn^{2+} ions. There is also evidence that Mn and Fe tend to co-precipitate in oxides, possibly because Fe oxide surfaces also catalyze the oxidation of Mn^{2+}. The products of the heterogeneous oxidation of Mn(II) at hematite, goethite, and albite surfaces are Mn(III)-bearing oxyhydroxides, predominantly β-MnOOH (feitknechtite) after 6 months at room temperature in aerated solutions containing 4.0-26.7 ppm Mn(II) at pHs ranging from 7.8 to 8.7 (*5*).

Mineralogy of Fe and Mn Oxides. Iron is very reactive in surface environments and both Fe(II) and Fe(III) can form stable compounds. Most of the Fe derives from the weathering of Fe(II)-bearing primary minerals (e.g., pyroxenes, amphiboles, biotites, and olivines) in the magmatic rocks of the Earth's mantle. The Fe released from minerals will be reprecipitated in the environment as various secondary oxide minerals (Table I), depending on environmental conditions (e.g., pH, Eh, temperature, and solution composition) (*6*). We only give a brief description of a few important Fe minerals occurring commonly in soils and sediments. Further information concerning Fe oxides in the environment can be found in review papers (*6-8*). Ferrihydrite has been shown to be common in environments and exists as very small, poorly ordered crystals (~2-5 nm in size) with a large surface area (159-720 $m^2\ g^{-1}$) (*9*), which makes it very important in controlling the fate and transport of organic and inorganic contaminants in surface and ground waters. Ferrihydrite forms in the environment by biotic or abiotic oxidation and subsequent hydrolysis of dissolved Fe(II). The biotic oxidation process is catalyzed by microorganisms such as *Gallionella, Leptothrix*, and algae (*6*). Ferrihydrite is the least stable of the Fe(III) minerals and will spontaneously transform to goethite. Goethite, a yellow-brown colored oxyhydroxide, is widespread in soils and sediments and is usually associated with one or more of the other Fe minerals such as hematite, maghemite, lepidocrocite, and ferrihydrite. Goethite precipitates from solution via nucleation-crystal growth process. Goethite formation is favorable at low temperature and high water activity. Lepidocrocite, less frequent than goethite in soils and sediments, has a bright orange color. The carbonate concentration has been demonstrated to determine the ratio of lepidocrocite to goethite formation. Higher carbonate concentrations suppress lepidocrocite formation in favor of goethite (*10, 11*). This may indicate the importance of biota in goethite vs. lepidocrocite

Table I. Iron oxide minerals formed in surface and subsurface environments

Mineral	*Formula*
Ferrihydrite	$Fe_5HO_8 \cdot 4H_2O$ (2)
Goethite	α-FeOOH
Lepidocrocite	γ-FeOOH
Feroxyhyte	δ'-FeOOH
Akaganeite	β-FeOOH
Hematite	α-Fe_2O_3
Maghemite	γ-Fe_2O_3
Magnetite	Fe_3O_4

formation. Hematite, a red-colored Fe oxide mineral, forms by structural rearrangement and dehydration of ferrihydrite, not by dehydration of goethite. Low water activity, low organic matter content, high temperature, and high rate of Fe release from rocks favor the formation of hematite. A recent study has demonstrated that hematite can result from microbial transformation of magnetite (*12*). Magnetite can be inherited from the parent rock or formed biologically in the environment by magnetotactic (*13, 14*) or iron-reducing bacteria (*15*). Magnetite can also be precipitated from the reaction of ferrihydrite (or any other reactive Fe(III) oxide) with dissolved Fe(II) (*6*).

Mn(II), originally released through weathering of igneous rocks in surface or subsurface environments, can be oxidized to Mn(III, IV), resulting in the formation of Mn(III, IV) (hydr)oxides, oxyhydroxides, or manganates (Table II). The manganates are Mn(III,IV) minerals that have highly negatively charged surfaces, open crystal structures, and contain a variety of charge-balancing, low valence cations (e.g., Na, Mg, Ca, Cu, Ni, etc.). A few Mn oxide minerals contain only Mn(II) but most are found to be compounds of Mn(IV) with some Mn(III). Since Mn(III, IV) minerals in soils and sediments are often poorly crystallized or amorphous, characterization is difficult and frequently incomplete with diffraction techniques (*16-18*). Recent studies using X-ray absorption fine structure (XAFS) spectroscopy enable specific species to be distinguished (*19-21*). Several comprehensive reviews on Mn mineralogy have been published (*18, 22-24*).

Table II. Manganese oxide minerals formed in surface and subsurface environments

Mineral	Formula
Ramsdellite	MnO_2
Hollandite (α-MnO_2)	$Ba_x(Mn^{4+}Mn^{3+})_8O_{16}$ ($x \approx 1$)
Pyrolusite (β-MnO_2)	MnO_2
Nsutite (γ-MnO_2)	MnO_2
Vernadite (δ-MnO_2)	MnO_2
Groutite (α-MnOOH)	MnOOH
Feitknechtite (β-MnOOH)	MnOOH
manganite (γ-MnOOH)	MnOOH
Hausmannite	$Mn^{2+}Mn^{3+}{}_2O_4$
Bixbyite (α-Mn_2O_3)	Mn_2O_3
Pyrochroite	$Mn(OH)_2$
Manganosite	MnO
Birnessite	$Na_{.7}Ca_{.3}Mn_7O_{14}\cdot 2.8H_2O$
Chalcophanite	$ZnMn_3O_7\cdot 3H_2O$
Coronadite	$Pb_x(Mn^{4+}Mn^{3+})_8O_{16}$ ($x \approx 1$-1.4)
Cryptomelane	$K_x(Mn^{4+}Mn^{3+})_8O_{16}$ ($x \approx 1.3$-1.5)
Lithiophorite	$LiAl_2(Mn^{4+}Mn^{3+})_3O_6(OH)_6$
Manjiroite	$Na_xMn_8O_{16}$
Romanechite	$Ba_{0.66}(Mn^{4+}Mn^{3+})_5O_{10}\cdot 1.34H_2O$
Todorokite	$(Ca, Na, K)_{0.3-0.5}(Mn^{4+}Mn^{3+})_6O_{12}\cdot 3.5H_2O$

Most Mn(III, IV) minerals have crystal structures that fall into one of three major groups: Chain, tunnel, and layer structures (*22-24*). The chain structures consist of cross-linked single or double rows of MnO_6 octahedra running parallel to the **c** axis (e.g., pyrolusite). The tunneled structures are formed from two or more cross-linked chains of octahedra running parallel to the **c** axis in minerals with tetragonal symmetry (e.g., hollandite) or parallel to the **b** axis in minerals having monoclinic symmetry (e.g., todorokite). The layer structures consist of layered sheet of Mn octahedra (e.g., buserite). While Burns and Burns (*22*) distinguish between chain and tunnel type Mn minerals, the primary difference between them is the tunnel diameter. For example, the layered manganate minerals, birnessite and buserite, are distinguished based on their characteristic *d*-value between their octahedral sheets of 7.3 Å and 10 Å, respectively. Buserite has an extra layer of water molecules lying between the octahedral sheets which give it a wider *d*-spacing. Dehydration of buserite results in the formation of birnessite.

Many studies have been made on the synthesis of Mn minerals, and the transformation of one form to another (*23*). Abiotic Mn(II) oxidation begins with the precipitation of Mn(III)-bearing oxides (e.g., Mn_3O_4) or hydroxides (MnOOH), which subsequently disproportionate slowly to form Mn(VI) oxides (e.g., MnO_2; Table III). The disproportionation reaction is believed to be the rate controlling step in the abiotic formation of MnO_2 (*25*). The oxidation of $Mn(OH)_2$ in alkaline suspension forms buserite, which on dehydration gives birnessite. Birnessite may transform to nsutite, and finally to pyrolusite, and large amounts of foreign ions prevent the formation of both nsutite and pyrolusite. Birnessite containing many foreign cations are readily transformed to hollandite type minerals upon heating. Tian et al. (*26*) recently reported the abiotic synthesis of mixed-valent semiconducting, catalytically active Mn oxides, which have mesoporous structures, high thermal stability, and catalyze the oxidation of cyclohexane and n-hexane in aqueous solutions. This group includes Mn_2O_3 and Mn_3O_4 in which the crystal structures are based on various arrangements of MnO_6 octahedra.

Table III. Mn(II) oxidation reactions.

Reaction	*Oxidation State of Product*
1) $Mn^{2+} + {}^1/_2O_2 + H_2O \rightarrow MnO_2 + 2H^+$	+4
2) $Mn^{2+} + {}^1/_4O_2 + {}^3/_2H_2O \rightarrow MnOOH + 2H^+$	+3
3) $3Mn^{2+} + {}^1/_2O_2 + 3H_2O \rightarrow Mn_3O_4 + 6H^+$	+2.67
4) $Mn_3O_4 + 2H^+ \rightarrow 2MnOOH + Mn^{2+}$	+3
5) $Mn_3O_4 + 4H^+ \rightarrow MnO_2 + 2Mn^{2+} + 2H_2O$	+4
6) $2MnOOH + 2H^+ \rightarrow MnO_2 + Mn^{2+} + 2H_2O$	+4

Surface chemistry. Fe and Mn oxides together with aluminum oxides are important mineral phases in soils, sediments, and waters where they usually occur as particle coatings or as discrete particles. They are very reactive components of natural environments because of their small particle size, large surface area, and variable surface charge, which are responsible for their ability to affect the fate, transport, and bioavailability of a variety of heavy metals and organic compounds in the environment. Knowledge of the surface area of a solid is more important than the mass in understanding and interpreting its ion sorption properties. There are two categories of methods for surface area measurements: 1) physical determination of the size and morphology of solid particles and 2) measurement of the adsorption of gas (e.g., N_2) or solute molecules (e.g., dye) having known dimensions and interpretation of the resulting data with a particular adsorption model (*27*). The adsorption techniques are practically used to determine specific surface areas of oxides since the physical methods cannot be applied routinely. Specific surface areas of some synthetic Fe and Mn oxides measured by N_2 adsorption (BET) are listed in Table IV. Ferrihydrite and Mn oxides have much higher surface areas than most of the others in Table IV.

Table IV. Specific surface area, site density, and zero point of charge of synthetic Fe and Mn oxides

Mineral	*Surface area* ($m^2 g^{-1}$)	*Site density* (*sites/nm²*)	*ZPC*	*References*
Ferrihydrite	200-300	0.1-0.9 moles per mole Fe	7.9-8.2	(*9*)
			9.4	(*28*)
Goethite	43-81			(*29-31*)
				(*32, 33*)
		2.6-16.8	7.7	(*34-36*)
			9.4	(*28*)
Hematite	6-116			(*37*)
	18-31			(*38*)
		5-22	8.4	(*34-36*)
			8.5	(*28*)
Lepidocrocite	108			(*39*)
Magnetite	5			(*40*)
			6.5	(*35, 36*)
			7.1	(*28*)
Pyrolusite			3.7	(*28*)
Nsutite	150			(*41*)
δ-MnO_2	300			(*42*)
	206	18	2.3	(*43*)
	137	20.5	2.3	(*44*)

The surface charge of Fe and Mn oxides develops from surface functional groups (hydroxyl groups) capable of complex formation with inorganic and organic ions. Some important parameters such as surface site density and zero point of charge (Table IV) are used to characterize the surface chemical properties of Fe and Mn oxides. Surface site density refers to the number of surface functional groups (sites) per unit of surface area and is usually expressed as sites per squared nm (sites nm^{-2}). A high surface site density for most of the Fe and Mn oxides suggests a high ion adsorption capacity. The zero point of charge (ZPC) is the pH value at which the net surface charge density of an oxide mineral is zero. The net surface charge is positive when a suspension pH < ZPC, while negative when the pH > ZPC. The ZPC of Mn oxides is significantly lower than that of the Fe oxides, suggesting they will be more effective scavengers of cations at low pH. Conversely, the Fe oxides should have greater anion adsorption capabilities at low pH.

Solubility. The processes of dissolution and precipitation of Fe and Mn oxides control the mobility of Fe and Mn in soils and sediments and their availability to organisms. Dissolution of Fe and Mn oxides is a function of pH. Fe oxides decrease in solubility in the order: Ferrihydrite (amorp) > maghemite > lepidocrocite > hematite > goethite. The solubility of Mn oxides decrease in the order: pyrochroite > hausmannite > bixbyite > manganite > birnessite > nsutite > pyrolusite.

Adsorbing species such as organic compounds can accelerate the dissolution of Fe and Mn oxides, which is referred to as ligand-promoted dissolution. Generally, ligands that bind strongly to a given metal in solution have been found to accelerate the dissolution of the oxide of the same metal. For example, the dissolution rate of Fe(III) oxides is enhanced markedly by oxalate and ascorbate that form surface chelates with the oxide surfaces (*45, 46*). Mn oxides are dissolved by a variety of simple aromatic and nonaromatic compounds having structures similar to organic materials in natural waters (*47-50*).

Role of Fe and Mn oxides in the fate and transport of contaminants in surface and subsurface environments. Fe and Mn oxides, widespread in soils and sediments and having small particle sizes and large surface areas, play a critical role in the fate and transport of toxic metals and organic compounds in the environment through sorption and redox reactions. Fe and Mn oxides adsorb a wide range of inorganic anions and cations and organic compounds. Numerous studies have demonstrated that sorption processes are important in natural systems. In rivers, sorption processes can control the dissolved concentrations of solutes, give rise to high concentrations of toxic metals in the bed load, and affect the discharge rates of solutes into estuarine and coastal marine systems. In lakes and oceans, sorption processes regulate the flux of elements from the water column to sediments and affect the availability and toxicity of trace elements to phytoplankton. The uptake of sorbed metals by filter feeding and burrowing organisms is a potential environmental risk in sediments. In soils and aquifers, sorption processes can retard the transport of solutes and result in secondary repositories, i.e., regions where contaminants are accumulated downgradient from a source, from which they can be released in response to a change in the geochemical environment.

Iron and Mn oxides serve as active reactants in abiotic redox transformations of toxic metals and organic compounds in soils, sediments, and waters. Myneni et al. (*51*), using X-ray absorption fine structure spectroscopy, demonstrated that Se(VI) reduces to Se(IV) and Se(0) in the presence of Fe(II,III) oxide called green rust. The Se(VI) transformation rates are within the range of those reported from laboratory and field studies on Se speciation on soils and sediments. The authors indicated that similar green rust-mediated abiotic redox reactions are likely to be involved in the mobility of other trace elements (e.g., Cr(VI) and As(V)) in the environment. Mn oxides are the strongest oxidizing agent encountered other than oxygen. At pH 7 they are considerably more oxidizing than Fe oxides. They promote the oxidation of As(III), Co(II), Cr(III), Pu(III), and possibly other trace metals (*52*). Mn oxides also catalyze the degradation of organics (*47-50*) and the formation of humic substances and organic N complexes (*52*).

Microbial Fe(III) and Mn(VI) Reduction. Although Fe(III) and Mn(IV) can be abiotically reduced to Fe(II) and Mn(II) by organics (e.g., organic acids and aromatic compounds) and inorganics (e.g. sulfide, U(IV)), microbial enzymatic reduction of these two elements is the major mechanism in most soils and sediments (*53*). Many microorganisms have been found to possess the capability of reducing Fe(III) and Mn(IV). Among them are *Pseudomonas* sp., *Shewanella putrefaciens, Desulfovibrio* sp., *Geobacter metallireducens*, *Desulfuromonas acetoxidans, Thiobacillus thiooxidans*, *T. ferrooxidans*, *Sulfolobus acidocaldarius*, and *Aquaspirillum magnetotacticum.* Microbial Fe(III) and Mn(IV) reduction requires direct cell-oxide contact, by which electrons transport from cells to the Fe or Mn oxide surfaces. The microbial reduction of Fe(III) and Mn(IV) is found to be most significant in anoxic environments since most Fe(III) and Mn(IV) reducers do not reduce Fe(III) and Mn(IV) in the presence of O_2. Nitrate inhibits Fe(III) but generally not Mn(IV) reduction in soils and sediments. Microorganisms can reduce both Fe(III) and Mn(IV) simultaneously. However, the zones of net Fe(III) and Mn(IV) reduction are generally separated in time or space, in part because when Mn(IV) is available, Fe(II) is rapidly oxidized back to Fe(III) through a strictly chemical reaction.

Fe and Mn precipitating microorganisms

Diversity of Fe(II) and Mn(II) oxidizing microorganisms. A variety of microorganisms, including bacteria, fungi, algae and protozoa, are capable of bringing about the oxidation and precipitation of Fe and Mn (*54*). These organisms are ubiquitous in a wide range of environments such as soils, sediments, and marine and fresh waters, essentially wherever Fe and Mn deposits occur. Because of this ubiquity these microorganisms are believed to strongly influence the cycling of Fe and Mn and the other trace elements that co-precipitate or adsorb to the oxides (*55*) (Figures 1 and 2). Together with the Fe and Mn reducing microorganisms, the Fe and Mn oxidizing bacteria have a profound influence on the biogeochemical cycles of Fe and Mn.

Iron and Mn precipitation generally occurs extracellularly by processes that are considered to be "biologically induced" (*56*). The products of these processes tend to be the mineral forms predicted from thermodynamic considerations based on Eh, pH and metal concentrations. For Fe, "biologically controlled" mineralization processes occur in some bacteria and result in the intracellular formation of different types of minerals. For example, the magnetotactic bacteria synthesize magnetosomes that contain a single magnetic-domain crystal of a magnetic Fe mineral that are bounded by a membrane (*57*). The magnetic minerals generally are magnetite or greigite. Other bacteria are known that form ferrihydrite intracellularly.

In most cases of Fe and Mn precipitation, there is no obvious benefit to the microorganisms. However, the fact that such a diverse group of organisms promotes Fe and Mn precipitation suggests that there must be some selective advantage for the organisms. In some cases, there is obvious benefit to the organisms. For example, in the magnetotactic bacteria the magnetosomes are used to orient the organisms in the magnetic field and allow them to swim to their preferred microaerobic or anoxic niches, usually in the oxic-anoxic transition zone. In other bacteria, the energy available from the oxidation of Fe(II) or Mn(II) may be used metabolically by the bacteria for growth or maintenance energy.

The diversity of bacteria that catalyze the oxidative precipitation of Fe and Mn is large. In general, these processes cannot be considered a phylogenetic trait. Organisms that oxidize and precipitate Fe are found in both the Domain Bacteria and in the Domain Archaea, while bacteria that oxidize Mn are found in two major branches within the Domain Bacteria. The so-called "iron bacteria" include both bacteria that grow at the expense of Fe(II) oxidation (the chemolithotrophs) and well as those that accumulate or deposit iron on their cell walls, sheaths, or appendages (Table V). The best examples of organisms that obtain energy through Fe(II) oxidation are the acidophilic Fe(II)-oxidizing bacteria, such as *Thiobacillus ferrooxidans*, that are associated with mining operations (sulfide ores or coal) and promote the acidification of the environment. These organisms have pH optima for growth of between pH 2-4 and are fairly well characterized at the physiological and biochemical level relative to other iron bacteria. They are completely capable of supplying all their energetic needs through Fe(II) oxidation and can derive all their cellular carbon from CO_2 via the key enzyme in the process, ribulose-1,5-bisphosphate carboxylase/oxygenase, the primarily CO_2-fixing enzyme found in autotrophic plants and in aerobic chemolithoautotrophic bacteria. The other Fe bacteria are neutralophilic and have growth optima between pH 6-8. At least some members of this group seem to obtain energy through Fe(II) oxidation (*58-60*), but the function of Fe oxidation and precipitation for the other members is unknown. Most of the Fe bacteria require molecular O_2 for Fe(II) oxidation. Recently, however, anoxygenic phototrophic bacteria have been shown to oxidize Fe(II) anaerobically using the electrons from Fe(II) for photosynthesis (*61, 62*). Anaerobic Fe(II) oxidation coupled to nitrate reduction under autotrophic conditions has been observed in several strains of bacteria including *Thiobacillus denitrificans* and *Pseudomonas stutzeri* (*63*), and a novel thermophilic Archaea, *Ferroglobus placidus* (*64*).

The Mn(II) oxidizing bacteria all grow and oxidize Mn(II) around neutral pH (6-8). Like the Fe(II) oxidizers, the Mn(II) oxidizers deposit Mn(III,IV) oxides on their surfaces. Many Fe and Mn oxidizers belong to the same genera (Table V) and, in fact, all Mn(II) oxidizers are also Fe(II) oxidizers, either by virtue of the fact that they catalyze Fe(II) and Mn(II) oxidation enzymatically or because the Mn oxides formed by

Table V. Examples of iron and manganese bacteria

Iron Bacteria	Manganese Bacteria
Acidophiles (pH 2–4)	All are neutralophiles (pH 6-8)
Thiobacillus ferrooxidans	Common Gram-negatives
Leptospirillum ferrooxidans	*Pseudomonas*
Sulfolobus	*Aeromonas*
Neutralophiles (pH 6–8)	*Oceanospirillum*
Common Gram-positives	*Vibrio*
Arthrobacter	Common Gram-positives
	Arthrobacter
	Bacillus
Sheathed bacteria	Sheathed bacteria
Leptothrix	*Leptothrix*
Clonothrix	*Clonothrix*
Crenothrix	
Budding or appendaged bacteria	Budding or appendaged bacteria
Hyphomicrobium	*Hyphomicrobium*
Pedomicrobium	*Pedomicrobium*
Metallogenium	*Metallogenium*
Gallionella	

Mn(II) oxidizers oxidize Fe(II) (Figures 1 and 2). Although there is some evidence that Mn(II)-oxidizing bacteria can obtain some energy for growth (*65-67*), Mn chemolithoautotrophy has not yet been demonstrated (*68, 69*). One Mn(II)-oxidizing bacterium (marine strain SI95-9A1) has been shown to possess the genes for ribulose-1,5-bisphosphate carboxylase/oxygenase (*70*). Although no enzyme activity could be detected in the strain, the genes are fully functional in *Escherichia coli* (*70*). Criteria to establish Mn chemolithoautotrophy in a bacterium have been discussed by Nealson et al. (*68*), however, in no case have all these criteria been fulfilled.

A number of benefits other than energy that microbes may receive by catalyzing Mn(II) oxidation have been proposed. The oxidation of Mn(II) may be a mechanism for metal detoxification, or to help protect the cell from toxic oxygen species (*71*). The metal-encrusted surfaces of Mn(II)-oxidizing bacteria may also provide increased resistance to ultraviolet radiation, predation and viral attack (*72*). Since Mn oxides may serve as alternate electron acceptors for anaerobic growth, Mn oxide accumulation may serve as a mechanism to store a useful electron acceptor for later use (*73, 74*). Finally, because Mn oxides are strong oxidizing agents that oxidatively degrade humic substances to low molecular weight organic compounds, it has been proposed that Mn(II) oxidation serves as a mechanism for the Mn(II)-oxidizing bacteria to indirectly obtain carbon for growth (*75*). Recent work has suggested that some marine Mn(II)-oxidizing bacteria may indeed obtain carbon and energy via this mechanism (*76*).

Mechanisms of Fe and Mn oxidation/precipitation. Fe and Mn oxidation can be directly or indirectly catalyzed by microorganisms. Direct oxidation consists of enzymatic or cell-associated processes, whereas indirect oxidation occurs through alterations in the environmental conditions that enhance abiotic (chemical) reactions. For example, indirect oxidation would occur through changes in the Eh and pH conditions in the environment around the cell. This could be brought about through the production of oxidants or bases, or consumption of acids. Another indirect mechanism of possible importance in the case of Fe precipitation is the utilization of iron-binding organic ligands, such as organic acids or siderophores, leading to the release of the Fe from the complex which would allow it to precipitate. Some of the "iron bacteria" may not be able to oxidize Fe(II) at all, but rather accumulate the oxidized forms of iron on their cell surface via adsorptive mechanisms (*54, 71*).

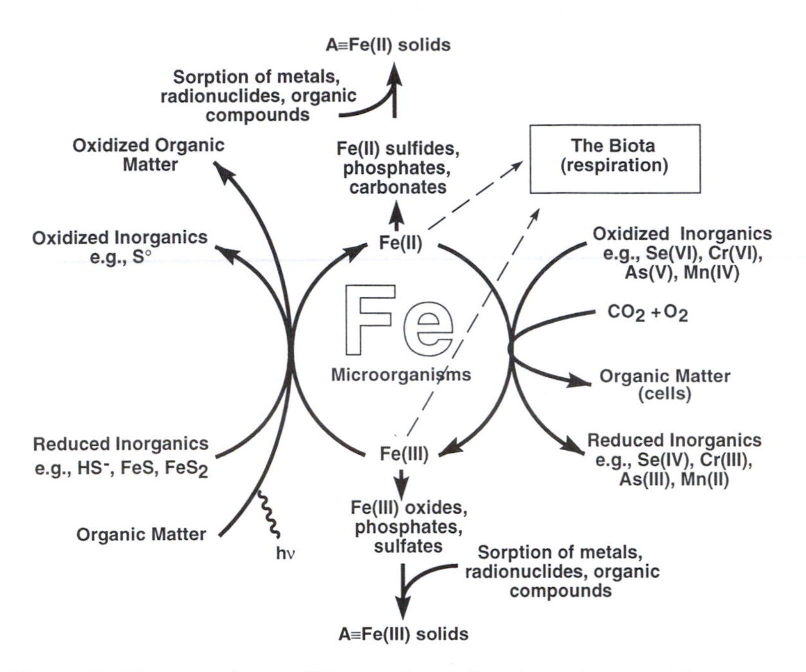

Figure 1. The central role of Fe transformations in environmental processes.

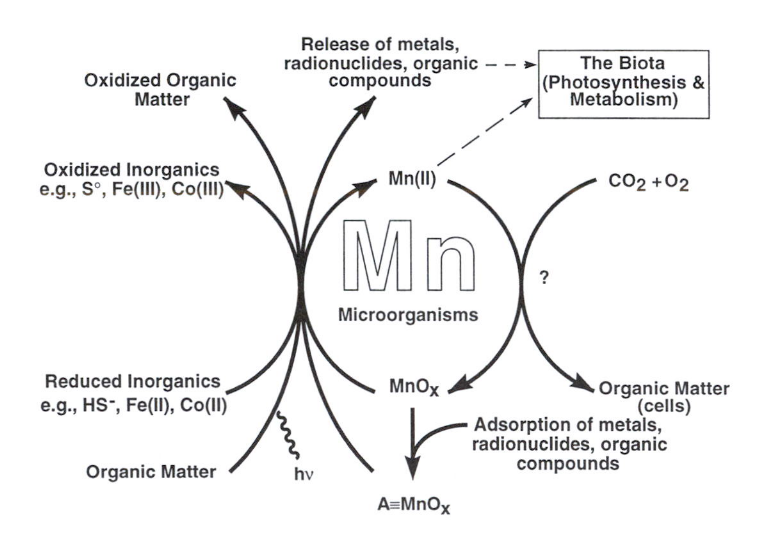

Figure 2. The central role of Mn transformations in environmental processes.

The biochemical mechanisms of Fe(II) oxidation by acidophiles and the current state of knowledge of iron oxidation by other iron bacteria and have been described by Ehrlich (*67*). Bacterial Mn(II) oxidation is described in more detail below; the reader is also referred to more detailed presentations of this topic (*67, 69*).

Environments with active Fe and Mn deposition

Fe and Mn precipitating organisms are widely distributed in nature and occur wherever soluble Fe and Mn species occur, from marine and freshwaters, to sediments, soils and desert varnish (*54*). They are most conspicuous in environments where soluble species are abundant, especially in the transition zone between anoxic and oxic conditions. Under anoxic conditions, reduced Fe(II) and Mn(II) accumulate and diffuse into the overlying oxygenated environment where metal oxidation occurs. These environments can be oxic/anoxic transition zones that occur in the water columns of stratified lakes, fjords, or marine basins, in a variety of freshwater and marine sediments, or in groundwater systems with high organic loads. If such environments exist in the presence of light, anoxygenic phototrophic bacteria can oxidize Fe(II) under completely anaerobic conditions. Even in oxic environments, reduced species of Fe(II) and Mn(II) can be generated through photochemical reactions (*77, 78*). Thus, another niche for Fe and Mn precipitating bacteria is the air-water interface where these organisms develop as surface films (*54*) or on the surfaces of rocks where they participate in the formation of rock varnishes.

The marine *Bacillus* sp. strain SG-1: A model system for studies of bacterial Mn(II) oxidation

The Mn(II)-oxidizing bacterium that has probably been studied in greatest detail with regard to mineralogy, genetics, and biochemistry is the marine *Bacillus* sp. strain SG-1 (Figure 3). Strain SG-1 is a sporeformer that was isolated from a Mn(II)-oxidizing enrichment culture from a marine sediment sample (*79*). Because in SG-1 it is the mature, free spores that oxidize Mn(II) on the spore surface (*80*) and thus growth does not have to be sustained, this organism has proved to be an excellent model system for studies of the microbiology and associated geochemistry of Mn(II) oxidation. This section describes some of the new insights that have been gained from the study of this organism.

Mineralogy of bacterially-produced Mn oxides. The oxidation states of synthetic Mn oxides formed in the laboratory at neutral pH tend to be low, around 3 (*5, 81, 82*). Since Mn oxides formed in many natural waters through microbial activities have a considerably higher oxidation state (≥3.4) (*83-87*), it is tempting to speculate that abiotic oxidation does not result in the same oxidation products as those found in nature (*88*).

Using SG–1 spores for the microbial production of Mn oxides, Hastings and Emerson (*89*) observed the formation of Mn(IV) minerals in seawater (pH 7.5) containing moderate (2 μM) concentrations of Mn(II). Hausmannite (Mn_3O_4) appeared to initially form, but with time there were gradual increases in the oxidation state. These results were consistent with Hem and Lind's model (*90*) that Mn(IV) minerals form by a two step process involving first the formation of Mn_3O_4 which subsequently disproportionates under low Mn(II) concentrations (*89*).

Mandernack et al. (*91*) extended the studies of Hastings and Emerson (*89*) and more systematically examined the Mn oxidation state and mineralogy of the Mn oxides formed in low ionic strength buffer and in natural seawater under various environmentally-relevant temperatures and Mn(II) concentrations. Experiments were conducted with SG–1 spores in HEPES buffer and HEPES buffered seawater, at pH 7.4-8.0, in 10 μM to 10 mM Mn(II), and at 3°C, 25°C, and 50-55°C. After two

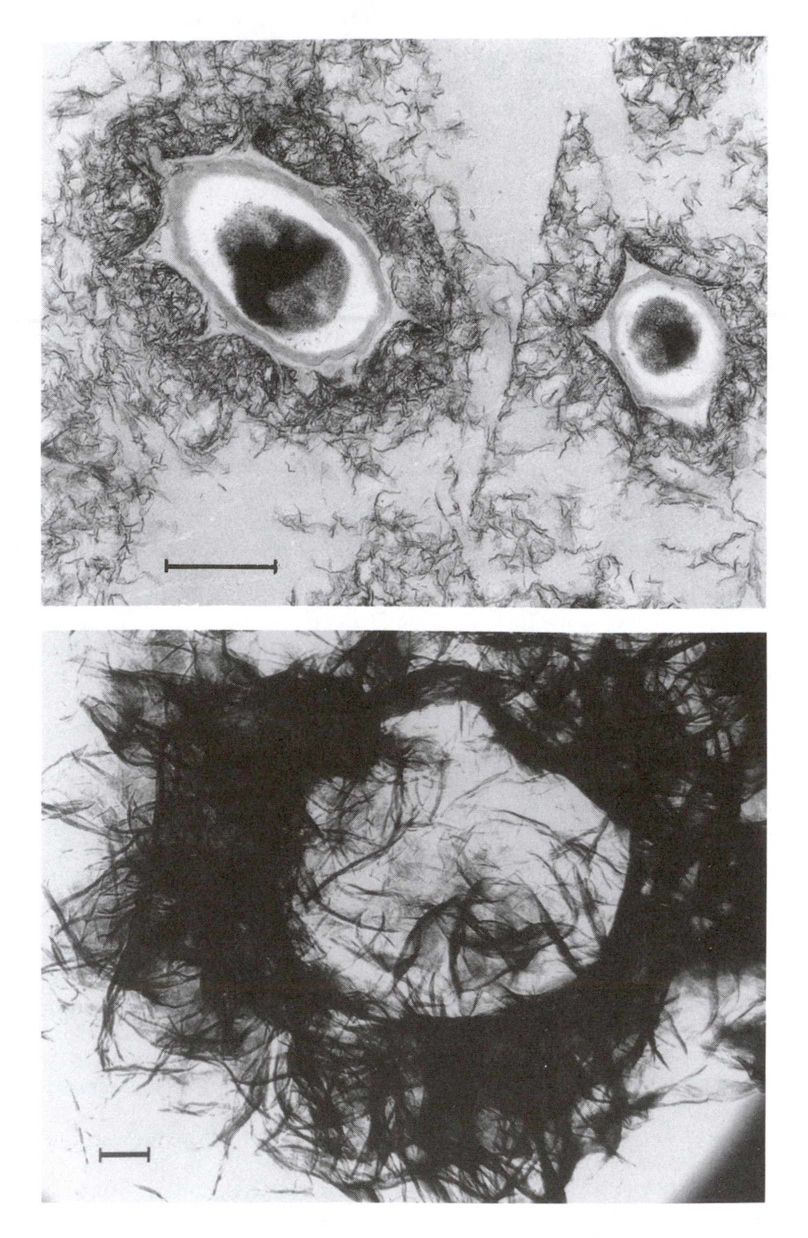

Figure 3. Top: Transmission electron micrograph of a thin section of the metal-oxidizing spores of the marine *Bacillus* sp. strain SG–1. The spores are coated with manganese oxides. Bar = 0.5 μm. Bottom: A todorokite-like manganese oxide precipitate produced by SG–1 as observed directly in the transmission electron microscope. The precipitate has been extracted to remove the organic material of the spore. Bar = 0.1 μm. (Reproduced with permission from ref. *123*. Copyright 1995 Plenum Press.)

weeks at high Mn(II) concentrations or high temperature, mixed phases of lower valence state minerals (Mn_3O_4; feitknechtite, β-MnOOH; and manganite, γ-MnOOH) formed in both seawater and buffer. At lower Mn(II) concentrations Mn(IV) minerals were observed. In low ionic strength buffer the Mn(IV) minerals most often resembled sodium buserite, as evidenced by collapse of a 10 Å to 7 Å phase upon drying. In seawater, both buserite and noncollapsible 10 Å manganates formed. In general, the Mn oxidation states of the minerals formed in seawater were higher than those formed in the buffered medium. In many cases Mn(IV) mineral formation occurred at pH and Mn(II) concentrations where the lower valence state minerals (Mn_3O_4, MnOOH) are thermodynamically stable with respect to disproportionation. These results suggested that SG–1 spores are capable of oxidizing Mn(II) to Mn(IV) without the formation of lower valence state solid phase intermediates.

Oxygen isotopic ($\delta^{18}O$) studies of chemically and microbially produced Mn oxides provided more direct evidence that Mn(II) is oxidized directly to Mn(IV) by SG–1 as well as another marine Mn(II)-oxidizing bacterium (*92*). Theoretically, the maximum amount of oxygen in Mn oxides that is derived from O_2 can be estimated from the stoichiometry of the Mn(II) oxidation reactions (see Table III). Thus, for oxidation reactions involving the formation of lower (II,III) Mn oxide intermediates, the maximum amount of oxygen that could be derived from O_2 is 25%. Since no further O_2 is incorporated upon disproportionation of these oxides to Mn(IV), the maximum signal from O_2 in Mn(IV) oxides formed by this two step pathway is 25%. In contrast, Mn(IV) oxides formed directly (Reaction 1 in Table III) could have up to 50% of the oxygen derived from O_2. Hausmannite (Mn_3O_4), synthesized either chemically or microbially, did not incorporate oxygen atoms from molecular O_2. Since disproportionation of Mn_3O_4 also would not incorporate oxygen atoms from O_2 (see reaction 5 in Table III), a Mn(IV) oxide formed by this pathway would have a $\delta^{18}O$ signature solely reflecting the water in which it formed. In contrast, the Mn(IV) oxides produced by both SG–1 and the Mn oxidizer showed that 40-50% of the oxygen in the Mn(IV) minerals is derived from O_2, consistent with the oxidation of Mn(II) to MnO_x by reaction 1 (Table III). These results suggest that Mn(II) oxidation by these bacteria occurs via a metal-centered oxidation in which Mn(II) is directly attacked by dissolved O_2 (or an intermediate derived from O_2) and electron transfer proceeds via an inner-sphere mechanism, as has been proposed on theoretical grounds (*93, 94*).

Environmental studies lend support to the laboratory studies that suggest that microbial Mn(II) oxidation results in the formation of Mn(IV) oxides without lower valence intermediates. Through analysis of a lake sediment, Wehrli et al. (*95*) found that the sediment is dominated by vernadite (δ–MnO_2), an X-ray-amorphous Mn(IV) oxide, and contained no reduced forms of Mn oxides (Mn_3O_4), hydroxides [($Mn(OH)_2$], or oxyhydroxides (MnOOH), as revealed by X-ray absorption spectroscopy. They concluded that the direct oxidation of Mn(II) to Mn(IV) by microorganisms provides an efficient transfer of oxidizing equivalents to the sediment surface. More recently Friedl et al. (*86*) examined the Mn oxide microbially formed at the oxic/anoxic boundary in the water column of a eutrophic lake. These authors identified the Mn mineral by extended X-ray absorption fine structure (EXAFS) spectroscopy as H^+–birnessite. The rapid formation of particulate Mn with high Mn oxidation state (> 3) above the oxic/anoxic interface of lakes and fjords also supports the direct Mn(II) → Mn(IV) mechanism (*83-85, 87*). Measurements of the $\delta^{18}O$ composition of Mn oxides in a freshwater ferromanganese nodule from Lake Oneida, NY indicate that 50% of the oxygen atoms in the Mn oxides are derived from O_2 providing evidence that they were microbially formed by the direct pathway without lower Mn oxidation state solid phase intermediates (*92*).

Genetics and Biochemistry. Early studies with purified SG-1 spores (*80*) showed that Mn(II) oxidation by the spores was inhibited by heat or various inhibitors, including mercuric chloride, sodium azide, and potassium cyanide. This work was suggestive of the involvement of a metalloprotein in Mn(II) oxidation. Ultrastructural studies of the spores by transmission electron microscopy showed that the Mn oxide was deposited on the outermost ridged layer of the spore (Figure 3, *72*). Indeed, de Vrind et al. (*73*) localized the Mn(II) oxidation activity to the outer spore layers and, like the whole spores, Mn(II) oxidation was inhibited by the same inhibitors. Oxygen was demonstrated to be required for Mn(II) oxidation (*73, 88*) and measurements of the consumption of O_2 and the production of protons during Mn(II) oxidation demonstrated that the stoichiometry for the Mn(II) oxidation reaction was consistent with reaction 1 in Table III (*73*)). Recent studies have shown that Mn(II)-oxidizing activity is localized in the exosporium (a membrane-like layer found external to the proteinaceous spore coats, compositionally distinct from the spore coats, but not present in all species). In SG–1 the exosporium is the ridged layer (Figure 3) which can be removed from the spores by passage through a French pressure cell (*95*). These studies, taken in total, suggested that a protein present in the exosporium is responsible for Mn(II) oxidation.

Studies of the biochemistry of another Mn(II)-oxidizing bacterium, *Leptothrix discophora*, have demonstrated the feasibility of identifying Mn(II)-oxidizing proteins by polyacrylamide gel electrophoresis (PAGE) and staining for Mn(II) oxidation activity by incubating the gels in Mn(II)-containing buffer (*96, 97*). By this approach a brown protein band from *L. discophora* with an apparent molecular weight of 110 kDa could be detected in the gel. A similar approach was attempted with SG–1 and although a high molecular weight Mn(II)-oxidizing protein band (~205 kDa) could occasionally be seen in the gels (*98*), recovery of activity from experiment to experiment was inconsistent.

A major breakthrough in the study of Mn(II) oxidation by SG–1 spores came through genetic studies (*99, 100*). van Waasbergen et al. (*99*) used transposon mutagenesis to obtain mutants of SG–1 that lost the ability to oxidize Mn(II) but still produced mature spores. In transposon mutagenesis, a transposon (transposable genetic element) is inserted randomly into different loci in the chromosome of the organism. The insertions disrupt the nucleotide sequence of the chromosome, thus mutating and/or inactivating genes at the location of the insertion or downstream from the insertion point in the case of an operon (genes that are co-transcribed). The transposon also serves as a tag for easy identification of the mutated gene. Twenty-seven non-oxidizing mutants in SG–1 were isolated, 18 of which were localized within one region of the chromosome (termed the Mnx region). This region of DNA was sequenced and found to be a series of seven genes which appear to be an operon (the *mnx* operon, Figure 4). Upstream of the first gene in the operon, regulatory sequences were present that suggested that the genes would be transcribed by the mother cell (the cell within which the endospore is forming) during sporulation. In fact, expression of the *mnx* genes occurred concomitant with onset of sporulation. TEM of two of the

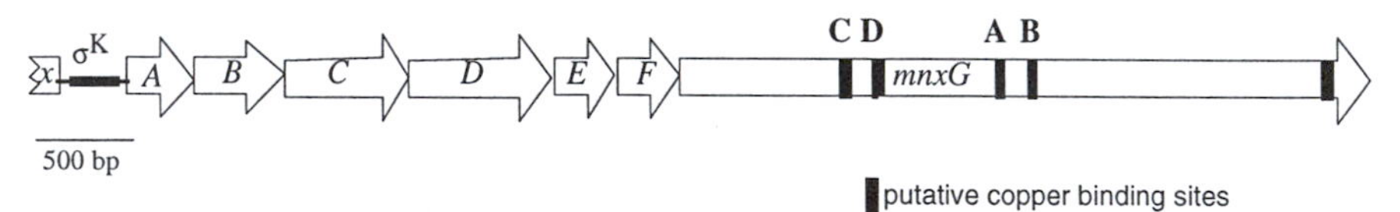

Figure 4. The organization of the Mnx operon of *Bacillus* sp. strain SG–1 based on DNA sequence analysis. *mnxG* is the gene that codes for the putative Mn(II)-oxidizing protein which shows significant similarity to the multicopper oxidase family of proteins, particularly in the regions of copper binding (boxed areas). The amino acid sequences of the copper binding sites designated with the letter A–D are shown in Figure 5. σ^K represents the site upstream of the operon where RNA polymerase binds to initiate RNA synthesis (DNA transcription). (Reproduced with permission from ref. *68*. Copyright 1997 Mineralogical Society of America.)

mnx mutants showed slight alterations in the outermost spore layers suggesting that one or more of the genes in the operon are involved in the deposition of the spore coat structure or code for a spore coat or exosporium protein (*100*). The mutants were unaffected with regard to germination and resistance properties as compared to wild-type SG–1 spores.

Sequence analysis of the *mnx* genes have provided some insights into the mechanism of Mn(II) oxidation. Two of the *mnx* genes, *mnxC* and *mnxG*, code for putative proteins (MnxC and MnxG, respectively) that have some sequence similarity to other proteins in the sequence databases, while the remaining genes do not. The MnxG protein has regions of sequence similarity with a family of proteins known as multicopper oxidases (Figure 5), proteins in which multiple copper ions serve as redox cofactors for the oxidation of a variety of different substrates (*101, 102*). In the three-dimensional protein structure, the amino acids, histidine, cysteine, and methionine, that make up each type of copper center, come into close contact and form a copper binding site. There is a great diversity among the family of multicopper oxidases and they occur in a wide range of organisms from bacteria and fungi to plants and animals. Generally they are involved in the oxidation of organic substrates, however, there are two multicopper oxidases that are involved in Fe(II) oxidation (Fet3 from yeast and human ceruloplasmin) (*103-106*) and one (CopA) that helps some bacteria resist the toxic effects of copper. Members of the multicopper oxidase family that are involved in oxidizing organic compounds include laccase (from plants and fungi) and ascorbate oxidase (from cucumber & squash).

Three types of copper ions based on their spectroscopic properties can be distinguished in multicopper oxidases. The type 1 or 'blue' copper is usually involved in a one electron oxidation of a substrate. The type 1 copper then transfers the electron

A

MnxG	**527**	**M**	**H**	**I**	**H**	**F**	**V**
MofA	**304**	**I**	**H**	**L**	**H**	**G**	**G**
Asox	94	I	H	W	H	G	I
Lacc	78	V	H	W	H	G	L
Hcer	119	F	H	S	H	G	I
CopA	99	I	H	W	H	G	I
Fet3	8	M	H	F	H	G	L
			2		3		

B

MnxG	**572**	**F**	**F**	**H**	**D**	**H**	**L**
MofA	**384**	**W**	**Y**	**H**	**D**	**H**	**T**
Asox	137	F	Y	H	G	H	L
Lacc	121	W	Y	H	S	H	Y
Hcer	178	I	Y	H	S	H	I
CopA	140	W	Y	H	S	H	S
Fet3	52	W	Y	H	S	H	T
				3		3	

C

MnxG	**281**	**H**	**V**	**F**	**H**	**Y**	**H**	**V**	**H**
MofA	**1174**	**H**	**P**	**V**	**H**	**F**	**H**	**L**	**L**
Asox	480	H	P	W	H	L	H	G	H
Lacc	508	H	P	I	H	K	H	G	N
Hcer	994	H	T	V	H	F	H	G	H
CopA	542	H	P	I	H	L	H	G	M
Fet3	341	H	P	F	H	L	H	G	H
		1			2		3		

D

MnxG	**334**	**H**	**C**	**H**	**L**	**Y**	**P**	**H**	**F**	**G**	**I**	**G**	**M**
MofA	**1279**	**H**	**C**	**H**	**I**	**L**	**G**	**H**	**E**	**E**	**N**	**D**	**F**
Asox	542	H	C	H	I	E	P	H	L	H	M	G	M
Lacc	585	H	C	H	I	A	S	H	Q	M	G	G	M
Hcer	1039	H	C	H	V	T	D	H	I	H	A	G	M
CopA	590	H	C	H	L	L	Y	H	M	E	M	G	M
Fet3	411	H	C	H	I	E	W	H	L	L	Q	G	L
		3	1	3				1					1

Figure 5. Amino acid sequence alignment of copper-binding sites in MnxG, MofA, and other multicopper oxidases. The letters A–D correspond to the copper binding sites shown in Figure 4. Abbreviations: Asox, ascorbate oxidase (cucumber and squash); Lacc, laccase (fungi); Hcer, human ceruloplasmin; CopA, a copper resistance protein from *Pseudomonas syringae*; and Fet3, an iron oxidizing/transport protein in yeast. The amino acid residues conserved among the different proteins are shaded and the copper-binding residues are numbered according to the type of copper they potentially help coordinate. (Reproduced with permission from ref. *68*. Copyright 1997 Mineralogical Society of America.)

to the type 2 and type 3 coppers ions which form a cluster within the protein and are involved in the binding and reduction of molecular oxygen.

$$\text{Substrate} \xrightarrow{e^-} \text{Type 1} \xrightarrow{e^-} \{\text{Type 2} + (\text{Type 3})_2\} \xrightarrow{e^-} O_2.$$

Multicopper oxidases and cytochrome oxidases are the only proteins known to catalyze the four-electron reduction of O_2 to H_2O. The multicopper oxidases possess a distinctive subdomain structure (*106*). Laccase, ascorbate oxidase, and Fet3 have three domains while human ceruloplasmin has six. In all the multicopper oxidases there is significant internal homology among the subdomains suggesting they all arose from a common gene ancestor by gene duplication (*101, 106*).

MnxG share most similarity with the multicopper oxidases in the regions of the copper binding sites (Figure 5) and based on the size of the protein and the subdomain structure, MnxG is most similar to human ceruloplasmin. In humans, ceruloplasmin appears to function both in the mobilization of Fe(II) from tissues and oxidation of Fe(II) which facilitates Fe(III) uptake by transferrin in the bloodstream. There are at least three compelling facts that point to MnxG as the Mn(II)-oxidizing protein: 1) Azide, an inhibitor of Mn(II) oxidation in SG–1 as well as in environmental samples (*79, 107, 108*), inhibits multicopper oxidases by bridging the type 2 and type 3 copper atoms (*109*); 2) Mn(II) oxidation in SG–1 is enhanced by the addition of small amounts of Cu(II) (*100*); and 3) using MnxG-specific antibodies, the MnxG protein has recently been localized to the exosporium, i.e., on the spore surface (*95*).

These results beg the question, is the real function of MnxG to oxidize Mn(II)? If so, what function does Mn(II) oxidation serve the spores? Alternate possible functions of MnxG include 1) helping to cross-link the outermost spore coat or exosporium, 2) by analogy to Fet3 and ceruloplasmin, MnxG may function in iron acquisition or storage, and 3) a mechanism to resist toxic chemicals or oxidants. In SG-1, it has been suggested that the Mn oxides may function as electron acceptors during germination under anaerobic conditions (*72, 110*) and thus SG–1 spores oxidize Mn(II) in order to create a store of Mn oxide. Clearly, understanding the true role of MnxG *in vivo* will require further investigation.

Recently, the amino acid sequence of the Mn(II)-oxidizing protein from *L. discophora*, MofA, was reported (*111*). The most striking result is that this protein, like MnxG, also appears to be a multicopper oxidase (Figure 5). Although MofA has more similarity to two other multicopper oxidases, phenoxazinone synthase and bilirubin oxidase (*111, 112*), the copper binding regions are highly conserved (Figure 5). Thus, we now have two very similarly functioning "Mn oxidases" (assuming the copper centers are involved in the oxidation of Mn^{2+}) in two phylogenetically unrelated Mn(II)-oxidizing bacteria. It is thus tempting to speculate on a universal role of copper in bacterial Mn(II) oxidation (*68*). In fact, the partial sequence of a gene involved in Mn(II) oxidation in *Pseudomonas fluorescens* suggests that the putative protein also has at least 2 of the copper binding sites found in the multicopper oxidases (*113*).

The second SG–1 gene in the Mnx operon whose putative protein, MnxC, is similar to something in the database, contains a signal peptide, suggesting that the protein is associated with a membrane. This protein has also been recently localized to the exosporium (Francis and Tebo, unpublished). MnxC is similar to a number of other cell surface proteins of unknown function and redox-active proteins. The region of similarity between MnxC and these other proteins is in a region of about 20 amino acids that includes a pair of cysteines that are separated by three other amino acids (a C-x-x-x-C motif). This motif is similar to a thioredoxin motif (C-x-x-C) suggesting MnxC might be involved in forming disulfide bonds between cysteines. However, the C-x-x-x-C motif is identical to that found in two yeast mitochondrial membrane proteins, SCO1 and SCO2. These proteins are known to play a role in assembly of cytochrome oxidase. Recently, the two cysteines were shown to bind copper and thus it was hypothesized that SCO1 and SCO2 are involved in the uptake of copper by the mitochondrion or the delivery of copper to cytochrome oxidase (*114*). In a similar fashion, MnxC may function to deliver copper to MnxG or to take up copper during sporulation.

Surface chemistry of SG–1 spores. Recently the surface chemistry and Cu(II) adsorption properties of SG-1 spores have been characterized (*115*). The specific surface area of freeze-dried SG-1 spores was 6.3–6.9 m^2 g^{-1} of spores as determined by a BET surface area analyzer or calculated from spore dimensions determined in the electron microscope. A value of 74.7 m^2 g^{-1}, however, was obtained with wet spores using a methylene blue adsorption method. Since BET analysis required freeze drying whereas the dye adsorption method uses wet spores, the difference in the surface area suggests that the spore surface is porous rather than smooth and that the porous structure collapses after freeze drying. The surface exchange capacity measured by the proton exchange method was found to be 30.6 μmol m^{-2}, equal to a surface site density of 18.3 sites nm^{-2}, based on the methylene blue adsorption surface area. The SG-1 spore surface charge characteristics were obtained from the acid-base titration data. The surface charge density varied with pH, with the point of zero charge (PZC) being 4.5. The surface was dominated by negatively charged sites, believed to be primarily carboxylate but also phosphate groups. Copper adsorption by SG-1 spores was rapid and complete within minutes and was lower at lower pH, increasing with pH. The high surface area and surface site density, comparable to Fe and Al mineral colloids, give SG-1 spores a high capacity for binding Cu(II) and other metals on their surface. Whether Cu(II) adsorption is due to nonspecific binding or specific binding related to the presence of MnxC or MnxG on the spore surface is yet to be determined.

Interactions with other metals. Mn(II)-oxidizing bacteria have the potential to mediate the cycling of a variety of other elements (Figure 2). This may occur either through the direct interaction of the Mn(II)-oxidizing protein with other metals or via interactions with the Mn oxides products. A variety of environmental investigations have suggested that other metals and lanthanides, e.g., cerium and cobalt, are acted upon by Mn(II)-oxidizing bacteria (*72, 82, 116-118*). In addition to Mn(II), SG–1 is capable of oxidizing Co(II) (*119, 120*) and Fe(II) (Tebo and Edwards, unpublished). Co(II) oxidation is affected similarly by environmental parameters such as pH, temperature, and metal concentration, as for Mn(II) oxidation. In fact, mutants in Mn(II) oxidation (*mnx* mutants) also do not oxidize Co(II), suggesting that the same enzyme is responsible for both oxidations. This same phenomenon has been observed for the manganese peroxidase enzyme of white rot fungi (*121*). As described above for copper, SG–1 spores bind a variety of other metals in the absence of Mn oxides. Metals such as cadmium, zinc, and nickel all seem to be taken up via non-specific adsorptive mechanisms. The metal binding and oxidizing capabilities and the physical properties of SG–1 spores suggest they have utility for metal removal and recovery applications (*122*).

Conclusions

During the past five years significant progress has been made toward gaining a better mechanistic understanding of bacterial Mn(II) oxidation. Perhaps most important are the realizations that copper may play a key role as a cofactor in Mn oxidation and that the Mn oxidases are related to the multicopper oxidase family of proteins. The results of measurements of the isotopic composition of the Mn oxide products formed through microbial activity are consistent with the molecular and genetic studies.

The development of a better understanding of both Fe and Mn biomineralization processes will lead to new insights into the biogeochemical cycles of Fe and Mn and to the possible biotechnological applications of these organisms for new material syntheses (e.g., catalysts or adsorbents) or for metal and radionuclide removal and recovery (e.g., in bioremediation and water treatment).

Acknowledgments

Funding for research on SG–1 has come from grants from the National Science Foundation (OCE94-168944 and MCB94-07776), the University of California Toxic Substances Research and Teaching Program, the National Sea Grant College Program (National Oceanic and Atmospheric Administration, U.S. Department of Commerce, #NA66RG0477 Project R/CZ 123, through the California Sea Grant College System), and the California State Resources Agency. K. Mandernack, Y. Lee, L. van Waasbergen, K. Casciotti, C. Francis, and D. Edwards have all made significant contributions to the study of *Bacillus* sp. strain SG-1.

Literature Cited

1. Stumm, W.; Morgan, J. J. *Aquatic Chemistry. Chemical Equilibria and Rates in Natural Waters;* John Wiley and Sons: New York, NY, 1996.
2. Towe, K. M.; Bradley, W. F., *J. Colloid Interface Sci.* **1967,** *24*, 384-392.
3. Chukhrov, F. V.; Zvyagin, B. B.; Gorshkov, A. I.; Ermilova, L. P.; Balashova, V. V., *Izvest. Akad. Nauk. SSSR, Ser. Geol.* **1973,** *4*, 23-33.
4. Russell, J. D., *Clay Minerals* **1979,** *14*, 109-114.
5. Junta, J. L.; Hochella Jr., M. F., *Geochim. Cosmochim. Acta* **1994,** *58*, 4985-4999.
6. Schwertmann, U.; Fitzpatrick, R. W. In *Biomineralization Processes of Iron and Manganese: Modern and Ancient Environments;* Skinner, H. C. W.; Fitzpatrick, R. W., Eds.; Catena Supplement 21; Catena Verlag, Cremlingen-Destedt, 1992, pp 7-30.
7. Schwertmann, U. In *Iron in soils and caly minerals;* Stucki, J. W.; Goodman, B. A.; Schwertmann, U., Eds.; D. Reidel Publishing Co., Dordrecht, 1988, pp 203-250.
8. Schwertmann, U.; Taylor, R. M. In *Minerals in soil environments (2nd ed.);* Dixon, J. N.; Weed, S. B., Eds.; Soil Science Society of America, Madison, 1989, pp 379-438.
9. Dzombak, D. A.; Morel, M. M. *Surface complexation modeling;* John Wiley & Sons: New York, NY, 1990.
10. Schwertmann, U.; Fitzpatrick, R. W., *Soil Sci. Soc. Am. J.* **1977,** *41*, 1013-1018.
11. Carlson, L.; Schwertmen, U., *Clay Minerals* **1990,** *25*, 65-71.
12. Brown, D. A.; Sherriff, B. A.; Sawicki, J. A., *Geochim. Cosmochim. Acta* **1997,** *61*, 3341-3348.
13. Bazylinski, D. A.; Moskowitz, B. M. In *Geomicrobiology: interactions between microbes and minerals;* Banfield, J. F.; Nealson, K. H., Eds.; Reviews in Mineralogy; Mineralogical Society of America, Washington, D.C., 1997, Vol. 35, pp 181-224.
14. Stolz, J. F. In *Biomineralization Processes of Iron and Manganese: Modern and Ancient Environments;* Skinner, H. C. W.; Fitzpatrick, R. W., Eds.; Catena Supplement 21; Catena Verlag, Cremlingen-Destedt, 1992, pp 133-146.
15. Liu, S. V.; Zhou, J.; Zhang, C.; Cole, D. R.; Gajdarziska-Josifovska, M.; Phelps, T. J., *Science* **1997,** *277*, 1106-1109.
16. Manceau, A.; Combes, J. M., *Physics and Chemistry of Minerals* **1988,** *15*, 283-295.
17. Mandernack, K. W.; Post, J.; Tebo, B. M., *Geochim. Cosmochim. Acta* **1995,** *59*, 4393-4408.

18. Dixon, J. B.; Skinner, H. C. W. In *Biomineralization Processes of Iron and Manganese: Modern and Ancient Environments;* Skinner, H. C. W.; Fitzoatrick, R. W., Eds.; Catena Supplement 21; Catena Verlag, Cremlingen-Destedt, 1992, Vol. 21, pp 31-50.
19. Manceau, A.; Gorshkov, A. I.; Drits, V. A., *Amer. Mineral.* **1992a,** *77*, 1133-1143.
20. Manceau, A., *Amer. Mineral.* **1989,** *74*, 1386-1389.
21. Manceau, A.; Gorshkov, A. I.; Drits, V. A., *Amer. Mineral.* **1992b,** *77*, 1144-1157.
22. Burns, R. G.; Burns, V. M. In *Marine Minerals;* Burns, R. G., Ed. Mineralogical Society of America, 1979, pp 1-46.
23. McKenzie, R. M. In *Minerals in soil environments;* Dixon, J. B.; Weed, S. B., Eds.; Soil Science Society of America, Madison, 1989, pp 439-466.
24. Post, J. E. In *Biomineralization Processes of Iron and Manganese: Modern and Ancient Environments;* Skinner, H. C. W.; Fitzpatrick, R. W., Eds.; Catena Supplement 21; Catena Verlag, Cremlingen-Destedt, 1992, pp 51-74.
25. Ehrlich, H. L., *Chem. Geol.* **1996,** *132*, 5-9.
26. Tian, Z. R.; Tong, W.; Wang, J. Y.; Duan, N. G.; Krishnan, V. V.; Suib, S. L., *Science* **1997,** *276*, 926-930.
27. Davis, J. A.; Kent, D. B. In *Mineral-water interface geochemistry;* Hochella Jr., M. F.; White, A. F., Eds.; Reviews in mineralogy; The Mineralogical Society of America, Washington, D.C., 1990, Vol. 23, pp 177-260.
28. Sverjensky, D. A.; Sahai, N., *Geochim. Cosmochim. Acta* **1996,** *60*, 3773-3797.
29. Zhang, P. C.; Sparks, D. L., *Soil Sci. Soc. Am. J.* **1989,** *53*, 1028-1034.
30. Goldberg, S.; Forster, H. S.; Heick, E. L., *Soil Sci. Soc. Am. J.* **1993,** , 704-708.
31. Manning, B. A.; Goldberg, S., *Soil Sci. Soc. Am. J.* **1996,** *60*, 121-131.
32. Tejedor-Tejedor, M. I.; Anderson, M. A., *Langmuir* **1990,** *6*, 602-611.
33. Hawke, D.; Carpenter, P. D.; Hunter, K. A., *Environ. Sci. Technol.* **1989,** *23*, 187-191.
34. James, R. O.; Parks, G. A., *Surface and Colloid Science* **1982,** *12*, 119-216.
35. Zelazny, L. W.; He, L. M.; Vanwormhoudt, A. In *Methods of soil analysis. Part 3. Chemical methods;* Sparks, D. L., Ed. SSSA, Inc., Madison, WI, 1996, pp 1231-1253.
36. Goldberg, S., *Advances in Agronomy* **1992,** *47*, 233-329.
37. Colombo, C.; Barron, V.; Torrent, J., *Geochimica et Comochimica Acta* **1994,** *58*, 1261-1269.
38. Breeuwsma, A.; Lyklema, J., *J. Colloid Interface Sci.* **1973,** *43*, 437-448.
39. Parffit, R. L.; Atkinson, R. J.; Smart, R. S. C., *Soil Science Society of America Proceedings* **1975,** *39*, 837-841.
40. Blesa, M. A.; Maroto, A. J. G.; regazzoni, A. E., *J. Colloid Interface Sci.* **1984,** *99*, 32-40.
41. Posselt, H. S.; Anderson, F. J.; Weber Jr., W. J., *Environ. Sci. Technol.* **1968,** 2, 1087-1093.
42. Morgan, J. J.; Stumm, W., *Journal of Colloid Science* **1964,** *19*, 347-359.
43. Yao, W.; Millero, F. J., *Environ. Sci. Technol.* **1996,** *30*, 536-541.
44. Godtfredsen, K. L. Ph.D. Dissertation, The Johns Hopkins University, Baltimore, MD (1992).
45. Zinder, B.; Furrer, G.; Stumm, W., *Geochim. Cosmochim. Acta* **1986,** *50*, 1861-1869.
46. Deng, Y., *Langmuir* **1997,** *13*, 1835-1839.
47. Stone, A. T.; Morgan, J. J., *Environ. Sci. Technol.* **1984,** *18*, 450-456.
48. Stone, A. T.; Morgan, J. J., *Environ. Sci. Technol.* **1984,** *18*, 617-624.
49. Stone, A. T., *Environ. Sci. Technol.* **1987,** *21*, 979-988.

50. Stone, A. T., *Geochim. Cosmochim. Acta* **1987,** *51*, 919-925.
51. Myneni, S. C. B.; Tokunaga, T. K.; Brown Jr., G. E., *Science* **1997,** *278*, 1106-1109.
52. Huang, P. M. In *Rates of Soil Chemical Processes;* Sparks, D. L.; Suarez, D. L., Eds.; Soil Science Society of America, Inc., Madison, 1991, pp 191-230.
53. Lovley, D. R., *Advances in Agronomy* **1994,** *54*, 175-231.
54. Ghiorse, W. C.; Ehrlich, H. L. In *Biomineralization Processes of Iron and Manganese: Modern and Ancient Environments;* Skinner, H. C. W.; Fitzpatrick, R. W., Eds.; Catena Supplement 21; Catena Verlag, Cremlingen-Destedt, 1992, pp 75-99.
55. Jenne, E. A. In *Trace Inorganics in Water;* American Chemical Society, Washington, D.C., 1967, pp 337 - 387.
56. Lowenstam, H.; Weiner, S. *On Biomineralization;* Oxford University Press: New York, NY, 1989.
57. Hallbeck, L.; Pederson, K., *J. Gen. Microbiol.* **1991,** *137*, 2657-2661.
58. Hallbeck, L.; Stahl, F.; Pederson, K., *J. Gen. Microbiol.* **1993,** *139*, 1531-1535.
59. Emerson, D.; Moyer, C., *Appl. Environ. Microbiol.* **1997,** *In press*
60. Widdel, F.; Schnell, S.; Heising, S.; Ehrenreich, A.; Assmus, B.; Schink, B., *Nature* **1993,** *362*, 834-836.
61. Ehrenreich, A.; Widdel, F., *Appl. Environ. Microbiol.* **1994,** *60*, 4517-4526.
62. Straub, K. L.; Benz, M.; Schink, B.; Widdel, F., *Appl. Environ. Microbiol.* **1996,** *62*, 1458-1460.
63. Hafenbradl, D.; Keller, M.; Dirmeier, R.; Rachel, R.; Rossnagel, P.; Burggraf, S.; Huber, H.; Stetter, K. O., *Arch. Microbiol.* **1996,** *166*, 308-314.
64. Kepkay, P.; Nealson, K. H., *Arch. Microbiol.* **1987,** *148*, 63-67.
65. Ehrlich, H. L.; Salerno, J. C., *Arch. Microbiol.* **1990,** *154*, 12-17.
66. Ehrlich, H. L. *Geomicrobiology;* Marcel Dekker: New York and Basel, 1996.
67. Nealson, K. H.; Tebo, B. M.; Rosson, R. A., *Adv. Appl. Microbiol.* **1988,** *33*, 279-318.
68. Tebo, B. M.; Ghiorse, W. C.; van Waasbergen, L. G.; Siering, P. L.; Caspi, R. In *Geomicrobiology: interactions between microbes and minerals;* Banfield, J. F.; Nealson, K. H., Eds.; Reviews in Mineralogy; Mineralogical Society of America, Washington, D.C., 1997, Vol. 35, pp 225-266.
69. Caspi, R.; Haygood, M. G.; Tebo, B. M., *Microbiology* **1996,** *142*, 2549-2559.
70. Ghiorse, W. C., *Ann. Rev. Microbiol.* **1984,** *38*, 515-550.
71. Emerson, D. Ph.D. dissertation, Cornell University, Ithaca (1989).
72. Tebo, B. M. Ph.D. dissertation, University of California, San Diego (1983).
73. de Vrind, J. P. M.; de Vrind-de Jong, E. W.; de Voogt, J.-W. H.; Westbroek, P.; Boogerd, F. C.; Rosson, R. A., *Appl. Environ. Microbiol.* **1986,** *52*, 1096-1100.
74. Sunda, W. G.; Kieber, D. J., *Nature* **1994,** *367*, 62-64.
75. Tebo, B. M.; Edwards, D. B.; Kieber, D. J.; Sunda, W. G. ASLO Aquatic Sciences Meeting, Santa Fe, New Mexico, February 10-14, 1997.
76. Waite, T. D.; Morel, F. M. M., *Environ. Sci. Technol.* **1984,** *18*, 860-868.
77. Sunda, W. G.; Huntsman, S. A.; Harvey, G. R., *Nature* **1983,** *301*, 234-236.
78. Nealson, K. H.; Ford, J., *Geomicrobiol. J.* **1980,** *2*, 21-37.
79. Rosson, R. A.; Nealson, K. H., *J. Bacteriol.* **1982,** *151*, 1027-1034.
80. Hem, J. D., *Geochim. Cosmochim. Acta* **1981,** *45*, 1369-1374.
81. Murray, J. W.; Dillard, J. G.; Giovanoli, R.; Moers, H.; Stumm, W., *Geochim. Cosmochim. Acta* **1985,** *49*, 463-470.
82. Tebo, B. M.; Nealson, K. H.; Emerson, S.; Jacobs, L., *Limnol. Oceanogr.* **1984,** *29*, 1247-1258.
83. Tipping, E.; Thompson, D. W.; Davison, W., *Chem. Geol.* **1984,** *44*, 359-383.
84. Tipping, E.; Jones, J. G.; Woof, C., *Arch. Hydrobiol.* **1985,** *105*, 161-175.

85. Friedl, G.; Wehrli, B.; Manceau, A., *Geochim. Cosmochim. Acta* **1997,** *61*, 275-290.
86. Kalhorn, S.; Emerson, S., *Geochim. Cosmochim. Acta* **1984,** *48*, 897-902.
87. Tipping, E., *Geochim. Cosmochim. Acta* **1984,** *48*, 1353-1356.
88. Hastings, D.; Emerson, S., *Geochim. Cosmochim. Acta* **1986,** *50*, 1819-1824.
89. Hem, J. D.; Lind, C. J., *Geochim. Cosmochim. Acta* **1983,** *47*, 2037-2046.
90. Mandernack, K. W.; Post, J.; Tebo, B. M., *Geochim. Cosmochim. Acta* **1995,** *59*, 4393-4408.
91. Mandernack, K. W.; Fogel, M. L.; Usui, A.; Tebo, B. M., *Geochim. Cosmochim. Acta* **1995,** *59*, 4409-4425.
92. Luther III, G. W. In *Aquatic Chemical Kinetics;* Stumm, W., Ed. John Wiley & Sons, Inc., New York, NY, 1990, pp 173-198.
93. Wehrli, B. In *Aquatic Chemical Kinetics;* Stumm, W., Ed. John Wiley & Sons, Inc., New York, NY, 1990, pp 311-336.
94. Wehrli, B.; Friedl, G.; Manceau, A. In *Aquatic Chemistry: Interfacial and Interspecies Processes;* Huang, C. P.; O'Melia, C. R.; Morgan, J. J., Eds.; American Chemical Society, Washington, D.C., 1995,.
95. Francis, C. A.; Casciotti, K. L.; Tebo, B. M. In *Annual Meeting of the American Society for Microbiology* . Miami, FL, May 4-8, 1997.
96. Adams, L. F.; Ghiorse, W. C., *J. Bacteriol.* **1987,** *169*, 1279-1285.
97. Boogerd, R. C.; de Vrind, J. P. M., *J. Bacteriol.* **1987,** *169*, 489-494.
98. Tebo, B. M.; Mandernack, K.; Rosson, R. A. In *Abstracts, Annual Meeting of the American Society for Microbiology* . Miami Beach, FL, 1988, pp. 201.
99. van Waasbergen, L. G.; Hoch, J. A.; Tebo, B. M., *J. Bacteriol.* **1993,** *175*, 7594-7603.
100. van Waasbergen, L. G.; Hildebrand, M.; Tebo, B. M., *J. Bacteriol.* **1996,** *178*, 3517-3530.
101. Ryden, L. G.; Hunt, L. T., *Journal of Molecular Evolution* **1993,** *36*, 41-66.
102. Huber, R., *Angew. Chem. Int. Ed. Engl.* **1989,** *28*, 848-869.
103. Kaplan, J.; O'Halloran, T. V., *Science* **1997,** *271*, 1510-1512.
104. Stearman, R.; Yuan, D. S.; Yamaguchi-Iwai, Y.; Klausner, R. D.; Dancis, A., *Science* **1996,** *271*, 1552-1557.
105. Eide, D.; Guerinot, M. L., *ASM News* **1997,** *63*, 199-205.
106. Solomon, E. I.; Sundaram, U. M.; Machonkin, T. E., *Chem. Rev.* **1996,** *96*, 2563-2605.
107. Rosson, R. A.; Tebo, B. M.; Nealson, K. H., *Appl. Environ. Micrbiol.* **1984,** *47*, 740-745.
108. Tebo, B. M., *Deep-Sea Res.* **1991,** *38*, S883-S905.
109. da Silva, J. J. R. F.; Williams, R. J. P. *The Biological Chemistry of the Elements;* Clarendon Press: Oxford, 1991.
110. de Vrind, J. P. M.; Boogerd, F. C.; Jong, E. W. d. V.-d., *J. Bacteriol.* **1986,** *167*, 30-34.
111. Corstjens, P. L. A. M.; de Vrind, J. P. M.; Goosen, T.; de Vrind-de Jong, E. W., *Geomicrobiol. J.* **1997,** *14*, 91-108.
112. Siering, P. L.; Ghiorse, W. C., *Geomicrobiol. J.* **1997,** *14*, 109-125.
113. Brouwers, G.-J.; de Vrind, J. P. M.; Corstjens, P. L. A. M.; Westbroek, P.; de Vrind-de Jong, E. W. In *Abstracts, Geological Society of America Annual Meeting* . Salt Lake City, UT, 1997.
114. Glerum, D. M.; Shtanko, A.; Tzagoloff, A., *J. Biol. Chem.* **1996,** *271*, 20531-20535.
115. He, L. M.; Tebo, B. M., *Appl. Environ. Microbiol.* **1998,** *64*, In Press.
116. Moffett, J. W., *Limnol. Oceanogr.* **1994,** *39*, 1309-1318.
117. Moffett, J. W., *Geochim. Cosmochim. Acta* **1994,** *58*, 695-703.
118. Moffett, J.; Ho, J., *Geochim. Cosmochim. Acta* **1996,** *60*, 3415-3424.

119. Tebo, B. M.; Lee, Y. In *Biohydrometallurgical Technologies;* Torma, A. E.; Wey, J. E.; Lakshmanan, V. L., Eds.; The Minerals, Metals, & Materials Society, Warrendale, PA, 1993, Vol. I, pp 695-704.
120. Lee, Y.; Tebo, B. M., *Appl. Environ. Microbiol.* **1994,** *60*, 2949-2957.
121. Souren, A. W. M. G., *Geochim. Cosmochim. Acta* **1998,** *62,* in press.
122. Tebo, B. M. In *Genetic Engineering–Principles and Methods;* Setlow, J. K., Ed. Plenum Press, New York, 1995, Vol. 17, pp 231-263.

Author Index

Balko, Barbara A., 301
Bargar, John R., 14
Bertsch, P. M., 282
Bickmore, B. R., 37
Brooks, Scott C., 358
Brown, Gordon E., Jr., 14
Burkhard, Caroline, 265
Casey, William H., 244
Chang, Fang-Ru Chou, 88
Comans, Rob N. J., 179
Feitz, Andrew J., 374
Fendorf, Scott, 358
Gates, W. P., 282
Goldberg, S., 136
Greathouse, Jeffrey A., 88
Grundl, Timothy J., 2
Haderlein, Stefan B., 342
He, L. M., 393
Hiemstra, T., 68
Hochella, M. F., Jr., 37
Huang, Weilin, 222
Hug, Stephan J., 374
Hunter, D. B., 282
Jardine, Phillip M., 358
Kemner, K. M., 282
Leboeuf, Eugene J., 222
Luther, George W., III, 265
Maurice, Patricia A., 57
Nordin, Jan, 244
Parks, George A., 14
Pecher, Klaus, 342
Peterson, Maria L., 323
Phillips, Brian L., 244
Pignatello, J. J., 204
Rakovan, J. F., 37
Refson, K., 88
Rochette, Elizabeth A., 358
Rosso, K. M., 37
Rufe, E., 37
Ruppel, David T., 265
Scheckel, Kirk G., 108
Scheidegger, André M., 108
Scherer, Michelle M., 301
Skipper, N. T., 88
Sparks, Donald L., 2, 108
Sposito, Garrison, 88
Strawn, Daniel G., 108
Su, C., 136
Suarez, D. L., 136
Taylor, David L., 358
Tebo, B. M., 393
Towle, Steven N., 14
Tratnyek, Paul G., 301
VanRiemsdijk, W. H., 68
Waite, T. David, 374
Weber, Walter J., Jr., 222
White, Art F., 323

Subject Index

A

Adsorption
- accumulation of substance at interface between solid and bathing solution, 108, 110
- inorganic species, 4–5

Adsorption mechanisms. *See* Oxyanion adsorption mechanisms on oxides

AFM. *See* Atomic force microscopy (AFM)

Alumina, formation of Co-Al hydroxide precipitates, 22–23

Aluminum, importance in immobilizing heavy metals, 126

Aluminum oxides. *See* Oxyanion adsorption mechanisms on oxides

Aluminum/nickel mixed-cation hydroxide phases. *See* Metal sorption at mineral–water interface; Nickel sorption case study

Anionic surface species, sorption to aluminum and silicon oxide materials, 5

Aqueous metal species reduction. *See* Reduction of aqueous metal species on Fe(II)-containing oxide surfaces

Arsenate
- attenuated total reflectance–Fourier transform infrared (ATR–FTIR) spectrum of solutions of Na_2HAsO_4 at pH 5 and 8, 162*f*
- diffuse reflectance infrared Fourier transform (DRIFT) difference spectrum of As(V) sorbed on am-$Fe(OH)_3$ at pH 5 and 8, 163*f*
- electrophoretic mobility (EM) of am-$Fe(OH)_3$ suspensions, 160*f*
- pressure-jump relaxation kinetics, 172
- sorption on oxides, 159–164
- *See also* Oxyanion adsorption mechanisms on oxides

Arsenic, adsorption kinetics, 170, 172

Arsenite
- diffuse reflectance infrared Fourier transform (DRIFT) difference spectrum of As(III) sorbed on am-$Al(OH)_3$ at pH 5 and 8, 168*f*
- DRIFT difference spectrum of As(III) sorbed on am-$Fe(OH)_3$ at pH 5 and 8, 166*f*
- sorption on oxides, 164–167
- *See also* Oxyanion adsorption mechanisms on oxides

Atomic force microscopy (AFM)
- advantages and difficulties, 57–58
- characterization of sorption processes, 45, 47–50
- fundamental principles, 57–58
- in situ imaging of samples, 3
- most common scanning probe microscope (SPM), 38
- real time clay/solution interaction studies, 43–45
- study of dissolution rates and mechanisms, 38–43
- TappingMode™ AFM (TMAFM), 58–59
- TMAFM phase imaging technique, 60, 62*f*
- *See also* Reactivities of environmental particles

Auto-inhibition of oxide mineral reductive capacity toward Co(II)EDTA
- association of $Co(II)EDTA^{2-}$ with Fe- and Mn-oxides, 359-360
- cobalt oxidation state affecting stability of Co-EDTA complexes, 359
- dynamic flow experiments, 361

experimental success dependence on MnO_2 preparation, 368–369
ferrihydrite-coated silica preparation, 361
iron-oxide studies, 369
manganese solids speciation by X-ray absorption near-edge structure (XANES) spectroscopy, 363, 368
manganese XANES spectra of model compounds, 365*f*
manganese XANES spectra of pyrolusite reacted with continuous pulse of $Co(II)EDTA^{2-}$, 367*f*, 368
measuring surface changes in Mn and Fe by X-ray absorption fine structure spectroscopy, 360
methods and materials, 360–363
migration of radionuclides away from low-level waste disposal facilities, 358–359
model manganese minerals first-derivative XANES spectra, 366*f*
oxidation of $Co(II)EDTA^{2-}$ to $Co(III)EDTA^{-}$ by ferrihydrite-coated silica under dynamic flow conditions, 369
oxidation of $Co(II)EDTA^{2-}$ to $Co(III)EDTA^{-}$ by pyrolusite-coated silica under dynamic flow conditions, 363, 364*f*
pyrolusite-coated silica preparation, 360–361
reduced effluent concentrations of $Co(II)EDTA^{2-}$ and $Co(III)EDTA^{-}$ as function of time and pore volume of MnO_2 coated silica bed, 364*f*
solid-phase preparation of Mn- and Fe-(hydr)oxide minerals, 360–361
spatial distribution of Mn_2O_3 on β-MnO_2, 368
specific affinity of cobalt for Mn oxides
spectral contribution of manganese minerals in β-MnO_2 reacted with $Co(II)EDTA^{2-}$ for varying flow periods, 369*t*
XANES spectroscopy procedures, 362–363

B

Bacillus sp. strain SG-1 for Mn(II) oxidation. *See* Microbially mediated oxidative precipitation reactions
Biotic reaction mechanisms, ranging from mineral grains to weathering on global scale, 7
Boron
absorption by iron oxide, 151
ATR–FTIR (attenuated total reflectance–Fourier transform infrared) difference spectra of am-$Al(OH)_3$ with and without boron, 150*f*
ATR–FTIR difference spectra of am-$Fe(OH)_3$ with and without boron, 152*f*
ATR–FTIR spectra of boric acid at pH 7 and 11, 149*f*
coordination and speciation on mineral surfaces, 148–153
electrophoretic mobility (EM) measurements for am-$Al(OH)_3$ and am-$Fe(OH)_3$, 151, 153
See also Oxyanion adsorption mechanisms on oxides; Tetraphenylboron degradation at hydrated smectite surfaces
Brønsted acidity
effect of strong intermolecular hydrogen bonding on acidity of waters, 254
effect on rates of exchange of oligomeric oxygens, 254
influencing inductive or catalytic role of protons, 258

C

Carbonate
adsorption on oxides, 140–148
ATR–FTIR (attenuated total reflectance–Fourier transform infrared) difference spectra of

aluminum oxides in presence of carbonates, 142, 147*f*
ATR–FTIR difference spectra of carbonate ions at varied pH, 141*f*
See also Oxyanion adsorption mechanisms on oxides
Catalysis, distinction from induction, 256, 258–259
Cesium sorption. *See* Radiocesium sorption on illite and natural sediments
Charge distribution (CD) MUlti SIte Complexation (MUSIC) model
describing pH dependent sulfate adsorption on goethite and competition between sulfate and phosphate, 83
extension of MUSIC model, 79–83
multi component interaction, 83
See also Metal (hydr)oxides; MUlti SIte Complexation (MUSIC) model
Chemical weathering. *See* Dissolution rates and mechanisms
Chromate
adsorption on oxides, 167, 169
kinetics of adsorption on oxides, 172
See also Oxyanion adsorption mechanisms on oxides
Chromium plating facility, waste water contamination site, 336, 338–340
Clay/solution interaction studies, real time
interest in mica weathering, 43
methods and techniques for AFM imaging, 44–45
TappingModeTM AFM (TMAFM), 44–45
TMAFM image of dispersed clay on polished sapphire substrate, 45, 46*f*
Cobalt ions. *See* Sorption studies
Cobalt(II)EDTA complexes. *See* Auto-inhibition of oxide mineral reductive capacity toward Co(II)EDTA
Contaminant degradation
general picture of mechanism in metal-oxide-water systems, 317
treatment with zero-valent metals, 315, 317
See also Photocatalytic degradation of trace contaminants
Contamination of aqueous and terrestrial environments with metals and semi-metals, 108
Corrosion science, unifying models of oxide-water interface, 317
Crystallography, chemical composition of ideal crystal planes, 69

D

Degradation of tetraphenylboron at hydrated smectite surfaces. *See* Tetraphenylboron degradation at hydrated smectite surfaces
Degradation of trace contaminants. *See* Photocatalytic degradation of trace contaminants
Desorption models. *See* Model for sorption and desorption of organic contaminants in soils and sediments
1,3-Dichlorobenzene. *See* Natural organic matter (NOM) as sorbent of organic compounds
Diffuse reflectance infrared Fourier transform (DRIFT)
limitation for adsorption studies, 138
method for oxyanion speciation and binding, 137–138
Dissociation of oligomers, reactivity trends, 253–255
Dissolution and hole-filling mechanisms, natural organic matter as sorbent of organic compounds, 205–207
Dissolution and mineral precipitation, advancing of modern spectroscopic techniques, 5
Dissolution rates and mechanisms

AFM image of phlogopite surface after pretreatment, acid etch, and rinse, 40*f*
comparison to conventionally measured rates, 41, 43
phlogopite surface after 24 hours in DI water, 41, 42*f*
phlogopite surface after second etching, 41, 42*f*
surface preparation technique, 39, 40*f*
use of atomic force microscopy (AFM), 38–39
Dissolved Mn(III) complexes and Mn(IV) species. *See* Mn(III) complexes and Mn(IV) species, reactivity with reductants
Distributed reactivity model (DRM), quantifying nonideal behavior, 224–225
DRIFT. *See* Diffuse reflectance infrared Fourier transform (DRIFT)

E

EDTA (ethylenediaminetetraacetic acid) complexes. *See* Auto-inhibition of oxide mineral reductive capacity toward Co(II)EDTA
Electron paramagnetic resonance (EPR), in situ imaging of samples, 3
Electron transfer, heterogeneous, 6
Electrophoretic mobility (EM)
distinguishing between inner- and outer-sphere complexes, 138–139
evaluating possible mixed reaction mechanisms, 139
surface speciation data, 139
See also Oxyanion adsorption mechanisms on oxides
Environmental contaminants, treatment with zero-valent metals, 315, 317
Environmental particles. *See* Reactivities of environmental particles
Environmental scanning electron microscopy (ESEM), in situ imaging of samples, 3
EPR (electron paramagnetic resonance), in situ imaging of samples, 3
Ethylenediaminetetraacetic acid (EDTA) complexes. *See* Auto-inhibition of oxide mineral reductive capacity toward Co(II)EDTA
Extended X-ray absorption fine structure (EXAFS)
effect of Fe reduction on clay structure, 296–298
inner- versus outer-sphere bonding, 137

F

Fate and transport of metals, 108, 109*f*
Fe(II)-containing oxide surfaces. *See* Reduction of aqueous metal species on Fe(II)-containing oxide surfaces
Fe oxide minerals
central role of Fe transformations in environmental processes, 402*f*
examples of iron and manganese bacteria, 401*t*
field evidence of heterogeneous metal reduction, 336, 338–340
iron oxide minerals formed in surface and subsurface environments, 395*t*
kinetics of Fe(II) oxidation, 394–395
mechanisms of Fe and Mn oxidation/precipitation, 401, 403
microbial Fe(III) reduction, 399
mineralogy of Fe oxides, 395–397
precipitating microorganisms for Fe and Mn, 399–403
role of Fe and Mn oxides in fate and transport of contaminants, 399
solubility, 398
surface chemistry, 397–398
See also Microbially mediated oxidative precipitation reactions

Fe(III) reduction. *See* Tetraphenylboron degradation at hydrated smectite surfaces
Fe(II)/Fe(III) systems for pollutant reduction. *See* Pollutant reduction in heterogeneous Fe(II)/Fe(III) systems
Ferrihydrate, arsenate adsorption, 172
Ferrihydrite, phosphate adsorption, 169–170
Ferrous iron, reactivity in heterogeneous aqueous systems. *See* Pollutant reduction in heterogeneous Fe(II)/Fe(III) systems
Fourier transform infrared spectroscopy (FTIR)
  in situ imaging of samples, 3
  See also Oxyanion adsorption mechanisms on oxides
Fulvic and humic acids
  abundant components in soils and natural waters, 61
  supportive of dual-mode concept, 215
  TappingMode™ AFM (TMAFM) imaging of fulvic acids, 61, 63–64

G

Geochemistry
  improved understanding on all time scales, 4
  unifying models of oxide-water interface, 317
Geosorbents
  general behavior for range of, 236–237
  sorption and desorption, 233, 235–237
Gibbsite. *See* Nickel sorption case study
Glass transition temperature, phase transition in synthetic organic polymers, 228–229
Glassy-rubbery phase transition of polymers, non-idealities in sorption of organic contaminants to natural organic matter (NOM), 3
Goethite
  AFM observations of surface reactions, 47, 48*f*, 49*f*
  inner-sphere adsorption mechanism for phosphate, 139
  phosphate adsorption, 169–170
  sorption of selenate, 153–154
  studies of Ni, Zn, and Cd sorption, 117
  TappingMode™ AFM (TMAFM) amplitude and phase images along edge of goethite-bacteria cluster, 64*f*
Gouy–Chapman double layer model, variable charge character of metal (hydr)oxides, 71–72
Grazing-incidence XAFS studies, Co(II) and Pb(II) sorption on single crystal α-Al_2O_3 and TiO_2, 23–24
Grindsted aquifer, formation and regeneration of reactive Fe(II) surface sites, 353, 355

H

Hematite, arsenite adsorption, 170, 172
Heterogeneous electron transfer
  exchange from surface bound Fe^{2+} to reducible pollutants, 6
  mechanism with iron, 6
  progress in understanding of electron transfer across mineral surfaces, 6
  surface passivation, 6
Heterogeneous Fe(II)/Fe(III) systems. *See* Pollutant reduction in heterogeneous Fe(II)/Fe(III) systems
Hole-filling and dissolution mechanisms, natural organic matter as sorbent of organic compounds, 205–207
Humic acid
  abundant components in soils and natural waters, 61
  calorimetric investigation of glass transition for dry and water-wet samples, 230*f*
  polymer-like behavior, 229
  supportive of dual-mode concept, 215

Hydrolytic processes at oxide mineral surfaces
atomic-force microscope image showing monomolecular steps at dislocation on calcite surface, 247*f*
comparison of surface sites and dissolved complexes, 248–252
dehydration, rate of water exchange, 249, 252
formalism of structural inorganic chemistry, 245
ligand-directed labilization, 255–256, 262
low-symmetry solids like Bayerite, 248, 250*f*, 251*f*
mechanisms of dissolution, 262
retreating monomolecular step on [100] surface of oxide mineral with rock salt structure, 249*f*
stoichiometry of surface complex, 262
structural similarities and dissimilarities between minerals and dissolved oligomers, 248
variation in hydrolytic rates with OH/Cr(III) ratios in monomers and oligomers, 246*t*
See also Interfacial kinetics
Hydrophobic effect, thermodynamic driving force for sorption, 206
Hydroxylapatite, study of Pb uptake, 60–61

I

Illite. *See* Radiocesium sorption on illite and natural sediments
Ilmenite, redox reactions, 324–327
Induction, distinction from catalysis, 256, 258–259
Inner-sphere adsorption, inference from zero point of charge shift, 137
Inner-sphere complexes, without water molecules, 138
Inorganic species, sorption of, 4–5
Interfacial kinetics
dissolution or growth processes, 245
general rules for understanding, 252–255
kinetic rate expressions, 252–253
ligand-directed labilizations, 255–256
non-equilibrium and time-critical, 244
proton catalysis and proton induction, 256, 258–259
rate coefficients for ligand-induced dissolution of bunsenite for set of ligands, 260, 261*f*
rate of exchange of waters between inner-coordination sphere of Ni(II) and bulk solution, 257*f*
rates of dissolution of isostructural, molecular solids such as orthosilicate minerals, 261*f*
rates of protonation-deprotonation reactions for simple oxide solids in electrolyte solutions via relaxation spectroscopy, 259
reactivities of different oxide minerals, 259–260
reactivity changes of sites in inner-coordination sphere by stable ligands, 255–256
reactivity trends for dissociation of oligomers, 253–255
reactivity trends with adsorbed ligands, 260, 262
solute adsorption versus metal detachment, 259
See also Hydrolytic processes at oxide mineral surfaces
Interfacial processes in environment
diversity of surface-mediated reactions, 3
evolving spectroscopic techniques, 3
principles from unrelated fields, 3
underlying themes, 2–3
Ion adsorption to mineral surfaces
chemical composition of ideal crystal planes by crystallography, 69
description of ion binding in model, 70

important phenomenon in many industrial applications, 68
pH dependence of ion binding, 70
series of reaction equations for ion binding description, 69
study of variable surface charge/potential of metal (hydr)oxides, 69
surface speciation of adsorbed phosphate on goethite by in situ spectroscopy, 70
See also Metal (hydr)oxides
Ion-exchange theory, radiocesium binding in sediments, 180–181
Iron, electron exchange of surface bound Fe^{2+} to reducible pollutants, 6
Iron(II)-containing oxide surfaces. *See* Reduction of aqueous metal species on Fe(II)-containing oxide surfaces
Iron-oxides
field evidence of heterogeneous metal reduction, 336, 338–340
oxidation of $Co(II)EDTA^{2-}$ to $Co(III)EDTA^{-}$ by ferrihydrite-coated silica under dynamic flow conditions, 369
See also Auto-inhibition of oxide mineral reductive capacity toward Co(II)EDTA; Oxyanion adsorption mechanisms on oxides; Reduction reactions at metal-water interface, role of oxides
Iron(III) reduction. *See* Tetraphenylboron degradation at hydrated smectite surfaces
Ising model approach, interaction between neighboring sites for proton affinity, 75–76

K

Kaolinite
studies of Co(II) sorption, 22–23
See also Nickel sorption case study
Kinetic studies. *See* Radiocesium sorption on illite and natural sediments
Kinetics and mechanisms of metal-water interfacial reactions
heterogeneous electron transfer, 6
microbially mediated reactions, 7
mineral precipitation and dissolution, 5
photochemical reactions, 7
sorption of inorganic species, 4–5
sorption of organic species, 5
spectroscopic/microscopic tools, 3–4
underlying themes, 2–3
See Interfacial kinetics
K-montmorillonite. *See* Montmorillonite-water systems

L

Lead (Pb), AFM study of Pb uptake by hydroxylapatite, 60–61
Lead (Pb) ions. *See* Sorption studies
Leptothrix discophora for Mn(II) oxidation. *See* Microbially mediated oxidative precipitation reactions
Ligand-directed labilizations
effect on mechanisms of dissolution, 262
reactivity changes of other sites in inner-coordination sphere, 255–256
Linear-free energy relations (LFER), predicting rate coefficients for complicated processes, 245
Linear partitioning model
quantifying sorption, 223-224
sorption of organic compounds, 5
Lithium-montmorillonite. *See* Montmorillonite-water systems

M

Magnetite, redox reactions, 324–327

Manganese complexes. *See* Mn(III) complexes and Mn(IV) species, reactivity with reductants
Manganese hydrous oxides
strong control on environmental distribution of cobalt, 359
See also Auto-inhibition of oxide mineral reductive capacity toward Co(II)EDTA
Mechanisms of metal-water interfacial reactions. *See* Kinetics and mechanisms of metal-water interfacial reactions
Metal (hydr)oxides
adsorption isotherms of Cd and PO_4 on goethite on double logarithmic scale, 78, 81*f*
adsorption of cation at positively charged metal (hydr)oxides, 78
adsorption of Cd on goethite in presence of PO_4 as function of pH, 84*f*
adsorption of oxyanions or weak organic acids, 78
adsorption of proton by broken SiO bond of silica, 72, 74*f*
adsorption of proton in aluminum hydroxide, 73, 74*f*
charge distribution (CD MUSIC) model, extension of MUSIC, 79–83
charging of silica as function of pH for various electrolyte concentrations, 73, 74*f*
comparison of charging of two homogeneous metal (hydr)oxide surfaces, 75, 77*f*
interfacial charge distribution, 79–83
ion binding, 78–85
Ising model approach for interaction between neighboring sites for proton affinity, 75–76
multi component interaction, 83
MUlti SIte Complexation (MUSIC) model, 73
pH dependence of ion binding, 78–79
phosphate surface complex structures on goethite in relation to charge distribution value by modeling, 80, 82*f*
pristine point of zero charge (PPZC or PZC) values, 70–71
proton affinity, 75–78
proton in FeOOH structure asymmetrical over O and OH, 76, 77*f*
refined MUSIC concept, 76, 78
schematic representation of inner sphere complex formation at surface, 79–80, 81*f*
spectroscopic technique to assess surface speciation, 79
Stern–Gouy–Chapman double layer model, 71–72
surface structure and sites, 72–75
two-step protonation reaction for two-pK model, 71–72
variable charge character, 70–78
Metal oxide-water interfaces, reaction mechanisms
conclusions, 29, 34
experimental details, 16–18
goniometer in grazing-incidence (GI) XAFS, 18
grazing-incidence XAFS studies of Co(II) and Pb(II) sorption on single crystal α-Al_2O_3 and TiO_2, 23–24
powder XAFS studies of Co(II) sorption on γ-Al_2O_3, α-Al_2O_3, α-SiO_2, and TiO_2 (rutile), 18–23
powder XAFS studies of Pb(II) sorption on α-Al_2O_3, α-Fe_2O_3, 23
procedures to prepare powder and single crystal sorption samples, 17*f*
single-crystal sorbent sample preparation, 17*f*, 18
surface complexation modeling (Co(II) on γ-Al_2O_3), 24–29
XAFS powder sorbent sample preparation, 16, 17*f*
See also Surface complexation modeling

Metal sorption at mineral–water interface
adsorption isotherms, 110
diffusion mechanisms, 114, 117–118
diffusion processes in natural materials, 116*f*
equilibrium models, 110, 112
fractional adsorption of Co to hydrous Fe-oxide (HFO) as function of pH and HFO-Co aging time, 114, 116*f*
fractional adsorption of Pb to HFO as function of pH and HFO-Pb aging time, 114, 115*f*
kinetic and molecular approaches, 112–131
kinetics of Ni sorption on pyrophyllite, kaolinite, gibbsite, and montmorillonite, 114, 115*f*
macroscopic sorption experiments not definitive mechanistic proof, 117–118
metal reaction effects on processes in subsurface environment, 108, 109*f*
nickel sorption on clay minerals (case study), 119–131
non-in-situ and in-situ molecular spectroscopic methods, 112–113
nucleation product types proposed, 118–119
outer- versus inner-sphere complexes effecting rate and reversibility of reactions, 113–114
radii of metal cations, 121*t*
schematic of surface complexes formed between inorganic ions and hydroxyl groups of oxide surface, 111*f*
sites of varying energy states, 118
studies of Ni, Zn, and Cd sorption on goethite, 117
surface precipitation and polynuclear surface complex formation, 118–119
time scales of metal sorption reactions, 113–114
X-ray absorption fine structure (XAFS) to study metal reaction mechanisms, 113
See also Nickel sorption case study
Metal species, aqueous, solubility, sorption, and transport in environment, 323
Metallurgical corrosion literature, zero valent iron remediation, 3
Metals, fate and transport of, 108, 109*f*
Microbial attachment to mineral surfaces
phase imaging of attachment features, 64–65
TMAFM amplitude and phase images along edge of goethite-bacteria cluster, 64*f*
Microbially mediated oxidative precipitation reactions
amino acid sequence alignment of copper-binding sites in MnxG, MofA, and other multicopper oxidases, 407*f*
amino acid sequence of Mn(II)-oxidizing protein from *Leptothrix discophora*, MofA, 408
central role of Fe transformations in environmental processes, 402*f*
central role of Mn transformations in environmental processes, 402*f*
diversity of Fe(II) and Mn(II) oxidizing microorganisms, 399–401
environments with active Fe and Mn deposition, 403
examples of iron and manganese bacteria, 401*t*
Fe and Mn precipitating microorganisms, 399–403
genetics and biochemistry of Mn(II) oxidation, 405–409
geochemistry of Fe and Mn and their minerals, 394–399
interactions with other metals, 409
iron oxide minerals formed in surface and subsurface environments, 395*t*

kinetics of Fe(II) and Mn(II) oxidation, 394–395
L. discophora bacterium for Mn(II) oxidation, 406
manganese oxide minerals formed in surface and subsurface environments, 396*t*
marine *Bacillus* sp. strain SG-1, model system for bacterial Mn(II) oxidation, 403–409
mechanisms of Fe and Mn oxidation/precipitation, 401, 403
microbial Fe(III) and Mn(IV) reduction, 399
microorganism participation in Fe and Mn mineralization processes, 393
mineralogy of bacterially-produced Mn oxides, 403–405
mineralogy of Fe and Mn oxides, 395–397
Mn(II) oxidation mechanism insights by sequence analysis of *mnx* genes, 407–408
Mn(II) oxidation reactions, 397*t*
organization of Mnx operon of *Bacillus* sp. strain SG-1 based on DNA sequence analysis, 406*f*
role of Fe and Mn oxides in fate and transport of contaminants in surface and subsurface environments, 399
solubility controlling mobility of Fe and Mn and availability to organisms, 398
specific surface area, site density, and zero point of charge of synthetic Fe and Mn oxides, 398*t*
surface chemistry of Fe and Mn oxides, 397–398
surface chemistry of SG-1 spores, 409
todorokite-like manganese oxide precipitate by SG-1 by TEM, 404*f*
transmission electron micrograph (TEM) of thin section of metal-oxidizing spores of marine *Bacillus* sp. strain SG-1, 404*f*

Microbially mediated reactions
catalyzing prokaryotic bacteria, 7
progress driven by advances in molecular biology, 7
Microcystin-LR blue-green algal toxin. *See* Photocatalytic degradation of trace contaminants
Microscopic and spectroscopic tools, overview of, 3–4
Mineral precipitation and dissolution, advancing of modern spectroscopic techniques, 5
Mixed cation surface precipitates, efficient scavenging mechanism for trace metal ion removal, 4
Mn(III) complexes and Mn(IV) species, reactivity with reductants
acidity enhancement of reaction, 276
decomposition rate constant for Mn(III) malonate with increasing pH, 268, 269*f*
empirical rate law for reaction of oxalate with soluble and polymeric MnO_2, 278–279
experimental analytical methods, 267–268
experimental preparation of Mn(III) complexes, 266
experimental reaction procedures, 266–267
Mn(III) complexes with sulfide, 270, 271*f*
MnO_2 precipitation induced by NaCl and seawater, 272*f*
polymeric and soluble MnO_2 stability, 270
polymeric MnO_2 and oxalate reactivity, 270, 272–278
reaction scheme for MnO_2 reduction, 276, 278
representative initial rate data to determine reaction order for reactants, 274*t*

representative kinetic studies at pH 4, 5, and 6 for MnO_2 reaction with oxalate, 273*f*, 275*f*
significant oxidants of organic matter in environment, 265
soluble Mn(III,IV)compounds at interfaces of oxic and anoxic zones and in suboxic zones, 266
stability of Mn(III) complexes (internal metal-ligand redox reactions), 268–270
UV-vis spectra for polymeric MnO_2, polymeric MnO_2 with oxalate, and Mn(oxalate), 277*f*

Mn oxide minerals
Bacillus sp. strain SG-1, model for bacterial Mn(II) oxidation, 403–409
central role of Mn transformations in environmental processes, 402*f*
examples of iron and manganese bacteria, 401*t*
genetics and biochemistry with purified SG-1 spores, 405–409
kinetics of Mn(II) oxidation, 394–395
manganese oxide minerals formed in surface and subsurface environments, 396*t*
mechanisms of Fe and Mn oxidation/precipitation, 401, 403
microbial Mn(VI) reduction, 399
mineralogy of bacterially-produced Mn oxides, 403–405
Mn(II) oxidation reactions, 397*t*
Mn(II)-oxidizing bacterium *Leptothrix discophora*, 406
precipitating microorganisms for Fe and Mn, 399–403
role of Fe and Mn oxides in fate and transport of contaminants, 399
solubility, 398
surface chemistry, 397–398
See also Microbially mediated oxidative precipitation reactions

MnO_2, polymeric and soluble. *See* Mn(III) complexes and Mn(IV) species, reactivity with reductants

Model for sorption and desorption of organic contaminants in soils and sediments
analogy between soil organic matter (SOM) and synthetic organic polymers, 225, 228–229
aqueous-phase sorption and desorption isotherms for rubbery cellulose and poly(phenyl methacrylate), 233, 234*f*
calorimetric investigation of glass transitions of soil-derived humic acid, 229, 230*f*
desorption behavior, 231, 233
distinct sorption/desorption examples of Canadian peat and Lachine shale, 233, 236
distributed reactivity model (DRM) to quantify non-ideal behavior, 224–225
dual reactive domain model (DRDM), 229, 231
general behavior of range of geosorbents, 236–237
hysteresis index, 233
linear partitioning sorption model, 223–224
measuring sorption under non-equilibrium conditions by phase-distribution relationship (PDR) approach, 225
PDR coefficient changes for sorption of phenanthrene by EPA-23 sediment as function of log time, 226*f*
phenanthrene sorption and desorption data for Canadian peat and Lachine shale, 235*f*
polymer-like behavior of natural humic acid, 229
reversible phenanthrene sorption and desorption data for silica gel-40, 239*f*

schematic illustration of domain types associated with soil or sediment, 227*f*
similarities of sorption and desorption behavior between SOM and synthetic organic polymers, 229, 231–233
sorption and desorption by natural geosorbents, 233, 235–237
sorption by exposed inorganic mineral domains, 237–238
sorption process and observations, 222–223
three domain model for characteristic changes in PDR parameters, 225, 238
time-dependent linear distribution coefficients for sorption by α-Al_2O_3, silica gel-40, and silica gel-100, 239*f*
time-dependent PDRs for sorption of phenanthrene by EPA-23 sediment, 226*f*
utility of DRDM for describing sorption in rubbery and glassy matrices, 231, 232*f*

Modeling
glassy and rubbery states for natural organic matter (NOM), 5
surface reactions, 4
See also Surface complexation modeling

Molybdenum (Mo)
adsorption kinetics, 170
DRIFT difference spectra of am-$Fe(OH)_3$ with and without Mo solutions, 158*f*
sorption on oxides, 157–159
See also Oxyanion adsorption mechanisms on oxides

Montmorillonite
inorganic reductants capable of reducing, 6
polyhedral representation showing linkage of tetrahedral and octahedral sheets, 296*f*
X-ray absorption near edge structure (XANES) analysis of Lewis acid oxidative degradation of tetraphenylboron (TPB) on Na-exchanged montmorillonite, 295, 296*f*
See also Nickel sorption case study

Montmorillonite-water systems
counterion solvation, 94, 96–100
influence of exchangeable cations, 88–89
inner-sphere and outer-sphere surface complexes, 92–94
inner-sphere configurations in Li-beidellite, 93*f*
interactions of clay surfaces with water dipoles and interlayer cations, 88–89
interlayer molecular structure for smectite-water systems, 89
interlayer water structure, 100–102
local coordination structure of water near Li^+, Na^+, and K^+, 94, 96
Monte Carlo (MC) simulations, 91–92
outer-sphere configurations in Li-hectorite, 93*f*
physical modeling based on quantum mechanical calculations, 90
radial distribution functions M^+-O (M = Li^+, Na^+, and K^+) in one-layer hydrates of montmorillonite, 96, 97*f*
radial distribution functions of O-H in one-layer montmorillonite hydrates, 100, 101*f*
radial distribution functions of O-H in two-layer montmorillonite hydrates, 102, 103*f*
radial distribution functions of O-O in one-layer montmorillonite hydrates, 100, 101*f*
radial distribution functions of O-O in two-layer montmorillonite hydrates, 102, 103*f*
sampling strategy, 91–92

self-diffusion coefficients for three interlayer cations in low-order hydrates of montmorillonite, 96, 97*t*
simulation methodology, 90–92
synergistic relationship between experiment and modeling, 89
trajectories of adsorbed Li^+ and hydrating water molecules, 95*f*
trajectories of K^+ on siloxane surfaces of one-layer hydrate of K-montmorillonite, 98*f*
trajectories of K^+ on siloxane surfaces of two-layer hydrate of K-montmorillonite, 99*f*
V structure of adsorbed water, 88–89
water molecules organizing in response to cation-water interactions, 102, 104
water structure network formation critical in early stages of MC simulation, 92
Mössbauer spectroscopy, in situ imaging of samples, 3
MUlti SIte Complexation (MUSIC) model
proton affinity of surface complexes, 75–78
structural approach in silica case, 73
surface complexes site definition, 154–155

N

Na-montmorillonite. *See* Montmorillonite-water systems
Natural organic matter (NOM) as sorbent of organic compounds
carbon dioxide as alternative characterization probe, 214
competitive sorption, 212–214
contribution of hole-filling to total sorption calculated by Freundlich slope method for 1,3-dichlorobenzene, 217*f*
dissolution and hole-filling mechanisms, 205–207
dual-mode model, 205–207
evidence of nanoporosity in NOM, 214
evidence supporting dual-mode model, 207–215
Freundlich parameters for selected additional systems, 210*t*, 211*t*
high affinity of organic compounds, 204–205
hydrophobic effect as thermodynamic driving force, 206
importance of hole-filling, 215–216
influence of mechanism on sorption and desorption rates, 220
isotherm of atrazine in Cheshire fine sandy loam with and without prometon, 213*f*
isotherm of 1,3-dichlorobenzene in Cheshire fine sandy loam soil suspension as function of contact time, 208*f*, 209*f*
isotherms of 1,3-dichlorobenzene in whole Pahokee peat soil, humin, and humic acid, 219*f*
linear partitioning or solid-phase dissolution process, 205
nonlinear isotherms, 207, 212
role of structural components of NOM in dual-mode mechanism, 216, 218–220
schematic of dual-mode model , 208*f*
supporting evidence in literature, 215
Nickel sorption case study
enhanced dissolution of clay and oxide minerals, 131
existence of mixed-cation hydroxide phases, 125
formation kinetics of mixed Ni-Al hydroxide phases, 126
growth of mixed Ni/Al phases on mineral surfaces using scanning force microscopy (SFM), 126, 129*f*, 130*f*, 131

importance of Al in immobilizing heavy metals, 126
kinetics of Ni sorption on pyrophyllite, 121*f*
normalized, background-subtracted, and k-weighted XAFS spectra of Ni sorbed on pyrophyllite, kaolite, gibbsite, and montmorillonite, 122*f*
phenomena at mineral/liquid interface for mixed-cation hydroxide formation, 131
radial structure functions by forward Fourier transforms of XAFS spectra, 124*f*
radial structure functions of pyrophyllite samples reacted with Ni, 127*f*
sorption on clay minerals, 119–131
structural information from XAFS analysis, 123*t*
structural parameters for Ni/pyrophyllite from XAFS analysis, 128*t*
synthesis of mixed-cation hydroxide compounds by induced hydrolysis, 125
synthesis of mixed Ni/Al compounds, 125–126
Nitroaromatic compounds. *See* Pollutant reduction in heterogeneous Fe(II)/Fe(III) systems
Nontronite, color change during tetraphenylboron degradation, 293
Nuclear magnetic resonance (NMR), in situ imaging of samples, 3

O

Organic compounds, sorption of, 5
Organic contaminants in soils and sediments. *See* Model for sorption and desorption of organic contaminants in soils and sediments
Organic polymers. *See* Polymers, synthetic organic
Outer-sphere complexes, at least one water molecule, 138
Oxalate. *See* Mn(III) complexes and Mn(IV) species, reactivity with reductants
Oxidative precipitation reactions. *See* Microbially mediated oxidative precipitation reactions
Oxide mineral reductive capacity towards Co(II)EDTA. *See* Auto-inhibition of oxide mineral reductive capacity toward Co(II)EDTA
Oxide minerals, reactivities of various, 259–260, 261*f*
Oxide-water interface
composition of oxide, 304–305
oxide as coordinating surface for reduction of aliphatic halide, 311–315
oxide as physical barrier to mass transport, 305, 308
oxide as semiconductor, 308–311
See also Reduction reactions at metal-water interface, role of oxides
Oxides in reduction reactions. *See* Reduction reactions at metal-water interface, role of oxides
Oxyanion adsorption mechanisms on oxides
advantage of Fourier transform infrared (FTIR) spectrometers over older dispersive instruments, 137–138
arsenate sorption and surface configuration, 159–164
arsenic adsorption kinetics, 170, 172
arsenite sorption and surface configuration, 164–167
boron coordination and surface configuration, 148–153
carbonate speciation and surface configuration, 140–148
chromate adsorption, 167, 169
chromate adsorption kinetics, 172

distinguishing between inner- and outer-sphere adsorption mechanisms, 136–137
electrophoretic mobility measurements, 138–139
extended X-ray absorption fine structure spectroscopy (EXAFS), 137
inner-sphere adsorption, 137
kinetics of oxyanion adsorption, 169–172
methods possible with DRIFT and ATR spectroscopy, 137–138
molybdenum adsorption kinetics, 170
molybdenum sorption and surface configuration, 157–159
outer-sphere adsorption, 138
oxyanion identification and mode of bonding to oxide surface, 136–137
phosphate adsorption kinetics, 169–170
phosphate speciation and surface configuration, 139–140
selenate sorption and surface configuration, 153–155
selenite sorption and surface configuration, 155–157
selenium adsorption kinetics, 170
vibrational modes for various oxyanions, 143*t*, 144*t*, 145*t*, 146*t*

P

Pauling bond valence, available charge per bond, 72
Pauling electrostatic valence principle, bond valence sum, 21
Phlogopite
AFM images in studying dissolution rates and mechanisms, 40*f*, 42*f*
potassium release problem, 4
TappingMode™ atomic force microscope (TMAFM) image of dispersed clay on polished sapphire substrate, 45, 46*f*
Phosphate
drying of oxide suspension, 138
kinetics of adsorption on oxides, 169–170
magic angle spinning nuclear magnetic residence (NMR) experiments, 140
speciation and surface configuration on oxides, 139–140
See also Oxyanion adsorption mechanisms on oxides
Photocatalytic degradation of trace contaminants
adsorption isotherms for uptake of organic carbon on TiO_2 from solutions of pH 3.5 and 8.6, 382*f*
approach to quantitative modeling of semiconductor-catalyzed contaminant photodegradation, 384–386
assessing veracity of proposed conceptual model, 389–390
blue-green algal toxin microcystin-LR (MLR) structure, 377*f*
change in pseudo-first order rate constant of MLR removal as function of initial concentration, 381*f*
comparison of time dependence of toxin removal under strong and weak adsorption conditions, 387*f*, 388*f*
concentration of MLR adsorbed at pH 3.5 as function of solution concentration of toxin, 381*f*
effect of pH on MLR degradation, 376, 379
effect of toxin and algal exudate concentration on pseudo-first order rate constant for removal of toxin, 388*f*
effect of toxin and exudate concentrations, 389
effect of toxin concentration, 379–380
effect of varying affinity of MLR and algal exudate for TiO_2 surface sites on rate of removal of toxin, 387*f*
effect of varying extent of toxin and exudate adsorption, 386, 389

loss of MLR from solutions of various pH, 378*f*

MLR remaining in pH 6.4 over time in absence and presence of TiO_2 and light, 377*f*

MLR TiO_2-catalyzed photodegradation experimental approach, 375–376

oxidative degradation by semiconductor-catalyzed photolysis, 374

qualitative model of MLR removal from solution, 380, 383–384

quantitative modeling of hypothetical system, 386, 387*f*, 388*f*

rates of photodegradation as function of suspension pH, 378*f*

surface active compounds, 374–375

TiO_2-catalyzed photodegradation breakdown of blue-green algal toxin, 375, 377*f*

See also Semiconductor-catalyzed contaminant photodegradation

Photochemical reactions, conceptual model for system, 7

Pitting, dissolution at defect in oxide surface by aggressive anion, 305, 308

Point of zero charge (PZC), shift determination by electrophoretic measurements, 137

Pollutant reduction in heterogeneous Fe(II)/Fe(III) systems

calculated parameter values of isotherm fits of surface precipitation model for Fe(II) uptake by iron (hydr)oxides, 352*t*

comparison of tracer reactivity within aquifer and in model systems exhibiting only natural organic matter or surface-bound Fe(II) as reductants, 353, 354*f*

electron balance for reduction of 4-chloronitrobenzene (4-Cl-NB), 345*f*

evidence for participation of ferrous iron in reductive transformation of pollutants, 342, 344

factors controlling reactivity of ferrous iron, 344–352

Fe(II) bound to oxide surface altering reactivity with time, 348*f*

formation and regeneration of reactive Fe(II) surface sites, 353, 355

general scheme for reduction of pollutants by reactive Fe(II) surface sites and processes that regenerate such reductants, 354*f*

groundwater remediation using metal iron as bulk reductant, 355

interaction of Fe(II) with iron (hydr)oxides, 349, 352

isotherm fits of surface precipitation model for sorption of ferrous iron on iron oxides, 351*f*

pH dependence on ferrous iron reactivity, 344, 347

pseudo-first-order reaction rates with nitroaromatic compounds (NACs) and polyhalogenated aliphatic compounds (PHAs), 346*f*

rate constant for dibromodichloromethane reduction as function of total ferrous iron, 350*f*

reaction rates (normalized) for two PHAs as function of equilibrium time, 348*f*

remodeling of Fe(II) at iron (hydr)oxides with time, 347, 348*f*

representative redox couples of various organic pollutants and iron species, 343*f*

schematic representation of field injection experiment in anaerobic part of landfill leachate plume, 354*f*

significance of surface-bound Fe(II) for pollutant reduction, 352–355

surface density of Fe(II) species, 347, 349

type and reactivity of Fe(II) subject to environmental factors, 355

typical example of ferrous iron reactivity at ambient pH in aqueous solutions containing Fe(III), 345*f*
uptake kinetics of dissolved Fe(II) by different iron oxides in anoxic aqueous suspensions, 350*f*
uptake kinetics of dissolved Fe(II) by goethite in anoxic aqueous suspensions at different pH values, 351*f*
Polyhalogenated aliphatic compounds. *See* Pollutant reduction in heterogeneous Fe(II)/Fe(III) systems
Polymer literature, non-idealities in sorption of organic contaminants to natural organic matter (NOM), 3
Polymers, synthetic organic
analogy with soil organic matter (SOM), 225, 228–229
distinct forms of mechanical behavior, 228
glass transition temperature, 228–229
similarities to sorption and desorption of soil organic matter, 229, 231–233
utility of dual reactive domain model (DRDM) for describing sorption in rubbery and glassy matrices, 232*f*
Potassium-montmorillonite. *See* Montmorillonite-water systems
Precipitation reactions. *See* Microbially mediated oxidative precipitation reactions
Pyrite, new direction with UHV-based scanning tunneling microscope (STM) research, 51–54
Pyrophyllite. *See* Nickel sorption case study

R

Radiocesium sorption on illite and natural sediments
average fraction of exchangeable-^{137}Cs in sediments after NH_4-extractions, 197*t*
2- and 3-box models for cesium sorption, 187, 195–196
exchangeability of sediment-bound radiocesium, 197–198
introduction of radioisotopes into environment, 179–180
ion-exchange theory, 180–181
K_D-values (in situ) from different European sediments, 182*f*
kinetic data of Cs adsorption on K- and Ca-saturated illite, 190*f*, 191*f*
kinetic data of Cs adsorption on sediments from Ketelmeer and Hollands Diep, 192*f*, 193*f*
measurements of model predictions of Cs adsorption and desorption on K- and Ca-saturated illite, 184*f*, 185*f*
measurements of model predictions of Cs adsorption and desorption on sediments from Ketelmeer and Hollands Diep, 188*f*, 189*f*
mechanistic interpretation, 196
modeling kinetics, 183–196
parameter values of 2-box and 3-box kinetic models, 195*t*
radiocesium binding in sediments, 180–183
remobilization process in anoxic pore waters by ion-exchange, 180
reverse (remobilization) rates, 199–200
reversibility of radiocesium partitioning, 183–196
schematic representation and equations defining 2-box and 3-box kinetic models, 194*f*
selective sorption by micaceous clay minerals, 180–181
slow (reverse) migration from clay-mineral interlayers into solution, 198–199
slow uptake rates by illite, 199
Reaction kinetics. *See* Interfacial kinetics

Reactions at mineral surfaces. *See* Interfacial kinetics
Reactivities of environmental particles
atomic force microscopy (AFM) as tool, 57–60
phase imaging of microbial attachment features, 64–65
study of Pb uptake by hydroxylapatite, 60–61
TappingMode™ AFM imaging of fulvic acids, 61, 63–64
Real time clay/solution interaction studies. *See* Clay/solution interaction studies, real time
Reduction of aqueous metal species on Fe(II)-containing oxide surfaces
aqueous Fe(II) release from natural magnetite and ilmenite as function of time and pH, 325*f*
coupled half cell reactions, 324, 326
Cr(VI) loss from solution in presence of Kesterson magnetite, 337*f*
Cr(VI) uptake and reduction to Cr(III) using magnetite, 334, 335*f*
effects of aqueous metal concentrations on electrode potentials, 331, 334
electrochemical studies, 326–327
evidence for formation of passivation layer, 334, 336
experimental metal reduction using naturally weathered magnetite, 336
extent of Cr(VI) reduction by LA River magnetite, 337*f*
field evidence of heterogeneous metal reduction by Fe(II) oxides, 336, 338–340
heterogeneous reduction of aqueous transition metals, 327, 329–334
high resolution transmission electron micrograph (TEM) images of synthetic magnetite crystals, 335*f*
mechanisms and kinetics of decoupled reactions by dynamic polarization experiments, 326–327
oxidation of ilmenite to pseudorutile, 326
oxidation of magnetite to maghemite, 324, 326
potentiodynamic scans as functions of applied potential E and measured current *i* for natural magnetite and ilmenite electrodes in anoxic solutions at pH 3, 328*f*
potentiodynamic scans for magnetite electrode for anoxic solutions with Cr(VI) and V(V), 332*f*
redox reactions involving magnetite and ilmenite, 324–327
reductive capacities of oxide minerals, 334
role of homogeneous reactions, 330–331
role of surface passivation, 334–336
self induced potentials as functions of time and pH for natural magnetite and ilmenite electrodes, 328*f*
self induced potentials for magnetite electrodes in solutions with various metal cations, 333*f*
solubility, sorption, and transport of aqueous metal species in environment, 323
suppression of reductive dissolution reactions, 331, 332*f*
time trends showing rates of ferric, chromate, and vanadate reduction in ilmenite suspensions, 329*f*
Reduction reactions at metal-water interface, role of oxides
adsorbate competition for limited sites in surface complexation model (SCM), 312–315
chemical background of Fe^0–H_2O systems, 302
composition of oxide at oxide-water interface, 304–305
conceptual models of potential processes in reduction of chlorinated

aliphatics (RX→RH) at iron-oxide-water interface, 303*f*
differences in energy level of E_F(metal), E_F(oxide), and E^0(RX/R•), 310
effect of acetate, ascorbate, and catechol concentration on rate of CCl_4 dechlorination by Fe^0, 316*f*
effect of initial CCl_4 concentration on rate of CCl_4 reduction by Fe^0, 314*f*
electron transfer dependent on electron energy levels for semiconductors, 310
electron transfer through oxide by direct and resonance tunneling, 311
electronic properties of iron oxides, 309*t*
groundwater remediation applications producing mixed-valent oxides, 309
implications for environmental and engineering applications, 315, 317
little evidence for auto-catalysis in dechlorination by Fe^0, 308
mass action equations for surface complexation model, 314*t*
mass transport to interface, 302
mixed potential diagram illustrating controls on kinetics of corrosion at pitted, oxide-covered metal, 307*f*
non-competitive SCM, 313
oxide as coordinating surface for reduction of RX at surface of oxide film, 311–315
oxide as physical barrier to mass transport, 305, 308
oxide as semiconductor, 308–311
oxide-water interface, 304–315
pit initiation and oxide breakdown, 305
pitting and dissolution by aggressive anion, 305
pitting corrosion acceleration by aggressive anions such as chloride, 308
rate equation assuming steady-state population of precursor complex, 312
scanning electron micrograph showing precipitates on Fe^0 surface after long-term exposure to CCl_4, 306*f*
SCM model parameters for iron metal systems, 315*t*
simplified iron-oxide-solution system revealing controlling factors, 310
surface layer of corrosion products under environmental conditions, 301
Reductive capacity of oxide minerals toward Co(II)EDTA. *See* Auto-inhibition of oxide mineral reductive capacity toward Co(II)EDTA
Resonance tunneling, electron transfer through oxide, 311

S

Sandia National Laboratory, waste water contamination site, 338
Scanning probe microscopes (SPM)
atomic force microscope (AFM) most common, 38
characterization of local nature of heterogeneous processes, 37
new directions in geochemistry research, 38
original scanning tunneling microscope (STM) for use in ultra-high vacuum, 37–38
original SPM introduction, 37–38
See also Atomic force microscopy (AFM); Scanning tunneling microscope (STM)
Scanning tunneling microscope (STM)
in situ imaging of samples, 3
original use in ultra-high vacuum (UHV), 37–38
UHV-based STM for geochemistry research, 50–54
See also Ultra-high vacuum (UHV) scanning tunneling microscope (STM)
Selenate

diffuse reflectance infrared Fourier transform (DRIFT) spectrum of sorption on am-$Fe(OH)_3$, 156*f*
sorption on oxides, 153–155
See also Oxyanion adsorption mechanisms on oxides
Selenite
diffuse reflectance infrared Fourier transform (DRIFT) spectrum of sorption on am- $Fe(OH)_3$, 156*f*
sorption on oxides, 155–157
See also Oxyanion adsorption mechanisms on oxides
Selenium
adsorption kinetics, 170
sorption and desorption isotherms, 171*f*
Semiconductor
details of electron transfer, 310
electronic properties of iron oxides, 309*t*
iron oxides at oxide-water interface, 308–311
Semiconductor-catalyzed contaminant photodegradation
adsorption of toxin and exudate to TiO_2, 384–385
initial concentrations of toxins and exudate, 384
production of primary oxidizing species, 385
proportion of toxin and exudate adsorbed to TiO_2, 385*t*
quantitative modeling, 384–386
solution-phase radical reactions, 386
surface-located radical reactions, 385–386
See also Photocatalytic degradation of trace contaminants
Smectite surfaces. *See* Tetraphenylboron degradation at hydrated smectite surfaces
Sodium-montmorillonite. *See* Montmorillonite-water systems
Sorbate species. *See* Metal oxide-water interfaces, reaction mechanisms
Sorbents
effect on type of surface complexes formed, 20–22
See also Geosorbents; Natural organic matter (NOM) as sorbent of organic compounds
Sorption and desorption models. *See* Model for sorption and desorption of organic contaminants in soils and sediments
Sorption density, effect on type surface complex formed, 18, 20
Sorption of inorganic species
to mineral surfaces, 4–5
nature of anionic surface species, 5
Sorption of organic compounds
linear partitioning model, 5
model of natural organic matter (NOM) with glassy and rubbery states, 5
Sorption process characterization using atomic force microscopy (AFM)
AFM observations of reactions on goethite surfaces, 47, 48*f*, 49*f*
analytic breakthroughs advancing field of sorptive geochemistry, 45, 47
Mn sorptive mechanisms on hematite, goethite, and albite surfaces, 47
potential for further AFM contributions, 47, 50
technique for fixing clays to be imaged in fluid cell, 47
Sorption studies
aqueous Pb(II) on goethite and alumina surfaces in presence of chloride ion, 23
aqueous Pb(II) on hematite and goethite powders, 23
differences in crystal structure and bond-valence sums at surface oxygens, 21
dissimilarity of polyhedral dimensions, 21
effect of sorbent type on sorption complexes, 20–22

effect of sorption density on sorption complexes, 18–20
EXAFS spectra of Co(II) on γ-Al_2O_3 as function of surface coverage, 19*f*
formation of Co-Al hydroxide precipitates on alumina, 22–23
grazing incidence XAFS of Co(II) and Pb(II) sorption on single crystal α-Al_2O_3 and TiO_2, 23–24
powder XAFS of Co(II) on γ-Al_2O_3, α-Al_2O_3, α-SiO_2, and TiO_2 (rutile), 18–23
powder XAFS of Pb(II) on α-Al_2O_3, and α-Fe_2O_3, 23
second-neighbor cobalt atoms, 18, 20
surface bonding model, 24
XAFS study of aqueous Co(II) on alumina powders, 22–23
See also Metal sorption at mineral–water interface
Spectroscopic and microscopic tools, overview of, 3–4
Stern–Gouy–Chapman model, variable charge character of metal (hydr)oxides, 71–72
Stern layer capacitance, model fitting parameter, 71–72
STM. *See* Scanning tunneling microscope (STM)
Surface complexation models (SCMs)
adsorbate competition for limited oxide sites, 312–315
cobalt sorption reactions, 26
Co(II) on γ-Al_2O_3, 24–29
metal sorption at mineral–water interface, 110, 112
mono- and multinuclear complexes and precipitation (model M9), 29
monomeric complexes and precipitation alone (model M1), 28
mononuclear reactions, 26–27
multinuclear sorption, 28–29
precipitation and multinuclear complexes, 27–28
SCM with triple layer model (TLM) capacitances, 26
SCM/TLM simulation of Co uptake on γ-Al_2O_3, 30*f*, 31*f*
SCM/TLM sorption reactions and equilibrium constants, 27*t*
significance of models, 29
site saturation and effect surface area, 27
sorption density of N_{Co} in Co(II) sorption complexes on γ-Al_2O_3 as predicted by models M1 and M9, 32*f*
uptake data for sorption of Co(II) by γ-Al_2O_3, 25*f*
varying partition coefficient, K_d, with initial total cobalt concentration, 33*f*
XAFS results for sorption of Co(II) by γ-Al_2O_3, 25*f*
Surface-mediated reactions, diversity of, 3
Surface passivation
evidence for formation of passivation layer, 334, 336
experimental metal reduction using naturally weathered magnetite, 336, 337*f*
heterogeneous electron transfer, 6
reductive capacities, 334, 335*f*
Surface precipitation, sorption of inorganic species, 4–5
Surface reactions, modeling of, 4
Synthetic organic polymers. *See* Polymers, synthetic organic
Synthetic polymer literature, non-idealities in sorption of organic contaminants to natural organic matter (NOM), 3

T

Takovite. *See* Nickel sorption case study
TappingMode^{TM} atomic force microscopy (TMAFM)

exciting cantilever into resonance oscillation, 58–59
imaging of fulvic acids, 61, 63–64
little lateral force on imaged surface, 44–45
phase imaging of microbial attachment features, 64–65
phase imaging TMAFM, 60, 62*f*
study of Pb uptake by hydroxylapatite, 60–61
TMAFM image of dispersed clay on polished sapphire substrate, 45, 46*f*

Tetraphenylboron degradation at hydrated smectite surfaces
attenuated total reflectance infrared (ATR–IR) for minimization of water absorption, 285
bands of interest for quantitative determination of major mother and daughter products in reaction of TPB at clay surfaces, 286*f*, 287
Brønsted acidity and Lewis acidity predominating at mineral surfaces, 283
Brønsted degradation pathway of TPB, 289, 290*f*
cesium-containing liquid waste precipitation with sodium-tetraphenylboron (Na-TPB), 283
changes in average local structure of Fe in montmorillonite, 297
coupling of TPB oxidation to structural Fe(III) reduction in smectites, 293, 295
details of ATR–IR setup, 286*f*
different Fe–O bond lengths depending on oxidation state, 296*f*, 298
effect of Fe reduction on clay structure by extended X-ray absorption fine structure (EXAFS), 296–298
effect of water content on reaction kinetics, 292–293
effect of water content on TPB phenyl vibrational bands in presence of Na-exchanged montmorillonite preceding reaction, 294*f*
fate of organic contaminants in terrestrial and aquatic environments, 282–283
fate of TPB by IR analysis, 285–287
IR spectra of TPB degradation on Al-exchanged montmorillonite and quantitative plot measured from IR spectra, 291*f*
Lewis and Brønsted acid degradation pathways for TPB, 283, 285
Lewis degradation pathway of TPB, 287, 288*f*
multiple reaction pathways of TPB-catalyzed degradation at smectite surfaces, 284*f*
non-linear best fits of IR absorption to proposed pseudo-first order kinetics, 293, 294*f*
overlaid IR spectra tracing TPB degradation on Al-exchanged hectorite and plot of reaction measured from IR spectra, 290*f*
overlaid IR spectra tracing TPB degradation on Na-exchanged montmorillonite and quantified plot measured from IR spectra, 288*f*
primary reaction pathways for Lewis and Brønsted degradation of tetraphenylboron (TPB), 284*f*
reaction kinetics of Lewis and Brønsted pathways, 289, 292
simultaneous Lewis and Brønsted degradation pathway of TPB, 289, 291*f*
X-ray absorption near edge structure (XANES) as in situ probe for Fe(II)/Fe(III), 295, 296*f*

Three domain model, measuring sorption under non-equilibrium conditions, 225, 226*f*, 227*f*

Trace contaminants and degradation. *See* Photocatalytic degradation of trace contaminants

Transition metals, aqueous, heterogeneous reduction, 327, 329–334
Transport and fate of metals, 108, 109*f*
Tunneling of electrons, electron transfer through oxide, 311

U

Ultra-high vacuum (UHV) scanning tunneling microscope (STM)
geochemical research, 50–51
low energy electron diffusion (LEED) patterns after cleaning, 51, 54
technique to restore pyrite growth surfaces in-vacuum, 51
UHV STM image of cleaned pyrite growth face, 51, 53*f*
UHV STM image of in-vacuum fractured pyrite surface, 51, 52*f*

V

Vejen landfill, transformation of xenobiotic organic compound, 355

W

Waste water contamination sites, variable in Cr(VI)/Cr(III) ratios, 338, 340
Wyoming-type montmorillonite. *See* Montmorillonite-water systems

X

X-ray absorption near edge structure (XANES), in situ probe for Fe(II)/Fe(III), 295
X-ray adsorption fine structure (XAFS)
grazing-incidence (GI) XAFS probing sorbates, 15–16
grazing incidence XAFS of Co(II) and Pb(II) sorption on single crystal α-Al_2O_3 and TiO_2, 23–24
in situ imaging of samples, 3
powder studies of Co(II) sorption on γ-Al_2O_3, α-Al_2O_3, α-SiO_2, and TiO_2 (rutile), 18–23
powder XAFS of Pb(II) sorption on α-Al_2O_3 and α- Fe_2O_3, 23
second-neighbor atom contributions, 15
spatially resolved technique X-ray spectromicroscopy, 4
structural information on sorbates, 15
synchrotron-based XAFS in simple model sorption systems, 14–15
See also Sorption studies
X-ray adsorption near edge structure (XANES)
contaminant-induced reduction of structural Mn(IV) in oxide minerals, 370
in situ imaging of samples, 3
See also Auto-inhibition of oxide mineral reductive capacity toward Co(II)EDTA
X-ray spectromicroscopy, spatially resolved technique of XAFS, 4